Avionics:
Systems and Troubleshooting
A Practical Guide to Advanced Avionics

Second Edition

T.K. Eismin

Production Staff
Editor Rhonda Cooper
Lead Illustrator Amy Siever
Designer/Production Coordinator Roberta Byerly

International Standard Book Number 1-933189-21-5
ISBN 13: 978-1-933189-21-5
Order # T-AVSAT-0201

For Sale by: Avotek
A Select Aerospace Industries company

Mail to:
P.O. Box 219
Weyers Cave, Virginia 24486
USA

Ship to:
200 Packaging Drive
Weyers Cave, Virginia 24486
USA

Toll Free: 1-800-828-6835
Telephone: 1-540-234-9090
Fax: 1-540-234-9399

Second Edition
First Printing
Printed in the USA

www.avotekbooks.com

www.avotek.com

To my parents,
Nick and Mary Lou:

You showed me how to live, work hard, respect others, and enjoy life. My mother taught me how to love and be loved. My father taught me about technology and electronics. Thank you for all you have done.

Special thanks to:

Nona Schaler for all her love and support, her honesty, hard work and her help on this text.

Preface

This textbook on avionics presents functional concepts of avionics systems and thus its information is valuable to a variety of individuals. It should be noted that this is an advanced textbook and it is assumed the reader will have thorough knowledge of aviation electronics, preferably holding a Federal Aviation Administration (FAA) Airframe and/or Powerplant certificate, or a National Center for Aerospace & Transportation Technologies (NCATT) Aircraft Electronics Technician (AET) certification.

Avionics technicians can utilize *Avionics: Systems and Troubleshooting* to gain a better understanding of line troubleshooting and computer controlled aircraft. The Aviation Maintenance Technician (AMT or A&P mechanic) should study this text for insight into advanced avionics systems including troubleshooting. Avionics engineers and bench technicians can also gain a better perspective of the entire aircraft through the study of this text.

Educators can use this text as a "ready to use" course on advanced electronics/avionics. The materials presented in this book will bring your students well ahead of any FAA advanced electronics requirements and satisfy the needs for both corporate and air carrier technicians. A student study guide is available for this text that provides a variety of questions presented in assignment format. Each page is designed so the assignments can be turned in individually. The instructor can choose from a variety of questions and find all the answers on the instructor's CD-ROM. The CD-ROM also contains each figure presented in the text. Over 500 illustrations can easily be made into handouts, lecture notes, or copied into unit exams.

Textbooks, by nature, must be general in their overall coverage of a subject area. As always, the aircraft manufacturer is the sole source of operation, maintenance, repair, and overhaul information. Their manuals are approved by the FAA and must always be followed. You may not use any material presented in this or any other textbook as a manual for actual operation, maintenance, or repairs.

The author has to the best of his abilities tried to provide accurate, honest, and pertinent material in this textbook. However, as with all human endeavors, errors, and omissions can show up in the most unexpected places. If any exist, they are unintentional. Please bring them to our attention. ➜

Email us at comments@avotek.com for comments or suggestions.

Avotek® Aircraft Maintenance Series:
Introduction to Aircraft Maintenance
Aircraft Structural Maintenance
Aircraft System Maintenance
Aircraft Powerplant Maintenance

Other Books by Avotek®:
Aircraft Turbine Engines
Aircraft Corrosion Control Guide
Aircraft Structural Technician
Aircraft Wiring & Electrical Installation
Aviation Maintenance Technician Reference Handbook
Avotek Aeronautical Dictionary
Fundamentals of Modern Aviation
Light Sport Aircraft Inspection Procedures

Acknowledgements

Airbus Americas

Airbus S.A.S.

Airbus Service Company, Inc.

Air Transport Association, Inc.

Bendix/King by Honeywell

Beech Aircraft Corporation

Boeing Commercial Airplane Company

Cessna Corporation

Collins Commercial Avionics

Embraer S.A.

Garmin International, Inc.

Gulfstream Aerospace

Hawker Beechcraft Corporation

Honeywell, Inc.

Mark Hopkins

IFR Systems, Inc.

Northwest Airlines, Inc.

Rockwell International Corporation

Tim Ropp

Seattle Avionics

Technical Staff of Hanger Six

Cover photo: © Gulfstream Aerospace Corporation

Contents

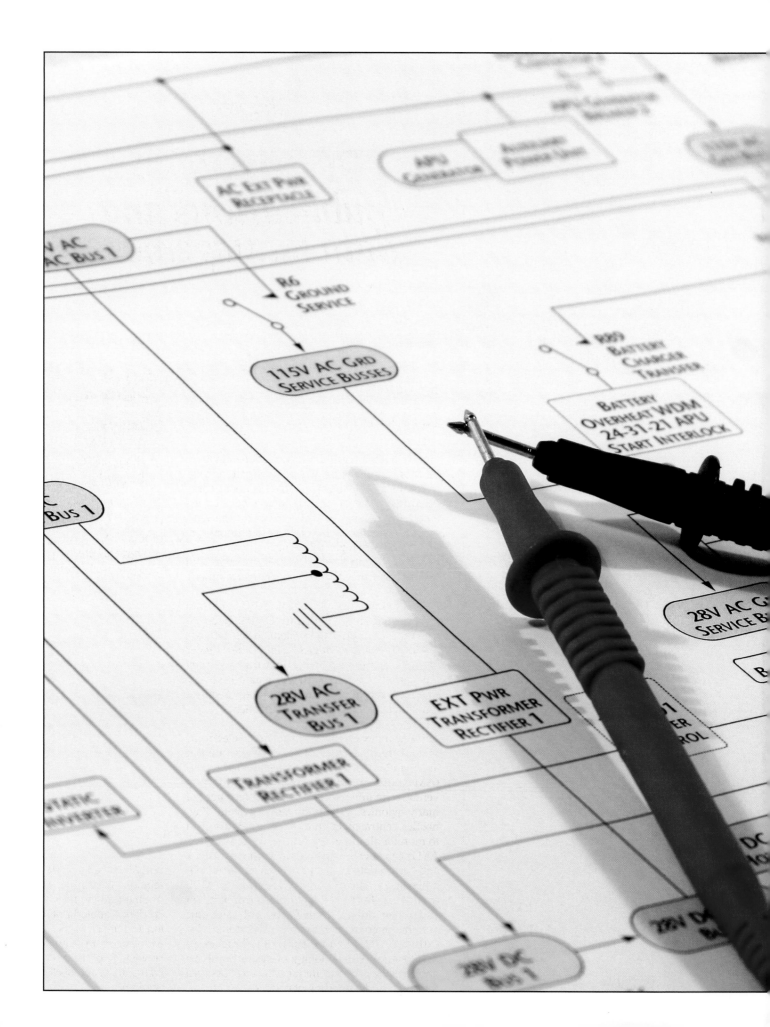

Chapter 1

TECHNICAL
publications and data management

Section 1

Introduction

For decades aircraft have been dependant on electronic systems for various navigation, communication, and flight control functions. These electronic systems have been termed *Avionics* (short for aviation electronics). Today the term avionics reaches far beyond the traditional systems. Today's high performance aircraft use avionics to control virtually every system from engine performance to passenger entertainment.

Contemporary avionics are quite different from those found on older aircraft. Digital electronics and computer-based systems have replaced analog devices. Modern avionics no longer employ electromechanical instruments; they have been replaced with solid-state displays. The most recent aircraft utilize integrated systems that share a variety of components.

This trend toward the increased use of electronic systems has created a shift in the way aircraft are maintained. *Aircraft maintenance technicians* (AMTs) are expected to perform routine preventative maintenance and line replacement of many avionics systems. To aid in system diagnostics, central maintenance computers are used to monitor almost every function of the aircraft. AMTs can now isolate faults from the flight deck and correct defects that previously required avionics specialists. Avionics specialists, or R&E (radio and electronics) technicians troubleshoot and repair only a small fraction of electronic systems onboard the aircraft. Therefore, it is extremely important for avionics technicians to understand the technology of aircraft systems and as equally important for AMTs to possess a thorough understanding of avionics.

Learning Objectives:

- *Technical Publications*
- *Industry Standards*
- *Component Location*
- *Data Management*

Left. Advanced aircraft require advanced troubleshooting skills. Efficient troubleshooting and repair starts with a good working knowledge of the aircraft's technical specifications.

A review of technical publications begins the textbook. Since all avionics systems require electrical power, technicians must have a good working knowledge of power systems and therefore a variety of corporate and transport category aircraft power distribution systems are reviewed. Common digital data formats, such as, ARINC 429, 629 and 664 are presented in great detail along with their related troubleshooting techniques. Complex avionics systems common to most modern aircraft are also presented. Integrated displays, autopilot/ autoflight, flight management, GPS, ACARS, passenger entertainment, and central diagnostic systems are all discussed, including system architecture, operation, and troubleshooting techniques for a variety of common avionics systems. *Avionics: System and Troubleshooting* provides a study of integrated electronic systems as they apply to flight line maintenance.

Section 2

Presentation of Technical Publications

Understanding how the systems work is the key to troubleshooting. Aircraft manufacturers provide technicians with documentation that describes how the systems work through numerous technical publications. The ability to use and interpret these manuals is the first step to good troubleshooting.

One of the biggest challenges for today's technician is to keep up with an overwhelming amount of paper work. As aircraft have become increasingly dependent on electronics for basic flight and navigation, the documentation for these systems has increased significantly. Major manufacturers, such as the Boeing Corporation, print literally tons of manuals covering inspection, maintenance, and repair of their avionics systems. These companies also incorporate a variety of electronic manuals distributed on CD-ROM, DVD and the Internet. The major advantage of this format is that it allows near instant access to various sections of the manuals using a personal computer. The speed of the CD-ROM format is exceptionally beneficial when performing maintenance tasks that require information from multiple sections of the aircraft manuals.

These electronic manuals help reduce the paperwork needed to keep information current. Today most airlines and many general aviation shops rely on digital technologies to store, retrieve, and view technical manuals.

Using the modern digital format for maintenance information also allows for rapid information updates. Manuals contained on CD can simply be updated by replacing that CD; a much simpler task than replacing pages in a paper manual. Another option is the transfer of updated material to the Internet. Using an internet download of technical information allows for literally instant updates of materials and is more environmentally friendly. With internet downloads there is no need to discard old CD-ROM discs or old pages of a paper manual.

Industry Standards

To help technicians in their quest to keep avionics systems airworthy, several standards exist. The most commonly used standards for aircraft maintenance publications are those developed by the *Air Transport Association* (ATA) and *General Aviation Manufacturers Association* (GAMA).

The ATA represents airlines, aircraft manufacturers, and various system manufacturers in an effort to ensure uniformity in many facets of the aviation industry and has helped gain standardization for reference materials and component locations. Although the ATA has its largest impact on transport category aircraft, many of the standards also apply to general aviation type aircraft.

GAMA has developed a standard for the organization of aircraft publications for general aviation aircraft similar to, and compatible, with ATA's standard. However, the GAMA version is more flexible to allow for the relatively small documents used by light, single-engine aircraft. The manufacturers of general aviation aircraft may choose between the GAMA or ATA standards. Many manufacturers of corporate type aircraft, such as the Gulfstream 550 or Cessna Citation, have chosen the ATA's standard even though they are actually general aviation aircraft. No matter which standard is chosen for the organization of technical publications, it is critical the technician become familiar with it as it will greatly reduce the time needed to find important technical information.

ATA iSpec 2200

The ATA's current standard for the organization of technical publications is *ATA iSpec 2200: Information Standards for Aviation Maintenance*. Established in May 2000, it contains the former ATA 100 and ATA 2100 specifications. Revised as needed, ATA iSpec 2200 provides standards for aircraft technical publications, such as maintenance manuals, minimum equipment lists, parts manuals, wiring diagrams, and computer based diagnostics, whether in paper or electronic formats. As noted above, ATA iSpec 2200 provides consistency between various manufacturers, suppliers, and airlines for the organization and display of aircraft publications. Therefore, technicians who work on a variety of aircraft, whether a Boeing B-777, Airbus A-380, or a Gulfstream G-550, need only become familiar with one system for the organization of publications and technical data.

The following is a brief overview discussing the organization of the more common technical publications an avionics technician will encounter in his or her career. Manuals will, in most cases, contain an introduction section, sometimes referred to as *Front Matter*, that will explain the manual's organization. This introductory section will also provide specific information that is pertinent to that manual and/ or aircraft model. For example, Sikorsky has incorporated the Illustrated Parts Catalog (IPC) into the aircraft maintenance manual on its S-92 models, but not in S-76C++ model manuals. For this reason, reading the introduction or front matter material is critical to setting the foundation for good troubleshooting.

Aircraft Manuals

Manuals, either paper or electronic, are designed to communicate information regarding the installation, operation, and maintenance of aircraft and related systems. The manuals most commonly used during avionics/ electronics troubleshooting and repair are the maintenance manuals, wiring diagram manuals, and parts manuals. Avionics technicians often use installation manuals, supplied by the component manufacturer/vendor, during initial system installation as well. In some cases, technicians will use the pilot's operation guide or the operator's manual during system inspection and/or troubleshooting.

The complexity of the manual system is usually a function of the type of aircraft. Light aircraft may contain only one, relatively simple, maintenance manual that also contains all wiring diagrams. Light aircraft typically use a separate parts manual or catalog. Large and/or complex aircraft typically employ a series of manuals, organized according to one of the previously discussed standards. In this text, ATA iSpec 2200 will be emphasized.

It should be noted that in this digital age the term manual still applies; however, the traditional paper manuals in many cases have now been replaced with computer based electronic formats. Whether speaking of paper or digital formats, this text will use the generic term manuals when discussing aircraft technical data.

Organization of chapters. As in its predecessor ATA 100, ATA iSpec 2200 separates the various aircraft systems into chapters. The current chapter and section numbers of the ATA iSpec 2200 are included in Appendix A of this text. Chapters consist of three two-digit numbers, for example 31-40-01. As shown Figure 1-2-1, the first two digits of the number designate the chapter, or specific aircraft system. For example, ATA Chapter 23-00-00 will contain information on the communications system. If the second and third sets of digits contain zeros, then that section contains information general to the overall system.

The second two digits represent a section within the given chapter or subsystem. ATA Chapter 23-20-00 represents materials covered in the communications chapter (23), VHF/UHF section (20). If the third set of digits contains zeros, then the information contained in this section is general to the overall subsystem (VHF/UHF). The first three digits of the chapter coding are specified within ATA iSpec 2200 and cannot be changed by the manufacturer or airline.

The last three digits of the code are typically assigned by the manufacturer or airline and bring further detail to the chapter. In Figure 1-2-1, ATA Chapter 23-21-01 designates the communications system, VHF radio, system number 1.

It should be noted that all numbers of a given chapter may not be used by a given manufacturer. In fact, ATA iSpec 2200 does not use all possible combinations for the first three digits of the code. This allows for flexibility in the standard and for future expansion of the systems.

First 2 Digits (Chapters)	Second 2 Digits (Section)	Third 2 Digits (Subject)	
23	00	00	Chapter digits 23 designated. All material under communications. (Chapter 23) -Assigned by ATA
23	20	00	The first section digit (2) designates all material under communication VHF or UHF. -Assigned by ATA
23	21	00	The second section digit (1) designates VHF systems only. -Assigned by manufacturer
23	21	01	Subject 1st and 2nd digit detail specific units within the designated chapter and section. (VHF com radio System 1) -Assigned by manufacturer

Figure 1-2-1. Visual representation of ATA iSpec 2200 breakdown of chapter designations

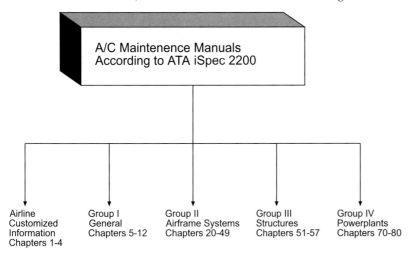

Figure 1-2-2. The five main sections of a maintenance manual are Customized Information, General Information, Airframe Systems, Structures and Powerplants.

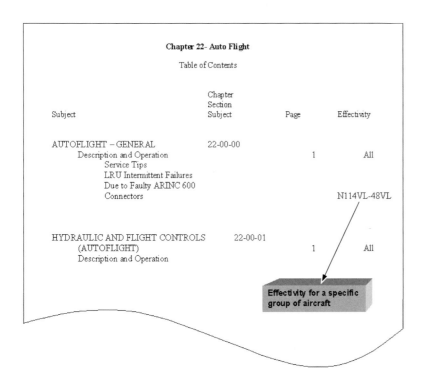

Figure 1-2-3. Effectivity column in a maintenance manual chapter index

Chapters can be further categorized into five main sections:

- Customized Information (Chapters 1-4), providing customer specific information

- General (Chapters 5-12), providing information and procedures for taxiing, towing, and servicing the aircraft

- Airframe Systems (Chapters 20-49), providing information and procedures on the such aircraft systems electrical, mechanical, pneumatic, and hydraulics

PAGE BLOCK TOPIC DESCRIPTION	
Page Block	Included Topic(s)
001-099	Description and Operation (D/O)
101-199	Troubleshooting (T) *Note: includes Electrical Schematics*
201-299	Maintenance Practices (MP)
301-399	Servicing (S)
401-499	Removal and Installation (R/I)
501-599	Adjustment and Test (A/T)
601-699	Inspection and Check (I/C)
701-799	Cleaning and Painting (C/P)
801-899	Approved Repairs (AR)

Table 1-2-1. Page block topic descriptions

- Structures (Chapters 51-57), providing information and procedures related to the wings, doors, fuselage, etc.

- Powerplant (Chapters 70-80), providing information and procedures for all engine related systems

Page blocks. Page blocks further categorize a chapter-section-subject of the maintenance manual. The term *page block* refers to a set of pages, such as 401-499. Each page block is consistent throughout the manuals and covers a given group of topics. The topics covered by the various page blocks are shown in Table 1-2-1. Within the manual, the actual page number is always comprised of three digits, such as 401. The page block number is located in the Table of Contents at the beginning of each chapter in the manual. See Figure 1-2-6.

The first page within a given block is always page XX1. For example, the first page under the topic Inspection/Check would be page number 601; the second page would be 602, etc. If there are more than 99 pages in any given block the next page will be numbered beginning with the letter A. For example, a sequence might be 601, 602... 698, 699, A600, A601, A602, etc.

List of effective pages. Each manual has a List of Effective Pages, which is a list of dates the pages became effective. The effective date also appears on the specific page. The dates should match to make sure the most up-to-date material is used.

Effectivity. Effectivity should not be confused with the *List of Effective Pages*. Effectivity is used to determine if the described maintenance is applicable to the aircraft. The aircraft serial number is most often used to determine effectivity, however as in Figure 1-2-3, the Description and Operation subsection service tip: LRU intermittent failures due to faulty ARINC 600 connectors applies only to aircraft with registration number N141VL to N148VL. The general information under the section Description and Operation applies to all aircraft covered by this maintenance manual.

Effectivity can be listed prior to the maintenance instructions, at the bottom left corner, or within the maintenance instructions. Read the introductory section of the manual to determine how effectivity is presented.

Effectivity can also be stated according to specific type of equipment installed on the aircraft, or specific aircraft serial, or registration number.

If the word *ALL* is listed under effectivity it designates that the corresponding pages apply to all the aircraft covered by that manual. A list of all aircraft arranged by serial number and/or registration number is included in the beginning of each manual.

Effectivity can also be designated using a flag note in diagrams or schematics. As shown in Figure 1-2-4, the flag is placed on the diagram in the effected area and the note is located in the lower left section of the page. In this example, flag number two denotes this configuration is for aircraft with serial numbers PA 261 to 277.

If the effectivity change is too large to contain in a simple note, the diagram or schematic will be labeled with a configuration number in the lower right corner of the page as shown in Figure 1-2-5. A configuration number designates a first, second, third, etc., page of the same diagram which has been modified for a given configuration of aircraft. This type of effectivity change occurs only in limited situations.

Cautions, warnings, and notes. Whenever using any manual pay particular attention to all cautions, warnings, and notes. A *warning* calls attention to any methods, materials, or procedures that must be followed to avoid injury or death. A *caution* calls attention to any methods, materials, or procedures that must be followed to avoid damage to equipment on the aircraft. A *note* is used to draw attention to a particular procedure that will make the task easier to perform.

Revisions. Manuals are revised as needed; always confirm that the most up-to-date version of the manual is used. Electronic updates are downloaded from the Internet or CD-ROMs are replaced, making updates simple. Paper manuals have to be manually updated and documented. Temporary revisions are used for items that require immediate release. In paper manuals, temporary revisions are printed on yellow paper and filed adjacent to the effected page. In electronic manuals, revisions are incorporated in the appropriate section of the

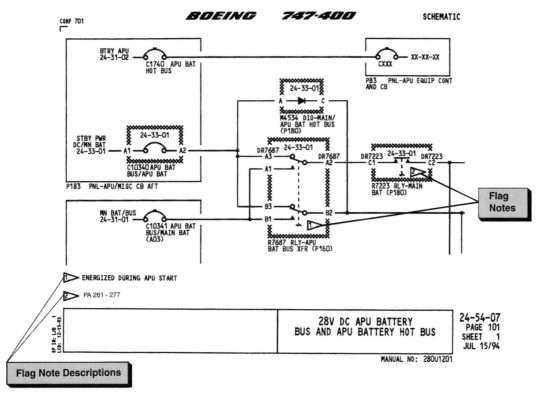

Figure 1-2-4. Flag notes on a schematic

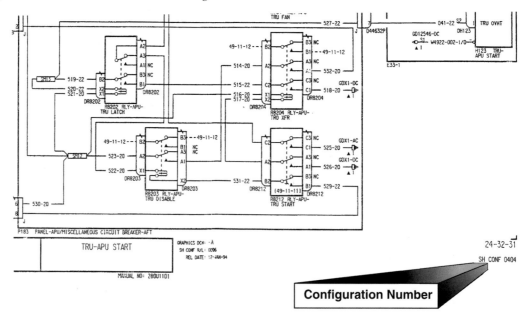

Figure 1-2-5. Schematic labeled for a specific configuration. Configuration number designates a page of the same diagram that has been modified

Courtesy of Boeing Commercial Airplane Company-for training purposes only

manual, for this reason always verify the current version and check for temporary revisions before beginning any work.

Having a thorough knowledge of how to use the manuals used in the maintenance and care of an aircraft is fundamental to troubleshooting aircraft systems. How manuals are organized has been reviewed, the following section will describe several specific manuals used in avionic/electronic troubleshooting.

Maintenance Manuals

Maintenance manuals provide information on installation, component location, maintenance, troubleshooting, testing, and inspection procedures. They provide specific information for flight line and hangar maintenance activities for both scheduled and unscheduled activities.

The description and operation section of a manual is often extremely helpful as an overview of the entire system under study. When in question about any system, start by reading the description and operation section. Typically, description and operation is the first subject for each section of a chapter (subject designator 00).

Since electronic systems are found on virtually every system of the aircraft, an electronic/avionics technician must become familiar with all sections of aircraft manuals. The airframe systems section (Chapters 20-49) likely contains a major portion of the electronics information. Chapter 20–Standard Practices

Airframe, often used by avionics technicians, contains important information, such as, wire splicing techniques, connector torques, and bundling information that is common to various areas of the aircraft. Other chapters containing important electrical system information include Chapter 22–Auto Flight, Chapter 23–Communications, Chapter 24–Electrical Power, Chapter 31–Indicating and Recording Systems, Chapter 34–Navigation, and Chapter 45–Central Maintenance Systems. The general group (Chapters 5-12) may also be of particular interest, since in many cases the aircraft are towed, taxied, or set into a particular configuration prior to maintenance.

The correct section of the maintenance manual can be found through the following procedures:

1. Identify the correct ATA chapter (example; Chapter 22–Auto Flight). If you are not familiar with the specific ATA chapter titles, a list is typically provided at the beginning of each manual.

2. Locate the table of contents for the chapter desired. The table of contents should be the first page(s) in that chapter.

3. Follow through the table of contents to find the specific system/procedure desired. It should be noted that sections are arranged numerically within the table of contents, and subjects are arranged alphabetically. For example, Autoflight - General (22-00-00) would be listed before Autoflight - Flight Director System (22-12-00).

4. If the correct chapter/section/subject number is already known, it is not necessary to reference the chapter table of contents. Simply go directly to the desired section within the manual.

An example of the Chapter 22 table of contents is shown in in Figure 1-2-6. Here it can be seen the top of the page shows the major heading (Chapter 22–Auto Flight) and the headings for each column. Column 1 (Subject) contains a description of the system. Column 2 (Chapter/Section/Subject) lists the ATA code. This ATA number is used to find the items in the maintenance manual. Column 3 (Page) and column 4 (Effectivity) contain more information which will further define the chapter.

Wiring Manuals

Wiring manuals are of particular importance to avionic technicians. Wiring system manuals have been developed containing diagrams, charts, lists, and all schematics needed to maintain the various electrical/electronic systems. For light aircraft, wiring diagrams and schematics may be contained in the maintenance manual; however, for complex aircraft wiring information is normally contained in a separate manual of one or more volumes.

ATA iSpec 2200 is the primary standard for both the transport category and the general aviation aircraft concerning the layout and design of wiring manuals. Figure 1-2-7 shows the general classifications of wiring manuals that are typically used by transport category aircraft. There are three broad categories of wiring manuals, schematics, diagrams, and lists. The various manuals may be produced as stand-alone publications or as part of other manuals.

Diagrams. Electrical diagrams are designed to provide an understanding of a system using line drawings and/or pictures of the various system components. There are two basic types of electrical diagrams used in modern aircraft manuals, block diagrams and wiring diagrams. *Block diagrams* define electrical circuits by separating a system or subsystem into functional blocks. Block diagrams typically show the various connections between subsystems divided into categories, such as, input and output connections, or analog and digital circuits. This type of configuration makes block diagrams especially useful in gaining an initial understanding of a given system. A typical block diagram is shown in Figure 1-2-8. *Wiring diagrams,* on the other hand, are typically very specific with details on wires, connectors, and pin numbers for a given system. Since wiring diagrams are so

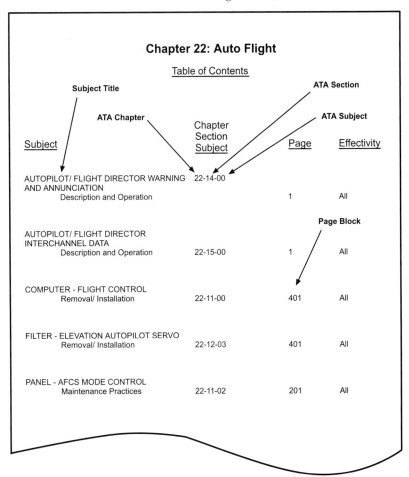

Figure 1-2-6. Example of chapter table of contents for a maintenance manual

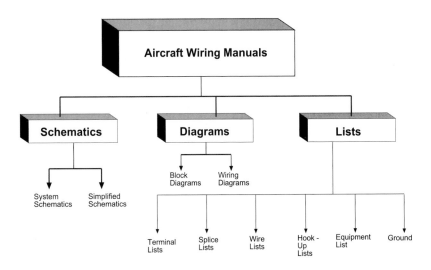

Figure 1-2-7. The general classification of wiring manuals typically used by transport category aircraft

specific, they typically have a limited scope, and in many cases, require several pages of wiring diagrams to detail an entire system or subsystem. Wiring diagrams are used during aircraft assembly, installation of systems, and detailed troubleshooting.

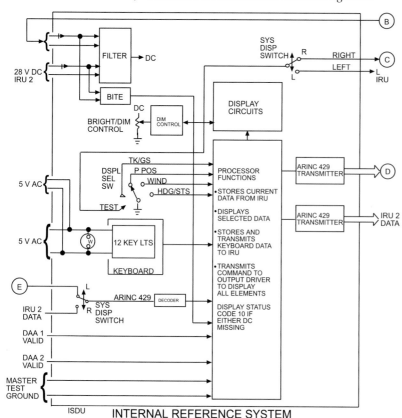

Figure 1-2-8. Typical system block diagram

Courtesy of the Boeing Commercial Airplane Company-for training purposes only

System schematics typically contain more depth than simplified schematics since they focus on a given system or subsystem. System schematics typically show all related components, connections, and wiring within a system. In many cases, system schematics will not show specific wire or connector numbers; that information is contained in the wiring diagrams.

The terms *schematic* and *wiring diagram* are often used interchangeably when referring to electrical/electronic systems. Technicians often use the term wiring diagram to refer to a schematic, and vice versa. Some manufacturers even seem to confuse the terms. For example, it is easy to find an aircraft system schematic in a wiring diagram manual. It is therefore important to become familiar with the publications for the specific aircraft maintained and the various types of wiring information provided.

Wiring manual numbering. Schematics and diagrams identified according to ATA iSpec 2200 have a code numbering system. The first four digits of the code are used to identify block diagrams and simplified system schematics. If a schematic has only four digits, it is covered again in more detail on another schematic. A fifth digit is added to identify all third level schematics. If the complete system is covered in one schematic, a -0 is used for the fifth digit. If the system requires two, three or more schematics, the fifth digit is -1, -2, -3, etc. For example, a simplified system schematic could be 31-22; a system schematic could be 31-22-01.

In many cases a given system or subsystem is too complex to include on one page (sheet) of the manual. The *sheet number* is used to designate the number of pages required for a given schematic or diagram. The number 33-22-00, Sheet 2 indicates the second page for the third level schematic where the entire system is shown on one schematic which requires two pages (sheets).

Schematics. Schematics are typically more general than wiring diagrams. Schematics are often used for identification of basic system components during troubleshooting. In some cases, there may be two levels of schematics provided by the aircraft manufacturer, simplified schematics and system schematics. *Simplified schematics* are wiring illustrations with intermediate depth and scope. These schematics typically show all components but do not detail all wiring. Simplified schematics are often used for elementary troubleshooting.

A320 AIRCRAFT WIRING LIST												
WIRE IDENTIFICATION					FROM TERMINATION A			FROM TERMINATION B			WIRING DIAGRAM MANUAL REFERENCE	EFFECT
NUMBER	COLOR IF NOT WHITE	TYPE/ GAUGE	LENGTH IN CM	ROUTE	LOCATION ZONE	FUNCTIONAL IDENTIFICATION NUMBER	CONNECTOR PIN NUMBER	LOCATION ZONE	FUNCTIONAL IDENTIFICATION NUMBER	CONNECTOR PIN NUMBER		
2373-1705		CF 20	215	1M	223	340RH1	5	223	2749VT	20G	23-73-52	ALL
2373-1706		CF 24	310	1M	223	300RH1	4	223	340RH1	4	23-73-52	ALL
2373-1707		CF 24	310	1M	223	300RH1	3	223	340RH1	3	23-73-52	ALL
2373-1708		CF 24	310	1M	223	300RH1	2	223	340RH1	2	23-73-52	ALL
2373-1709		CF 24	310	1M	223	300RH1	1	223	340RH1	1	23-73-52	ALL
2373-1710	B	PF 24		1M	223	300RH1	50	221	320RH1	2	23-73-54	ALL
2373-1710	R	PF 24		1M	223	300RH1	49	221	320RH1	1	23-73-54	ALL
2373-1712		CF 20	115	1M	221	320RH1	3		6681VN	B1	23-73-54	ALL
2373-1713		CF 24	15	1M			20	223	300RH1	20	23-73-54	ALL
2373-1714	B	PF 24		1M	223	300RH1	5	221	2791VC	P	23-73-53	ALL
2373-1714	R	PF 24		1M	223	300RH1	4	221	2791VC	N	23-73-53	ALL
2373-1716		CF 24	15	1M			3	223	300RH1	3	23-73-53	ALL
2373-1717	B	PF 24		1M	223	300RH1	7	221	2791VC	V	23-73-53	ALL
2373-1717	R	PF 24		1M	223	300RH1	6	221	2791VC	U	23-73-53	ALL
2373-1722		CF 24	150	1M	223	300RH1	45	223	2749VT	5E	23-73-50	ALL
2373-1723		CF 24	150	1M	223	300RH1	11	223	2749VT	16K	23-73-50	ALL

Figure 1-2-9. Typical master wire list

Courtesy of Airbus S.A.S.-for training purposes only

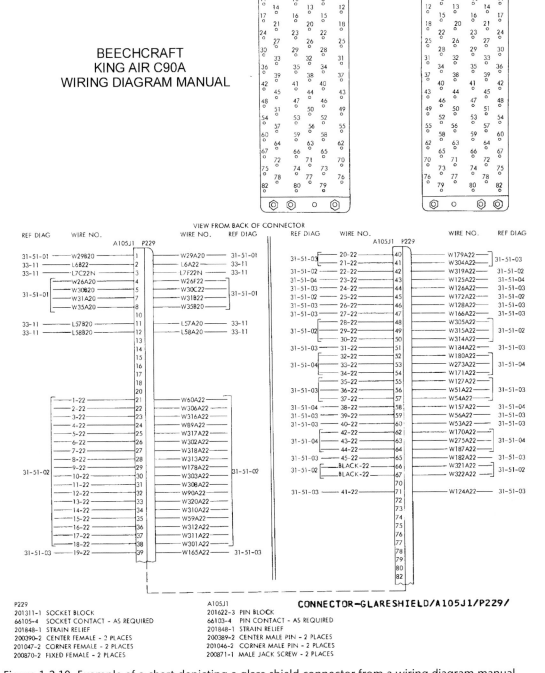

Figure 1-2-10. Example of a chart depicting a glare shield connector from a wiring diagram manual

Courtesy of Hawker Beechcraft Corporation-for training purposes only

In some cases, schematics are numbered corresponding to the maintenance manual page blocks. Schematics will always be in page block 101-199 (troubleshooting). Wiring diagrams are labeled with page numbers 1, 2, 3, etc. according to the number of pages (not sheets) for a given system or subsystem.

Lists and charts. A series of Aircraft Wiring Lists (AWL) contain very specific information for a given circuit. The Equipment list, Hook-up list, and Master Wire lists are all part of the air-craft wiring lists. The Equipment list contains information on various electrical/electronic components and references the appropriate aircraft schematic or diagram. The Hook-up list contains information on specific wire connections to plugs and receptacles, terminal blocks, splices, and ground points. The Master Wire list contains detailed information on wire size, types, and lengths. In some cases a separate splice list, ground list, and terminal list are included in the manuals to show further detail of the system.

2. Dead—end Shielding of Coaxial Cable, Single or Multiconductor Shielded Cables

A. Remove the outer jacket over the shield braid. Use extreme caution to avoid damage to shielding and inner conductors.

NOTE: A Reon R—720 cable jacket removal tool can be used on BMS 13—51 cable.

B. Fold the shielding back over the cable jacket 1/4 to 3/8 inch. See Figure 3.

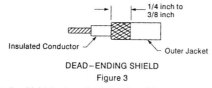

DEAD—ENDING SHIELD
Figure 3

C. Insulate the shield dead—end with Raychem RT—876 heat shrinkable sleeve. Select proper size sleeve to fit diameter of wire. See Subject 20—10—14.

As an option, insulate the shield dead—end with tape, using tape wrap procedures of Subject 20—30—12.

20—10—15 Page 4
MAR 01/94

Figure 1-2-11. Portion of a standard practices chapter of a wiring diagram manual   *Courtesy of the Boeing Commercial Airplane Company-for training purposes only*

A master wire list from an Airbus A-320 is shown in Figure 1-2-9. The heading on each wire list identifies the meaning of each column. The three major categories are Wire Identification, From Termination A, and From Termination B. The *wire identification* includes information about the wire number, color, manufacturer, length, and routing. The *From/To* termination identifies the location of each end of the wire, electrical component to which the wire is connected, the specific connector, and the pin for that wire. The right most columns show the wire diagram ATA number and effectivity for each wire. For the A-320, the wire number identifies the ATA chapter and section of the system for that wire. In 1-2-9, notice the first four digits of the wire number (left hand column) are identical to the ATA chapter and section (right hand column). Each of the wires on this list begins with 2373; indicating ATA Chapter 23–Communications, section 73–cabin inter-communications system.

Charts are used to depict specific component or wire locations, in the aircraft or on a sub-system, such as pins or sockets of a connector plug. An example of a chart depicting a glare shield connector is shown in Figure 1-2-10. Charts are contained in Chapter 91 of the wiring manuals.

Standard Practices

The procedures and practices used repeatedly during aircraft maintenance, troubleshooting, and repair are contained in the standard practices chapter of each manual. According to the ATA iSpec 2200, standard practices are located in Chapter 20. Chapter 20 of the maintenance manual contains procedures and practices that are common to a variety of maintenance operations. This chapter also contains practices that are common to a variety of wiring, electrical, and electronic systems.

Other pertinent information includes electrical bonding practices, soldering of electrical connectors, installation of undersize wire insulation sleeves, proper torquing of electrical connectors, and cleaning of electrical connections. Figure 1-2-11 shows an excerpt from a wiring diagram manual, standard practices chapter. Remember Chapter 20–Standard Practices contains common information. When a procedure cannot be found in other sections of the manual, be sure to refer to Chapter 20. The procedure you are looking for may be a standard practice.

Using wiring manuals. Many light aircraft wiring diagrams and schematics are often contained in the aircraft's maintenance manual. Larger aircraft wiring diagrams and schematics are contained in separate manuals, which are arranged by ATA chapter. A wiring diagram from a Beechcraft King Air is shown in Figure 1-2-12A. It should be noted that adjacent to each wiring diagram is the associated parts list as shown in Figure 1-2-12B. This system allows for quick identification of defective components. The King Air Wiring Diagrams Manual also contains an alpha-numeric index. This index allows the technician to identify a given wiring diagram using a specific part number. To find a given wiring diagram for this aircraft, the systems can be referenced using an appropriate part number or ATA code.

For transport category aircraft, the correct wiring diagram or schematic is often found using a given ATA code and chapter index. In some cases, specific schematics or diagrams are referenced by the maintenance manual. Wire numbers and part numbers can also be used to find their related schematic. The master wire list or equipment list could also be used to find the appropriate diagram. Figure 1-2-13A shows a portion of the wire list for a Boeing 747-400. As shown in the list, wire W880-M555-18 is located on diagram 38-32-11. Diagram 38-32-11 containing wire W880-M555-18 is shown in Figure 1-2-13B.

The wiring diagram manuals for Airbus S.A.S. aircraft are divided into three categories: Aircraft Schematic Manuals (ASM), Aircraft Wiring Manuals (AWM), and Aircraft Wiring Lists (AWL). The ASM contains simplified and moderately complex illustrations. The AWM contains the detailed electrical/electronic diagrams. By knowing a system function the correct AWM or ASM can be easily located using the alphabetical index. This is represented by the top left block, Access by Function, of Figure 1-2-14. The top right

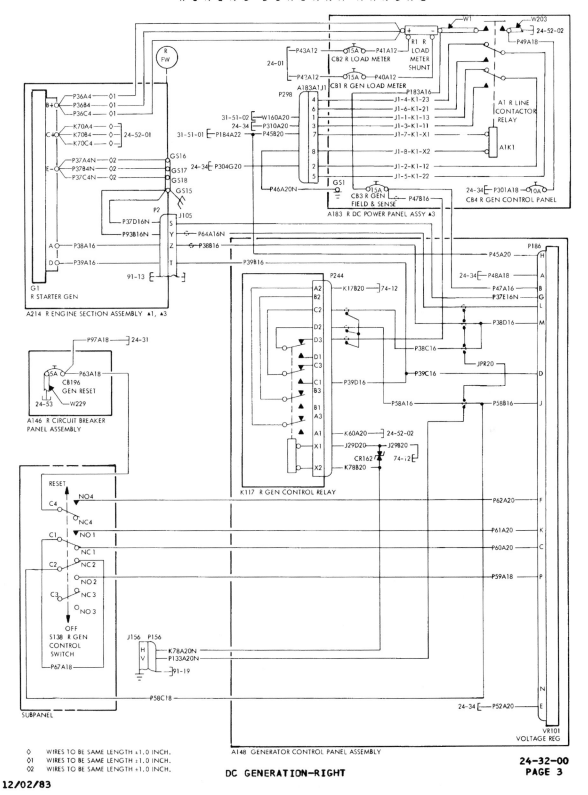

Figure 1-2-12A. Wiring diagram of DC generation circuit from a Beechcraft King Air

```
                        B E E C H C R A F T
                        K I N G   A I R   C 9 0 A
                    W I R I N G   D I A G R A M   M A N U A L
```

GAMA/ATA CODE & REF DES	PART NO.	DESCRIPTION 1 2 3 4 5 6 7	UNITS PER ASSY	U INSTL ZONE
32-00-A146		PANEL ASSY,RH CIRCUIT BREAKER/SEE CHAPTER 24-53	1	246
-CB196	7277-2-5	CIRCUIT BREAKER,GENERATOR RESET	1	246
-A146W229		. BUS BAR/SEE CHAPTER 24-53/.	1	246
-A148	109-364017-13	PANEL ASSY,GENERATOR CONTROL.	1	153
-CB162	100-361045-1	. TRANSZORB	1	153
-A148K117	K-D4A	. RELAY,RH GENERATOR CONTROL.	1	153
-A148P186	MS27473T20A16S	. PLUG,RH VOLTAGE REGULATOR	1	153
-A148P186	MS27506A20-2	. CLAMP,STRAIN RELIEF	1	153
-A148P244	S01048-8308	. PLUG,RH GENERATOR CONTROL RELAY	1	153
-VR101	101-364270-5	. REGULATOR,RH VOLTAGE.	1	153
-A183	90-364151-1	PANEL ASSY,RH DC POWER.	1	621
-A183A1	109-361030-9	. RELAY ASSY,RH LINE CONTACTOR.	1	621
-A183A1J1	211068-1	. . RECEPTACLE,RH LINE CONTACTOR RELAY.	1	621
-A183A1J1	66099-4	. . PIN,CONTACT	AR	621
-A183A1K1	K-D4A	. . RELAY,LH LINE CONTACTOR	1	621
-A183CB1	7277-2-15	. CIRCUIT BREAKER,RH GENERATOR LOAD METER	1	621
-A183CB2	7277-2-15	. CIRCUIT BREAKER,RH LOAD METER	1	621
-A183CB3	7277-2-15	. CIRCUIT BREAKER,RH GENERATOR FIELD AND SENSE. .	1	621
-A183CB4	7277-2-10	. CIRCUIT BREAKER,RH GENERATOR CONTROL PANEL. . .	1	621
-A183GS1	131270-3	. GROUND STUD	1	621
-A183GS1	AN960-10L	. WASHER.	1	621
-A183GS1	MS35338-43	. LOCKWASHER.	1	621
-A183GS1	MS21042L3	. NUT	1	621
-A183P298	208678-1	. PLUG,RH LINE CONTACTOR RELAY.	1	621
-A183P298	66105-4	. SOCKET,CONTACT.	AR	621
-A183P298	MS3187-16-2	. PLUG,SEALING.	AR	621
-A183R1	100-380007-3	. SHUNT,RH LOAD METER	1	621
-A183W1	101-364046-56	. BUS BAR	1	621
-A183W203	101-364046-111	. BUS BAR	1	621
-A214		ENGINE SECTION ASSY,RH.	1	420
-A214G1	50-369122-21	GENERATOR ASSY,RH STARTER	1	420
-A214G1	50-369122-13	GENERATOR ASSY,RH STARTER	1	420
-A214GS15		GROUND STUD/CONSISTS OF THE FOLLOWING/.	1	420
-A214GS15	MS35207-264	SCREW	1	420
-A214GS15	MS35338-43	LOCKWASHER.	2	420
-A214GS15	MS25082-3	NUT	1	420
-A214GS15	AN960-10L	WASHER.	4	420
-A214GS15	MS21042L3	NUT	1	420
-A214GS16		GROUND STUD/TYPICAL A214GS16,A214GS17 AND . . . A214GS18/CONSISTS OF THE FOLLOWING/	3	420
-A214GS16	AN5-7A	BOLT.	1	420
-A214GS16	MS35338-45	LOCKWASHER.	1	420
-A214GS16	AN960-516	WASHER.	1	420
-A214GS16	MS21042L5	NUT	1	420
-A214P2		PLUG,RH FIREWALL NO 2/SEE CHAPTER 91/	1	420
-J105		RECEPTACLE,RH FIREWALL NO 2/SEE CHAPTER 91/ . . .	1	621
-J/P156		CONNECTOR,RH CABIN GROUNDING/SEE CHAPTER 91/. . .	1	
-S138	13AT10S	SWITCH,RH GENERATOR CONTROL	1	245

```
        1   UNITED KINGDOM ONLY      CODES OF EFFECTIVITY
```

Figure 1-2-12B. Associated parts list for the DC generator of a Beechcraft King Air

Courtesy of Hawker Beechcraft Corporation -for training purposes only

block, Access by Wire Number, can be used if the correct wire number is known. In this approach the correct AWM or ASM can be located through the master wire list (Chapter 91). As shown in the center block of Figure 1-2-14, the FIN (described in the next paragraph) can be used to reference the correct AWM or ASM page(s). This is typically the fastest way to find a given wiring diagram.

Functional item numbers. The Functional Item Number (FIN) is a unique number given to each line replaceable unit on modern Airbus aircraft. It should also be noted that other aircraft manufacturers employ unique identification systems similar to the Airbus FIN, although the FIN will be the only system presented in this text. The FIN is located on a placard next

BOEING 747
WIRING DIAGRAM MANUAL

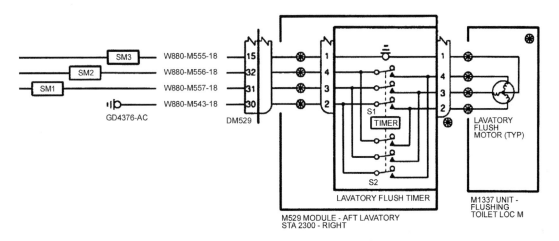

R E V	M O D	BUNDLE/WIRE/GA/CO	TY	FA	FT	/IN	DIAGRAM	EQUIP	TERM	TT	SP	EQUIP	TERM	TT	SP	EFFECTIVITY
		61B40880–WIRE BUNDLE–AFT LAVATORIES M AND N														
	W880–M	0555–18	UA	010	00		38–32–11	DM0529	15			SM00003	S			ALL
	–M	0556–18	UA	010	00		38–32–11	DM0529	32			SM00002	S			ALL
	–M	0557–18	UA	010	00		38–32–11	DM0529	31			SM00001	S			ALL
	–M	1021–22	UA	011	00		38–12–11	DM0528	4			SP04496	S			ALL
	–M	1022–22	UA	011	00		37–12–11	DM0528	18			SP04498	S			ALL
	–M	1023–22	UA	011	00		32–12–11	DM0528	19			SP04500	S			ALL
	–M	1038–20	UA	012	00		38–12–11	DM0528	8			GD00616	A..E			ALL

BUNDLE PART NUMBER AND TITLE
– – – – – – – – – – – – – – – – – – – LENGTH

W880– M
0555– 18

Figure 1-2-13A. Example of a typical wire list for a Boeing 747-400

Courtesy of the Boeing Commercial Airplane Company.

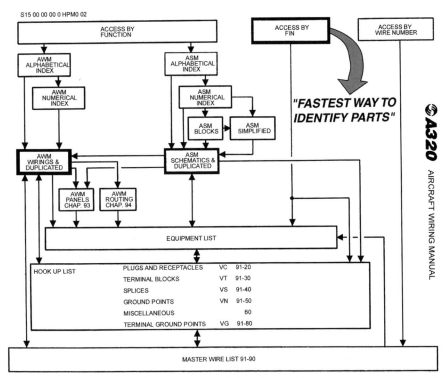

Figure 1-2-13B. A portion of the wiring diagram #38-2-11–in this figure the wire W880-M555-18 can easily be found

Courtesy of the Boeing Commercial Airplane Company -For training purposes only

Figure 1-2-14. Flow chart used to help access manuals for the A-320

Courtesy of Airbus S.A.S.

Figure 1-2-15. Example of a Functional Identification Number (FIN)

Courtesy of Airbus S.A.S.

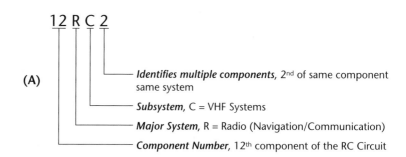

(A)

Identifies multiple components, 2nd of same component same system

Subsystem, C = VHF Systems

Major System, R = Radio (Navigation/Communication)

Component Number, 12th component of the RC Circuit

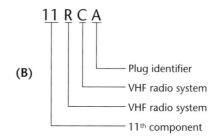

(B)

Plug identifier

VHF radio system

VHF radio system

11th component

Figure 1-2-16. Description of a typical FIN. (A) VHF radio component, (B) connector plug for the VHF radio system

ALPHABETICAL CODE FOR AIRBUS FUNCTIONAL ITEM NUMBERS	
System Identification Code (Major Category)	System Function
C..................................	FLIGHT CONTROL SYSTEMS
D..................................	DE-ICING
E..................................	ENGINE MONITORIING
F..................................	FLIGHT INSTRUMENTS
G..................................	LANDING GEAR HYDRAULIC
H..................................	AIR CONDITIONING
J..................................	IGNITION
K..................................	ENGINE CONTROL AND STARTING
L..................................	LIGHTING
M..................................	INTERIOR ARRANGEMENT
P..................................	DC POWER SUPPLY DISTRIBUTION
Q..................................	FUEL
RÖ..................................	RADIO (NAVIGATION AND COMMUNICATION)
SÖ..................................	RADAR NAVIGATION
TÖ..................................	SPECIAL ELECTRONICS
VÖ..................................	FICTICIOUS CIRCUITS
W..................................	FIRE PROTECTION AND WARNING
XÖ..................................	AC GENERATION AND DISTRIBUTION

Table 1-2-2. Alphabetical code for Airbus FINs   *Courtesy of Airbus S.A.S.*

to each component as seen in Figure 1-2-15. The FIN consists of a two letter code which defines the system or circuit of a specific Line Replaceable Unit (LRU). A numerical prefix or suffix is added to the code which gives each LRU a specific Functional Item Number. An example of a typical FIN is 12RC2.

There are 18 major categories of FIN alphabetical identifiers as shown in Table 1-2-2. The complete two letter code system is listed in the Appendix B of this text. Figure 1-2-16 shows a breakdown of a typical FIN. If the FIN represents a connector plug the suffix becomes a letter. The letter *A* designates the first plug in that system the letter *B* the second plug and so on.

The Airbus Functional Item Number is used to identify components throughout the entire series of aircraft manuals. In most cases, the FIN is the fastest way to identify a given manual page or to locate a given part number of an LRU. As seen in Figure 1-2-17, the FIN appears just below each LRU on a wiring diagram or schematic. In the center portion of this diagram, it can be seen that the flight control/ ELAC 2 push button switch has a FIN of 6CE2.

Wire identification systems. All aircraft wires that are three inches or longer must be labeled with a wire identification number. This number is typically comprised of an alphanumeric code based on the military standard MIL-W-5088. Figure 1-2-18 shows a typical wire code system. The wire code con-

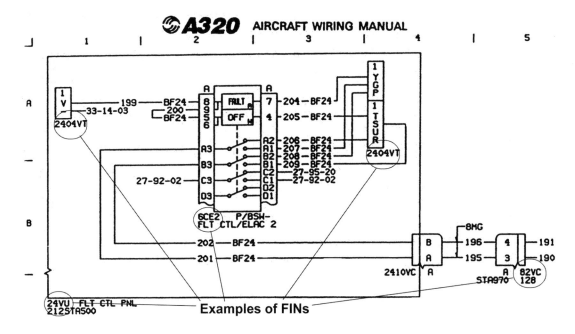

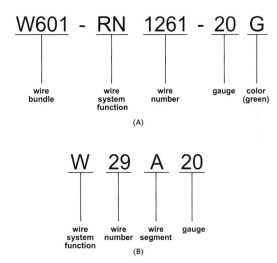

Figure 1-2-17. Example of various FINs

Courtesy of Airbus S.A.S.-for training purposes only

sists of a wire bundle number, circuit function letter(s), a wire number, segment letter, wire gauge, and color. It should be noted that wire bundle numbers are typically used on large aircraft only; segment letters are typically found on light aircraft. On some aircraft, portions of the wire number also correspond to the ATA chapter and section number for the system where the wire is installed. Each manufacturer has a specific wire code system. Be sure to read the introduction of the wiring manuals to determine the specific code for the aircraft.

Sensitive wires are considered critical to flight safety and must not be modified without specific manufacturer's approval. Sensitive wires are typically used for primary flight controls and other crucial circuits found on fly-by-wire computerized aircraft. Sensitive wires are always identified in the wire code and are often made with a distinct color marking. On the A-320 for example, an *S* at the end of the wire number (2792-1568-CF22-S) designates a sensitive wire. On this aircraft, there is a pink band located at every end of a sensitive wire. It should be noted, any work performed on sensitive wires (including a simple disconnect-reconnect of a connector) typically requires authorization of an inspector.

There are several different wire types used on modern complex aircraft. Two common specialty wires include: shielded cable and data bus cable. Wires also vary with different insulation or conductor types. Be sure to choose the correct type of wire during both installation and replacement. Typically wire type is defined

Figure 1-2-18. Samples of wire code systems for: (A) Complex aircraft, (B) Light aircraft

in the wire lists section of the aircraft manuals. Some companies identify the wire type and gauge on the wiring diagrams.

Home diagrams. Most aircraft components interact with other systems or subsystems throughout the aircraft. In many cases, a diagram or schematic must show these interrelations to explain the component in question. The home diagram of a component is the diagram where the component is shown in full detail. The home diagram will show all electrical connections and wires for that unit. A component is considered home only when it is shown in the ATA chapter, section and subject which correspond to that component. Components in there home diagram are

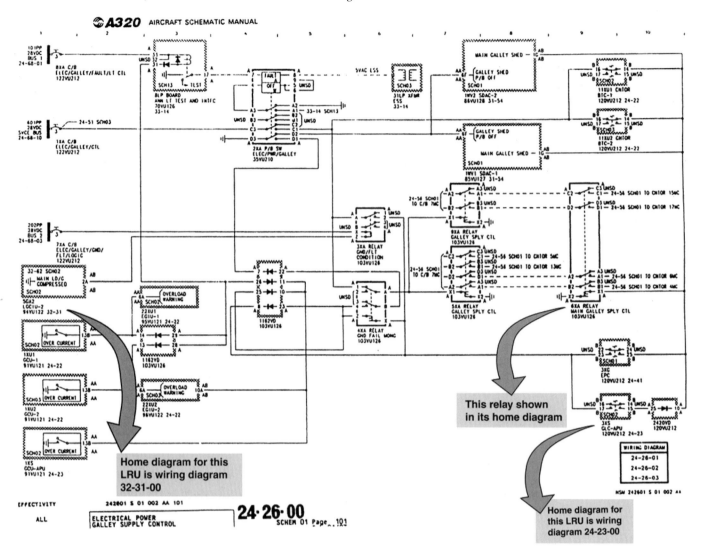

Figure 1-2-19. Electrical power galley supply control—note that this is a home diagram for relay 6XA and the home diagram for 3XS is located in wiring diagram 24-2

Courtesy of Airbus S.A.S.-for training purposes only

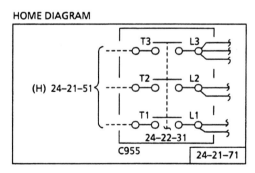

Figure 1-2-20. Example of relay diagram showing home diagram of wires

Courtesy of the Boeing Commercial Airplane Company -For training purposes only

enclosed in a solid bold line. In Figure 1-2-19, the 6XA relay, main galley supply control, (located on the right side of the schematic) is in its home diagram.

In most cases, a component which is not in its home diagram will not be shown in com-

plete detail. Components not in their home diagram are enclosed by a cross-hatched line. Adjacent to all components not in their home diagram is the ATA chapter where they can be found in detail (i.e., their home diagram). In Figure 1-2-19, the LGCIU-2 (located on the left side of the schematic) is not in its home diagram. The corresponding home diagram for the LGCIU-2 is located on schematic 32-31-00, page 101, as indicated on the bottom left side of the figure.

Individual wires may also be assigned a home diagram or schematic. If a wire is represented as a dotted line it is not in its home diagram. If a wire is not in its home diagram it will likely contain an (H) preceding the ATA chapter of the wire's home diagram. Figure 1-2-20 shows a relay where the wires from T1, T2, T3, are not located in their home diagram. Diagram number 24-21-51 must be referenced to find more detail concerning the wires to T1, T2, & T3.

Other Manuals

Most manufacturers of transport category aircraft offer a variety of specific manuals designed to be used for certain maintenance situations. The *Ramp Maintenance Manual* (RMM) is an abbreviated maintenance manual designed specifically for line maintenance and minor troubleshooting. The RMM contains most of the information a technician would need for minor repairs between flights. If additional information is needed, the RMM contains a cross references to other more specific manuals for that aircraft. The RMM often contains simplified electrical schematics and a reference to all wiring diagrams for each system.

Aircraft troubleshooting manuals (TSM) are designed specifically for system troubleshooting. These manuals contain block diagrams, electrical schematics, as well as, functional diagrams of mechanical, hydraulic, and pneumatic systems. The TSM is an excellent place to start when troubleshooting systems. The *Fault Isolation Manual* (FIM) is a manual that contains various repair strategies for given faults in a flow chart format. The FIM is typically used in conjunction with central maintenance computer systems found on modern aircraft. The computer identifies faults using a given code and that code is referenced through the fault isolation manual.

Section 3

Component Location

Each aircraft manufacture has a *component location system* for identifying the location of various assemblies, subassemblies, and components on the aircraft. Using this reference system the technician can quickly locate a specific wire, component, or connector. Per ATA iSpec 2200 and GAMA Specification 2, information for the component location system is in Chapter 6–Dimensions & Areas.

Aircraft have three reference planes known as the datum, the waterline, and the butt line. The *datum* is an imaginary vertical plane and the *waterline* is an imaginary horizontal plane, both are shown in Figure 1-3-1. In Figure 1-3-1A the reference planes are located so all measurements have a positive value. In Figure 1-3-1B the reference planes are located so that both positive and negative numbers are used. Measurements to the left of the datum and below the water line are negative. The *butt line* (BL) is a vertical plane that divides the aircraft

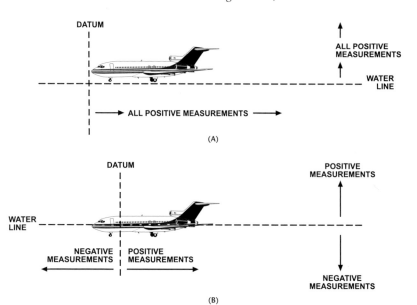

Figure 1-3-1. Examples of a datum and waterline: (A) All positive measurements, (B) Both positive and negative measurements

front to rear through the center of the fuselage Figure 1-3-3). Dimensions are typically given in inches; however, some foreign made aircraft use centimeters.

Stations

Fuselage stations (FS) are used to indicate locations longitudinally along the aircraft fuselage. Fuselage stations are measured from the datum reference plane. Fuselage station 00.00 is located at the datum. An example of fuselage stations for a typical corporate type aircraft are shown in Figure 1-3-2.

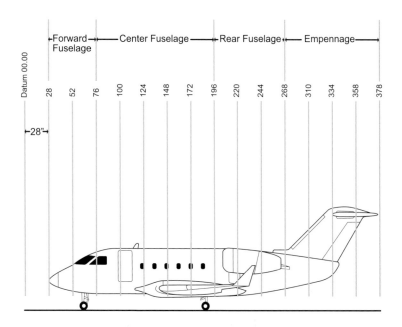

Figure 1-3-2. Example of fuselage stations on a corporate jet

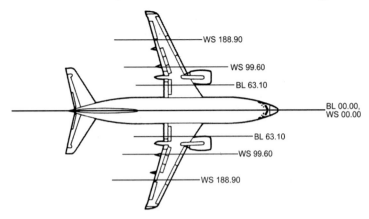

Figure 1-3-3. Butt lines and wing stations of a typical aircraft

Wing stations (WS) are used to indicate locations longitudinally along the wings of the aircraft. Wing station 00.00 is located in the center of the fuselage at the butt line. In Figure 1-3-3 the butt line numbers are used from station 00.00 to station 99.60. Outboard of station 99.60 the measurements are referred to as wing stations. On most aircraft WS numbers are used in wing areas and BL numbers indicate fuselage stations left and right of station 00.00. On all aircraft both WS and BL measurements have the same reference point, butt line 00.00.

Other stations such as nacelle stations (NS) or elevator stations (ES) are often used on larger

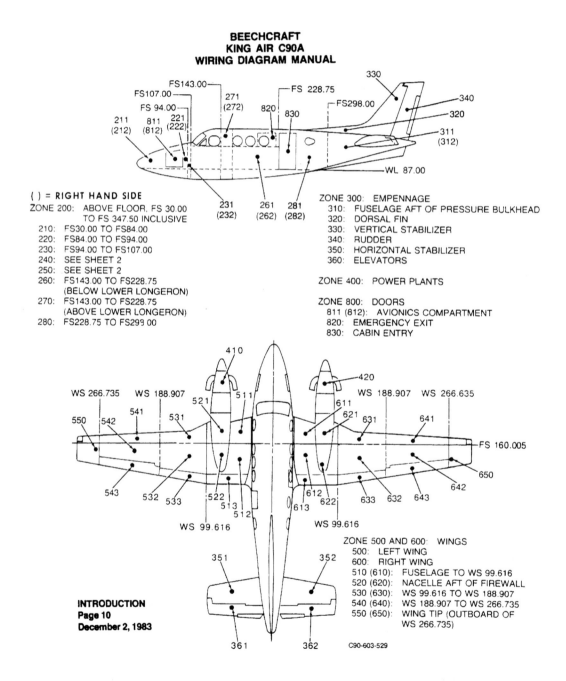

Figure 1-3-4. Zone locations for a typical aircraft

Courtesy of Hawker Beechcraft Corporation-for training purposes only

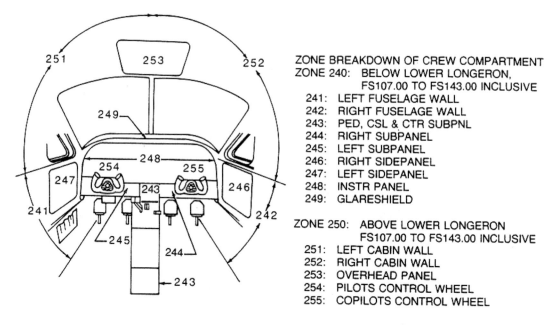

ZONE BREAKDOWN OF CREW COMPARTMENT
ZONE 240: BELOW LOWER LONGERON,
            FS107.00 TO FS143.00 INCLUSIVE
  241: LEFT FUSELAGE WALL
  242: RIGHT FUSELAGE WALL
  243: PED, CSL & CTR SUBPNL
  244: RIGHT SUBPANEL
  245: LEFT SUBPANEL
  246: RIGHT SIDEPANEL
  247: LEFT SIDEPANEL
  248: INSTR PANEL
  249: GLARESHIELD

ZONE 250: ABOVE LOWER LONGERON
            FS107.00 TO FS143.00 INCLUSIVE
  251: LEFT CABIN WALL
  252: RIGHT CABIN WALL
  253: OVERHEAD PANEL
  254: PILOTS CONTROL WHEEL
  255: COPILOTS CONTROL WHEEL

Figure 1-3-5. Zone locations of a typical crew compartment; odd numbers are on the left; even numbers are on the right

Courtesy of Hawker Beechcraft Corporation

aircraft to help technicians find components in specific areas of the aircraft. It should be noted that for any station on the aircraft the 00.00 reference will always be taken from one of the three main reference planes: the datum, the waterline, or the butt line.

Zones

A system of zoning, designed by the ATA, further assists in identifying component location on large aircraft. Figure 1-3-4 shows an example of the zone system used on a Beechcraft King Air C90A. Here it can be seen that most zone numbers on the left side of the aircraft begin with odd numbers and zones on the right begin with even numbers. The right/left side is determined when sitting inside the aircraft facing forward. For example, in fixed-wing aircraft the pilot sits in the left seat; the copilot sits in the right seat.

The zoning system is based on a three-digit code. The first digit is the major zone designator. The second nonzero digit is the subzone, and the third nonzero digit is a specific area within the subzone.

The major zone areas are:

100 Lower half of the fuselage to the rear of the pressure bulkhead (below the main cabin)

200 Upper half of fuselage to the rear of the pressure bulkhead

300 Empennage, including aft of the rear pressure bulkhead

400 Powerplants and struts or pylons

500 Left wing

600 Right wing

700 Landing gear and landing gear doors

800 Doors

900 Reserved for uncommon differences between aircraft types not covered in zones 100-800

Subzones (the second digit of the code) are used to designate specific stations or given areas within a major zone. Figure 1-3-4 shows zones 530 and 630 as between wing stations 99.616 and 188.907. Figure 1-3-5 shows a breakdown of the crew compartment of the King Air C90A. In this example, all three digits of the zone code are used to specify location.

To understand component location systems used during maintenance and troubleshooting, study the following examples. Three different aircraft that represent a cross section of the industry will be examined—the Beechcraft King Air, Airbus A-320, and the Boeing 747. Troubleshooting and repair of these systems require access to different manuals and often requires considerable time to simply find the defective component or wire on the aircraft.

BEECHCRAFT
KING AIR C90A
WIRING DIAGRAM MANUAL

GAMA/ATA CODE & REF DES	PART NO.	DESCRIPTION 1 2 3 4 5 6 7	UNITS PER ASSY	INSTL ZONE	USABLE ON CODE
21-00-A146		PANEL ASSY,RH CIRCUIT BREAKER/SEE CHAPTER 24/ .	1	246	
-CB188	7277-2-2	. CIRCUIT BREAKER,LH TORQUE METER	1	246	
-CB189	7277-2-2	. CIRCUIT BREAKER,RH TORQUE METER	1	246	
-CB234	7277-2-5	. CIRCUIT BREAKER,NO.1 INVERTER POWER SELECT. .	1	246	
-CB236	7277-2-5	. CIRCUIT BREAKER,NO.2 INVERTER POWER SELECT. .	1	246	
-CB212	7277-2-5	CIRCUIT BREAKER,INVERTER NO.1,26 VAC POWER. . .	1	512	
-CB213	7277-2-10	CIRCUIT BREAKER,INVERTER NO.1,115 VAC POWER . .	1	512	
-CB214	7277-2-5	CIRCUIT BREAKER,INVERTER NO.2,26 VAC POWER. . .	1	612	
-CB215	7277-2-10	CIRCUIT BREAKER,INVERTER NO.2,115 VAC POWER . .	1	612	
-CB241	7277-2-5	CIRCUIT BREAKER,NO.2 INVERTER CONTROL	1	612	
-CB243	7277-2-5	CIRCUIT BREAKER,NO.1 INVERTER CONTROL	1	512	
-CB245	7277-2-7	CIRCUIT BREAKER,NO.1 INVERTER POWER SELECT. . .	1	512	
-CB246	7277-2-7	CIRCUIT BREAKER,NO.2 INVERTER POWER SELECT. . .	1	612	
-CR148	100-361045-1	TRANSZORB,NO.1 INVERTER POWER RELAY	1	512	
-CR149	100-361045-1	TRANSZORB,NO.2 INVERTER POWER RELAY	1	612	
-CR176	100-361045-1	TRANSZORB,NO.1 INVERTER POWER SELECT RELAY. . .	1	512	
-CR177	100-361045-1	TRANSZORB,NO.2 INVERTER POWER SELECT RELAY. . .	1	612	
-CR207	1N4005	DIODE,INVERTER SELECT RELAY	1	222	
-F114	MDA5	FUSE/TYPICAL F114 AND F115/	2	222	
-XF114	4532	FUSEHOLDER.	1	222	
-GS160		GROUND STUD/TYPICAL GS160,GS161 & GS162/MADE. . UP FROM THE FOLLOWING/	3	222	
-GS160	131270-3	GROUND STUD/BLIND/.	3	222	
-GS160	MS21042L3	NUT	3	222	
-GS160	MS35338-43	LOCKWASHER.	3	222	
-GS160	AN960-10L	WASHER.	3	222	
-J186	1506-105	TEST JACK,115 VOLTS AC/BLUE/.	1	245	
-K106	50-380048-5	RELAY,AC INVERTER OUT WARNING LIGHT	1	222	
-K144	109-381002-1	RELAY,NO.1 INVERTER POWER SELECT.	1	512	
-K145	109-381002-1	RELAY,NO.2 INVERTER POWER SELECT.	1	612	
-K153	50-380048-1	RELAY,INVERTER SELECT	1	222	
-K154	109-381002-1	RELAY,NO.1 INVERTER POWER	1	512	
-K155	109-381002-1	RELAY,NO.2 INVERTER POWER	1	612	
-S136	MS24659-21A	SWITCH,INVERTER SELECT.	1	245	
	M81714/5-1	RAIL,MODULE	1	222	
-TB107	M81714/3AB1	BLOCK,NO 1 INVERTER POWER TERMINAL.	1	222	
-TB108	M81714/3AB1	BLOCK,NO 2 INVERTER POWER TERMINAL.	1	222	
-W300		HARNESS ASSY,INVERTER/SEE CHAPTER 24-22/. . . .	1	521/621	

Figure 1-3-6A. Beechcraft component description page   *Courtesy of Hawker Beechcraft Corporation-for training purposes only*

Identifying Component Locations

King Air C90A

For electronics or avionics technicians, locating aircraft components will often start at the wiring diagram. Assume that our task is to troubleshoot an inverter power system on a King Air C90A. The first step would be to find the wiring schematic in the wiring diagrams manual under ATA Chapter 24–Electrical power. As seen in Figure 1-3-6A, each Beechcraft wiring diagram is comprised of two pages, a component description page and a system schematic page.

Through study of the schematic and operation of the system, it is determined that two circuit breakers should be tested, CB236 and CB246. To locate these circuit breakers on the aircraft, simply find the circuit breaker number in the left hand column of the component descrip-

tion page (Figure 1-3-6B). Follow the component description line to the right and find the installation location in the Installation Zone column. Circuit breaker 236 is located in zone 246; CB246 is located in zone 612. Referring to the zone location charts (Figures 1-3-4 and 1-3-5), it is determined that CB236 is installed in the right side panel of the flight deck. CB246 is located in the right wing root, center section.

Airbus A-320

On transport category aircraft, locating components is often a difficult task due to the sheer size of the aircraft. On the A-320 there are over 3,000 panels and racks containing equipment, switches, and circuit breakers. Switches and circuit breakers are typically mounted on a panel. Line replaceable units (LRUs), such as the VOR receiver, autoflight computers, or EFIS display management computers are typically mounted in an electronics equipment rack.

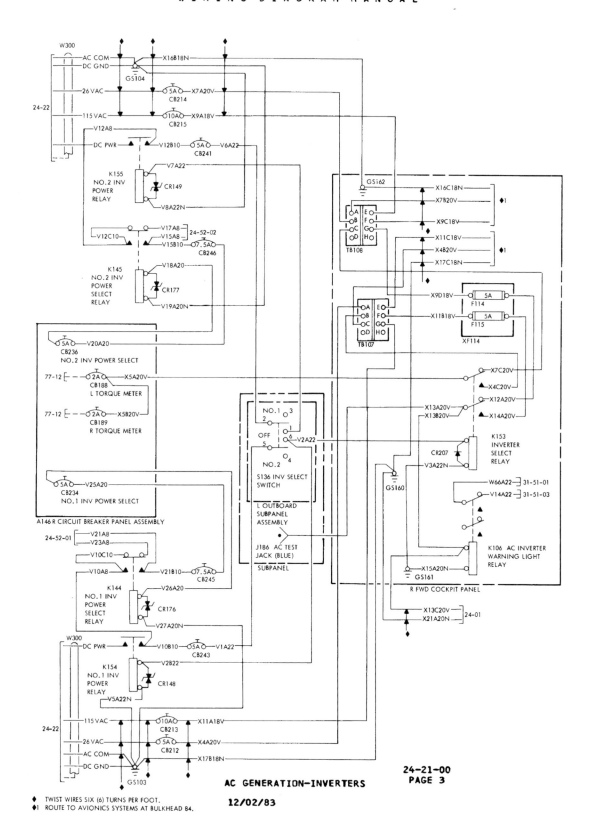

Figure 1-3-6B. Beechcraft wiring diagram

Courtesy of Hawker Beechcraft Corporation-for training purposes only

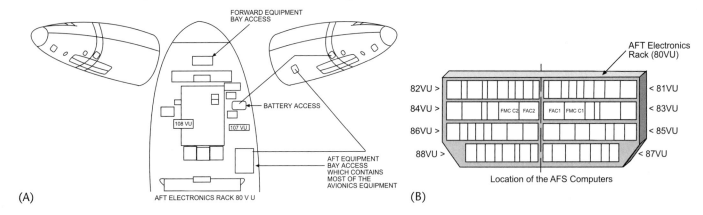

Figure 1-3-7. (A) Avionics equipment locations – equipment bay access, (B) Typical avionics equipment rack showing avionics equipment locations

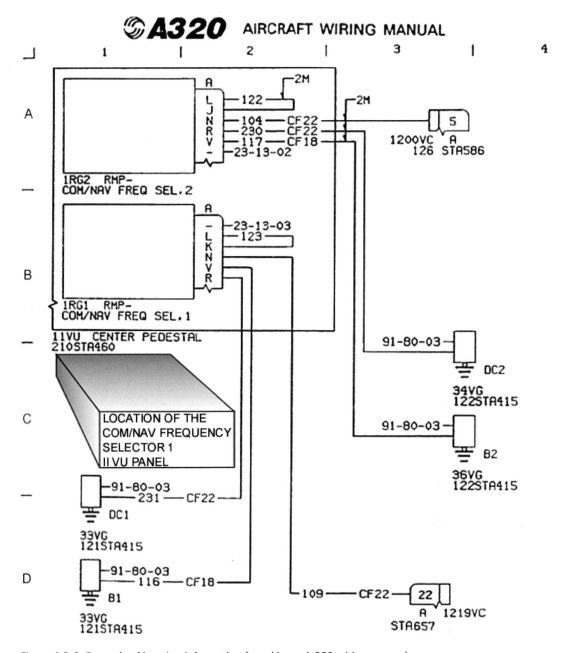

Figure 1-3-8. Example of location information found in an A-320 wiring manual

Courtesy of Airbus S.A.S.-for training purposes only

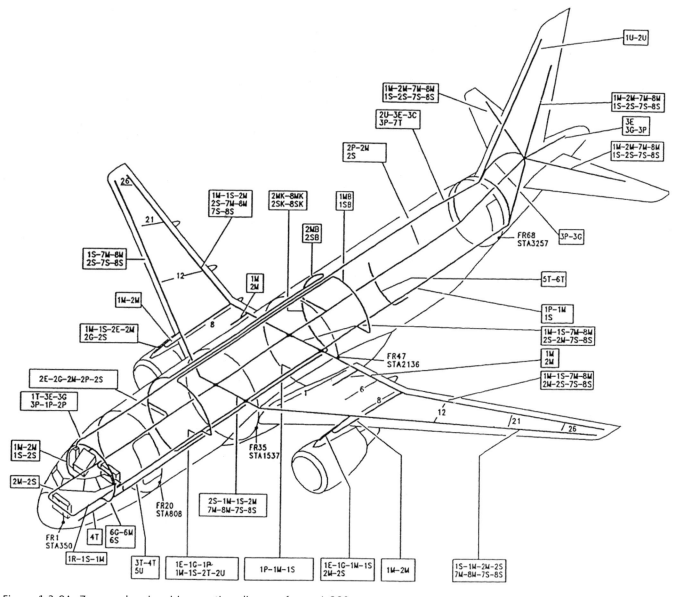

Figure 1-3-9A. Zone and main wiring routing diagram for an A-320

Courtesy of Airbus S.A.S.-for training purposes only

Airbus designates all panels and racks with a VU number (the aircraft station number), and/or the zone number. For example, the electrical power distribution control switches are located on the 35VU panel on the lower-overhead section of the flight deck. The Autoflight system computers are located in the aft electronics racks 84VU and 83VU. It should be noted that a panel or rack will always have removable components. For example, if an item contains several switches and those switches are not individually replaced (i.e., the complete item must be replaced as a unit) that item is considered an LRU, not a panel or rack.

Most of the avionics and computerized LRUs are located in the forward or aft equipment bays on the 90VU, 108VU, and 80VU electronic racks. See Figure 1-3-7A. These racks are further divided into individual shelves as seen in Figure 1-3-7B that shows the location of the autoflight computers.

The A-320 Aircraft Schematics Manuals (ASM) and Aircraft Wiring Manuals (AWM) identify the location of various electronics components using the VU number. As shown in Figure 1-3-8, the radio management panel-communication/navigation frequency selector number 1 (RMP-COM/NAV FREQ SEL.1) is located on the 11VU panel. The schematic states that 11VU is located on the center pedestal at zone 210 and fuselage station 460. It should be noted that all station numbers are measured in centimeters. Zone and station information also shows adjacent connector plugs and ground connections.

The wire routing charts for the A-320 will help the technician to locate wire bundles running through the aircraft. The AWM, Chapter 24 contains a wire zoning and routing diagram as shown in Figures 1-3-9A and 1-3-9B. This diagram can be extremely helpful when troubleshooting a defective wire.

⊘*A320* AIRCRAFT WIRING MANUAL

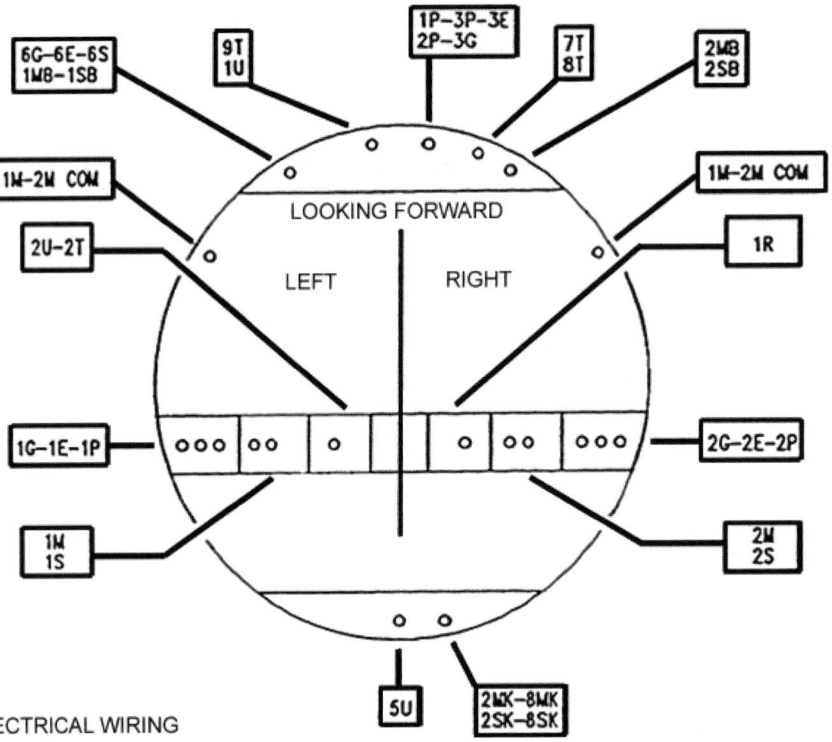

LOOKING FORWARD

LEFT   RIGHT

Label (around diagram)
1P–3P–3E / 2P–3G
6G–6E–6S / 1MB–1SB
9T / 1U
7T / 8T
2MB / 2SB
1M–2M COM
1M–2M COM
2U–2T
1R
1G–1E–1P
2G–2E–2P
1M / 1S
2M / 2S
5U
2MK–8MK / 2SK–8SK

ELECTRICAL WIRING

1G: GENERATOR 1 FEEDERS
2G: GENERATOR 2 FEEDERS
3G: APU GENERATOR FEEDERS
4G: BATTERY FEEDERS
5G: EXTERN AL POWER FEEDERS
6G: EMERGENCY AC
1P: POWER SUPPLY FROM AC
     AND DC BUS 1
2P: POWER SUPPLY FROM AC BUS 2
3P: APU START SUPPLY
1S: SENSITIVE SYSTEM 1
2S: SENSITIVE SYSTEM 2
6S: CONTROL AND MONITORING
     FOR EMERGENCY GENERATOR
7S: FLY BY WIRE (SENSITIVE)
8S: FLY BY WIRE (SENSITIVE)

1M: SYSTEM 1 NOT SENSITIVE
2M: SYSTEM 2 NOT SENSITIVE
6M: POWER SUPPLY FOR SECONDARY
     EMERGENCY CIRCUIT
7M: FLY BY WIRE (NOT SENSITIVE)
     GENERATOR 1 AND EMERGENCY
8M: FLY BY WIRE (NOT SENSITIVE)
     GENERATOR 2
1E: GENERATOR 1 ENERGIZATION
2E: GENERATOR 2 ENERGIZATION
3E: APU GENERATOR ENERGIZATION
6E: EMERGENCY EXCITATION

ELECTRONIC WIRING

1R: SYSTEM 1 RADIO COMMUNICATION
     AND RADIO NAVIGATION
1T: VHF 1
2T: VHF 2
3T: ATC 1, ATC 2, AND T. CAS
4T: DME 1 AND DME 2
5T: RADIO ALTIMETER 1
6T: RADIO ALTIMETER 2
7T: HF 1
8T: HF 2
9T: VHF 3

1U: VOR 1
2U: VOR 2
3U: GLIDE GP1-GP2
4U: LOCALIZER GP1-GP2
5U: MARKER
1R: ADF 1 & ADF 2

Page   1
Feb 01/93

94-00-03

EFFECTIVITY

ALL

ZONING AND ROUTING
MAIN WIRE ROUTING

Figure 1-3-9B. Zone and main wiring routing diagram for an A-320

Courtesy of Airbus S.A.S.-for training purposes only

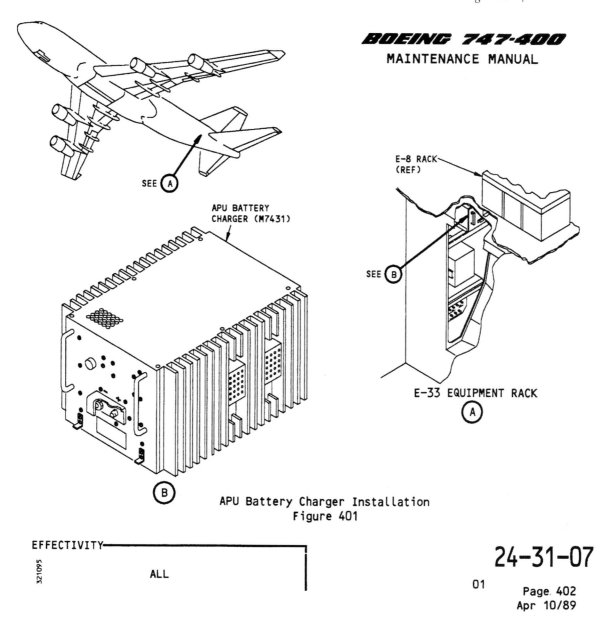

BOEING 747-400
MAINTENANCE MANUAL

APU BATTERY
CHARGER (M7431)

SEE Ⓐ

E-8 RACK
(REF)

SEE Ⓑ

E-33 EQUIPMENT RACK
Ⓐ

Ⓑ

APU Battery Charger Installation
Figure 401

EFFECTIVITY

321095

ALL

24-31-07

01

Page 402
Apr 10/89

Figure 1-3-10. Maintenance manual showing locations of APU battery charger

Boeing 747-400

Because the Boeing 747-400 wiring diagrams do not contain equipment location numbers, their component location system is somewhat more complicated than that of the Airbus system. In the Boeing system component locations for specific equipment are identified on the Equipment List. The Wiring List is used to find specific wires; the Splice List for any wire splices; the Terminal List for terminal strips; and the Hook-up List for connectors or pins within a connector. The wiring, splice, terminal, and hook-up lists are located in Chapter 91 of the maintenance publications. The equipment list is a separate document subdivided by ATA chapters. As seen in Figure 1-3-10, maintenance manuals for Boeing aircraft do a good job of identifying component

locations. In most cases however, troubleshooting and repair of a given system will require the use of several different manuals. In this case, the use of digital manuals often helps to simplify the process through the use of computerized search tools.

Figure 1-3-11 shows an excerpt from the B-747 equipment list. In this example, the equipment (listed under the EQUIP column) includes two different diode assemblies, and two information signs for the passenger compartment upper deck. In this case, the number used to identify each piece of equipment begins with an M. The M indicates the equipment is considered an electrical system component that is not listed in another specific category. Other equipment designators are listed in Table 1-3-1.

R M E O EQUIP V D	O P PART NUMBER T PART DESCRIPTION	USED ON	DWG VENDOR QTY DIAGRAM STATION– WL–BL	EFFECTIVITY
M01559	1 YHLZD–9 DIODE ASSY R965		V09922 1 21–25–12 PANEL–14–	ALL
M01559	2 69B40800–3 DIODE ASSY R965		V09922 1 21–25–12 PANEL–14–	ALL
M01562	60B50192 INFO SIGN–PASS UPR DK		V81205 1 33–24–21 530-370-R 50	ALL
M01564	60B50192 INFO SIGN–PASS UPR DK		V81205 1 33–24–21 550-370-R 50	ALL

MODEL 747 CUSTOMER XXX	REV DATE MAY 02/85	MANUAL D6–XXXXX	EQUIPMENT LIST VOLUME–1	SECTION M01500 PAGE 1

Figure 1-3-11. Example of equipment list for a Boeing 747   *Courtesy of the Boeing Commercial Airplane Company-for training purposes only*

Please reference Figure 1-3-11 while reading the following paragraph. The equipment list contains component part numbers, descriptions, a vendor number, and the quantity installed on the aircraft. The diagram number and aircraft effectivity are also shown on the equipment list. The equipment location is shown just below the vendor information. The location of M01562 is shown as fuselage station 530, water line 370 and butt line R (right) 50. All dimensions are given in inches. For items located on a specific panel, the panel number is substituted for the station, water line and butt line locations. In our example notice that Panel 14 is listed for the location of both diode assemblies.

When troubleshooting from an electrical schematic, it may not be necessary to refer to the equipment list to find the location of a component. Once a technician is familiar with the aircraft, many of the panels and equipment center locations become second nature. For example, when studying the electrical power standby system control and distribution schematic, Figure 1-3-12, it was determined that circuit breaker C80 should be checked. C80, the main battery charger circuit breaker, is found in the upper left corner of the schematic. The breaker is located on the panel (P6) as indicated on the schematic. A technician who is familiar with the B-747 would likely know that P6 is the main power circuit breaker panel located at the right side rear of the flight deck.

A SUMMARY OF THE EQUIPMENT DESIGNATORS FOR THE BOEING 747 EQUIPMENT LIST AND WIRING DIAGRAMS. *Note: Similar designators are used on other transport category. For specific designators be sure to see the aircraft manuals.*	
Equipment Designator	Equipment Category
A..	Anti-icing/de-icing
B..	Electronics Equipment: Accessories units, amplifiers, antennas, capacitors, computers, control units, directional gyros, filters, flux valves, ILS racks, music reproducers, receivers, SELCAL, transmitters, vertical gyros, and other electronics not listed in a specific category
C..	Circuit breakers, current limiters, and miscellaneous protective devices
D..	Connectors
E..	Equipment racks
F..	Fuel system components
G..	Generators and related components
GD..	Grounds (airframe)
J...	Junction boxes
K..	Command post mission system electronic modules
L..	Lights, lamps, and related assemblies
M...	Electrical systems: batteries, ballast assemblies, bells, chimes, control units, heaters, horns, lavatory assemblies, motors, phase adaptors, power units, pumps, strato light assemblies

Table 1-3-1. Summary of equipment designators for the Boeing 747 equipment list and wiring diagrams

Courtesy of the Boeing Commercial Airplane Company-for training purposes only

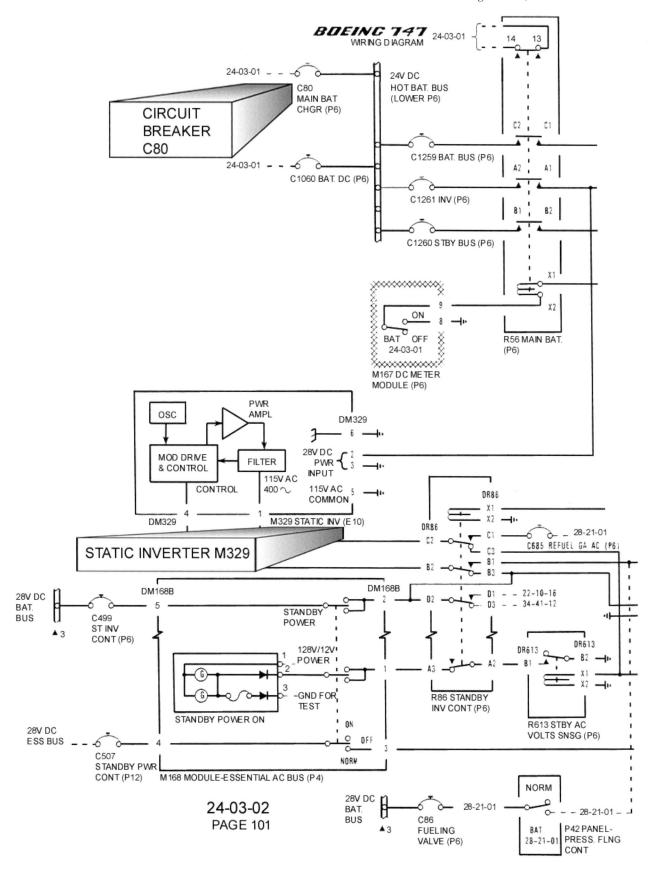

Figure 1-3-12. Portion of a standby system control and distribution schematic

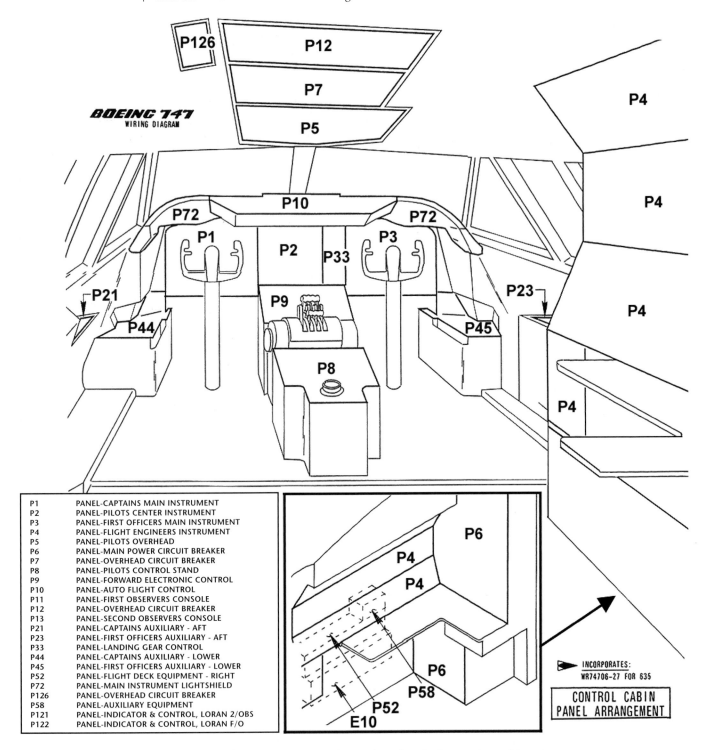

Figure 1-3-13. Example of Chapter 91 of the Wiring Diagrams Manual used to find control panel locations

Courtesy of the Boeing Commercial Airplane Company-for training purposes only

If needed, Chapter 91 (Charts) of the wiring diagrams manual can be used to find the P6 panel (Figure 1-3-13). This chart gives the locations of all flight deck panels. The exact circuit breaker (C80) can be identified by the label on the front side of the P6 panel (MAIN BATT CHGR), or circuit breaker C80 can be identified using Chapter 91. For example, Figure 1-3-14 shows the rear side of P6 to identify the location of each circuit breaker. It should be noted, this

Figure shows only a portion of P6; the entire panel requires eight pages of charts.

As with most transport category aircraft, the majority of the B-747 electronic/avionics equipment is located in various equipment racks found throughout the aircraft. On the schematic for a given system, the equipment rack identifier is located just below the component and designated by an *E* number. A

P number designates panels. To indicate a shelf on a rack, a dash and the shelf number is added, for example, E003-2 would represent equipment rack number 003, shelf number 2.

Figure 1-3-12 shows a static inverter on the left side of the diagram. Shown just below the bold outlines of the inverter is the identifier M329 STATIC INV (E10). M329 can be used to find the static inverter on the aircraft main equipment list. E10 identifies the inverter is located in the equipment rack E10. As seen from the chart in Figure 1-3-15, E10 is located on the copilot's side towards the rear of the flight deck. Chapter 91 also contains other charts showing the locations of various items, such as junction boxes, service interphone jacks, antennas, and coaxial cables. See Figure 1-3-16. Remember, Chapter 91 of the wiring diagram manual contains panel and rack locations.

Boeing 747 Component Locator Guide. The Component Locator Guide for the 747-400 aircraft is a five by eight inch manual of approximately 450 pages. The manual is designed for 747-400 line technicians and provides quick access to the locations of various components. The guide also provides a list of ATA chapters for maintenance or wiring manuals. There are three sections to the manual: the Index, Access/Area, and Recognition/Identification. Each section of the manual is color coded for quick reference. The example below will help to explain the use of the Component Locator Guide.

Assume the technician wishes to find the center auto pilot elevator servo. To locate the component, proceed as follows:

1. Determine the correct ATA chapter for the auto pilot servo (Chapter 22, Autoflight). A list of ATA chapters is included in the locator guide introduction.

2. In the yellow index pages find ATA 22. Next find the correct component in that chapter. See Figure 1-3-17.

3. Read the information in each column of the servo - elevator A/P, center to determine: quantity (1), access/area (STA2600, WL320, and LBL20), page showing the access diagram (28), page showing the recognition diagram (49), and ATA Chapter (22-12-01).

4. Turn to access page 28 (access pages are pink). This Figure shows the various access panels in the horizontal stabilizer (Figure 1-3-18A). The recognition page will show which panel provides access to the servo.

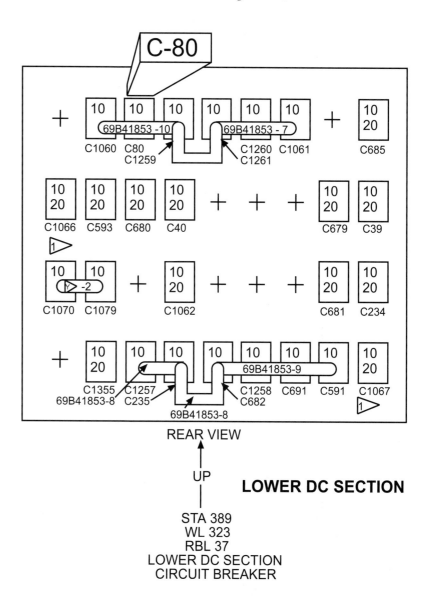

Figure 1-3-14. Diagram showing the rear of the P6 panel, main circuit breaker power

Courtesy of the Boeing Commercial Airplane Company-for training purposes only

5. Turn to recognition page 49 (recognition pages are blue). This will show that the center elevator servo is located behind panel 315 A (Figure 1-3-18B). This diagram also gives a basic drawing of the servo to help the technician recognize the unit.

6. Find Chapter 22-12-01 in the maintenance and/or wiring diagram manuals for the information needed to finish the needed repair.

The Component Locator Guide has provided the technician all the necessary information to locate the center elevator servo. Other LRUs can be found in a similar manner.

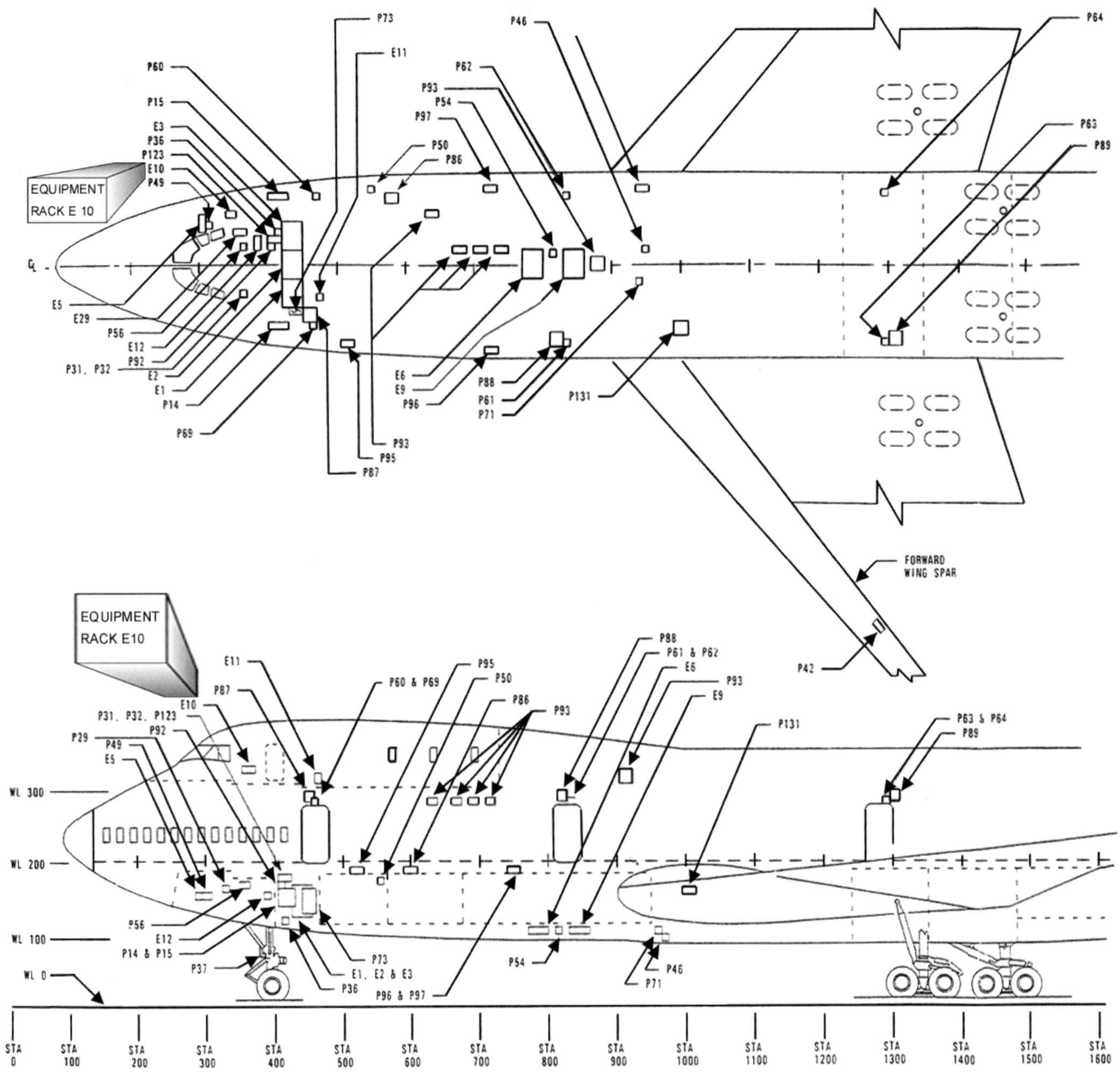

Figure 1-3-15. Panel and equipment locations diagram

E1	EQPT CTR-MAIN L	STA 437
E2	EQPT CTR-MAIN CTR	STA 437
E3	EQPT CTR-MAIN R	STA 437
E5	EQPT CTR-FWD R	STA 300
E6	EQPT CTR-FWD FUSLG	STA 790
E8	EQPT CTR-AFT L	STA 2322
E9	EQPT CTR-CTR FUSLG	STA 835
E10	EQPT CTR-UPPER R	STA 359
E11	EQPT CTR-UPPER L	STA 472
E12	BAT RACK-INS	STA 390
P1-P13	SEE CONTROL CABIN 91-06-11	
P14	CENTER-MAIN PWR-L	STA 430
P15	CENTER-MAIN PWR-R	STA 430
P21	SEE CONTROL CABIN 91-06-11	
P23	SEE CONTROL CABIN 91-06-11	
P27	PNL-E/E EQPT-MN OK-ZONE D	STA 1655
P28	PNL-E/E EQPT-MN OK-ZONE E	STA 2292
P29	PNL-E/E EQPT-FWD LWR LOBE	STA 370
P31	PNL-GS 1 ANT SW	STA 400

P32	PNL-GS 2 ANT SW	STA 400
P33	SEE CONTROL CABIN 91-06-11	
P36	PVL-EXT PWR RECP	STA 410
P37	PNL-CONT	STA 376
P42	PNL-PRESS FUELING CONT	STA 329
P44	SEE CONTROL CABIN 91-06-11	
P45	SEE CONTROL CABIN 91-06-11	
P46	PNL-POTABLE WATER SYS	STA 980
P49	PNL-FWD EQPT CTR UTIL-R	STA 310
P50	PNL-OXY SVCE SYS-PASS & CREW	STA 560
P52	SEE CONTROL CABIN 91-06-11	
P54	PNL-CTR EQPT	STA 810
P56	PNL-FWD EQPT	STA 370
P57	PNL-APU BAT-AUX	STA 2600
P58	SEE CONTROL CABIN 91-06-11	
P59	PNL-LWR CARGO EQPT-AFT	STA 1810
P60	PNL-1-ATT OVHD-R	STA 464
P61	PNL-2-ATT OVHD-L	STA 83
P121	SEE CONTROL CABIN 91-06-11	

Courtesy of the Boeing Commercial Airplane Company-for training purposes only

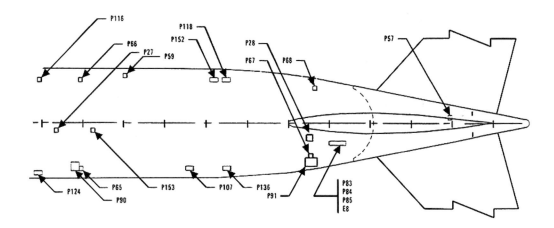

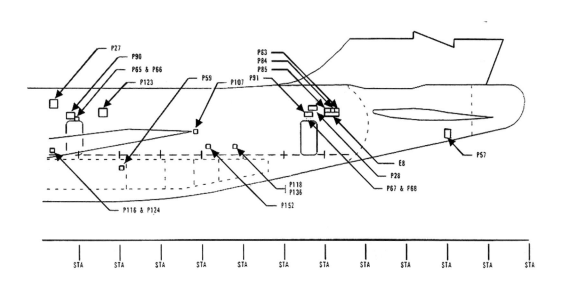

P122	SEE CONTROL CABIN 91-06-11		P83	PNL-APU-CMPNT	STA 2333
P123	PNL-INS BAT RLY	STA 400	P84	PNL-MISC CMPNT	STA 2323
P124	PNL-CB-MN DX CAR HOLG PWR	STA 1610	P85	PNL-AIT PRESS CMPT	STA 2310
P126	SEE CONTROL CABIN 91-06-11		P86	PNL-M-LWR CAR EQPT-FWD	STA 590
P131	PNL-ELEC-CTR WING-FR SPAR	STA 1000	P87	PNL-ENTRY DR 1 OVHD-L	STA 457
P136	PNL-CB-MN DK CAR HOLG PWR	STA 2050	P88	PNL-ENTRY DR 2 OVHD-L	STA 823
P152	PNL-CB-MN DK CAR HOLG PWR	STA 2025	P89	PNL-ENTRY DR 3 OVHD-L	STA 1305
P153	PNL-MN-DK CARGO	STA 1720	P90	PNL-ENTRY DR 4 OVHD-L	STA 1687
P62	PNL-2-ATT OVHD-R	STA 830	P91	PNL-ENTRY DR 5 OVHD-L	STA 2261
P63	PNL-3-ATT OVHD-L	STA 1295	P92	PNL-SPLIT SYS BRKR	STA 404
P64	PNL-3-ATT OVHD-R	STA 1295	P93	PNL-LTG CONT XFR	
P65	PNL-4-ATT OVHD-L	STA 1694	P95	PNL-XFR RLY-WP/GALY	STA 520
P66	PNL-4-ATT OVHD-R	STA 1694	P96	PNL-XFR RLY-WP/GALY	STA 760
P67	PNL-5-ATT OVHD-L	STA 2261	P97	PNL-XFR RLY-WP/GALY	STA 760
P68	PNL-5-ATT OVHD-R	STA 2261	P98	PNL-XFR RLY-WP/GALY	STA 960
P69	PNL-1-ATT OVHD-L	STA 464	P107	PNL-REMOTE IND CONT	STA 1950
P71	PNL-WATER INJ SVCE	STA 970	P116	PNL-CB-MN DX CAR HOLG PWR	STA 1610
P72	SEE CONTROL CABIN 91-06-11		P118	PNL-CB-MN DX CAR HOLG PWR	STA 2050
P73	PNL-FLT INSTR RLY	STA 446			

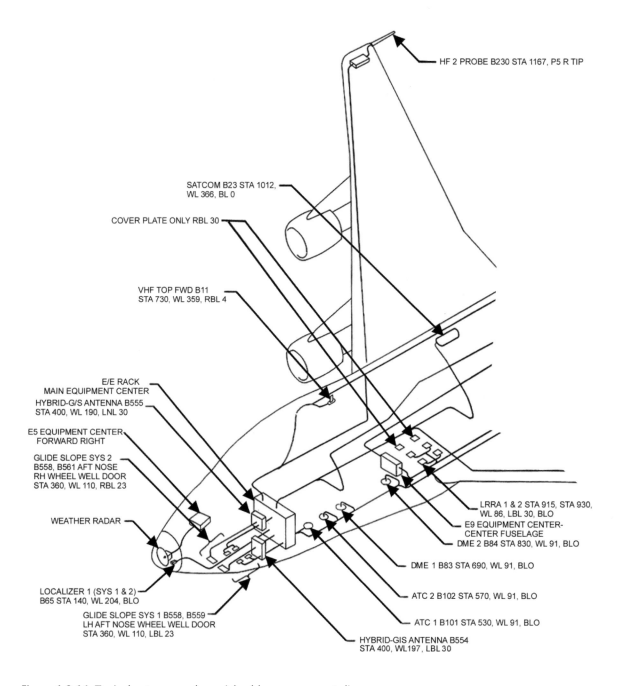

HF 2 PROBE B230 STA 1167, P5 R TIP

SATCOM B23 STA 1012, WL 366, BL 0

COVER PLATE ONLY RBL 30

VHF TOP FWD B11 STA 730, WL 359, RBL 4

E/E RACK MAIN EQUIPMENT CENTER

HYBRID-G/S ANTENNA B555 STA 400, WL 190, LNL 30

E5 EQUIPMENT CENTER FORWARD RIGHT

GLIDE SLOPE SYS 2 B558, B561 AFT NOSE RH WHEEL WELL DOOR STA 360, WL 110, RBL 23

WEATHER RADAR

LOCALIZER 1 (SYS 1 & 2) B65 STA 140, WL 204, BLO

GLIDE SLOPE SYS 1 B558, B559 LH AFT NOSE WHEEL WELL DOOR STA 360, WL 110, LBL 23

LRRA 1 & 2 STA 915, STA 930, WL 86, LBL 30, BLO

E9 EQUIPMENT CENTER-CENTER FUSELAGE

DME 2 B84 STA 830, WL 91, BLO

DME 1 B83 STA 690, WL 91, BLO

ATC 2 B102 STA 570, WL 91, BLO

ATC 1 B101 STA 530, WL 91, BLO

HYBRID-GIS ANTENNA B554 STA 400, WL197, LBL 30

Figure 1-3-16. Typical antenna and coaxial cable arrangement diagram

Courtesy of the Boeing Commercial Airplane Company-for training purposes only

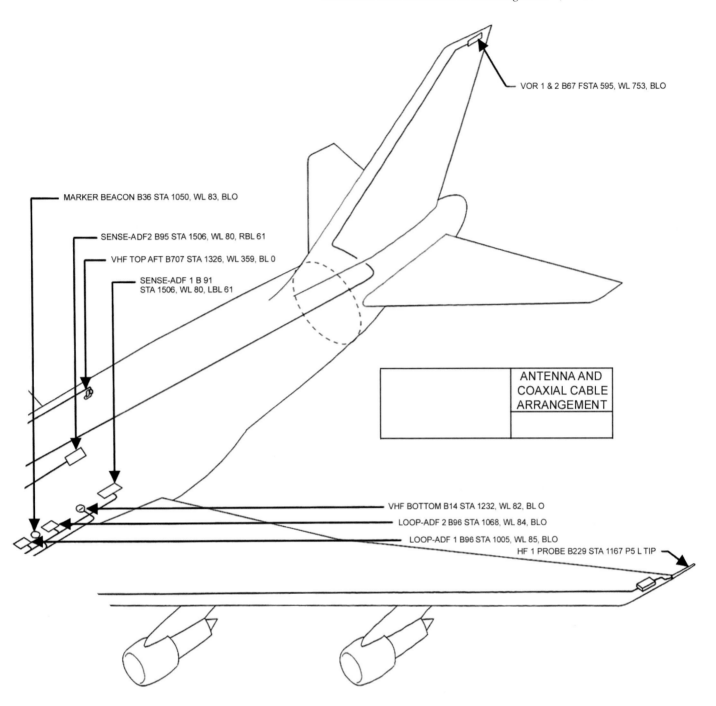

VOR 1 & 2 B67 FSTA 595, WL 753, BLO

MARKER BEACON B36 STA 1050, WL 83, BLO

SENSE-ADF2 B95 STA 1506, WL 80, RBL 61

VHF TOP AFT B707 STA 1326, WL 359, BL 0

SENSE-ADF 1 B 91
STA 1506, WL 80, LBL 61

ANTENNA AND
COAXIAL CABLE
ARRANGEMENT

VHF BOTTOM B14 STA 1232, WL 82, BL O

LOOP-ADF 2 B96 STA 1068, WL 84, BLO

LOOP-ADF 1 B96 STA 1005, WL 85, BLO

HF 1 PROBE B229 STA 1167 P5 L TIP

ATA 22	AUTOFLIGHT			ACCESS	RECOG-NITION	ATA 22
COMPONENT		QTY	ACCESS/AREA	(pink)	(blue)	CH/SEC
ACCELEROMETER - AFT MODAL SUPPRESSION, LEFT		1	BELOW MAIN DECK - STA 2260	8	46	22-21-03
ACCELEROMETER - AFT MODAL SUPPRESSION, RIGHT		1	BELOW MAIN DECK - STA 2260	8	46	22-21-03
ACCELEROMETER - FWD MODAL SUPPRESSION, LEFT		1	ABOVE NOSE GEAR WHEEL WELL	8	46	22-21-03
ACCELEROMETER - FWD MODAL SUPPRESSION, RIGHT		1	ABOVE NOSE GEAR WHEEL WELL	8	46	22-21-03
COMPUTER - FLIGHT CONTROL, CENTER		1	MEC - E1-3	22,42	8	22-11-01
COMPUTER - FLIGHT CONTROL, LEFT		1	MEC - E1-1	22,42	8	22-11-01
COMPUTER - FLIGHT CONTROL, RIGHT		1	MEC - E1-4	22,42	8	22-11-01
PACKAGE - ROLLOUT POWER CONTROL, CENTER		1	FR, FL, ZONE 324	31	47	22-13-04
PACKAGE - CNTRL LATERAL CONTROL, LEFT		1	WING LANDING GEAR WELL	25	48	27-11-08
PACKAGE - ROLLOUT POWER CONTROL, LEFT		1	FR, FL, ZONE 324	31	47	22-13-04
PACKAGE - CNTRL LATERAL CONTROL, RIGHT		1	WING LANDING GEAR WELL	25	48	27-11-08
PACKAGE - ROLLOUT POWER CONTROL, RIGHT		1	FR, FL, ZONE 324	31	47	22-13-04
PANEL - PASSENGER OXYGEN AND YAW DAMPER		1	FLIGHT DECK - P5	41	3	22-21-01
PANEL - AFCS MODE CONTROL		1	FLIGHT DECK - P10	41	2	22-11-02
SERVO - ELEVATOR A/P, CENTER		1	STA 2600, WL320, LBL 20	28	49	22-12-01
SERVO - ELEVATOR A/P, LEFT		1	STA 2600, WL320, RBL 12	28	49	22-12-01
SERVO - ELEVATOR A/P, RIGHT		1	STA 2600, WL320, RBL 22	28	49	22-12-01
SERVO - LATERAL A/P		1	WING LANDING GEAR WELL	25	48	22-13-01
SWITCH - AFCS GO-AROUND		4	FLIGHT DECK - PILOT'S CONTROL STAND	41	45	22-11-04
SWITCH - CAPTAIN A/P DISENGAGE, S282		1	FLIGHT DECK - CONTROL WHEEL	41	45	22-11-03
SWITCH - F/O A/P DISENGAGE, S283		1	FLIGHT DECK - CONTROL WHEEL	41	45	22-11-03

Figure 1-3-17. Chapter 22 of Boeing 747 Component Locators Guide

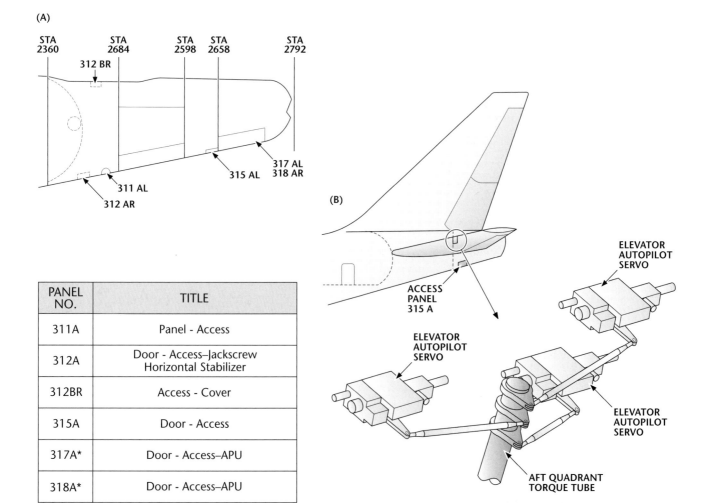

NOTE:
Panels with an "L" at the end (e.g. 317 AL) in the figure are on the left side of the airplane, and panels with an "R" at the end are on the right side. However, actual panel numbers on an airplane do not show and "L" or an "R."

Figure 1-3-18. Example of Boeing 747 Component Locators Guide: (A) access page, (B) recognition page

Section 4

Data Management: New Trends

The revolution in aircraft technical data management is a result of evolving aircraft technologies. The new generation of large transport aircraft incorporates digitally controlled systems and complex onboard computer networks. These electronic networks operate entire aircraft systems like brakes, engine bleed air, cabin environment, and pressurization. Even small piston aircraft used as fleet trainers, like the Cirrus SR20, have similar advanced technologies: Garmin G1000 glass cockpit and autopilot, satellite weather, on-screen approach plates, "synthetic vision" landing cameras, terrain warning systems, and flight data capturing systems. These systems monitor and store maintenance data on aircraft engine, airframe, and electrical system performance and faults. The computing power as well as knowledge required of these technical systems is significant. Data storage, retrieval, and display requirements for these technologies are enormous. In fact, many organizations have the option of actually storing data with outside vendors, including manufactures, as opposed to storing it on their in-house servers.

Maintenance Provider (In-House) Technical Data Systems

As modern aircraft have grown in complexity, electronic technical data needs have evolved as well. Even CD-ROM formats are no longer sufficient to support the multiple needs of today's aircraft technician.

Air carriers performing their own maintenance or contract maintenance providers who do it for them now build and house their own computer databases for technical documentation. They use their own Information Technology departments. The down side can be the effort required to keep specialized "home grown" networks and the labor cost of maintaining electronic data as time goes on.

OEM or Externally Managed Technical Data Systems

As another option, maintenance databases for technical documents on modern aircraft, such as the Boeing 787, are often main-

Figure 1-4-1. Online technical manual accessed in a web-based user format

Courtesy of Hangar of the Future Research Laboratory, Purdue University

Figure 1-4-2. Computers as tools in a modern maintenance operation

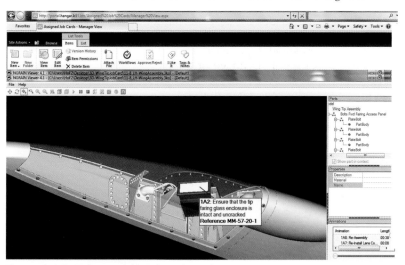

Figure 1-4-3. Enhanced 3D job task card

Courtesy of the Hangar of the Future Research Laboratory, Purdue University

Figure 1-4-4. Tablet PCs are useful for displaying and signing off on electronic job task cards

tained directly by the Original Equipment Manufacturer (OEM). They maintain secure online data access specific to an air carrier's fleet of aircraft. For a subscription cost or built into the purchase price, an OEM-managed technical data system provides a "one stop shop" technical data storage and retrieval services for an air carrier's maintenance provider. Such a system provides access to fault isolation manuals, maintenance documentation, and engineering diagrams contained in databases maintained by the OEM directly, but accessible to maintenance technicians from most standard computers. A technician can access necessary technical manuals or job task card instructions online in a web-based format as shown in Figure 1-4-1.

An air carrier can still adjust the OEM data to fit their particular aircraft mission and maintenance operations. This produces job task instructions specific to an air carrier's particular aircraft.

Boeing Aircraft Company's Airplane Health Management (AHM) system is an example of the move toward OEM-managed technical data. The major benefit of this system is, where in the past maintenance providers or air carriers were required to store and maintain multitudes of heavy paper-based aircraft maintenance, troubleshooting, wiring diagram and parts catalog manuals, these technical documents can now be accessed and maintained almost exclusively online and accessed as a web-based user interface if so desired.

Modern database servers have relatively limitless storage space and flexibility. The move toward either provider or OEM data management control systems provides ability to link separate manuals (for example a maintenance manual and a wiring diagram manual) using hyperlinked keywords. This offers rapid cross-referencing as well as important revision and document control improvements. With this system, an OEM for example, can make technical manual revisions and updates for all of the aircraft it supports instantaneously. This ensures control and distribution of up-to-date technical data changes for a manufacturer's supported aircraft worldwide. This approach makes it easier to maintain and "data mine" the vast amounts of information previously contained and stored in bulkier paper-based binders or on CD-ROMS with limited interactive capability.

The integration of portable electronic devices (laptops and hand held computers) into maintenance operations has also helped the front line technician, even when working on older model aircraft. Assistive computer devices as shown in Figure 1-4-2, are now considered a part of the modern technician's "tool box". Data once held on servers and accessible only by engineering support departments and desktop computers is now available planeside in real time, where maintenance and troubleshooting tasks are being performed.

With today's available computing power, aircraft technicians can rapidly access aircraft technical documentation and data much quicker. They can quickly assess structural references, view animated tutorials, and even order parts for a given maintenance task, using laptops, tablet PCs, and hand-held devices.

The Maintenance Hangar of the Future

At this writing, completely paperless maintenance systems are still in the process of emerging into standardized use by the industry, although some maintenance providers have begun to migrate to such systems. Many enterprise management applications and software packages already exist to support and manage paperless maintenance. However, today most facilities utilize maintenance data in a combination of paper-based and computer technologies.

Current research is exploring the use of enhanced online job task card delivery systems. These network-enabled task card systems will rapidly deliver work instructions in a realistic visual format to a technician's computer screen with intuitive, user controlled screen displays (Figure 1-4-3). They will display instructions and information with varying levels of detail depending on the user requirements and experience level. For example, requesting a 3D image of a part of the aircraft or the aircraft itself for better visualization and understanding of the job task.

In these new systems, aircraft job task cards can also be worked and signed off completely electronically (paperless maintenance). A sample of an electronic job task card being tested at Purdue University's Hangar of the Future research laboratory is shown in Figure 1-4-4. The future of hanger maintenance is still yet to be determined. However, the complete paperless aircraft maintenance facility is still sometime off in the future.

Chapter 2

POWER
distribution, digital and data bus systems

Section 1

Introduction

All aircraft and their avionics systems have similar characteristics. This chapter will present topics that are common to virtually every modern aircraft. To fully understand the operation of an avionics system, the technician must first become familiar with the aircraft's electrical power distribution system. Transport category aircraft power distribution systems are very complex and involve numerous computers for operation and control. This chapter will provide a review of power distribution systems for both simple and complex aircraft, with an emphasis on computerized transport category systems.

Modern avionics systems operate using a variety of computers and microprocessor circuits. These computer-based systems rely on digital data for communication between components. There are currently several formats used for transmission of digital data. The information presented in this chapter will help the avionics technician become familiar with commonly used data bus systems found on today's aircraft. A discussion on multiplexing and demultiplexing techniques and analog to digital converters is also reviewed in this chapter.

Many of the computer-based systems employ electronics that are sensitive to the discharge of static electricity. Standard techniques for safely handling these components will be presented.

Learning Objectives:

- *Power Distribution Systems*
- *Analog to Digital Conversion*
- *Flat-Panel displays*
- *Electrostatic Discharge Sensitive Equipment Safety*
- *Binary Numbering Systems*
- *ARINC Standards*
- *Digital Data Bus Systems*
- *Ethernet Communications*
- *Troubleshooting and Repair*

Left. The four-engined Airbus A-380 can carry more than 850 passengers and 20 crew. A fuel capacity in excess of 85,000 gallons allows for a range greater than 9,000 miles. Providing electrical power and data to passengers and aircraft is a challenge.

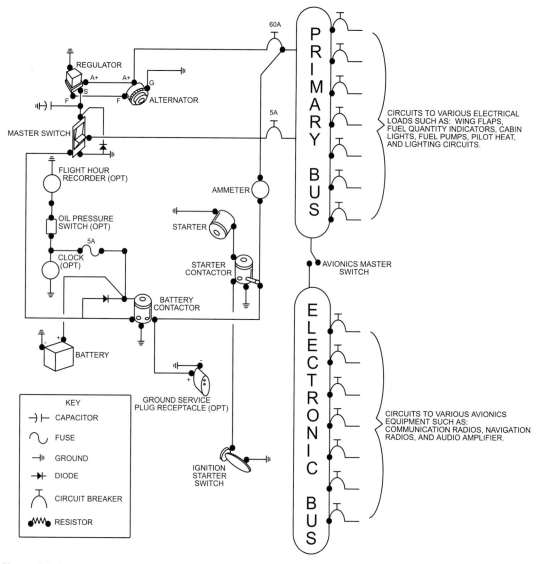

Figure 2-2-1. A simple power distribution system

For training purposes only

Section 2

Power Distribution Systems

An aircraft's power distribution system is comprised of one or more electrical distribution points. The distribution points are often made of a solid copper bar referred to as a *bus* or *bus bar*. In some cases, the distribution point is a simple terminal (stud) where several wires connect. On light, single-engine aircraft, there is often only one bus used to distribute the positive voltage of the electrical power. The negative voltage is connected to the various loads through the metal structure of the airframe. Any aircraft using the airframe to distribute negative voltage is considered to have a *single-wire electrical system*. A single-wire electrical system may also be referred to as a *negative ground system*. As aircraft become more com-plex so do their power distribution systems. Business-type and transport category aircraft contain several power distribution busses. It is very important for electronics technicians to understand the characteristics of a bus system in order to effectively troubleshoot various avionic components.

Bus Hierarchy

Since aircraft are designed with safety as a primary consideration, it stands to reason that electrical systems require special attention. Light aircraft that fly during night or poor weather conditions rely heavily on electronic equipment to provide a safe and comfortable flight. Modern transport category aircraft rely on electrical systems to power computers to operate virtually every system on the aircraft. On aircraft with more than one distribution bus, the busses are arranged into a hierarchy ranging from least to most critical. The *bus hier-*

archy is designed to ensure that the most critical electrical systems are the least likely to fail.

Light Aircraft Power Distribution

Figure 2-2-1 illustrates a simple power distribution system that has only a two-stage hierarchy. That is, the electrical systems may be powered by the engine-driven generator or by the aircraft's battery. Figure 2-2-2 shows the electrical load distribution of a Beechcraft King Air.

The King Air system is comprised of:

1. One main battery bus

2. One isolation bus

3. Two generator busses (left- and right-side)

4. Four dual fed busses

5. One 26VAC bus

6. Two sub-panel busses

7. One hot battery bus

The King Air has both direct and alternating current power systems. DC is used to power the majority of the aircraft systems. As seen in the upper left portion of the diagram in Figure 2-2-2, 26 VAC and 115 VAC power is used by various avionics systems. AC is supplied through one of two static inverters, labeled INV NO. 1 and INV NO. 2 on the schematic. During maintenance of these systems, be sure AC power is available from one of these static inverters.

The diagram for AC power distribution is shown in Figure 2-2-3. Here it can be seen that one or more components of the AC system can fail without loss of AC electrical power. For example, if the left power relay should fail the left inverter would not receive power. However, the right inverter system would still be operational and inverter power would be routed though the inverter select relay to the 26 VAC and 115 VAC busses.

The King Air system is designed to allow for more than one failure before critical systems are lost. The most critical bus on this aircraft is the hot battery bus (Figure 2-2-2). The hot battery bus receives DC power whenever a charged battery is installed in the aircraft. Notice that critical systems, such as fire protection and fuel pumps, are connected to the hot battery bus. The clock, RNAV memory, and stereo are also connected to the hot battery bus. It is advantageous that these systems remain powered even if the battery relay is turned off.

As their names imply, the left generator bus receives power from the left generator; the right generator bus receives power from the right generator (Figure 2-2-2). The isolation bus is used to parallel connect the system during normal operation. The isolation bus is connected to the left and right busses through a 325-amp isolation limiter. An isolation limiter is basically a large fuse used to protect the system. Figure 2-2-4 shows a detail of the left generator bus and the number 3 dual fed bus connection, Figure 2-2-2 shows an expanded view of this circuit. This type of power distribution system contains a series circuit with a 60A current limiter, a 70A diode, and a 50A circuit breaker. These three components are used to isolate each dual fed bus and protect the electrical system in the event of a catastrophic failure. Any given bus could short to ground effectively disconnecting the electrical components powered by that bus; however, the remainder of the electrical system will operate normally. For this protection to operate correctly the current limiter must be rated at a higher value than the circuit breaker (60A-current limiter, 50A-circuit breaker). The current limiter is designed to open only if the circuit breaker fails to protect the circuit (i.e., the current limiter is simply a backup for the circuit breaker.) The diode in the circuit is used to prevent any current from leaving the dual fed bus to power the left generator bus. Most modern corporate and commuter-type aircraft employ a split bus power distribution system similar to the one described for the King Air.

Transport Category Aircraft Power Distribution

On transport category aircraft, such as the Boeing 727, 737, 747, 777, and 787, and the Airbus A-320, A-340, and A-380, the power distribution systems produce thousands of watts of power and are quite complex. There are several busses and three or more AC generators supplying electrical power during flight. Direct current is supplied by the generators through *transformer rectifier* (TR) units, which convert AC power into DC power. The generators and distribution systems are computer controlled and automatically monitored for malfunctions and defects during flight.

The electrical system hierarchy for transport aircraft can be broken into two segments: the hierarchy for supplying power, and the hierarchy for the distribution busses. The typical power supply hierarchy might be arranged as shown in Figure 2-2-5. Here it can be seen that the two engine-driven generators are least critical and the static inverter would be considered the most critical AC power source.

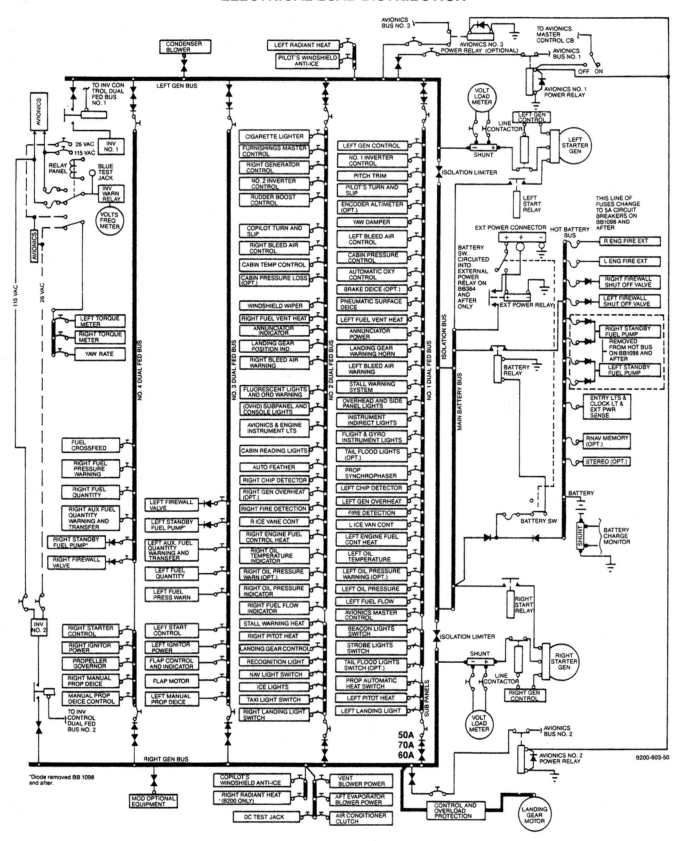

Figure 2-2-2. Beechcraft King Air C90A electrical load distribution

Courtesy of Hawker Beechcraft Corporation–for training purposes only

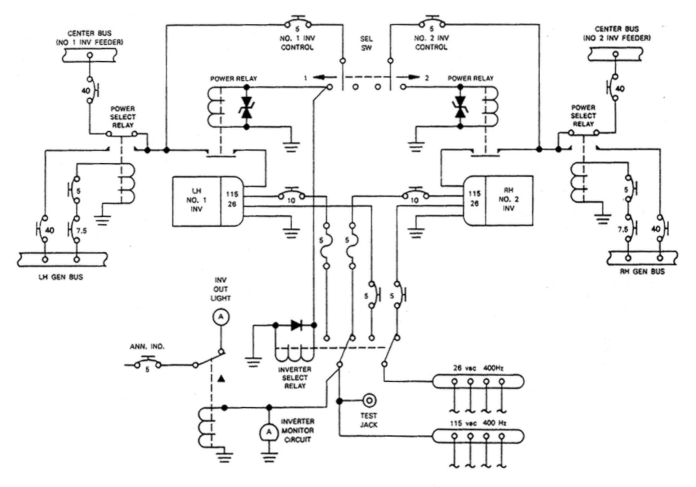

Figure 2-2-3. Beechcraft King Air C90A AC power distribution system   *Courtesy of Hawker Beechcraft Corporation–for training purposes only*

A typical bus hierarchy is shown in Figure 2-2-6. The diagram is divided vertically with AC shown on the left and DC on the right. The most critical busses are located on the bottom of the chart. Loss of these critical power distribution busses would most likely be catastrophic, therefore, they are typically fed by redundant power sources.

There are three basic types of power distribution systems found on transport category aircraft, the split bus system, the parallel system and the split-parallel system. The *split bus system* is used on most twin engine commercial aircraft such as the Boeing 737, 757, 767, 777, 787, and the McDonnell Douglas MD 80, and the Airbus Industries A-320, A-310 and A-380. In a split bus system each engine-driven generator supplies power to a specific bus (left or right). The left and right busses operate independently and must never be connected while both generators are supplying power.

In a *parallel electrical system*, the entire electrical load is equally shared by all of the working generators. Parallel AC power distribution systems are typically found on commercial aircraft containing three or more engines such as

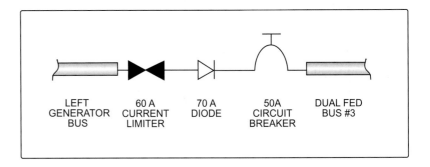

Figure 2-2-4. A typical circuit connecting two distribution busses

the Boeing 727, early 747s, the Lockheed L-1011, and the McDonnell Douglas MD-11, formerly the DC-10.

A *split-parallel power distribution system* is used on some modern four-engine aircraft such as the Boeing 747-400. A split-parallel system allows all generators to operate in parallel during normal conditions. In the event of a system malfunction, one or more generators may operate independently.

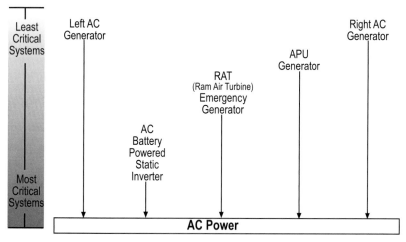

Figure 2-2-5. Power supply hierarchy

Split Bus Systems

The Airbus A-320 employs a state-of-the-art split bus power distribution system (Figure 2-2-7). Since this aircraft is typical of most twin-engine transport category aircraft, it will be presented here for the discussion on split bus power distribution. The A-320 system consists of four AC generators; two powered by the main engines, one powered by the auxiliary power unit (APU) and one emergency generator powered by a ram air turbine or RAT. The two main (engine-driven) generators are least critical and a failure of one generator would cause only momentary loss of non-critical systems, like galley power. Assuming the left generator failed, the flight crew would start the APU generator to assume all loads from the failed main generator. If the right main generator should also fail, the APU generator would supply all electrical power to the aircraft. All transport category aircraft with a split bus elec-

trical system utilize the APU to power a backup generator. Therefore, the APU must be capable of operation during normal flight. If the APU generator should fail, the ram air turbine (RAT) would be used to drive an emergency power generator. If that system should fail the aircraft would still have AC power available through the two static inverters. The static inverters receive their power from the aircraft's DC batteries.

Since the static inverter system is the most critical AC power source on the A-320, the system must be checked each night during routine maintenance. The static inverter system is tested using the aircraft's computer diagnostic system. If the test shows an inverter system failure, it must be repaired prior to the next flight.

The *ram air turbine* (RAT) is a device used on many twin-engine transport category aircraft to supply emergency hydraulic and electrical power. The RAT consists of a propeller driven hydraulic pump deployed from the fuselage in the event of a catastrophic hydraulic or electrical system failure (Figure 2-2-8). As long as the aircraft has airspeed greater than 80 knots, the RAT propeller will "wind mill" as it passes through the air stream. The turning propeller powers a hydraulic pump. The pressurized hydraulic fluid is routed to a hydraulic motor that is used to turn the emergency AC generator.

On the A-320 the emergency generator has a maximum output of 5 KVA at 115 VAC, 400 Hz. The main and APU generators have a maximum output of 90 KVA at 115 VAC, 400 Hz. Comparing these two values, it is easy to see that if the aircraft is operating on emergency generator power, the electrical system is

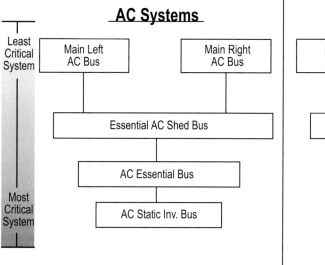

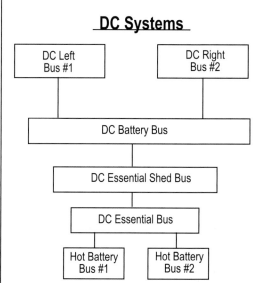

Figure 2-2-6. Bus hierarchy

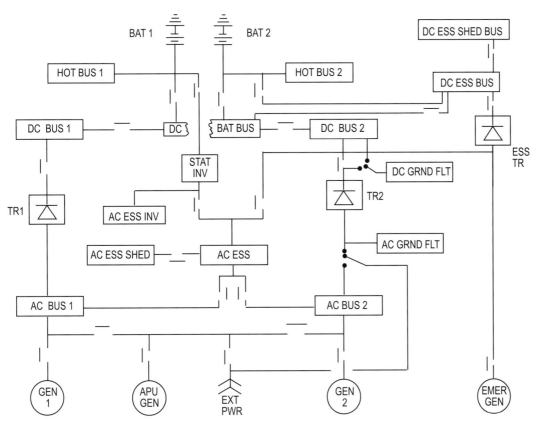

Figure 2-2-7. A split bus power distribution system for a transport category aircraft

Courtesy of Airbus S.A.S.–for training purposes only

extremely limited and the aircraft should land as soon as practical.

Parallel Bus Systems

The Boeing 727 is a three engine aircraft employing a parallel power distribution system. As seen in Figure 2-2-9, the system contains three main engine-driven generators and one APU-driven generator that is ground operable only. Each generator is connected to a paralleling, or synchronizing bus (sync bus), through its respective generator breaker (GB) and bus tie breaker (BTB). During normal flight, generators 1, 2 and 3 are connected to the sync bus; however, if a generator fails that generator is isolated by opening its GB. If AC bus number 1, 2 or 3 should short to ground, the related BTB and GB would open to isolate that bus.

On the B-727 AC bus number 1, 2 or 3 would power the least critical AC loads. DC bus number 1 or 2 powers the least critical DC loads. The next most critical systems would be powered by the essential (ESS) AC and DC busses as shown on the left side of Figure 2-2-9. The essential AC bus can receive power from any AC generator. The essential DC bus can be powered by either DC bus 1 or 2, or the essential TR unit. The standby busses located at the top of

Figure 2-2-8. A Ram Air Turbine (RAT) shown extended from the aircraft fuselage

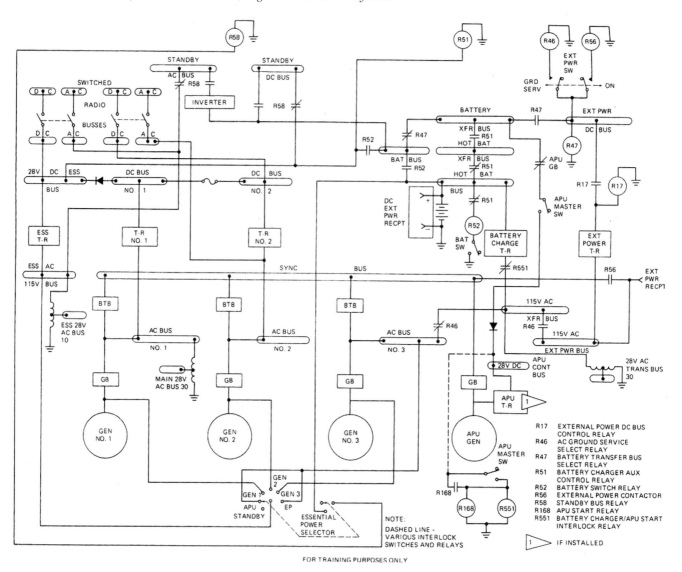

Figure 2-2-9. Boeing 727 power distribution system

Courtesy of the Boeing Commerical Airplane Company–for training purposes only

the diagram power the most critical electrical loads on the aircraft. The standby busses will receive power from the battery even if all three engine-driven generators fail.

Split-Parallel System

A split-parallel electrical power distribution system allows for flexibility in load distribution and yet maintains isolation between systems when needed. The *split system breaker* (SSB) can be seen in the simplified diagram of a split-parallel system (Figure 2-2-10). When closed, the split system breaker connects all four main-engine generators, thus paralleling the system. The system is operated with all generators paralleled during normal flight conditions. When open, the split system breaker isolates the right and left hand systems.

A split-parallel system is used on the Boeing 747-400 aircraft. As seen in Figure 2-2-11, this

system employs four engine-driven integrated drive generators (IDG), two auxiliary power units (APU) generators, and can accept two separate external power sources (EXT 1 and EXT 2). As seen in the system schematic, the four IDGs are connected to their respective AC busses through generator control breakers (GCB). The AC busses are paralleled through the bus tie breakers (BTB) and the split system breaker (SSB). When the SSB is open, the right system operates independent of the left. With this system, any generator can supply power to any load bus or any combination of the IDGs can operate in parallel. It should be noted that the term *integrated drive generator* (IDG) is used here to represent the AC generators. An IDG is simply an AC generator built into (integrated with) the same housing as the constant speed drive unit. A constant speed drive is a type of hydraulic transmission that turns the generator at a constant speed regardless of the engine RPM. The AC generator must rotate at a constant speed in order to maintain a constant output frequency of 400 CPS. Modern transport category aircraft employ integrated drive generators on all main engines.

Ground Handling and Ground Service Busses

Ground handling (GH) busses are used to power lighting and miscellaneous equipment for cargo loading, fueling, and cleaning the aircraft. The AC ground handling busses are powered by closing the ground-handling relay (GHR) to either the APU or EXT power. The DC ground handling busses receive power from the transformer rectifier (TR) shown in Figure 2-2-11, on the lower left side of the diagram. The GH busses are not powered during normal flight.

On the Boeing 747-400 the ground service (GS) busses are used to light the aircraft interior, power the main battery charger, and other miscellaneous systems required for maintenance and initial start-up of the aircraft. The ground service busses are controlled from the flight attendant station located at the number two left door of the aircraft. This switch energizes the

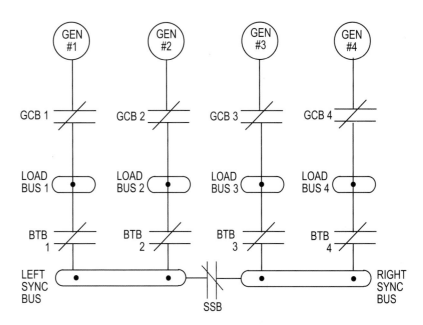

KEY: GCB = GENERATED CONTROL BREAKER
         BTB = BUS TIE BREAKER
         SSB = SPLIT SYSTEM BREAKER

Figure 2-2-10. A split-parallel power distribution system for a transport category aircraft

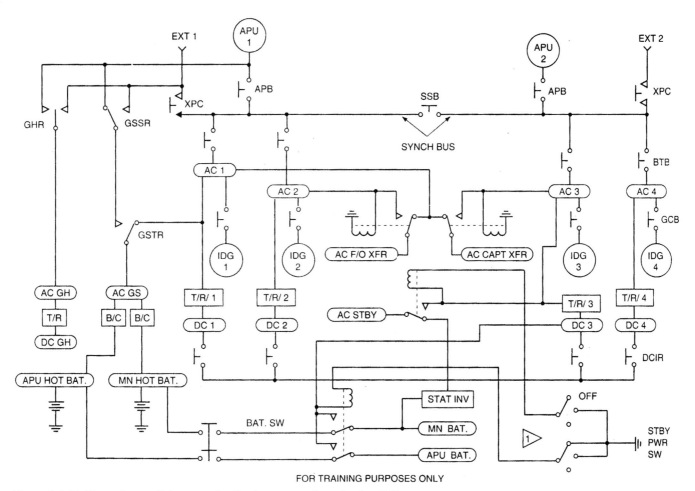

FOR TRAINING PURPOSES ONLY

Figure 2-2-11. The split-parallel power distribution system for a Boeing 747

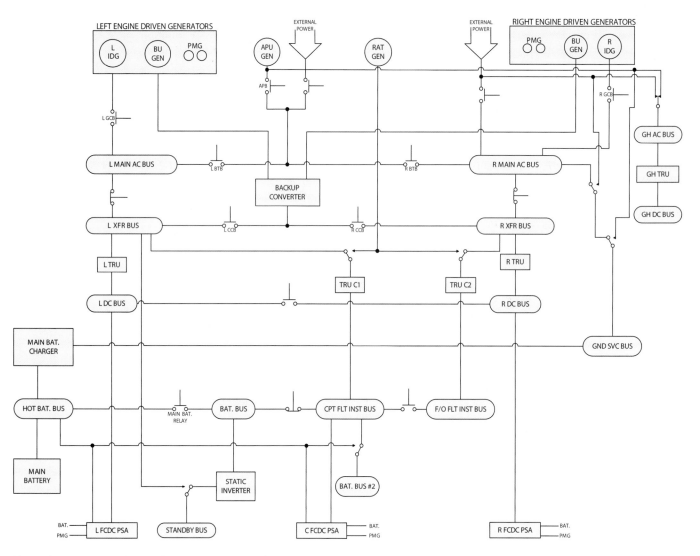

Figure 2-2-12. B-777 Electrical Power System Block Diagram. Key: HX - Heat Exchanger; XFR - Trans/Rectifier; SG - Starter Generator; RAT - Ram Air Turbine; P - Hydraulic Pump; M - Motor; NGS - Nitrogen Generation System; ECS - Environmental Control System; RPDM - Remote Power Distribution Module

ground service relay (GSR), which connects the GS bus to which ever is currently on line, the APU or EXT power.

No Break Power Transfer

The 747-400 uses an automated power distribution control system featuring a no break power transfer (NBPT). A no break power transfer means that the automated system can change the AC power source without a momentary interruption of electrical power. For example, when external power is being used and the aircraft is preparing to depart, the engines are started and the main generators are brought on line. During a NBPT, the generator control units monitor the power source currently on line (the external power) and the power source requested by the flight crew (an engine-driven generator). If the power requested is within specifications, both power supplies are paral-

leled for a split second and no power interruption occurs. If the requested power is out of limits the GCU tries to adjust the system and then connects the requested power to the busses. If the power systems cannot be adjusted to the correct tolerance for paralleling, the requested power source will be rejected or there will be a momentary power interruption.

During the power transfer, both sources are paralleled for a fraction of a second. This can only happen if the individual power sources are within extremely tight voltage, frequency, and phase limits. For example the 747-400 requires any difference between generators to be within 10 V, 6 Hz, and 90° maximum phase difference to complete a NBPT. If the NBPT fails it is most likely that one of the power sources is out of limits. This is particularly true when switching from ground power to aircraft power. Many ground power units are poorly regulated and cannot meet the specifications for the NBPT.

The Boeing 777 Electrical Power System

The B-777 is an aircraft heavily reliant on electrical power for basic flight operations. For this reason, this aircraft contains a total of seven engine-driven generators and one generator driven by the ram air turbine. The B-777 electrical power system is unique in several respects; therefore, a brief discussion of the system will be presented here. Since this is a twin engine aircraft, the basic AC power system is very similar to the traditional split bus power distribution system discussed earlier. This aircraft however, employs two additional backup AC generators, two DC generators specifically available for flight controls, and employs a modern electronic power control system. A simplified B-777 power distribution system diagram is shown in Figure 2-2-12. Please refer to this diagram during the following discussion. The two main AC generators are traditional integrated drive generators (IDG) driven by the two main engines. Each generator has an output of 115 VAC at 400 Hz with a maximum output of 120 KVA.

Since the B-777 is a fly-by-wire aircraft, each engine drives a backup generator used in the event of an electrical emergency. The two backup generators are variable speed constant frequency (VSCF) type units driven directly by each engine without the use of a constant speed drive. An electronic converter circuit, located in the main equipment bay, converts the variable frequency generator output into a usable 115 VAC, 400 Hz with a maximum output of 20 KVA. It should be noted that this output is considerably less than the main generators. The backup generators (BU Gen) are located in the upper right and left of the diagram (Figure 2-2-12). The converter circuitry (near the top center of the diagram) receives power from the BU generators, converts the power to usable AC and sends the current through two converter circuit breakers (the R-CCB or L-CCB) to power the respective right or left AC transfer bus. This backup power is available to the AC transfer busses during any autoland flight configuration, even if all other systems are operating normally. The backup generators also supply AC power whenever the main generator systems fail.

The backup generator assemblies also contain two permanent magnet DC generators that are used to power three flight control direct current (FCDC) power supply assemblies (PSA). The FCDC power supply assemblies are used during an emergency to send direct current (DC) to the flight control systems necessary for aircraft control. The power control assemblies are located along the bottom of the diagram in Figure 2-2-12. Here it can be seen that the FCDC PSA also receives battery current in the event all generators fail.

Like other split bus power distribution systems the B-777 employees a ram air turbine (RAT) to generate electrical power in case of a catastrophic power failure. The RAT is used to power standby systems through the transfer busses (L/R XFR BUS) with a maximum output power of 7.5 KVA. The output of the emergency generator is 115 VAC with a frequency between 392 and 510 Hz. The output frequency is determined by the speed of the RAT propeller. The generator attaches directly to the RAT strut/propeller assembly. Since this is a critical, emergency only system the RAT generator and GCU are operationally tested every 6,000 hours of aircraft flight time.

The power distribution system of the B-777 is controlled through a modern electronic load management system (ELMS). All electrical power goes through the ELMS for distribution throughout the aircraft. There are several ELMS power control panels containing the load switching devices that distribute power. ELMS receive data from various systems and flight deck control panels through an ARINC 629 data bus in order to monitor, distribute, and protect electrical power. The electrical load management system replaces complex relay logic circuit cards found on previous aircraft. It should be noted that ELMS does not control large current electrical loads. These heavy loads are controlled directly by the BPCU and backup generator load switching devices.

The B-777 relies on more computer technologies than any previous generation aircraft. Since computers operate mainly on direct current, this form of power is essential to flight safety. For this reason, the B-777 employs five transformer rectifier units (TRU). Two are fed from the traditional source, the R/L main AC buses. Two are fed from the R/L transfer buses, and one is fed from the ground handling AC bus. The B-777 power distribution diagram, Figure 2-2-12, shows the captain's and first officer's flight instrument busses are supplied power from several sources. These two busses are used to power critical flight instruments for both pilots and are considered essential for flight safety. During normal flight configurations the two instrument busses receive DC power from the transformer rectifier units (TRU) C1 or C2. The captain's and first officer's flight instrument busses can also be powered directly from the main battery bus or from battery bus number two. This configuration ensures extremely reliable DC power to the flight instrument busses.

115 VOLT ALTERNATING CURRENT

	VARIABLE FREQUENCY (370-770 HZ)	CONSTANT FREQUENCY (400 HZ)
Normal Operation	**4 Variable Frequency Generators (VFG)** Main Engine Drive (150KVA Output)	**4 Ground Power Units(GPU)** (90 KVA Output) **2 APU Generators** (120KVA Output)
Emergency Operation	**Emergency Generator** Ram Air Turbine (70KVA Output)	**Static Inverter (STAT INV)** Powered By DC System

28 VOLT DIRECT CURRENT

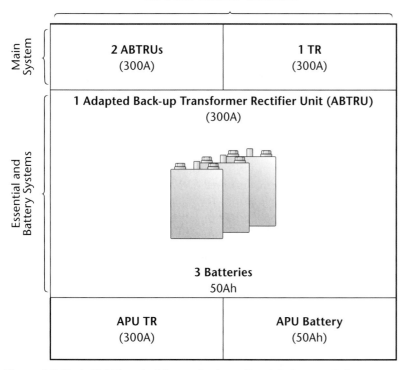

	2 ABTRUs (300A)	1 TR (300A)
Main System		
Essential and Battery Systems	**1 Adapted Back-up Transformer Rectifier Unit (ABTRU)** (300A) **3 Batteries** 50Ah	
	APU TR (300A)	APU Battery (50Ah)

Figure 2-2-13. A-380 Electrical Power Options: Top-AC, Bottom-DC

The More Electric Airplane

The future of aircraft electrical systems is still unclear, yet it is quickly becoming obvious that the *more electric airplane* will replace older designs and incorporate additional electrical and electronic systems. Conventional hydraulic and pneumatic systems will be replaced with electrically operated devices, complex mechanical systems will be replaced with solid state electronics, and many engineers predict the main turbine engines will someday be replaced with a total electric powerplant. The Airbus A-380 aircraft, introduced in 2007, has already incorporated more electrical systems than any other aircraft, and the Boeing 787, still under development at the time this text was written, promises a new level of electric airplane. The power distribution system found on these two aircraft will be discussed here and can be expected to shape the future of the more electric airplane.

The A-380 Electrical Power System

The A-380 is a large four-engine transport category aircraft designed using concepts that increase the use of electrical systems in order to decrease aircraft weight and increase efficiencies. The A-380 has six engine-driven AC generators; the main engines drive four generators and two are driven by the APU. There is one APU (auxiliary power unit) mounted in the tail of the aircraft that drives both APU generators. The APU can operate both on the ground and in flight. The four main generators are capable of a 150 KVA, 115 VAC output with a frequency range of 370 to 770 Hz. Generator frequency changes according to engine RPM; hence the name variable frequency generator (VFG). The two APU generators are capable of a 120 KVA, 115 VAC with a constant output frequency of 400 Hz. The APU speed is held constant to control AC frequency. The aircraft also employs a ram air turbine (RAT) and a static inverter system for emergency AC power. During normal operation the DC power system is fed through multiple transformer rectifier units, which change 115 VAC supplied by the generators into 28 VDC. There are three 50 amp-hour batteries for DC back up and a separate battery for APU starting. The various power supply units and their related functions are shown in Figure 2-2-13.

An aircraft electrical system of this complexity requires a sophisticated electrical power distribution system. In order to ensure safe operation during all flight conditions, the electrical power distribution systems are divided into two major categories and additional subsystems as shown in Figure 2-2-14. The main distribution system carries the majority of all electrical power for normal aircraft operations. The main distribution system is divided into two subsystems, primary and secondary. The primary distribution is controlled through the primary electrical power distribution center (PEPDC) located in the electrical equipment bay just below the flight deck. This load center is divided into side 1 and side 2 (left and right) and receives commands from the flight deck electrical panel and sensors that monitor the system throughout the aircraft. The PEPDC contains the components for power distribution, network management, and circuit protection. The PEPDC controls electrical power to the secondary distribution system, the emergency distribution system, and all electrical loads of 15 amps or greater.

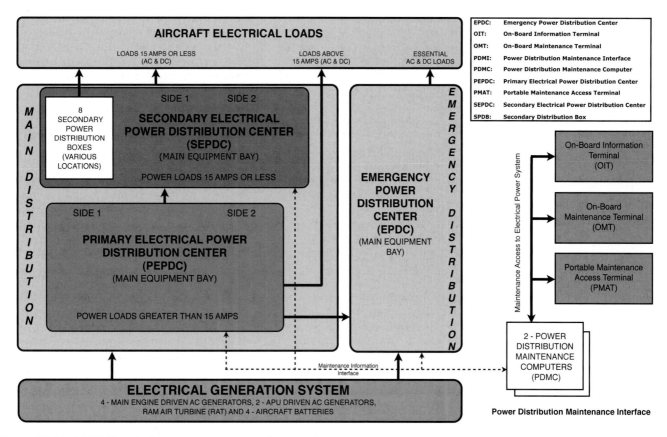

Figure 2-2-14. A-380 Electrical power system general categories

The secondary distribution system is controlled by two secondary electrical power distribution centers (SEPDCs) and eight secondary power distribution boxes (SPDB). The secondary system controls and protects electrical circuits that consume less than 15 amps of current. The two SEPDCs (one fed by primary distribution side 1 the other fed by side 2) are located in the electrical equipment center. The eight SPDBs are located in various sections of the aircraft and act as local distribution centers. This distribution concept helps to reduce long runs of wire from one central location, hence saving wiring complexity, weight, and of course money.

The emergency distribution system is controlled through the emergency power center (EPC) located in the equipment bay. As the name implies, equipment powered by the emergency system is the most critical to flight safety and must be the last to fail in the event of an emergency. Therefore, the EPC can receive power from several sources including all engine-driven generators, the RAT generator, and three nickel-cadmium batteries. The EPC also controls the operation of the aircraft's static inverter in order to convert emergency battery power into essential AC power.

The A-380 power distribution system operates similar to a typical split bus system in that each

generator is completely independent of the others. The AC power from one generator is never connected in parallel with other AC power and therefore, the phase relationship and frequency of each generator can operate in a wide range. This independent concept offers greater flexibility and simplicity allowing the A-380 to utilize variable frequency generators (VFGs) that do not require a constant speed drive (CSD) unit for frequency control, thus eliminating complex mechanical components. It is important to note that the output of a VFG will have a frequency between 370 to 770 Hz. In order to accommodate this range of AC frequencies, the power distribution system and AC loads must all be designed accordingly.

Figure 2-2-15 shows the main AC power sources along the top of the diagram. The main power sources include the four engine-driven variable frequency generators, APU generator A and B, and the four external power connections. The external power connectors are used to supply power to the aircraft while the aircraft is parked. Each external power connector can supply up to 90 KVA at 115 VAC with a constant frequency of 400 Hz. Each external power connector is associated with a given AC bus (for example, EXT. 1 feeds AC bus 1). During normal flight operations, the four main generators supply up to 150 KVA to their respective AC busses (AC bus 1 through 4). In the event of a main generator failure, APU generator A or B can supply power to one or more AC busses.

If all other generators fail the RAT generators (located at the top/center of the diagram) can be used to feed emergency power to the AC essential bus. The static inverter can also supply emergency AC power to the aircraft if all generators fail. The static inverter is fed by the DC power system and converts DC power to AC using electronic circuitry. The static inverter supplies only 2.5 KVA of 115 VAC power to the AC emergency bus to operate only essential AC systems needed to land the aircraft immediately.

The A-380 DC power distribution system is located on the bottom half of Figure 2-2-15. There are multiple sources for DC power; the aircraft's batteries, as well as the transformer rectifier units. Located along the very bottom of the diagram are four batteries that are the sole source of electrical power if there are no operable generators, or external power available. The batteries can be used for engine (APU) starting or during flight in an extreme emergency. During normal operations the AC systems feed three adaptive back-up transformer rectifier units (ABTRU), and two TRUs to power the DC loads. The ABTRUs and TRUs receive 115 VAC and employ transformers to step down the voltage to 28 VAC. The units then rectify the 28 VAC changing the power to 28 VDC.

Control and protection for both the AC and DC power distribution systems are accomplished through a series of contactors (Figure 2-2-15), which connect or disconnect the busses and power supplies. The electrical power contactors are controlled by pilot commands through the main electrical panel located on the flight deck. All pilot commands are sent to control units in the electronics equipment bay, which in turn sends commands to each contactor. These power control units also continuously monitor the system for faults and will automatically reconfigure electrical contractors to maintain safe power distribution.

Maintenance of the A-380 electrical power distribution system. The A-380 incorporates two power distribution maintenance computers (PDMC) that allow maintenance personnel access to, and control of electrical power components. During maintenance activities, technicians would typically use one of three access terminals for information and control of electrical power. The access terminals are known as: the *on-board maintenance terminal* (OMT), *on-board information terminal* (OIT), and the *portable maintenance access terminal* (PMAT). The OMT and OIT are permanently installed

on the aircraft, and as the name implies the PMAT is a portable unit used from various locations within the aircraft. Each of these access points is similar in design to a personal or laptop computer, allowing technicians direct access to information about electrical power systems during maintenance. The technician can also use these terminals to open, secure, tag, and close any electric system contactor while the aircraft is on the ground. All requests from any maintenance terminal are sent to the PDMCs, which in turn controls electrical components through the primary and secondary distribution systems located in the electrical equipment bay. Diagnostics for the power distribution system is collected, sorted, and stored through a central maintenance information system, which will be discussed later in this text.

The Boeing 787 Electrical Power System

At this time, since the aircraft is still under development, there is only limited information concerning the power distribution systems found on the B-787. However, the following discussions will present the major design concepts and various components of this revolutionary electrical power system. The B-787 is a twin engine aircraft designed to carry up to 330 passengers. The aircraft is designed with a strong emphasis on reliability and maintainability. In order to accomplish the goal of 30 percent lower airframe maintenance costs compared to similar aircraft, Boeing incorporated the philosophy of replacing traditional mechanical systems with electrical/electronic components. For example, the engine bleed air pneumatic starting system typically used on large turbine fan engines was replaced with an electric starter system. This is the first aircraft of this size to incorporate an electric motor to start the main turbine engines. Many of the other traditional hydraulic and pneumatic systems have also been converted to electrical systems on the B-787. These systems include cabin air conditioning, pressurization, motor-drive hydraulic actuators, and wing ice protection. The B-787 also incorporates an electrically powered nitrogen generation system used to create a nitrogen vapor for fuel tank safety. As the fuel tanks empty of liquid, the tank space is filled with nitrogen. This greatly lowers the risk of explosion.

The huge electrical demands of the B-787 require power generation unsurpassed by previous aircraft. The B-787 employs four main engine starter generators and two APU starter generators for a combined output of 1.45 megawatts of electrical power. Each main

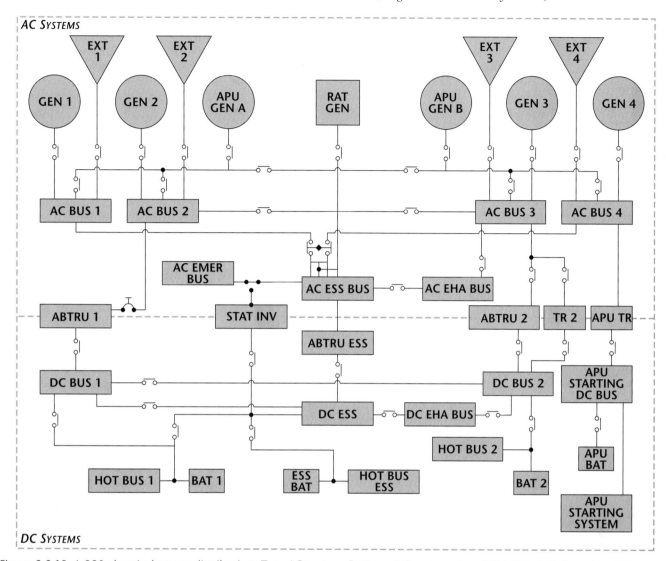

Figure 2-2-15. A-380 electrical power distribution: Top–AC system, Bottom–DC system. Key: STAT INV–static inverter, APU–auxiliary power unit, TRU–transformer rectifier unit, ABTRU–adapted backup transformer rectifier unit, EXT–external power, ESS–essential, GEN–generator, RAT–ram air turbine, EMER–emergency, EHA–electro-hydrostatic actuators

engine starter generator is capable of an output of 250 KVA and the APU generators can output 120 KVA each. It should be noted that these generators are also used for starting the engines and are therefore "starter generator" units that remain engaged with the engine gear box during both start and generation modes. During generation mode, the units produce an alternating current output with a variable frequency. The units are called variable frequency starter generators (VFSG). The VFSG units are similar in design to the variable speed constant frequency (VFSG) generators discussed earlier and found on the Airbus A-380. The output frequency of the B-787 generators depends on engine speed and has a range of 360 to 800 Hz. The main generators on the B-787 can output nearly twice the power as those found on the A-380 and the output voltage is 235 VAC. Using a voltage higher than traditional systems (235 VAC instead of 115 VAC) allows for reduced current flows, there-

fore reducing the size of feeder cables, electrical components, and control devices. This helps to reduce aircraft weight, increase reliability, and decrease costs.

During a normal main-engine start, both starter generators are powered in start mode for optimum performance. If one starter generator should fail, one unit can be used for engine starting but at a slower pace. Since the APU is a smaller engine and requires less power for starting, only one starter generator is set to start mode and that unit can easily be powered by the aircraft's battery. The output from the APU generators can be used for starting the main engines. The power from the left-side main generators can be used for starting the right engine, or vice versa. Ground power can also be used for engine starting.

In start mode, the starter generators run as synchronous motors controlled by high power

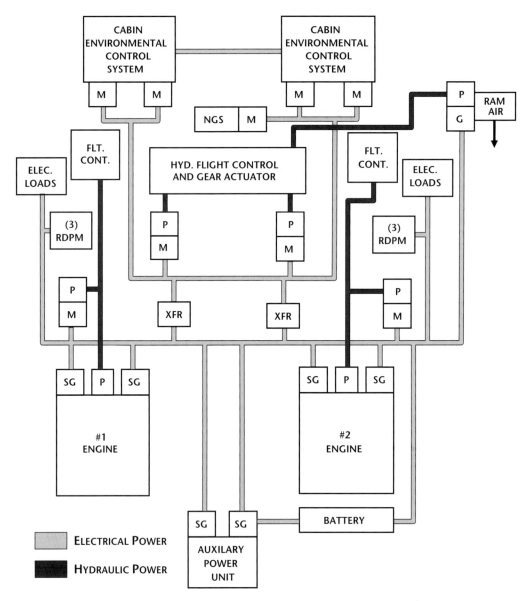

Figure 2-2-16. B-787 Electrical power distribution diagram

start converters. The start converters provide adjustable voltage and frequency for optimum start speed under different load conditions. Once any engine has reached a consistent run speed, the motor/generator controllers modify the circuitry so the starter generators now operate in current generating mode to power aircraft loads.

The six engine-driven generators found on the B-787 are each connected to a highly electronic power distribution system (Figure 2-2-16). The aircraft employs a hybrid system supplying four different voltage values: 235 VAC, 115 VAC, 270 VDC, and 28 VDC. The 115 VAC and 28 VDC power supplies most traditional systems. The 235 VAC and 270 VDC power is used mainly for high horsepower electric motors. The electric motors are used to replace systems traditionally powered by hydraulics and pneumatics. The 235 VAC power is sent to trans-

former rectifier units to create the 28 and 270 VDC power.

There are two electrical equipment bays found on the B-787 (one forward, one aft), which contain the majority of the electronic control circuitry needed for power distribution. The forward equipment bay distributes power to the majority of the electrical loads at 115 VAC or 28 VDC. The 235 VAC electrical equipment is supplied through the aft bay. In order to minimize the number of long distance power cables, there are 17 remote power distribution modules located in various places of the aircraft. The remote power distribution modules (RPDM) control current to over 1,000 low- and medium-sized electrical loads using advanced modular-design solid-state technologies. The RPDMs replace the traditional thermal circuit breakers and contactors found on older aircraft. Heavy loads, such as the two 88 horsepower electric motor-driven hydraulic pumps, are controlled by individual motor controllers, which handle up to 100 KVA each.

During generation mode, six generator control units (GCU) monitor and regulate the six generators (four main and two APU) and distribute power to the main bus system. Power from the four engine-driven generators supplies current to over 1,300 electrical loads through a complex network of distribution busses, power distribution modules, motor controllers and electrical contactors. During normal flight configuration, the four main engine-driven variable frequency starter generators supply all electrical power. In the event of a generator failure, there are two APU generators available, and in the event both main and APU engines fail, the aircraft deploys a ram air turbine (RAT) for emergency electrical power.

Powering the Aircraft for Maintenance Purposes

Aircraft have some means to provide electrical power to various systems without starting the main or APU engines. On light aircraft, the battery can be connected to the bus and used as a temporary power source for troubleshooting or maintenance of various electronic or avionics systems. Alternating current, if used, would be available through the aircraft's static inverter.

> **CAUTION:** *The battery is being discharged during this configuration. Always limit maintenance activities using battery power, and recharge the battery if needed.*

All corporate and transport category aircraft will accept external power for prolonged ground operations of electrical systems. Many smaller general aviation aircraft also have external power receptacles. Whenever practical use external power, not battery power, to supply electricity to the aircraft. When connecting external power to any aircraft, always refer to the operations or service manual if you are unfamiliar with the procedures. Always observe all safety precautions and be sure to warn other technicians working on the aircraft before you connect or disconnect external power. With the aircraft in certain configurations, application of external power can activate systems that could injure fellow technicians and/or damage the aircraft.

The following procedures are typical of those used to connect external power to the Beechcraft King Air. First, ensure the aircraft is in a safe configuration. Be sure no wiring or other systems are disassembled, which may cause damage if power is connected, and

Figure 2-2-17. Installation of a typical external power connection

inform all technicians working on the aircraft that external power will soon be connected. Second, place the battery switch in the *on* position (the aircraft's battery must be at least 20 V). Third, turn the avionics master off. Lastly, connect the external power plug to the aircraft, being sure the plug is seated firmly into the socket. Figure 2-2-17 shows the external power socket and related connections for a King Air. Like most aircraft, the polarity of the external power plug must be correct for the external power relay to close.

The King Air maintenance manual states several precautions that must be observed when connecting external power, briefly stated they are:

1. If unknown, determine the polarity of the external power supply plug.

2. Turn off all radio/avionics equipment and both engine generators before connecting external power. Typically the avionics master switch can be used to disconnect all avionics simultaneously.

3. A battery of at least 20 volts must be in the aircraft and the battery master switch must be *ON* prior to connecting external power. This allows the normally closed avionics master relay to energize and disconnect the avionics bus.

4. The external power supply must be capable of the correct amperage and voltage when used for engine starting purposes. For the Super King Air 200, a continuous 300 amperes at 24 to 30 volts and 1,000 amperes for 1 second must be available.

5. Proper ground connections are needed to ensure voltage stability at high amperage.

6. The aircraft battery may be damaged if the voltage applied is in excess of 30 volts.

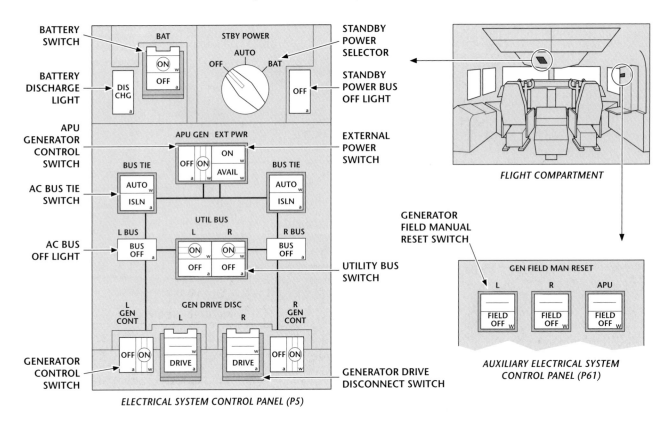

Figure 2-2-18. Electrical control panels for a Boeing 757

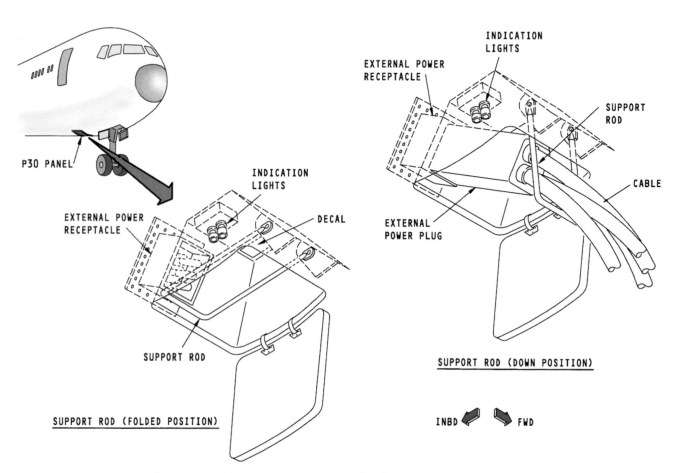

Figure 2-2-19. Typical external power panel for a transport category aircraft

Courtesy of the Boeing Commercial Airplane Company

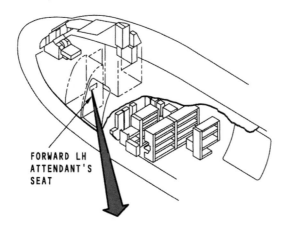

GROUND SERVICE ON LIGHT

ILLUMINATED (WHITE) - GROUND SERVICE
BUS IS POWERED DIRECTLY FROM EXTERNAL
POWER OR THE APU GENERATOR.

EXTINGUISHED - GROUND SERVICE
BUS IS UNPOWERED OR POWERED FROM THE
RIGHT MAIN AC BUS.

Figure 2-2-20. Forward flight attendant's panel and ground service switch   *Courtesy of the Boeing Commercial Airplane Company*

Most of the above suggestions apply not only to the King Air, but also to all general aviation corporate aircraft.

On transport category aircraft external power typically supplies 115 VAC, 400 Hz. The 28 VDC is made available through one of the aircraft's transformer-rectifier units, which converts the incoming 115 VAC to 28 VDC. Figure 2-2-18 shows the electrical system control panel used to activate electrical power on a Boeing 757 aircraft. This is the P5 panel located on the center overhead panel of the flight deck.

The P5 panel controls all power distribution functions for the aircraft using alternate action and momentary contact type switches. The alternate action (*IN* or *OUT*) switches are latched to the last operated position. A mechanical flag located inside the switch indicates the switch position, such as, *ON* or *AVAIL*.

Connecting External Power

To connect external power to this aircraft the following steps must be taken.

1. Use caution. Be sure it is safe to connect electrical power to the aircraft. If other maintenance is being performed, certain systems may be powered inadvertently when applying external power. Or, if portions of the electrical system are disconnected for maintenance, connecting external power may create a short to ground and damage the electrical system and/or the aircraft structure.

2. The external power cord must be plugged into the aircraft external power panel located near the aircraft's nose wheel (Figure 2-2-19). If the connected power is within specifications, the AC CONNECTED light on the external power panel and the external power switch available (AVAIL) light on the P5 panel will illuminate. Also, the ground handling bus will automatically receive power. The ground handling bus will supply power to necessary lighting for minor maintenance and cleaning of the aircraft.

3. Locate the forward flight attendant panel (P21) just inside the aircraft entrance (Figure 2-2-20). By depressing the ground service switch on P21, the ground service bus will be powered by external power. This bus supplies equipment that must be powered for both air and ground modes. Figure 2-2-21 shows that the ground service bus is connected to the right AC bus during flight and to external power or APU, if selected, during ground operations.

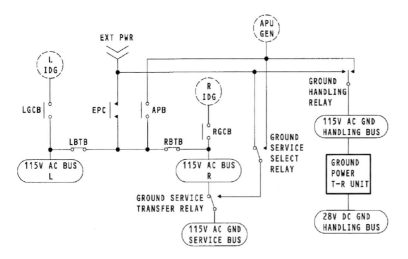

Figure 2-2-21. AC external power distribution

Courtesy of the Boeing Commercial Airplane Company

4. If it is necessary to power the left or right AC busses, depress the external power switch on the P5 panel, Figure 2-2-18. This will connect external power to the aircraft bus ties. If the bus tie switches are latched in the *AUTO* position, the electrical power will automatically connect to the AC busses. If the bus ties are in the isolation mode, depress the appropriate bus tie switch to activate its associated bus.

Digital Circuits and Basic Computer Operations

The basic building blocks for any computer system are silicon wafers manufactured to contain millions of microscopic transistors. These transistors are combined with resistors, diodes, and other electronic components to create logic gates, which are massproduced into a single unit, known as a microprocessor. A typical microprocessor (Figure 2-2-22) is a collection of complex digital circuits, which perform specific functions. In general, microprocessors can be thought of as miniature computers, such as a hand-held calculator. Aircraft computers contain one or more microprocessors and the related control circuitry needed for the operation of the system or subsystems. For the most part all computers can be broken into four basic elements: (1) central control unit (CCU), (2) arithmetic logic unit (ALU), (3) central processing unit (CPU), and (4) memory.

The central control unit (CCU) connects each section of the computer to coordinate control of all activities of that system. The CCU is typically designed to operate in steps, or stages according to the software programming for that computer. The CCU completes one task then moves to the next, and eventually completes each operation through a series of smaller steps. The CCU receives its instructions from the central processing unit (CPU). Computer technologies have continued to shrink in size with each new generation; with this miniaturization comes the ability to combine the operations and functions of certain systems. Therefore, the CCU and CPU may be contained within the same element of a given computer and the terms become difficult to distinguish from one another. These two computer elements (CPU and CCU) often share similar functions and in some cases share electronic circuitry.

Other elements of a typical computer worth noting are the arithmetic logic unit (ALU), the clock (or timer), and the memory. The ALU performs the various calculations necessary for manipulating the binary digits that eventually become the outputs for that computer. All complex computers perform thousands of calculations using thousands of digits (binary ones and zeros) to complete their given task. These calculations are performed by the ALU. Each computer must adhere to a strict timing format in order to complete each task; therefore, each computer must contain some type of clock circuitry. The clock or synchronizer allows all sections of the computer to operate in sync, so all operations are performed in the proper sequence and within the correct time limits. Computer memory comes in two basic types, volatile and nonvolatile. Non-volatile memory is a permanent memory which is not lost if the computer looses power. Nonvolatile memory must provide information for the basic operations of the microprocessor/computer, and may be used for the storage of other data as well. Volatile memory is used as a "note pad," or short-term storage of data needed during the manipulation of numbers. Volatile memory is lost whenever the computer power is termi-

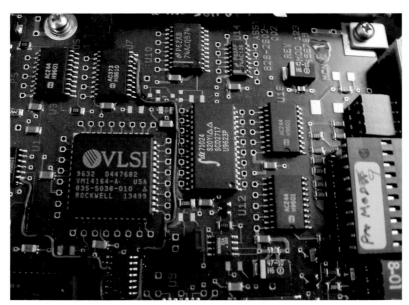

Figure 2-2-22. Photo of a typical microprocessor

nated or battery backups deplete their energy. On most aircraft computer systems both volatile and nonvolatile memory are contained on solid-state integrated circuits.

Two additional computer elements typically contained outside the main computer circuitry are the digital data bus(ses) and the system peripheral device(s). The digital data bus is a pathway for the transmission of data between computer subsystems. Smaller, self-contained, computers may contain an internal data bus built into the unit, installed onto a printed circuit card, or even built into the microprocessor circuitry. In more practical terms, most aircraft computers are connected to multiple subsystems, which are all connected through one or more digital data busses. In most aircraft, the data bus consists of a pair of 24 gauge copper wire twisted together and shielded to protect the digital signal from interference. Some of the latest aircraft employ fiber optic cable to transmit the digital signal to the various computer systems, however, traditional copper wire is still used for the majority of digital data transmissions. Fiber optic cables will be covered later in this text.

A peripheral device is a component that allows the computer to communicate with humans and/or other electronic devices. Typical peripheral devices include a keyboard, used to input data into the system; or an LCD display on the instrument panel, used to display output data from the computer to the pilots. In many cases a peripheral device is another aircraft LRU or component that operates transparent to the pilots. An example of this type of peripheral is the electronic engine control servos. A servo is a device that receives an instruction signal, often from a computer, and performs a mechanical function, such as adjusting engine fuel flow. The servo is considered a peripheral since it receives information from the computer and performs a mechanical function.

Multiplexing and Demultiplexing

Many avionics systems receive data inputs from a variety of sensors or subsystems located throughout the aircraft. In order for these remote devices to communicate with the avionics' computer circuitry, the input data must be received in the correct format. Most digital data is transmitted in a *serial* form, that is, only one binary digit at a time. However, many subsystems or individual sensors output data in *parallel* form. Therefore, some means are needed to convert parallel data into serial data and vice versa.

As discussed earlier in this chapter, transmission of data in serial form means each binary

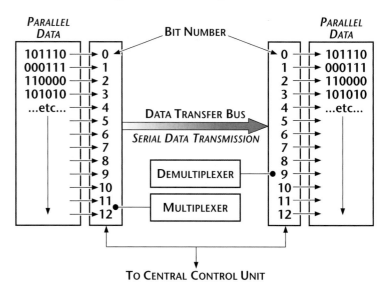

Figure 2-2-23. Parallel data converted to serial data, transmitted, and converted back to parallel data

digit is transmitted for only a very short time period. After one bit of information is sent, the next bit follows; this process continues until all the desired information has been transmitted. This type of system can be thought of as *time-sharing*, because each transmitted signal shares the wires for a short time interval. The time required to transmit the data is also known as *bandwidth*.

Parallel data transmission is a continuous-type transmission requiring two wires, or one wire and one ground, for each signal to be sent. Parallel transmission receives its name because each circuit would be wired in parallel with respect to the next circuit. One pair of transmitting wires may be used to handle enormous amounts of serial data. If the information signals were transmitted in parallel form, hundreds of wires may be required to perform a similar task.

Serial data transmission requires less wire than a parallel system; however, an interpretation circuit is needed to convert all parallel data to serial-type information prior to transmission. The device that converts parallel data into serial form is called a multiplexer or (MUX) circuit. The unit used to convert serial data into parallel form is called a demultiplexer or (DEMUX) circuit. As illustrated in Figure 2-2-23, parallel data is sent to a multiplexer, where it is converted to serial data and sent to the data transfer bus. The data transfer bus is a two-wire connection between the multiplexer and the demultiplexer. The demultiplexer receives the serial data and reassembles it into parallel form. In this example, the byte 10100 is being received by the multiplexer in parallel form. Starting at the top and working down, the multiplexer transmits each digit individu-

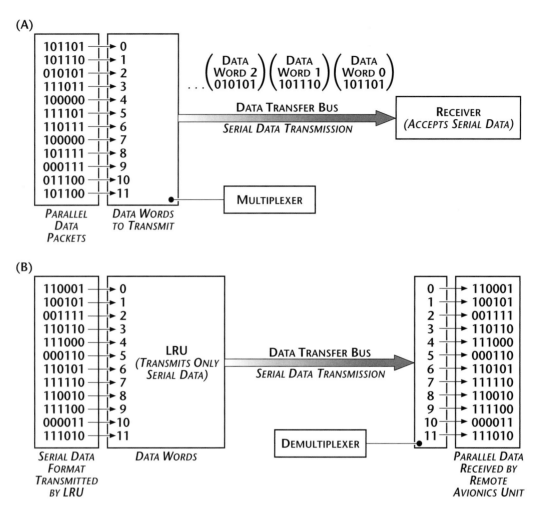

Figure 2-2-24. Serial and parallel data transmissions: (A) Multiplexer located in the transmitter, (B) demultiplexer located in the receiver

ally. Bit number 0 is the first to be transmitted. Bit number 1 is the next digit transmitted, bit number 2 next, and so on. This system repeats until all the parallel data are converted to serial form and individually connected to the data transfer bus. The demultiplexer receives a serial data input from the transmission bus and reassembles it into a parallel form. The output of the demultiplexer is identical to the input of the multiplexer (10100).

The circuitry of Figure 2-2-23 contains both a multiplexer and demultiplexer. This arrangement is used only when both the transmitter and receiver require parallel data formats. If the transmitter uses parallel data and the receiver accepts serial data, only a multiplexer would be used (Figure 2-2-24A). If the LRU transmits serial data and the receiver requires parallel data a demultiplexer would be used (Figure 2-2-24B). In most cases, the data sent to the multiplexer is more than one single data bit. In this case, a series of data packets (data words) are sent to each input connection of the multiplexer. Figure 2-2-24 shows the multiplexer inputs as data words containing six bits each.

These packets of data are then sent, one bit at a time, through the serial bus to the receiver. The multiplexer converts the incoming data into a series of data packets with each bit lasting only a fraction of a second. Each data packet is then transmitted to the receiver through the serial bus. The receiver software contains the program information needed to reassemble the data into useful information.

On systems that share a serial data bus, some means of control must be used to coordinate the MUX/DEMUX arrangement. As shown in Figure 2-2-23, a CCU (central control unit) can be used to coordinate the transmission and reception of data. This control is essential to ensure all serial data is transmitted and received at the proper time intervals. This system of serial data transmission may seem somewhat complex; however, the alternative, parallel data transmission would require one wire for each data bit to be transmitted. Since thousands of bits of information are transferred among various airborne systems, serial data transmission techniques are the obvious choice for modern aircraft.

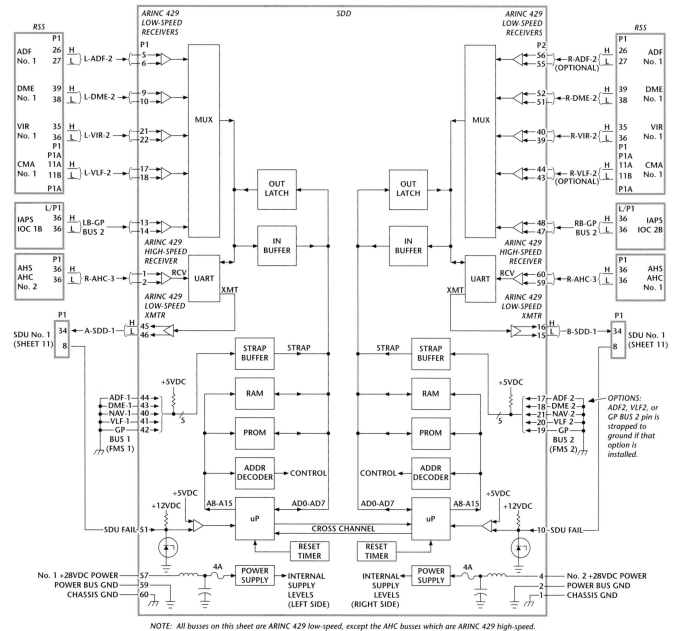

Figure 2-2-25. Example of multiplexers used within an LRU

The use of multiplexers and demultiplexers is only necessary when a change from serial to parallel (or vice versa) is required. In many cases, serial data is transmitted to another component, which can "read" serial data. In this case, no change in format is required.

When multiplexers or demultiplexers are required for operation of a system, they are typically contained within another unit. For example, Figure 2-2-25 (top left and top right side of diagram) shows the data from the RSSs (radio sensor system) to the SDD (sensor display driver) is sent directly into a multiplexer (MUX) circuit. In troubleshooting this system, the line technician may never know the MUX circuit exists. If the internal multiplexer fails,

the line technician may detect a defective SDD, the bench technician, on the other hand, could determine the MUX circuit inside the LRU was inoperative.

MUX/DEMUX Systems for In-Flight Entertainment

On many transport category aircraft, MUX/DEMUX circuits are used for the control of passenger entertainment. Each passenger has a control (installed in the armrest of the seat) used to select the audio and video for in-flight entertainment. In this case, multiplexers are often stand-alone units and can be replaced individually by the line technician. Each indi-

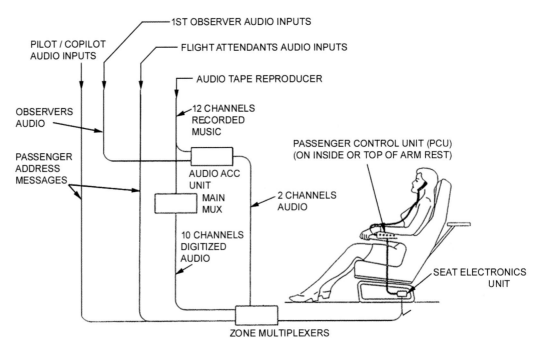

Figure 2-2-26. Multiplexers used in a typical passenger entertainment system

vidual armrest control would transmit parallel data. A group of controllers would be sent to an area multiplexer where the data is assembled and transmitted to the passenger entertainment control center, or main multiplexer. Data from the entertainment control center is transmitted in serial form and demultiplexed for a group of seats.

On some aircraft, the multiplex system controls passenger entertainment, passenger reading lights, and flight attendant call lights. The MUX/DEMUX units are often located beneath a group of seats or in the headliners above the seats. Figure 2-2-26 shows a diagram of the passenger entertainment system on a Boeing 767. Notice, this system contains both zone and main multiplexers.

Section 3

Analog to Digital Converters

As the popularity of digital avionics grows, many aircraft employ both analog and digital systems. This creates communication problems because pure digital systems are incapable of reading analog data and vice versa. To allow communication between analog and digital systems, analog to digital and digital to analog circuits were developed. Often referred to as *A/D* or *D/A converters*, these circuits can be found on virtually any system employing both analog and digital devices. Many sensors on modern aircraft create only analog outputs. A temperature monitor is a typical example of a device creating an analog signal. This output must then be converted to a digital format in order to communicate with modern computerized control systems.

A/D converters utilize microprocessor circuitry to convert a given analog signal into a digital format. A/D converters are typically designed for a specific purpose. That is, the conversion circuitry must be programmed to analyze a given analog input and produce a specific digital output. Figure 2-3-1 shows the conversion of an analog signal from a temperature transducer. This A/D converter must be programmed to accept the analog signal with voltage limits of 0 to +20 VDC and output a digital signal of +5 VDC for logic 1, and 0 VDC for logic 0.

Like multiplexers, A/D converters are most often found incorporated into the circuitry of various LRUs. For example, the inputs to a digital autopilot may include some analog signals. These signals would be converted to a digital format inside the autopilot computer (LRU). Since the outputs from the digital autopilot to the servomotors are analog signals, the autopilot circuitry must also incorporate a D/A converter.

For the most part, a line technician would not replace a specific A/D or D/A converter. These units are typically incorporated into a given LRU. If the output signal from an LRU is not

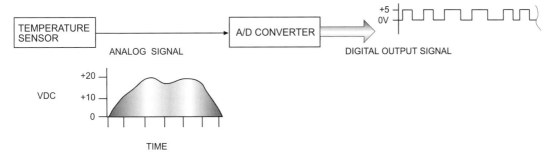

Figure 2-3-1. Conversion of an analog signal to a digital signal

correct, replace the unit, including the A/D and D/A converter.

Section 4

Flat-Panel Displays

Over two decades ago, electronic instruments were introduced for display of aircraft systems data. New electronic instruments, commonly referred to as EFIS (electronic flight instrument systems), are replacing traditional electromechanical instruments. The electronic instrument displays consisted of a CRT (cathode ray tube) similar to a typical television or computer monitor. These newer aircraft were said to have a *glass cockpit*, describing the lack of traditional instruments and the glass face of the CRT. The CRT revolution helped to improve both crew awareness and system reliability. Unfortunately, CRT displays are relatively heavy, large, and consume a lot of power. The many benefits of the CRTs made it very popular. However, the industry had long since looked for a replacement. The answer came in the form of the flat panel display.

A *flat-panel display* (FPD) is a solid-state device used to display various formats of video information. There are several versions of flat panel displays currently available and each type is lighter, smaller, consumes less power, and produces less heat than a typical CRT display. Early FPDs had several limitations. Color FPDs were hard to manufacture and had limited color range. Early FPDs had a slow response time and poor resolution. When viewed from an angle, the display seemed to go blank, and in general these displays were poorly lit. For years, flat-panel displays have been widely used on laptop computers. Computer designers accepted the limitations of the flat panel display since they consume very little power, a very important consideration for portable computers. In fact, it would be safe to say that laptop computers would be virtually impossible without the development of flat-panel displays.

Today the technology exists to produce a flat panel display, which overcomes previous limitations. Modern FPDs equal the brightness, resolution, and update speeds of a typical CRT. Improved production techniques have enhanced reliability and lowered the cost of FPDs. Flat-panel displays are now the display of choice for new aircraft instrument systems. A modern flat panel display is shown in Figure 2-4-1. As the name implies, this unit is less than 2 inch thick. Replacing CRT displays with flat panels provides extra room behind the instrument panel and creates more options in flight deck design. Virtually every aircraft instrument manufacturer has committed to flat-panel displays. The Boeing 777 was the first transport category aircraft to employ all flat panels for primary flight and systems data. Now FPDs are being installed into light aircraft business jets, commuter aircraft, and helicopters. Even

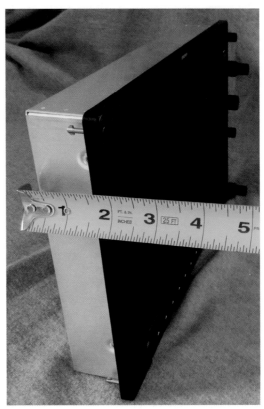

Figure 2-4-1. A typical LCD Flat Panel

Figure 2-4-2. Example of Flat Panel Display in a light aircraft instrument panel

older aircraft can take advantage of flat panel displays. Units, which are virtually drop-in replacements, are being developed for aircraft employing older traditional instruments.

There are several types of flat panel displays currently available or under development that show promise for use in future aircraft. Today the *liquid crystal display* (LCD) is used on many aircraft in both monochrome and full-color formats. In the future, technological developments will bring improvements to LCDs and introduce other types of flat panels.

Liquid Crystal Displays

One of the most popular flat panel displays, LCDs, have been used on aircraft for several years. The Boeing 756 and 767 were two of the first large aircraft to employ LCDs. On these aircraft, LCDs are used as backup instruments for engine data. These early displays were one color, poorly lit and had a limited viewing angle. Newer displays are full color, backlit, and have a wide viewing angle. The Boeing 777 employs the new technology full color LCD

displays. Even light aircraft, like the Cirrus SR-20, have an instrument panel containing LCDs (Figure 2-4-2).

LCD Theory of Operation

Liquid crystals are a substance that exhibits the properties of both a liquid and a solid. As light passes through a liquid crystal the light follows the alignment of the molecules of the crystal. A 19th-century botanist named Friedrich Reinitzer first discovered liquid crystals. In the 1960s it was discovered that an electrical charge could be used to align the crystal molecules and the crystals could be used to control light. Since the early 1970s, liquid crystals have been used for various types of displays.

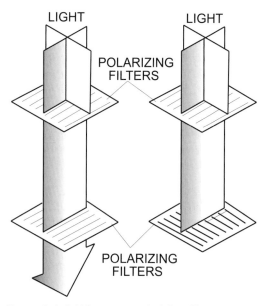

Figure 2-4-4. When two polarizing filters are arranged parallel – light passes through, but when arranged perpendicular – no light passes

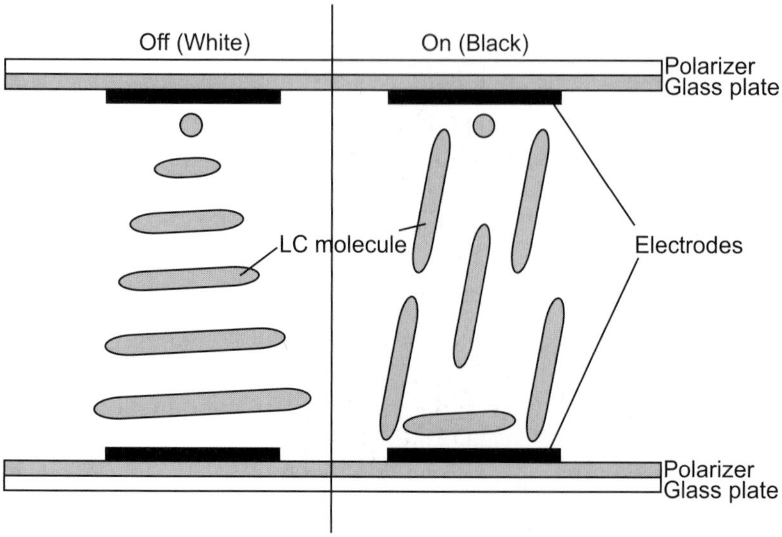

Every liquid crystal display contains several key elements layered together. As seen in Figure 2-4-3, the LCD uses two polarizing filters on the outside layers. Just inside the polarizer are two finely grooved glass plates. Inside the glass plates are the LCD electrodes, and, finally the liquid crystal material is located between the electrodes.

A polarizer is a light filter made of extremely fine lines. The filter could be thought of as an incredibly small blind used to control light through a window. As seen in Figure 2-4-4, if two polarizing filters are arranged perpendicular, they block light from traveling through the filters.

Within the LCD layers, just inside the polarizer, are two grooved glass plates. The grooved

Figure 2-4-3. Diagram of a typical liquid crystal display pixel

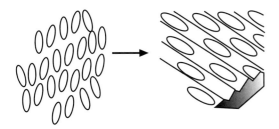

Figure 2-4-5. Liquid crystal molecules aligned with the grooved glass

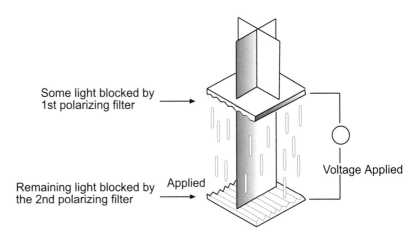

Figure 2-4-6. When a voltage is applied to the LCD electrodes, the liquid crystal molecules align

plates are used to align the liquid crystal molecules. As seen in Figure 2-4-5, liquid crystal molecules naturally tend to align themselves with the grooves in the glass plate. These glass plates are often referred to as alignment plates. The crystal molecules twist light entering the top plate as it passes through the crystal. By applying a voltage to the crystal, the alignment of the molecules can be controlled as seen in Figure 2-4-6.

When the LCD layering is complete, light is controlled by the two polarizing filters and by the liquid crystal molecules. Figure 2-4-7 shows the operation of a basic LCD. When there is no voltage applied to the LCD, light enters the top polarizer, twists through the liquid crystal, and passes through the bottom polarizing filter. When a small voltage is applied to the LCD electrodes, the crystal molecules align and the polarizing filters block light. In a typical LCD, the glass alignment plates and the electrodes are combined into one element. Only a portion of the light passes through the LCD. LCDs always block some light and hence create design challenges.

The previous discussion focused on one liquid crystal element commonly referred to as a *pixel*. In order to form a complete display with high resolution, thousands, or even millions of the individual pixels are grouped together. Each *pixel* is a distinct LCD acting independently of all other pixels in that display. Each pixel creates one dot on the flat panel display. The dots are used to form letters, numbers, and video images. A typical digital watch employs an LCD display with relatively large pixels. If one examines the watch display using a magnifier, the individual pixels can easily be identified (Figure 2-4-8). The number of pixels per square inch determines the display *resolution*. A higher resolution creates a clearer image. Unfortunately, the higher the resolution, the smaller the pixel and the more difficult the display is to produce. To obtain the resolution necessary for a typically full color primary flight display, millions of pixels are required.

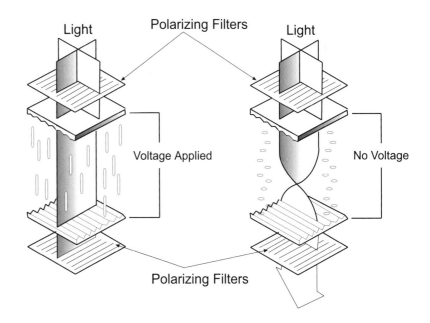

Figure 2-4-7. Operation of a basic LCD pixel

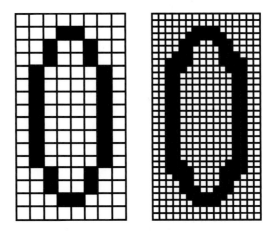

Figure 2-4-8. Pixel arrangement for a low- and high-resolution display

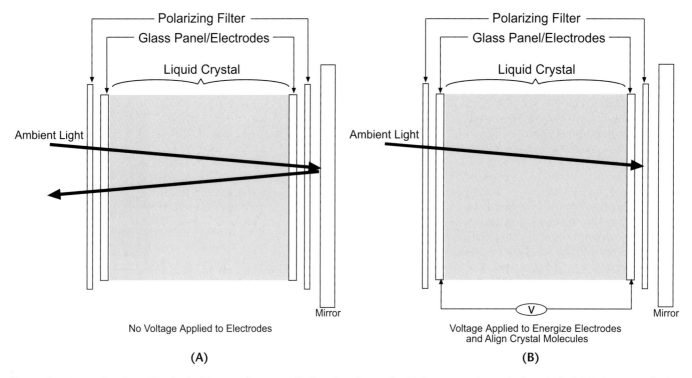

Figure 2-4-9. A reflective LCD pixel: (A) No voltage applied to the electrodes-light passes through the pixel, (B) Voltage applied to electrodes-light is blocked by the polarizing filters

Controlling the LCD

There are two basic methods used to control the operation of an LCD, static and dynamic drives. A *static drive control system* is used to control simple displays containing only a few pixels. The static drive system was first introduced on early versions of flat panel displays. Early displays had a very slow response time and were not well suited for displays requiring rapid change. Static drives basically consisted of a separate wire to each given pixel in the display. Static drives required so many wires that large displays containing thousands of pixels are impractical.

A *dynamic LCD drive* employs a system that shares wires to the various pixels on the display. The wiring is typically set up into a grid system that allows for large displays with fewer electrical connections. Today, the most common dynamic drives are the active matrix and the passive matrix. A *passive matrix LCD* employs a grid of conductors with pixels located at each intersection of the grid. Current is simultaneously sent through two conductors on the grid to control the illumination of each pixel. An *active matrix* has a transistor located at each pixel intersection, requiring less current to control the illumination of the pixel. Since less current is used, each pixel can be switched on and off more frequently and the screen response time is greatly improved. Active matrix displays are often referred to as *thin film transistor* (TFT) displays. Active matrix LCDs are used for air-

craft instrument and other displays requiring a fast update rate. Passive matrix LCDs are suitable for calculators and other displays that do not require a rapidly changing image. Passive matrix LCDs typically found on aircraft are used for secondary instruments or simple data displays.

Transmissive vs. Reflective

LCDs can be illuminated using reflected ambient light or by passing light from a dedicated source through the liquid crystal. *Reflective LCDs* employ a mirror mounted to the rear of the liquid crystal layers (Figure 2-4-9). Ambient light enters the front of the display, passes through the crystal, is reflected off the mirror, and passes once again through the crystal. This area of the LCD appears *illuminated*. When the crystal electrodes are energized, the light cannot pass through the crystal/polarizer layers and the mirror reflects no light. This portion of the display appears dark. The major advantage of a reflective LCD is its compact size and low power requirements. A reflective display, however, is difficult to see in some lighting conditions and is totally useless without ambient light or an additional light source.

Transmissive displays. Transmissive displays are LCDs containing a dedicated light source mounted to the rear of the display. Transmissive FPDs are much brighter and easier to view than reflective displays. Unfortunately, transmissive displays are less efficient since they require energy to power the light source. The most common light source is known as a *cold cathode fluorescent tube* (CCFT). A CCFT is a highly efficient fluorescent tube that weaves back and forth across the back of the FPD to provide a bright even light. Most of the FPDs used on aircraft are transmissive displays. Figure 2-4-10 shows a typical transmissive color display.

Color LCDs. One of the more common LCDs employs a grayscale display. Blocking light through various pixels produces the LCD image. The active pixel looks gray, or black, since no light travels through that pixel. In most cases, this type of display relies on room light for illumination. Grayscale displays are commonly used for calculators and watches. Due to their poor quality, grayscale displays have limited use on aircraft. This type of LCD may take on virtually any color if a filter is added to the display. Amber is a common display color and is often referred to as a monochrome display.

Color LCDs are now the flat panel display of choice for most aircraft applications. Color FPDs are much more complex to produce since they require three pixels (red, blue, and green) to replace one pixel of a grayscale display. Most color FPDs contain an active matrix control system and require an additional lighting source for illumination. Figure 2-4-10 shows a cross sectional view of a typical color LCD. The display is illuminated using a light source located at the rear of the LCD. There are three color filters located on the viewing side of the FPD. As light passes through the LCD it also passes through the color filters. Each color pixel is con-

trolled independently allowing the display to achieve a full color range.

Viewing angle. Early FPDs had very limited viewing range. As the FPD was viewed from the side, the image would fade and lose contrast. For several years this problem made FPDs impractical for primary aircraft displays. Recently engineers have developed thinner LCDs and LCDs with a dual domain alignment layer. The thin glass alignment layer (Figure 2-4-10) is modified to direct light into a wide viewing angle. This process creates a type of curved lens on each pixel. Modern aircraft displays have a wide viewing angle, however, they are costly to manufacture.

Touch screen technologies. Many flat panel displays employ touch screen technologies. The display surface detects the presence and location of a touch to its surface, allowing for interaction with the device or system, without the use of additional controls, such as a keyboard or mouse. Touch screens have become commonplace on portable electronic devices like cellular phones, however, touch screens are in limited use on aircraft and only for non-critical systems.

During flight, turbulence could make the use of a touch screen difficult; because there are no distinct levers, buttons, or switches, the excess motion may cause the operator to make an incorrect selection. This situation limits aircraft use of touch screens to ground use or non-critical functions, such as passenger entertainment.

There are a variety of touch screens currently available using different technologies. In each case, the display is made of several layers. Each

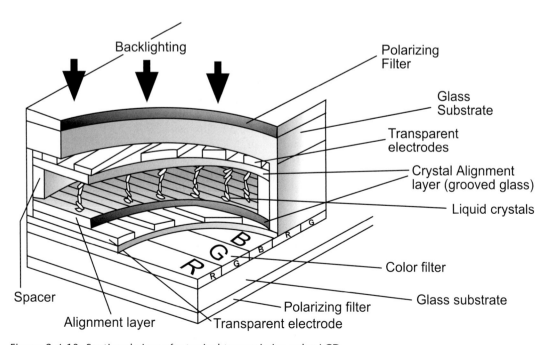

Figure 2-4-10. Sectional view of a typical transmissive color LCD

layer is integrated with an interactive display such as an LCD. The touch screen component of the display is housed on the outermost layers of the panel. The two most common technologies measure a change in resistance or capacitance whenever the display is touched.

A resistive touch screen panel contains two thin electrically conductive layers separated by a narrow gap. When an object, such as a finger, presses on the panel's outer surface the two metallic layers become connected at that point. This causes a change in the electrical current traveling through the layers. The change in electrical current is processed using digital circuitry to determine the location of each touch.

A capacitive touch screen panel consists of an insulator, for example glass coated with a transparent conductor film such as Indium Tin Oxide. The human body is also an electrical conductor, thus touching the surface of the screen results in a distortion of the screen's electrostatic field measurable as a change in capacitance. The change in capacitance is then sent to the controller for processing. Unlike resistive touch screens, one cannot use a capacitive touch screen through most types of electrically insulating material, such as gloves. Other touch screens may use surface acoustic waves or infrared light and sensor technologies to detect touch.

Repair and Maintenance of FPDs

For the most part, FPDs require very little maintenance. Cleaning the display should be done on a regular basis using a soft cloth. Be careful not to scratch the display. In the event an aircraft FPD fails, it is typically replaced, not repaired. However, replacing aircraft FPDs is extremely costly, so many manufactures and secondary repair facilities are developing methods to repair FPDs.

The most common defect seen with FPDs is one or more inactive pixels. An inactive pixel will always be set to the *ON* condition or always set to *OFF*. This will create a small black spot on the display or create a pixel that is always illuminated. Most aircraft FPDs have a tolerance for defective pixels. A typical primary flight display can have up to 15 defective pixels before the display must be replaced. As of this time, repairing a defective pixel is not possible.

If the entire FPD fails to operate, check all wiring and connectors to the display. This is an external repair and can often be accomplished in the field. If one or more horizontal or vertical lines appear on the display, the video driver, display-controller, or related electrical connections have failed. The FPD can still be repaired, however, it is a difficult task best left to specialists.

The back lighting of the FPD may also become defective. With age, the CCFT will dim naturally creating a poorly lit display. The display illumination can be tested using a light meter. Although it is rare, the CCFT may fail completely. A defective CCFT can be replaced given the right equipment and a lot of practice.

Section 5
Electrostatic Discharge Sensitive Components

Over the past decade, avionics systems have gone through a transformation of improved performance and reliability. This transformation was mainly the result of the extensive use of microelectronics in computer circuits. Microelectronic devices, such as integrated circuits and microprocessors, incorporate semiconductor material measuring only a few millionths-of-an-inch thick. These subminiature structures are extremely vulnerable to damage from static electricity. Devices made of CMOS (complementary metal oxide semiconductor) components are especially vulnerable to static discharge.

Components that are electrostatic discharge sensitive are commonly referred to as ESDS parts. Since these components are becoming popular on all types of avionics equipment, the avionics technician must be aware of the potential dangers from static electricity. It is estimated that over five billion dollars in microelectronics and PC boards are scrapped annually due to damage from static electricity. Thousands of warranty repairs and unnecessary failures are created in avionics shops annually. Awareness and proper precautions for ESDS components can increase productivity, customer satisfaction, and improve air safety.

Electrostatic Discharge

Electrostatic discharge is the discharge, movement of electrons, created when any material containing a static charge comes in contact with a material containing a different or neutral charge. A static charge is present when a material possesses an excess or deficiency of electrons. A neutral charge is present when a material possesses an equal number of electrons and protons. Any time two components of different static charge get close enough together, a static discharge will occur. Typically, a distance of 10 mm or less is required to produce a discharge of static electricity generated due to normal maintenance activities.

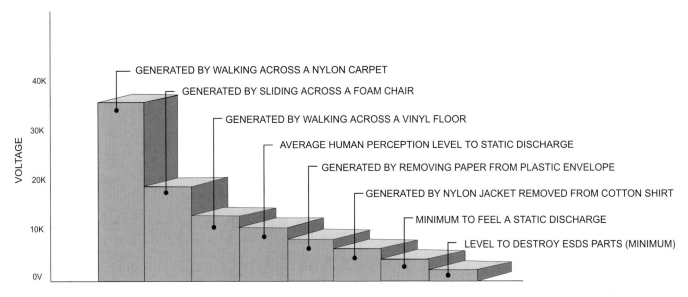

Figure 2-5-1. Graph showing various static voltage levels

Production of a Static Charge

Friction between two different materials produces the majority of static electricity. As two dissimilar materials rub together, there is often a transfer of electrons from one surface to another. The amount of electron transfer is mainly a function of the type of materials making contact, the amount of friction between the two, and the relative humidity of the surrounding air. Static electricity can also be generated on some materials, such as plastic, by subjecting them to heat or pressure.

Most everyone has experienced the production of a static charge while walking across a nylon carpet with plastic sole shoes. In this situation approximately 35,000 volts can be generated if the relative humidity is 20 percent or less. Normally, you do not notice the static charge until you come in contact with another object, such as a doorknob. At that time you experience the discharge of the static electricity to the doorknob which was previously stored on your body. This type of discharge will damage ESDS parts.

The minimum perception level to feel a static electrical discharge is typically 3,000 volts. For a discharge to be heard it requires approximately 5,000 volts; and the discharge cannot be seen if less than 10,000 volts. The current during discharge is extremely low and produces only minor discomfort to humans, however, a static discharge of as little as 100 volts can damage certain electronic components. Figure 2-5-1 shows an example of the voltage generated by common materials found in a maintenance facility. It is imperative that technicians realize that almost any situation can create enough static charge to damage sensitive components.

Identifying ESDS Components

In order to protect components from static discharge one must first identify ESDS parts and then take the necessary precautions. Technicians must be aware of the various types of labels that have been designed to identify sensitive components. The most common labels are shown in Figure 2-5-2. One or more of these labels should be placed on all ESDS parts. The label should be easily visible to anyone who might come in contact with the component. The identification labels must be used on components installed in the aircraft, as well as components found in shipping, storage, or in a repair facility.

ESDS parts come in all shapes and sizes. The LRUs found on virtually any modern aircraft

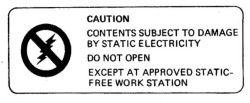

Figure 2-5-2. Common labels used to identify ESDS parts

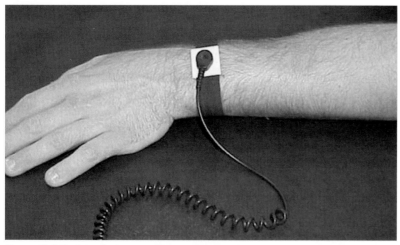

Figure 2-5-3. A typical grounding wrist strap

are most likely electrostatic sensitive and should not be handled without proper precautions. Each equipment rack containing a sensitive component must be labeled with the appropriate decal. Other components that are often vulnerable to static discharge include computer cards or circuit boards, and certain integrated circuits. These components are most vulnerable during removal and installation, as well as shipping, handling, and bench repair.

Protecting Components

The basic premise for the protection of all ESDS parts is to prevent a static charge from entering that component. This is accomplished by:

1. Electrically neutralizing anyone or anything coming in contact with or near the part, or

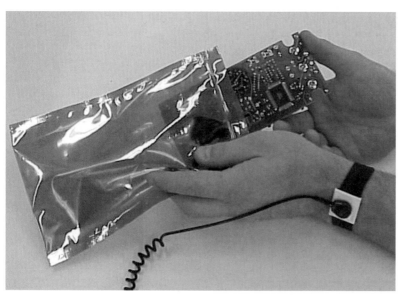

Figure 2-5-4. Placing a circuit card into an antistatic bag

2. Electrically insulating the sensitive part from potential static discharge.

A charged body does not need to contact the ESDS component to damage it. If a sensitive component is close enough, less than approximately a half inch to the static charge, the part can be damaged.

Neutralization is typically accomplished by connecting all personnel and equipment to an electrical ground source. Prior to working with an ESDS part the technician must connect a grounding strap to bare skin and ensure its security. A wrist strap connected to ground through a flexible wire is typically used for this purpose (Figure 2-5-3). All test equipment, including bench tops, must also be grounded. If equipment installed in the aircraft is labeled ESDS, always connect a wrist strap to the aircraft's ground prior to touching any component. A grounding socket is often provided in the aircraft's electrical equipment bay for this purpose.

CAUTION: *Always use an approved wrist strap. Commercially available straps contain a resistor in series with the ground circuit to protect the user from electrocution if contact is accidentally made with a hot wire.*

To electrically insulate a component, place all sensitive parts in the appropriate conductive container. These containers are often specially designed boxes for specific line replaceable units or conductive bags used for smaller items such as cards or integrated circuits. Once an item has been placed in its protective container, be sure it is labeled with an ESDS decal.

Technicians should be aware that there are two types of protective bags currently available:

1. The pink antistatic polyethylene bag is used to protect components from generating a static charge caused by movement while stored inside the bag

2. The dark gray antistatic bag will protect the component from static fields outside the bag, static discharge current, and static fields generated inside the bag (Figure 2-5-4). The gray colored bag offers the most protection and should be used under most circumstances.

If in doubt, ask your manufacturer for more details concerning the correct protective container. Once the ESDS component is placed inside the correct protective container, the unit is safe from damage caused by static electricity and may be safely handled by ungrounded personnel.

Other means to protect components include the use of antistatic caps that should be placed over all connections of sensitive LRUs, and metal clips or conductive foam used to short together the leads of individual components or cards. Whenever using any protective device be sure to inspect it for defects or unacceptable wear.

At the workstation, the ESDS components should not be set directly onto the tabletop. The workstation should be equipped with either a conductive or static dissipative mat. The *conductive mat* may be used whenever a component is serviced in a power off condition. If the component is operated while placed on the conductive-type mat, the mat itself may bridge the connections between any components and cause component failure. In the case where components are operated while in contact with the mat surface, a *static dissipating mat* must be used. Be sure to check that your workstations are equipped with the correct type of protection.

Nonconductors at the workstation are also a source of problems when working around ESDS parts. If plastic tools or pens, for example, come into contact with sensitive component damage may occur. Nonconductive plastic does not dissipate its charge easily. A grounded technician holding that tool is not enough. Examine your work habits, and if you find nonconductors used at workstations, the best solution is to install an ionization air blower. An ionization air blower will safely delete any static charge formed on nonconductors. It is also helpful to maintain a relatively high humidity level in all work areas. High humidity helps to dissipate static charges before they become a problem.

Types of Static Damage

Damage from static discharge can take two forms. They are commonly referred to as hard and soft failures. A *hard failure* caused by static discharge will create an immediate system defect. A *soft failure* caused by static discharge may only injure the component and cause erratic operations and/or eventual failure. A component that experienced a soft failure is often said to be *wounded*. A wounded part may pass all bench and/or preflight tests and still fail during flight. Wounded parts account for approximately 90 percent of all static damage. Often wounded parts cycle from aircraft to shop several times before the fault can be detected and repaired.

Modern avionics equipment contains a variety of ESDS parts. Every technician must be aware of the problems involved with static electricity and take the necessary precautions. The defect(s) caused by a static discharge to a sensitive component will not be visible to the naked eye; however, the results could be catastrophic. When in doubt, assume the component is ESDS and take the necessary precautions.

Section 6
Binary Numbering Systems

Computers understand only two distinct values: *binary 1* and *binary 0*. These ones and zeros become a voltage translating to on (1) and off (0) inside the computer circuitry. For example, binary one might equal +10 volts DC and binary 0 equal 0 volts DC. In order for a computer to perform the various functions necessary for it's operation, these two values (1/0) must be grouped into words or patterns that take on different meanings. A group of ones and zeros such as 00101001 might represent the number 29 or this same combination might represent a low oil pressure warning from engine number 2, or some other meaning. The software of the computer must be programmed to understand the correct numbering system in order to encode or decode data and therefore perform the correct function.

Although there are hundreds of codes used to convert binary bits into meaningful data, a few common numbering systems are used by most computer engineers when designing software. These numbering systems can be thought of as the language used by that system made up of a very small alphabet comprised of only two symbols (0 and 1). Just like various languages (German, French, Spanish, and English) are all basically comprised of the same 26 letters of the alphabet, all digital numbering systems are comprised of two characters: 1 and 0. Of course, each aircraft computer must know what language is being used at any given time in order to operate properly. It is also important that avionics technicians become familiar with these code systems since it is often required to decode or encode data in order to troubleshoot or repair a modern digital aircraft.

For most aircraft computers there are three common numbering systems employed during data transmission and processing: hexadecimal notation (base 16), binary coded decimal (base 10), and octal notation (base 8). Each of these numbering systems must be represented by various combinations of binary bits to form a base 16, 10 or 8 number. Pure binary (without grouping of bits) can also be used for computer operations, but the grouping of the binary bits greatly increases com-

DECIMAL VALUE	BINARY CODED DECIMAL (BCD) 4-BIT GROUPS
0	0000
1	0001
2	0010
3	0011
4	0100
5	0101
6	0110
7	0111
8	1000
9	1001

Table 2-6-1. Binary Coded Decimal (BCD) values

16 POSSIBLE COMBINATIONS OF A 4-BIT GROUP	BCD NUMBER (10 POSSIBILITIES)	DECIMAL EQUIVALENT OF THE BCD VALUE (0-9)
0000	0000	0
0001	0001	1
0010	0010	2
0011	0011	3
0100	0100	4
0101	0101	5
0110	0110	6
0111	0111	7
1000	1000	8
1001	1001	9
1010		
1100	These values do not exist in the BCD numbering system	Values greater than 9 do not exist as a single digit in the decimal numbering system
1101		
1110		
1111		

Table 2-6-2. BCD cannot use all combinations of the four bit groups

BCD uses a group of four bits to represent each digit of a decimal number. For example $9_{(base\ 10)}$ = $1001_{(BCD)}$. The computer simply looks at four bits each time it needs to determine the decimal value. Once again this grouping process helps to speed computer operations when dealing with large quantities of data interchanged between the inputs and outputs of a computer system. One might notice the four bits used to represent the value of nine in BCD are identical to the bits used in the binary representation of nine (1001). BCD and plain binary are very similar, but remember that each digit of the BCD code is always comprised of four bits. For example the number $3_{(base\ 10)}$ = $0011_{(BCD)}$ and $3_{(base\ 10)}$ is equal to 11 in binary. In other words $3_{(base\ 10)}$ = $0011_{(BCD)}$ = $11_{(base\ 2)}$

It should be noted that all numbering systems used with computer technologies have their advantages and disadvantages, and BCD is no exception. One main advantage of BCD is that humans can understand it since it uses binary bits to represent base ten numbers. However, BCD does not make good use of all four bits of the four-bit code. Remember that BCD uses a group of four bits to represent each digit of a decimal number. Table 2-6-2 shows that BCD cannot use some combinations of the four bit groups. Since no value higher than a nine exists in the base ten system, six combinations of the four bit group are unusable. These unusable data bits slow the computer program through the manipulation of bits, which contain no useful information.

Octal notation is a binary representation of an octal number. Octal numbers are base eight and are therefore comprised of eight different symbols (0, 1, 2, 3, 4, 5, 6 and 7). Octal notation is comprised of a series of three-bit groups. Each bit within the group is a binary digit limited to a 1 or a 0. Table 2-6-3 shows a relationship of decimal, octal and octal notation values. Notice the table shows that the decimal values of 8 and 9 as not applicable, meaning that no symbol (digit) can exist in a base eight numbering system for these values. Base eight numbers use symbols 0 through 7. If an octal number must represent a value greater than seven, a second digit must be added. Although octal notation is generally more difficult for humans to understand than binary coded decimal (BCD), it is a useful tool for many computer programs since every possible combination of the three-bit code is utilized. There are no unused bit combinations.

To decode or encode an octal notation number, it may become necessary to convert between decimal, octal and/or octal notation. Remember octal notation is a group of three binary bits used to represent an octal number. Octal is a base eight number comprised of seven differ-

puter speed since information can be manipulated in groups. This text will provide a brief explanation of the aforementioned numbering systems, however, if extensive computer engineering/programming is anticipated, it is recommended that further study of these binary codes be conducted.

Binary coded decimal (BCD) is a binary representation of a base ten number. This code uses groups of four bits to represent ten symbols. We know these symbols as zero, one, two, three...seven, eight, and nine (0,1,2,3,4,5,6,7,8,9). However, using BCD the computer sees these nine symbols as different combinations of four binary bits (Table 2-6-1).

ent symbols. The following example will convert 74110 to an octal (base 8) number, and then to octal notation (a three-bit representation of octal). Follow Figure 2-6-1 during this discussion of number conversion. It is common convention to indicate the base of the number in a subscript immediately following the number, such as, 741_{10}. This indicates that 741 is a base ten number.

To determine the octal notation code of the decimal number 741_{10} complete the following steps:

- Step 1: Divide the decimal number by 8
- Step 2: Record the quotient and the remainder
- Step 3: Divide the previous quotient by 8,
- Step 4: Record the quotient and remainder
- Step 5: Repeat this process until the quotient is zero
- Step 6: Convert each remainder into a 3-digit (triad)
- Step 7: Record each triad starting from the bottom moving upward to determine the octal notation code

In order to convert from an octal notation number into a decimal equivalent it is important to

DECIMAL VALUE	OCTAL VALUE	OCTAL NOTATION 3-BIT GROUPS
0	0	000
1	1	001
2	2	010
3	3	011
4	4	100
5	5	101
6	6	110
7	7	111
8	N/A	N/A
9	N/A	N/A

Table 2-6-3. Decimal, octal, and octal notation values

know the value of each three-bit group. Table 2-6-4 shows the decimal equivalents of the first five triad groups.

To convert an octal number to a decimal value follow the steps in Table 2-6-5. The octal number 653 (six, five, three; not six hundred fifty-three) is equal to 427 decimal. To make this conversion, first identify each of the octal digits as shown in step 1. Next, determine the exponential power for each of the octal digits. Third, determine the decimal equivalent of 8^N for each digit. Fourth, multiply the value of each col-

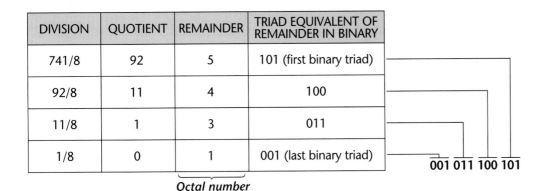

DIVISION	QUOTIENT	REMAINDER	TRIAD EQUIVALENT OF REMAINDER IN BINARY
741/8	92	5	101 (first binary triad)
92/8	11	4	100
11/8	1	3	011
1/8	0	1	001 (last binary triad)

Octal number from bottom up

001 011 100 101

$741_{10} = 1345_8 = 001\ 011\ 100\ 101$ (octal notation)

741 (base ten) = 1345 (base 8) = 001011100101 (octal notation)

Figure 2-6-1. Conversion of Decimal 741 to octal and then to octal notation

3-digit group (Triad)	5th	4th	3rd	2nd	1st
Power of 8 for each group	8^4	8^3	8^2	8^1	8^0
Decimal value of base 8 number	4,096 ($4,096 = 8^4$)	512 ($512 = 8^3$)	64 ($64 = 8^2$)	8 ($8 = 8^1$)	1 ($1 = 8^0$)

Table 2-6-4. Decimal equivalents of octal triads

		3rd DIGIT MOST SIGNIFICANT DIGIT	2nd DIGIT	1st DIGIT LEAST SIGNIFICANT DIGIT
STEP 1	Octal Number (653)	6	5	3
STEP 2	Power of 8	8^2	8^1	8^0
STEP 3	Decimal value of 8^N	64 ($64 = 8^2$)	8 ($8 = 8^1$)	1 ($1 = 8^0$)
STEP 4	Total the decimal value of this digit	384 (6×64)	40 (5×8)	3 (3×1)
STEP 5	Sum the decimal equivalents	$384 + 40 + 3 = 427_{10}$		

Table 2-6-5. Octal to decimal conversion

umn by the octal digit for that column. Lastly, sum the decimal value for each column. This example shows that $653_8 = 427_{10}$.

The following example will show the conversion of the octal notation number 001 010 100 001 into its decimal equivalent. Follow the information in Table 2-6-6 during this discussion.

- Step 1: Record each three-bit triad into its appropriate group
- Step 2: Determine the decimal equivalent of the three-bit triad

- Step 3: Record the exponential power of the octal digit for each column
- Step 4: Determine the decimal value of each octal column
- Step 5: Multiply the triad value with by decimal value of the base eight numbers
- Step 6: Sum the decimal equivalents determined in step 5. This total is the decimal value of the original octal notation number. Therefore 001 010 100 001$_{(octal\ notation)}$ is converted into decimal number 673$_{(base\ 10)}$.

Hexadecimal notation is a binary representation of a hexadecimal number. Hexadecimal notation is comprised of a series of four-bit

	TRIAD GROUP	4th DIGIT MOST SIGNIFICANT DIGIT	3rd DIGIT	2nd DIGIT	1st DIGIT LEAST SIGNIFICANT DIGIT
STEP 1	3-digit octal triad	001	010	100	001
STEP 2	Decimal equivalent of triad	1	2	4	1
STEP 3	Power of 8	8^3	8^2	8^1	8^0
STEP 4	Decimal value of base 8 number	512 ($512 = 8^3$)	64 ($64 = 8^2$)	8 ($8 = 8^1$)	1 ($1 = 8^0$)
STEP 5	Decimal equivalent of octal group	(1×512) 512	(2×64) 128	(4×8) 32	(1×1) 1
STEP 6	Sum the decimal equivalents of each octal group	$512 + 128 + 32 + 1 = 673_{10}$			

Octal notation 001 010 100 001 = 673_{10}

Table 2-6-6. Conversion of octal notation number 001 010 100 001 to decimal 673

groups. Each bit within the group is a binary digit limited to a 1 or 0. Grouping of data allows computers to store, manipulate, and perform calculations faster than if each bit was handled individually. The hexadecimal numbering system is commonly used for programming of various aircraft computers and all technicians should become familiar with this system. Table 2-6-7 shows the relationships of decimal, hexadecimal, and hexadecimal notation values.

The hexadecimal numbering system is comprised of 16 symbols (1, 2, 3, 4, 5, 6, 7, 8, 9, A, B, C, D, E and F). In these systems it can be seen that A through F are used to represent the last 6 values. This is necessary because hexadecimal numbers must be comprised of 16 different symbols with no repeats. For example, the number 11 (base 10) is actually comprised of two repeated symbols 1 and 1. In the hexadecimal numbering system the number 11 does not exist; that value would be represented by the symbol B. It should be noted that hexadecimal notation and binary coded decimal are both systems comprised of four-bit groups. However, only in the case of hexadecimal notation do all combinations of the four-bit group represent specific symbols. Hexadecimal notation therefore, makes more efficient use of the four-bit group than does the binary coded decimal system. Hexadecimal notation is typically the numbering system of choice for many airborne digital systems.

Figure 2-6-2 shows how to convert a decimal number into a hexadecimal number. To convert the decimal number, first divide by 16 (because hexadecimal is a base 16 number) and record the quotient and the remainder. Next, divide the quotient by 16 and record the new quotient and the new remainder. Repeat this process until the quotient is zero. The values of the hexadecimal number are derived from the calculated remainders. The most significant digit is the last remainder calculated and the least significant digit is the first remainder calculated. In the first example, decimal number 324 is converted to hexadecimal number 114 ($324_{10} = 144_{16}$). Keep in mind; the hexadecimal value would be called "one, four, four" (not one hundred forty-four). The value of one hundred or forty-four does not exist in the hexadecimal numbering system.

In the second example of Figure 2-6-3, the decimal value 412 is converted to 14C hexadecimal ($412_{10} = 19C_{16}$). This conversion is made following the same steps as outlined above, however, in this case, the value C must be substituted for the decimal value of 12 as 12 does not exist in the hexadecimal numbering system.

To convert a hexadecimal number to a decimal value follow the steps in Table 2-6-8. The hexa-

DECIMAL VALUE (BASE 10)	HEXADECIMAL VALUE (BASE 16)	HEXADECIMAL NOTATION (4-BIT GROUP)
0	0	0000
1	1	0001
2	2	0010
3	3	0011
4	4	0100
5	5	0101
6	6	0110
7	7	0111
8	8	1000
9	9	1001
10	A	1010
11	B	1011
12	C	1100
13	D	1101
14	E	1110
15	F	1111

Table 2-6-7. Decimal, hexadecimal, and hexadecimal notation values

Convert 324_{10} to a hexadecimal value:

$$\frac{324}{16} = 20 + remainder\_4 \text{—(Least significant digit)}$$

$$\frac{20}{16} = 1 + remainder\_4$$

$$\frac{1}{16} = 1 + remainder\_1 \text{ (Most significant digit)}$$

Therefore: $324_{10} = 144_{16}$

Figure 2-6-2. Decimal 324 to hexadecimal conversion

Convert 412_{10} to a hexadecimal value:

$$\frac{412}{16} = 25 + remainder\_12 \text{—(Least significant digit)}$$

$$\frac{25}{16} = 1 + remainder\_9$$

$$\frac{1}{16} = 1 + remainder\_1 \text{ (Most significant digit)}$$

Therefore: $412_{10} = 19C_{16}$

Note: 12 does not exist in HEX...12=C

Figure 2-6-3. Decimal 412 to hexadecimal conversion

decimal number 653 (six, five, three; not six hundred fifty-three) is equal to 1,619 decimal. To make this conversion first identify each of the hexadecimal digits as shown in step 1. Next, determine the exponential power for each of the hexadecimal digits. Third, determine the decimal equivalent of 16_N for each digit. Fourth, multiply the value of each column by the hexadecimal digit for that column. Lastly, sum the decimal value for each column. In this example $653_{16} = 1,619_{10}$.

In the example shown in Table 2-6-9, the conversion from hexadecimal to decimal is handled in a similar manor. In this example $FA2_{16} = 4,002_{10}$. Remember the F is substituted for 15, and A is substituted for 10.

Remember, hexadecimal notation is simply the binary representation (using four-bit groups) of each of a hexadecimal digit. An example of a conversion from hexadecimal to hexadecimal notation is shown in Table 2-6-10. In this example, $2F9_{(hexadecimal)} = 001011111001_{(hexadecimal\ notation)}$.

The example in Table 2-6-11 represents a complete conversion of a hexadecimal notation value to a hexadecimal value and then into a decimal value. Notice the values of Table 2-6-10 and Table 2-6-11 are the same. It takes time and practice to master an understanding of the various numbering systems presented in this text. Study the

tables presented here, and then simply substitute values and practice, practice, practice.

Section 7

ARINC Standards

ARINC Incorporated is a global corporation consisting of various U.S. and international airlines and aircraft operators and their subsidiaries. The company provides services related to a variety of aviation communication and navigation systems. Both U.S. and international aircraft and component manufacturers have accepted ARINC technical standards. ARINC also provides standardization guidelines for engineering and development of both software and hardware systems for a variety of military and civilian aviation electronic systems.

One of the most important facets of ARINC, from a technician's point of view, would have to be the standards set for various avionics systems. As the avionics industry was begin-

		3rd DIGIT MOST SIGNIFICANT DIGIT	2nd DIGIT	1st DIGIT LEAST SIGNIFICANT DIGIT
STEP 1	Hex Number (653)	6	5	3
STEP 2	Power of 16	16^2	16^1	16^0
STEP 3	Decimal value of 16^N	256 $(256 = 16^2)$	16 $(16 = 16^1)$	1 $(1 = 16^0)$
STEP 4	Total the decimal value of this digit	1,536 (6 x 256)	80 (5 x 16)	3 (3 X 1)
STEP 5	Total the decimal values	$1,536 + 80 + 3 = 1,619_{10}$		

Table 2-6-8. Hexadecimal to decimal conversion

		3rd DIGIT MOST SIGNIFICANT DIGIT	2nd DIGIT	1st DIGIT LEAST SIGNIFICANT DIGIT
STEP 1	Hex Number (FA2)	F	A	2
STEP 2	Power of 16	16^2	16^1	16^0
STEP 3	Decimal value of 16^N	256 $(256 = 16^2)$	16 $(16 = 16^1)$	1 $(1 = 16^0)$
STEP 4	Total the decimal value of this digit	3,840 (15 x 256)	160 (10 x 16)	2 (2 X 1)
STEP 5	Total the decimal values	$3,840 + 160 + 2 = 4,002_{10}$		

Table 2-6-9. Hexadecimal to decimal conversion

ning to emerge, it became clear that some type of commonality was needed to ensure systems were compatible. As avionics became even more interactive, the need for commonality grew. ARINC has documented characteristics and specifications or standards for a multitude of communication and navigation systems. Most of these standards are focused on transport category avionics; however, many of the general aviation avionics systems also adhere to ARINC specifications.

Individual system specifications and their related interfaces are addressed by ARINC. ARINC equipment specifications include: (#579) airborne VOR receivers, (#578) airborne ILS receivers, (#594) ground proximity warning systems, (#566) VHF communication transceivers, (#542) digital flight data recorders, (#741) aviation satellite communications, (#738) air data and inertial reference systems, (#728) analog and discrete data converter systems, (#725) electronic flight instrument (EFI), (#724) aircraft communication and reporting systems, and (#708) airborne distance measuring equipment. As discussed below, ARINC is also responsible for three of the most common data bus standards used on transport category aircraft (ARINC 429, 629 and 644).

Section 8
Digital Data Bus Systems

Many aircraft currently employ avionics systems that operate using digital electronics. Although these systems range in complexity from a simple radio receiver/transmitter to a complete autoflight system, they all adhere to some type of data bus standard. Adherence to an established standard is extremely important to ensure commonality of data transfer. A *data bus standard* can be thought of as a common language that allows several different units to communicate. Without a data bus standard, communication between various avionics systems would be impossible.

Several digital data bus standards are currently in use on modern aircraft. Agencies such as ARINC Incorporated and the General Aviation Manufacturers Association (GAMA) set standards for various data bus systems. The three commonly used ARINC data bus standards are: ARINC 429, ARINC 629 and ARNC 664. In industry, these standards are often referred to by their number only; for example, ARINC 664

	3rd DIGIT MOST SIGNIFICANT DIGIT	2nd DIGIT	1st DIGIT LEAST SIGNIFICANT DIGIT
Convert $2F9_{16}$ into equivalent 4-bit groups	2	F (F = 15)	9
Hex notation 4-bit group	0010	1111	1001

Hexadecimal $2F9_{16}$ = Hexadecimal Notation 001011111001

Table 2-6-10. Conversion of hexadecimal number to hexadecimal notation

		3rd DIGIT MOST SIGNIFICANT DIGIT	2nd DIGIT	1st DIGIT LEAST SIGNIFICANT DIGIT
HEXADECIMAL NOTATION VALUE *(as determined in Table 2-6-10)*		001011111001		
STEP 1	4-bit group	0010	1111	1001
STEP 2	Hex value of 4-bit group	2	F (F = 15)	9
STEP 3	Power of 16	16^2	16^1	16^0
STEP 4	Decimal value of 16^N	256 $(256 = 16^2)$	16 $(16 = 16^1)$	1 $(1 = 16^0)$
STEP 5	Decimal value of this digit	512 (2 x 256)	240 (15 x 16)	9 (9 x 1)
STEP 5	Total the decimal values	$512 + 240 + 9 = 761_{10}$		

In other words: $001011111001_{(hexadecimal\ notation)} = 2F9_{16} = 761_{10}$

Table 2-6-11. Hexadecimal notation to decimal conversion

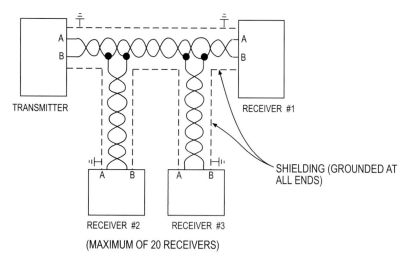

Figure 2-8-1. Typical data bus cable connections for a polarity sensitive bus

might be called simply 664. This text will refer to these standards in both configurations. ARINC 429 is a one-way data bus also known as Mark 33 Digital Information Transfer Systems (DITS). ARINC 629 is a faster, two-way, multi-transmitter data bus system found on the B-777. ARINC 429, 629, and/or 664 data bus systems are found on virtually all transport category aircraft that employ digital systems. ARINC 429 data bus systems are also found on many corporate and commuter aircraft. Several manufacturers employ proprietary data bus systems designed specifically for use by that manufacturer only. These proprietary data bus systems are typically considered trade secrets and used only within a given system where that manufacturer designs all components. Whenever systems must communicate between varieties of LRUs (designed by many different manufacturers), a standard ARINC data bus is typically employed.

The *ASCB (Avionics Standard Communication Bus)* and the *CSDB (Commercial Standard Digital Bus)* are two data bus systems used extensively on corporate-type general aviation aircraft, such as, the Beechcraft King Air, Cessna Citation, or Falcon 20. The ASCB format is used solely on Sperry or Honeywell avionics equipment. Sperry and Honeywell have since merged and are currently referred to solely as Honeywell. The CSDB format is used on aircraft that employ systems manufactured by Collins Avionics. Collins is a subsidiary of Rockwell International and manufactures avionics for both general aviation, mainly corporate and commuter aircraft, and transport category aircraft. Manchester data is also common to many corporate- and commercial-type aircraft and is adaptable to a variety of formats. In many cases two or more data bus standards may be used on the same aircraft. It is extremely important that all avionics technicians become familiar with these data bus specifications if they intend to repair and maintain digital systems.

A point of clarification: a data bus is typically thought of as the wires that transmit digital signals between two LRUs, not the wiring inside the LRU. The data bus standard applies only to the data transferred on the external data bus, not to the data manipulated inside the LRU. For example, a system may employ the ARINC 429 data bus standards for communication between the flight management computer (FMC) and the thrust management computer (TMC). However, inside the FMC and TMC the data may be manipulated in a completely different format as specified by the LRU software. Each data bus is dedicated to a specific data format. For example, a technician will never find a data bus that transmits ARINC 429 and then changes format to transmit ARINC 629. These two different formats would require two different data bus cables.

A common digital bus is a pair of solid copper wires, approximately 24-gauge. Each wire is coated with a thin plastic insulation, twisted together, wrapped in a foil shield, and covered again with insulation. It is important that the shield be grounded at each end of the cable. Many data bus systems are polarity sensitive and it is therefore very important that the two conductors never be switched. That is, wire A must always be connected to terminal A; wire B to terminal B (Figure 2-8-1). The digital signals can be measured between wires A and B using an oscilloscope or data bus analyzer.

Manchester II Coding

Both general aviation and transport category aircraft often incorporate a Manchester data bus. The *Manchester II code*, often simply referred to as Manchester, is a serial digital data format that incorporates a voltage change in each data bit. This means: a) logic 1 begins with a positive voltage and changes to zero (or negative) voltage halfway through the bit time period, and b) logic 0 begins with a zero (or negative) voltage and transitions to a positive voltage half way through the bit time period. This type of format allows the Manchester code to be self clocking. A self-clocking data bus carries a timing signal with each data bit. A self-clocking data bus helps to ensure the correct timing of all digital systems connected to that bus. Figure 2-8-2 shows a typical Manchester code with a 10 Kbits/sec transmission rate. As can be seen in this diagram, binary 1 equals +12V and binary zero equals 0V. The time period for each bit is 100 μsec and each bit is divided in half (50 μsec). The first half of the bit contains the data and the second half of the bit contains a clock pulse. A Manchester code can be adapted to a variety of different word configurations, voltage levels, and transmission speeds. In some cases, Manchester is used by

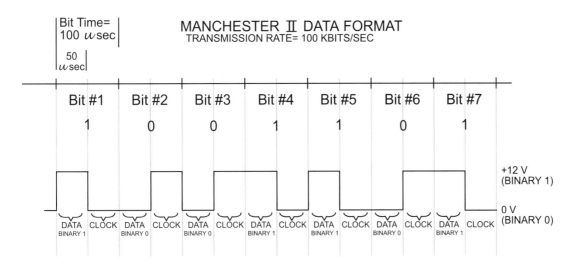

Figure 2-8-2. A typical Manchester digital signal format

the processing LRU and the data is converted to a different format for transmission on the data bus. For transmission between two LRUs, Manchester will always be transmitted on a dedicated Manchester data bus.

RS-232

RS-232 stands for *recommended standard number 232* and is another data bus standard used on modern aircraft systems, such as the Garmin G-1000 integrated flight display system. The RS-232 data bus is also a common interface between various units of avionics test equipment and standard personal computer devices. This standard defines a serial data format, which can be transmitted with full duplex capabilities; full duplex allows for simultaneous transmission and reception of data. RS-232 signals are represented by voltage levels with respect to the system ground. The exact voltage levels used to represent binary 1 and binary 0 may change between various systems; typically, however, voltage levels with as little as three volts and as high as 15 volts are acceptable. RS-232 is a bipolar bus where a negative voltage (-3 to -15 Vdc) indicates an *OFF* or 1-state. This is known as the *MARK* condition. Positive voltage (+3 to +15 Vdc) indicates an *ON* or 0-state typically called the *SPACE* condition. On some modern computers using RS-232, the negative voltage is not used and a simple zero voltage is considered the Mark.

The RS-232 standard requires a common ground between the data terminal equipment (DTE) and the data communications equipment (DCE). These terms are used to define the pin connections for the device and the direction of signals transmitted or received on those pins. Typically the main computer element of the system is the DTE device and other

peripheral components are the DCE units. The connector used for termination of the data bus cable is typically one of two configurations: a 25-pin D-sub connector, or a 9-pin D-sub connector. The sub stands for subminiature since this connector is relatively small and the connector-mating surface is D-shaped. D-sub connectors are also available in larger pin configurations, although they are less common.

A D-sub connector contains two or more parallel rows of pins or sockets surrounded by a D-shaped metal shell Figure 2-8-3. The metal shield provides mechanical support, shielding against electromagnetic interference from outside the cable and ensures correct orientation of the mating connectors. The socket's outer metal shell fits tightly inside the plug's shell and connects to the overall shielding of the cables. This creates an electrically continuous screen covering the whole cable and connector system. This protective shielding is particularly beneficial for aircraft use since signal interference in avionic systems can create a hazard to flight safety; since the transmit and receive wires are located within the same cable, interference may be generated from within the RS-232 data bus itself. This is particularly true with long cables or high-speed data transfer.

Figure 2-8-3. A typical D-Sub Connector: (A) Plug, (B) Sockets

25 PIN CONNECTOR ON DTE DEVICE (PC CONNECTION)

Male RS232 DB25

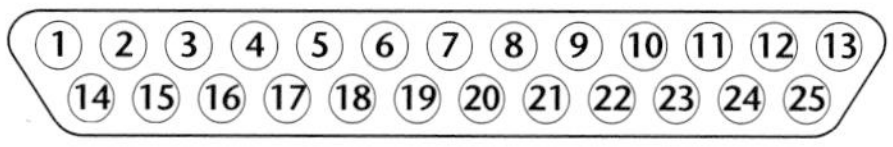

① ② ③ ④ ⑤ ⑥ ⑦ ⑧ ⑨ ⑩ ⑪ ⑫ ⑬
⑭ ⑮ ⑯ ⑰ ⑱ ⑲ ⑳ ㉑ ㉒ ㉓ ㉔ ㉕

Pin Number	Direction of Signal
1	Protective Ground
2	Transmitted Data (TD) Outgoing Data
3	Received Data (RD) Incoming Data
4	Request to Send (RTS) Outgoing Flow Control
5	Clear to Send (CTS) Incoming Flow Control
6	Data Set Ready (DSR) Incoming Handshaking Signal
7	Signal Ground Common
8	Carrier Detect (CD) Incoming Signal
20	Data Terminal Ready (DTR) Outgoing Handshaking Signal

9 PIN CONNECTOR ON DTE DEVICE (PC CONNECTION)

Male RS232 DB9

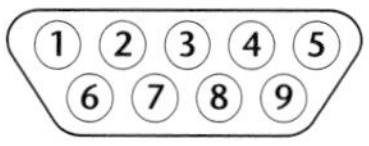

① ② ③ ④ ⑤
⑥ ⑦ ⑧ ⑨

Pin Number	Direction of Signal
1	Carrier Detect (CD)
2	Received Data (RD) Incoming
3	Transmitted Data (TD) Outgoing
4	Data Terminal Ready (DTR) Outgoing Handshaking Signal
5	Signal Ground Common Reference Voltage
6	Data Set Ready (DSR) Incoming Handshaking Signal
7	Request to Send (RTS) Outgoing Flow Control
8	Clear to Send (CTS) Incoming Flow Control

Figure 2-8-4. Contacts for a typical RS-232 25-pin and 9-pin connector

The specific pin connections of the D-sub connectors are spelled out in the RS-232 standard. This text will provide a brief explanation of one configuration for the 9 and 25-pin connectors. Keep in mind that variations exist in different systems; always refer to the appropriate maintenance data for specific connector details. As seen in Figure 2-8-4 each pin has a specific function, which is used to control data flow and ensure proper transmission. The RTS (request to send) and CTS (clear to send) are two pins/wires used for flow control. The RTS is a computer output asking if the receiver is ready for communication (data). The CTS is a return signal "answering" the request and approving (or not) the transmission of data. This process is often referred to as hand shaking. Remember no data will be sent until the CTS is set to the correct state (Mark or Space defined by the system's software). This concept becomes necessary since the receiving device may not be able to process data fast enough. This control can be thought of as a stop sign controlled by the receiver whenever the data buffer is full. For further control the DTR (data terminal ready) and DSR (data set ready) are available and may be used on some systems.

Keep in mind that although D-sub connectors are introduced here in the section on RS-232 data transfer, they are commonly used to transmit a variety of electrical signals not just RS-232. The rectangular shape and variety of pin configurations of the D–sub connector along with the protective outer shielding of the metal back shell make the D-sub connector ideal for a variety of applications. Many avionic systems found on modern aircraft employ D-sub connectors. To mention a few, D-sub connectors found may be used to transmit ARINC 429, RS-232, CSDB, and discrete and/or analog signals to a given LRU. Many avionic systems employ D-sub connectors with 26, 44, or 62 pins/contacts.

Avionics Standard Communication Bus (ASCB)

The *ASCB (avionics standard communication bus)* is a bidirectional data bus operating at 0.667 MHz. This system uses multiple bus controllers that coordinate bus transmission activities. Figure 2-8-5 shows the structure of a typical ASCB system. In this example, there are three bus controllers and two parallel busses providing the redundancy needed to ensure flight safety. During operation only one bus controller is active at any given time. The other two controllers are ready in active standby.

The subsystems transmit data only by command of the bus controller. The bus controller contains the software necessary to "call" various subsystems and coordinate data transmission in the correct sequence. The bus controller sends a request for messages on both busses. This request will have the specific address of the subsystem that the controller wishes. The subsystem will then respond to the request with its own message.

Data format. All ASCB data is transmitted at 0.667 MHz in a standard digital format using positive logic. ASCB uses a non-return to zero (NRZ) format for each data bit. That is, a signal of +5 V is used for a binary one, 0 V for binary zero, and the data fills the entire bit time. There is no change in voltage level during each bit period for timing purposes. A signal loss during transmission of approximately 0.5 V (allowing +4.5 V to reach the receiver) is typically acceptable. A greater signal loss will most likely cause an error in data transmission (Figure 2-8-6). An error in data transmission can easily be seen using an oscilloscope or data bus analyzer.

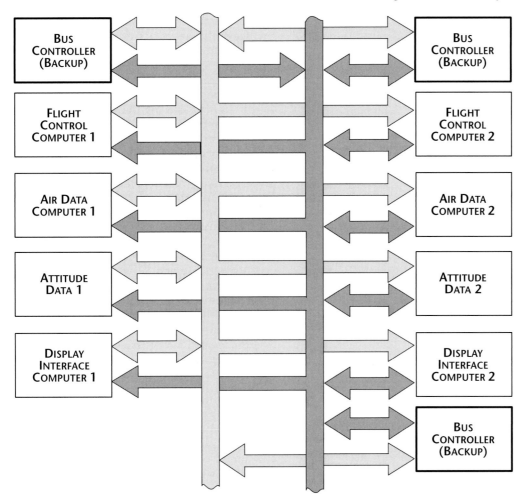

Figure 2-8-5. A typical Avionics Standard Communication Bus (ASCB) structure

The data from any subsystem will be transmitted using a specified format programmed into the transmitting LRU. The software of the receiving LRU will decode the data accordingly. All ASCB data transmissions will contain an address at the beginning of the message. Any receiver on the bus can choose to accept or ignore a given message through analysis of the address. Figure 2-8-7 shows the format for a typical bus controller request and the transmission of information by a subsystem. Notice the information content transmitted by the subsystem is shown in the box *IRS Data*. The information transmitted before and after the actual data is used for error checking and bus control functions.

Sequencing data. The request for data from the bus controller will be transmitted in a given sequence and at specific time intervals according to the controller software. Not all subsystems will be accessed during each controller request. Critical systems will be accessed more frequently than less critical systems. For example, the bus controller may send a request for data 40 times a second (40 Hz). This means that requests to a specific subsystem can be made at rates such as 40, 20, 10, 5, etc., times per sec-

ond. The ASCB request/answer time periods are known as *bus frames*. The total time allowed for a given frame must be less than the update period (40 times a second, or 25 milliseconds for the previous example).

At the beginning of each bus frame the bus controller transmits a frame-start message and

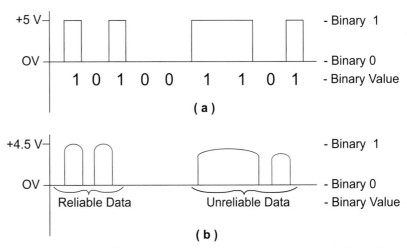

Figure 2-8-6. ASCB data transmission: (A) Accurate data transmission, +5 V=Binary 1, (B) Inaccurate data transmission, data signal less than +4.5 V

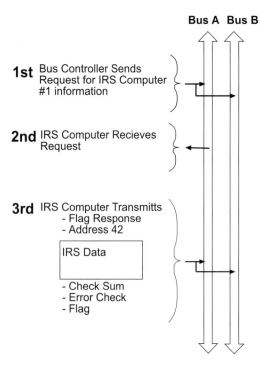

Figure 2-8-7. Typical ASCB data bus controller configuration

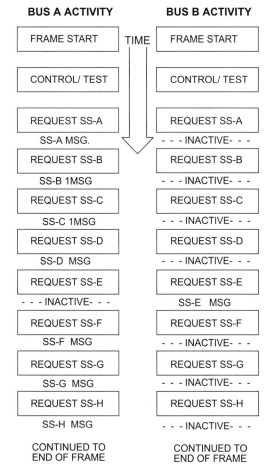

SS= Subsystem
MSG= Message from subsystem
INACTIVE= No message on that bus at this time
(i.e.- that subsystem transmits on opposite bus only)

Figure 2-8-8. Example of ASCB bus activity

a control/test message. The frame start message is a wake-up call telling all subsystems that a new frame is starting. After the control test message, the controller requests specific subsystems to transmit data. Each subsystem is requested to respond individually as shown in Figure 2-8-8. At the end of the bus frame, the controller waits for the specific time period and initiates the next bus frame sequence. With this configuration, the bus controller coordinates all bus activities. The failure of any single transmitter will not effect the operation of others on the bus.

Data bus cable. The typical data bus cable used for ASCB transmission is two-wire, 24-gauge, shielded cable. Figure 2-8-9 shows some of the ASCB bus specifications as discussed in this paragraph. The maximum bus cable impedance is 125Ω ± 2Ω, and maximum capacitance is 12 ± 2 picofarads. The end of each wire pair is terminated using a 127Ω, ±1/4Ω, nonconductive metal film resistor. The avionics equipment is connected to the bus through a stub bus and a transformer/coupler assembly. The maximum bus length is 150 feet, and the maximum stub length, between bus and avionics equipment, is 36 inches. All bus cable shielding must be terminated to ground at each end of the cable. A schematic of a typical bus coupler is shown in Figure 2-8-10.

Commercial Standard Digital Bus (CSDB)

The CSDB (commercial standard digital bus) is a one-way data bus system between one transmitter and a maximum of 10 receivers. CSDB is used in Collins general aviation electronics equipment and is certified on a large number of aircraft systems. CSDB adheres to a specification standard RS-422A established by the Electronics Industries Association, and has also been recognized as a standard aircraft data bus by the General Aviation Manufacturers Association (GAMA). The CSDB data bus cable is a twisted pair of shielded wire, which can be up to 150 feet long.

Data formats. The CSDB standard allows for operation in one of three forms: continuous repetition, noncontinuous, and burst transmissions. During *continuous repetition transmission*, the data is consistently updated at given intervals. Continuous repetition data is typically used for primary data, such as aircraft attitude. For *noncontinuous data transmission* the information is updated at a given rate only when that data is available. Noncontinuous data is typically used for secondary data, such as test data. *Burst data* is information that is meant for a one-time transmission intended to announce a specific action, such as capture of the glide slope signal. Reception and transmission of the three different data forms is a function of the software within a given system. All three types of data may be transmitted on the same bus. CSDB data is transmitted on a NRZ (non-return to zero) format where logic state 1 exists when bus line A is positive with respect to line B. Logic state 0 exists when bus line B is positive with respect to line A. The CSDB system operates at a high speed of 50 Kbits/sec or 12.5 Kbits/sec for low speed.

Data frames. The CSDB data transmission is divided into frames. Each frame is a fixed time interval for a given bus. The length of each frame is a function of the maximum update rate required for that system. Each frame begins at the start of a synchronization (sync) block and ends at the start of the next sync block. The sync block consists of a fixed number of bytes assigned to that bus. A typical sync block byte is the hexadecimal number "A5". As can be seen in Figure 2-8-11, each frame contains one or more message blocks.

Message blocks. The CSDB message block is a specific serial message, consisting of a defined number of bytes. Each message is divided into an address byte, a status byte, and one or more data bytes (Figure 2-8-11). The data bytes are transmitted with the least significant (LS) byte first and the most significant (MS) byte last. Each byte consists of eight bits of data along with a start bit, a parity bit, and a stop bit. Odd parity is used for CSDB transmissions.

ARINC 429

An ARINC 429 data bus is a one-way communication link between a single transmitter

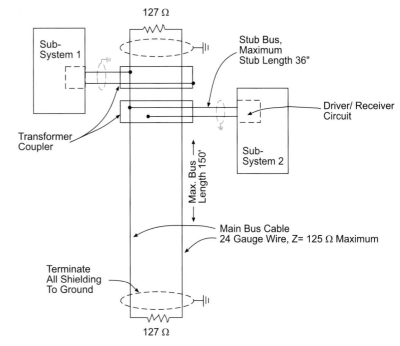

Figure 2-8-9. ASCB data bus specifications   *For training purposes only*

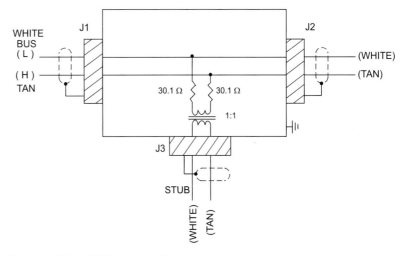

Figure 2-8-10. ASCB bus coupler

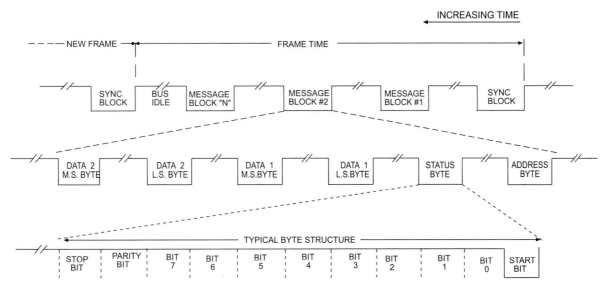

Figure 2-8-11. The CSDB data format consists of bus frames, message blocks, bytes, and bits

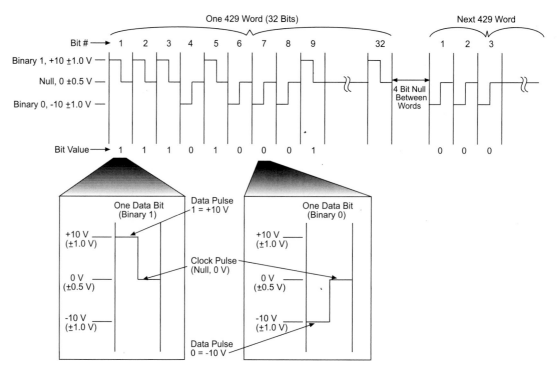

Figure 2-8-12. Digital signal for an ARINC 429 data bus

and one or more receivers. The 429 data bus standard is by far the most common data bus format in use today on civilian aircraft. The ARINC 429 system provides for the transmission of 32 bits of information in each data word. A data word is sometimes referred to as a byte. Each word is separated by four bits of bus silence known as a four-bit null. A null is a signal, which is equal to neither binary 1 nor binary 0. For an ARINC 429 transmitter the nominal voltage values are as follows: a digital 1 requires a signal of +10 ± 1.0 V; a digital 0 requires a -10 ±1.0 V signal. The null signal is equal to 0.0 ± 0.5 V. Figure 2-8-12 shows the basic digital signal format for an ARINC 429 transmitter. As seen in the expanded view of this diagram, each bit actually consists of two voltage values, the data signal (+10 volts or -10

volts) and the null signal (0 volts). The null signal is always transmitted immediately after the data and makes up one half of the bit.

The null signal is used for timing, or synchronizing, the data bus. Each 429 transmitter contains an internal clock; the timing signal is then transmitted through the data bus to each receiver via the null voltages. This type of data transmission is referred to as *RZ (return to zero)*. Using RZ format allows ARINC 429 to be "self clocking."

The ARINC 429 signal is also a bipolar format. Bipolar means the data signal actually reverses polarity (two-polarity) when it changes from binary 1 to binary 0. Binary 1 is equal to +10 volts; binary 0 equals –10 volts. Since the data bus consists of two shielded wires, reversing polarity of these wires changes the data signal from binary 1 to binary 0 (Figure 2-8-13). The ARINC 429 signals use a bipolar, return-to-zero format. This means the bus voltage must reverse polarity to change from binary 1 to binary 0 and each bit must return to zero to provide the synchronizing pulse.

There can be a maximum of 20 receivers for any one ARINC 429 transmitter. Each receiver must have an input impedance of 12,000 Ω or greater to keep signal loss to a minimum. The transmitter impedance should be 75 ±5 Ω matching the impedance of the data bus cable, which should also be approximately 75 Ω.

An ARINC 429 receiver will recognize a binary 1 as any value +6.5 to +13 V; and

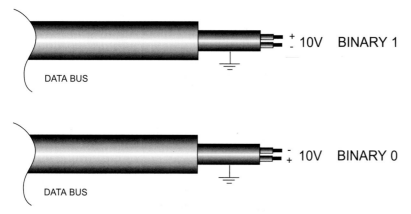

Figure 2-8-13. Typical ARINC 429 data bus cable. Note that the signal polarity reverses for binary 1 and binary 0

(A) BCD, DISCRETE Formats

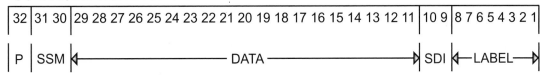

(B) BNR Format

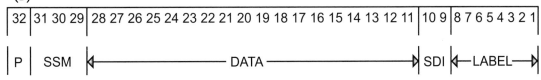

(C) AIM Format

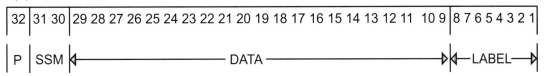

Figure 2-8-14. ARINC 429 data word format: (A) BCD - Discrete, (B) Binary, (C) AIM

a binary 0 as -6.5 to -13 V. The null voltage value must be +2.5 to -2.5 V to be accepted by the receiver. Values outside of these limits are often caused by poor electrical connections or faulty LRUs. The receiver has a greater voltage tolerance than the transmitter (transmitter tolerance: binary 1 = +10 V ± 1 V, binary 0 = -10 V ± 1 V, null = 0 V ± 0.5 V). The receiver must have this greater tolerance to allow for any signal loss, noise, or other forms of distortion that may occur between the transmitter and the receiver. The use of an oscilloscope or data bus analyzer can be used to detect excessive signal distortion. Oscilloscopes and data bus analyzers will be discussed later in this chapter.

The ARINC 429 standard allows for the transmission of digital data in two speeds. The majority of the digital systems transmit ARINC 429 data at low-speed (12.0 to 14.5 Kbits/sec). The high-speed bus is typically reserved for systems requiring large quantities of data, which is frequently updated. For example, the flight management computer would operate at high speed. The high-speed bus transmits at a frequency of 100 Kbits/sec. At high-speed, each bit lasts only 10 μsec, at low speed each bit lasts 70 to 83 μsec. To eliminate timing problems, the ARINC 429 system never mixes high-speed and low-speed data on the same bus. It should be noted low-speed has a range (12.0 to 14.5 Kbits/sec) the exact speed is determined by the transmission software design.

429 Data Word Formats

One of four data word formats must be used to conform to the ARINC 429 standards: *binary data* (BNR), *binary-coded decimal* (BCD), *acknowledgment ISO alphabet maintenance* (AIM), or *discrete*. As illustrated by Figure 2-8-14, the ARINC 429 standard assigns the first eight digits of a byte as the word label, and bit number 32 is a parity bit. The data field is comprised of digits 11 through 29 for BCD, and discrete word formats. The data field is comprised of digits 11 through 28 for the BNR word format. The data field is bits 9 through 29 for the AIM format. The sign-status matrix (SSM) includes bit numbers 30 and 31 for the BCD, AIM, and Discrete word formats. The SSM includes bit numbers 29, 30 and 31 for the BNR word format. Digits 9 and 10 are a *source-destination identifier* (SDI) for all formats except AIM. The AIM format does not use a SDI (bits 9 and 10 are part of the data).

To eliminate confusion, remember each of the four ARINC 429 formats (BNR, BCD, AIM and discrete) is comprised of a 32-bit word. Each word has the same basic structure as shown in Figure 2-8-14. However, the SSM may vary in size and the SDI may be eliminated for certain word formats.

The *SDI* serves as the address for the 32-bit word. That is, the SDI identifies the source or destination of the word. Any receiver connected to that bus receives all information sent to a common

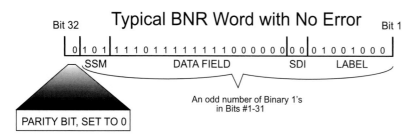

Typical BNR Word with No Error

Figure 2-8-15. The ARINC 429 parity bit is bit number 32. The parity bit is always set to allow for an odd number of binary 1s in the complete ARINC 429 data word (bits 1-32)

number 11 through 28 for BNR, bits 11 through 29 for BCD and Discrete, and bits 9 through 29 for AIM. This data is the actual message, which is to be transmitted. For example, an airspeed indicator may transmit the binary message 0110101001. Translated to decimal, this could mean 425, or airspeed of 425 knots. In some cases, all bits of the data field are not needed. Pad bits "fill in" any portion of the data field, which are not used. Pad bits are always set to binary 0. The letter *P* is often used to represent pad bits.

The sign-status matrix (SSM) provides information, which might be common to several peripherals. Typically bit number 29 is used for common information, such as north, south, plus, minus, right, left, etc. Bits 30 and 31 are used to transmit information concerning system status. For example, if bit 30 and 31 are set to 00 it could mean failure warning, indicating a malfunction in the transmitter system. The specific meaning of the SSM digits may vary between word formats. The ARINC 429 specification provides details as to the SSM meanings.

serial bus. Each receiver accepts only that information labeled with the correct SDI. An SDI set to binary 00 represents an All Call situation. In other words, if the SDI is set to 00 all receivers connected to the bus will "listen" to the transmitted data. If the SDI is set to 01, 10, or 11 only receiver(s) assigned that specific code (SDI) will "listen" to the transmitted data.

The information, or data, of an ARINC 429 transmission must be contained within bits

Label #	Parameter name	Units	Range	Significant Digits	Resolution
206	Computed Airspeed	Knots	1024	14	0.0625
207	Max. Allowable Airspeed	Knots	1024	12	0.25

(A)

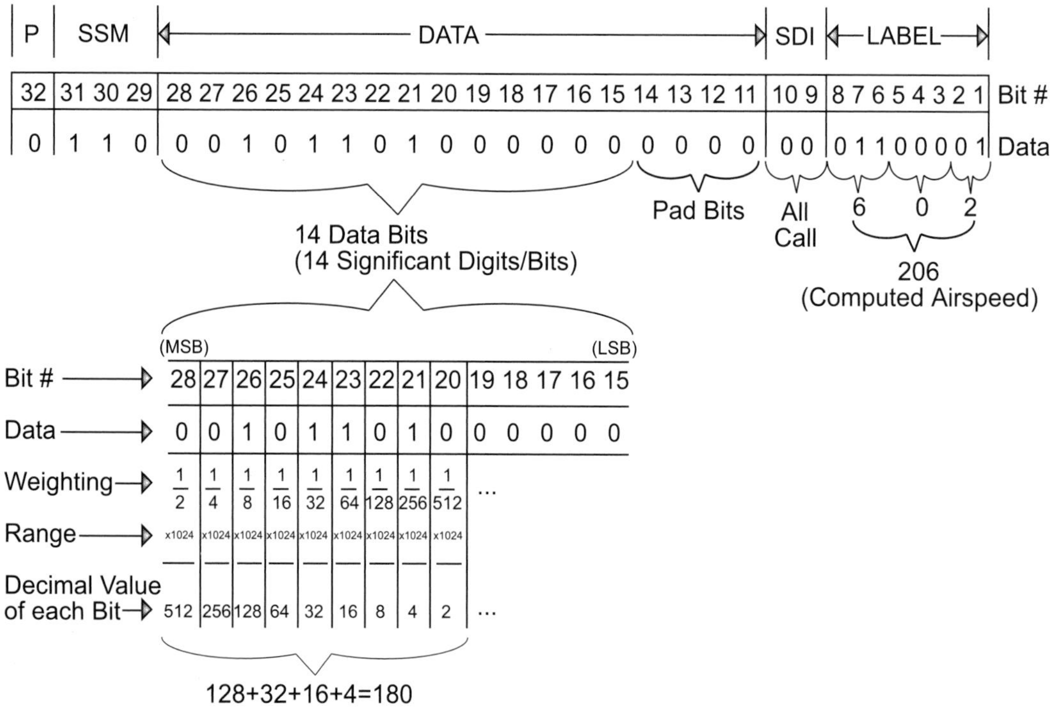

Figure 2-8-16. Decoding an ARINC 429 Binary word data field: (A) ARINC 429 specification table for binary words 206 and 207, (B) Analysis of the data field

LABEL	EQPT. ID (HEX)	PARAMETER NAME	UNITS	RANGE (SCALE)	SIG. DIG.	RESOLUTION	MIN. TRANSIT INTERVAL (msec)2	MAX. TRANSIT INTERVAL (msec)2
112	002	Runway Length	Feet	20480	11	10	250	500
	0A1	Selected EPR		4	12	0.001	100	200
	0A1	Selected N_1	RPM	4096	12	1	100	200
	0BB	Flap Lever Position-Left	Deg/180	+/-180	18	0.000687	80	160
114	002	Desired Track	Deg/180	+/-180	12	0.05	100	200
	029	Brake Temperature (Left Inner L/G)	Deg C	2048	11	1	100	200
	02F	Ambient Pressure	PSIA	32	14	0.002	100	200
	0BB	Flap Lever Position-Right	Deg/180	+/-180	18	0.000687	80	160
	0CC	Wheel Torque Output	Lb./ft.	16384	12	4	50	100
	10A	Selected Ambient Static Pressure	PSIA	1.5-20.0	11	0.016	100	500
	10B	Selected Ambient Static Pressure	PSIA	1.5-20.0	11	0.016	100	500

Table 2-8-1. ARINC 429 binary data word definitions

The ARINC 429 parity bit allows the receiver to detect errors that may have occurred during data transmission. The parity bit does not check the accuracy of the data itself; it only verifies accurate transmission on the data bus. ARINC 429 uses odd parity. This means that properly transmitted data will always contain an odd number of binary 1s in data bits 1 through 32 (Figure 2-8-15). Just prior to transmitting data onto the bus, the LRU's transmitter circuitry "counts" the number of binary 1s in data bits 1 through 31. If there is an even number, the parity bit is set to 1; if there is an odd number the parity bit is set to 0. This ensures the data leaving the transmitter always has an odd number of 1s in bits 1 through 32. When the data is received by an LRU, the circuitry "counts" the binary 1s. The receiver will only consider the data valid if there is an odd number of binary 1s in data bits 1 through 32. Therefore, if one or more data bits should be lost during transmission the receiver will most likely detect a fault due to inaccurate parity.

The receiver also performs a "reasonableness check" designed to detect any unreasonable information. For example, if the airspeed data changed from 205 knots to 12 knots in a fraction of a second, the receiver would ignore that information since it is unreasonable. The reasonableness check along with the parity bit ensures that the receiver will ignore all defects in data transmission. If the defect continues, a system fault flag will be displayed on the instrument panel to inform the pilot. The defect may also be recorded in the system's nonvolatile memory to be used later for maintenance purposes.

Word Labels and Equipment Identifiers

Once a receiver accepts a word from the bus it must determine how to decode the data field. To do this, the receiver "reads" the word label. Knowing the label allows the receiver's software to access the correct program to decode the data. As shown in Figure 2-8-16, the word label is contained in bits 1 through 8 of each data word. There are 256 possible combinations of word labels in the ARINC 429 code. The ARINC 429 standard assigns specific names to each of the possible 256 labels. Each label is assigned a specific word format and it cannot change. As mentioned earlier, the four available formats are BNR, BCD, AIM, and discrete.

It should be noted that ARINC 429 was first developed in the late 1970s. At that time there was a limited use of digital systems onboard aircraft and 256 label combinations was more than adequate. In the following years, a digital revolution took place and a multitude of systems were developed. Soon, 256 label combinations were not enough to cover all the digital systems.

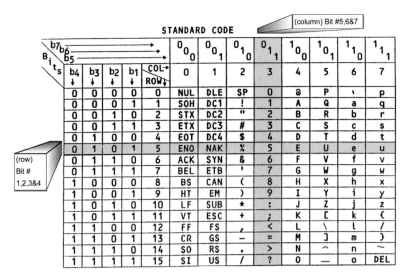

Figure 2-8-17. ARINC 429 ISO alphabet

At that time, equipment identifiers were added to most of the labels. The *equipment identifier* is a subcategory of the ARINC 429 label providing a greater variety of label combinations. The equipment identifier is a hexadecimal code assigned by ARINC. Table 2-8-1 shows a portion of the BNR data standards for labels 112 and 114. Here it can be seen that label 112 has 4 different equipment identifiers, and each identifier employs different units, range, significant digits, resolution, and transmission intervals.

The equipment identifier is not transmitted as part of the 429 data word. Since ARINC 429 employs a one-way data bus with only one transmitter, the software in each receiver "knows" what equipment identifier is assigned to each label. This situation is valid only if the transmitter and the receiver are both programmed for identical label/identifier combinations. Only one equipment identifier can be used for any given label on any given 429 bus. During aircraft operations the label is enough to identify the specific word configuration because the system software "knows" the assigned equipment identifier. However, when using a data bus analyzer (discussed later in this chapter), the equipment identifier may become critical. Since the data bus analyzer can read all combinations of a given ARINC 429 label, the user must "tell" the analyzer what equipment identifier is used when receiving or transmitting a given label. Once programmed with the correct label/identifier combination the analyzer can transmit or receive data in the correct format. Of course the equipment identifier is also important during system engineering.

Decoding the Word Label

The label data (bits 1 through 8) is comprised of two three-bit groups and one two-bit group (Figure 2-8-17). Bits 1 and 2 comprise the most significant digit of the word label, and bits 6, 7 and 8 comprise the least significant digit of the label. To determine the decimal value of the label, simply convert each binary group into its octal equivalent. The word label is an octal number so each digit can be one of eight characters (0, 1, 2, 3, 4, 5, 6, and 7). No value higher than 7 is possible for any digit of the word label. Also remember, the binary values are set in reverse order. For example, bit number $8 = 2^0$, bit number $7 = 2^1$, and bit number $6 = 2^2$ (Figure 2-8-18). In this example, bit number 8 is set to binary 1, bit 7 is set to 0, and bit 6 is set to 0; this group is therefore equal to a octal value of 1. This is the least significant group. Bit numbers 3, 4 and 5 are each set to 0 and therefore the octal value of this group is 0. Bit number 2 is set to 0 and bit 1 is set to 1 therefore the most significant group is set to an octal value of 2. In this example, the word label is 201. According to the ARINC 429 specifications, 201 is the label for DME distance.

Binary Data

Refer to Figure 2-8-16 during the following discussion on binary data. The octal notation code of the first eight bits is read as described above to achieve the word label. The bits 011/000/01 represent 206 (602 reversed). According to the ARINC 429 code, 206 represents computed airspeed (Figure 2-8-16A). The SDI label of 00 indicates transmission of this data to all receivers connected to the serial bus. The data field for label 206 is measured in units of knots, has a range (scale) of 1024, has 14 significant digits, and has a resolution of 0.0625. The range determines the maximum and minimum values for a given label. In this case the minimum computed airspeed is 0 knots and the maximum would be 1024. (Although we know no transport category aircraft can travel at 1024 knots, this value represents the maximum value that can be transmitted using label 206.) Since a minimum range value is not stated, it is assigned a value of zero. The number of significant digits determines how many bits will be used for the data field. For label 206 there are 14 bits of data. Since the BNR format allows for 18 bits in the data field, there will be 4 pad bits (bit numbers 11-14). As discussed earlier, all pad bits are set to zero. The resolution for a given label defines the smallest unit of data that can be transmitted. The resolution for label 206 is 0.0625 knots.

To decode the data segment of a BNR word a weighting factor must be determined for each bit in the data field. This is accomplished by assigning the most significant bit (MSB) a value of one half, the next digit one fourth, the next digit one eighth, the next one sixteenth, and so on down to the least significant bit (LSB) as seen in Figure 2-8-16B. Each weighting factor is then multiplied by the maximum value of the range. In this example, the weighting factor is multiplied by 1024. Multiplying the weighting factor by the range determines the decimal value for each bit. To determine the value of the data field, simply add the decimal values for all data bits assigned digital 1. In this example, bit numbers 26, 24, 23, and 21 are binary 1; all other data bits are binary 0. Therefore, the value of the data field is 128 (decimal value of bit number 26) + 32 (decimal value of bit number 24) + 16 (decimal value of bit number 23) + 4 (decimal value of bit number 21) = 180. The computed airspeed for the example in Figure 2-8-16 is 180 knots.

In this example (Figure 2-8-16), only a portion of the data field was used. The receiver "knew" the extent of the data field by the label code through internal software programming. Once the receiver analyzes the label, the computer recalls information about that data and how it will be presented. This allows the computer software to make a determination as to how the data and SSM should be read.

In Figure 2-8-16 the SSM digits 29, 30, and 31 are binary 0, 1, and 1, respectively. According to the 429 specifications, a SSM of 011 for label 206 represents a normal operation of plus value data. The parity bit (number 32) is a 0, which denotes an odd number of binary 1s in the transmitted word. No error is present according to the parity bit.

Binary Coded Decimal

The binary coded decimal (BCD) word format uses bits 11 through 29 for the data field and allows only bits 30 and 31 for the SSM. The data field is divided into four, four-bit groups and one three-bit group as shown in Figure 2-8-19A. The least significant digit is found in the four-bit group on the far right of the data field. In this example, DME is transmitting a distance of 150.71 NM. Once again, the label was used by the receiver's software to determine how the data would be analyzed. In Figure 2-8-19B, it can be seen that the ARINC 429 standard defines the unit's range or scale, significant digits, direction, and resolution for each label. In this example, the units are nautical miles (NM), the range is -1 to 399.99, there are five significant digits, and a resolution of 0.01. The resolution is used to determine where to place the decimal point after the data field has been decoded.

By studying Figure 2-8-19A it can be seen that bits 11 through 14 create the least significant digit of the data. In this example, bit number 11 is set to binary 1 and bits 12 through 14 are binary 0. This is decoded to be decimal value 1. Remember, bits 11 through 14 form a four-bit group to create the least significant digit of the data. Bits 15-18 form the next four-bit group (data bits 0111 are decoded as decimal value 7). This process is repeated for each four-bit group, 19 through 22 and 23 through 26. The most significant digit is formed by three bits of the data field (bits 27 through 29). In this example, the data field is decoded to be the digits 1, 5, 0, 7,

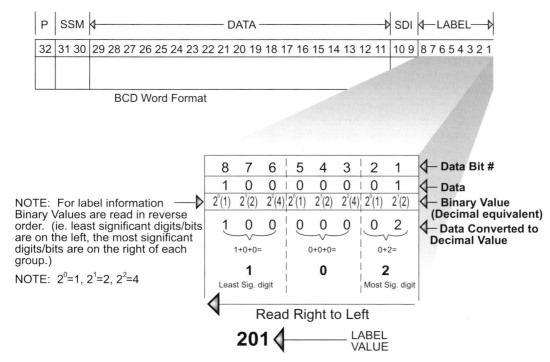

Figure 2-8-18. Decoding an ARINC 429 word label

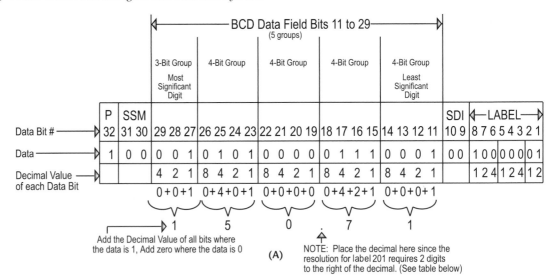

Figure 2-8-19. Decoding an ARINC 429 Binary Coded Decimal (BCD) data field: (A) Example of transmitted data, (B) Specifications for label 200, 201, and 230

and 1 (15071). To locate the decimal point in the data, simple reference the 429 specifications. In this case, the specification shows a resolution of 0.01. Therefore, the decoded data must contain two digits to the right of the decimal point. The decoded data is therefore equal to 150.71 nautical miles.

AIM Word Format

The AIM (acknowledgment, ISO-alphabet, maintenance) format is used for systems that require large amounts of data transfer. ISO is an acronym for the International Standards Organization. This group has identified a matrix that provides graphic, numerical, and alphabetic symbols for a group of binary digits. The ISO-alphabet used for ARINC 429 data uses a seven-bit code for each character. The values of bits 1 through 4 determine the *matrix*

row; bits 5 through 7 determine the *matrix column*. In Figure 2-8-17, the binary bits 0110101 represent the number five.

NOTE: b_1 *is shown on the right;* b_7 *is shown on the left.*

As seen in Figure 2-8-20, the data field for an AIM word is comprised of three, seven-bit groups. Each of these seven-bit groups can be decoded using the ISO-alphabet matrix (Figure 2-8-17). Since this format does not provide much data on each 32-bit word, a complete message may take several words. The ARINC 429 specification allows for a maximum of 127 individual words In the AIM format.

The ARINC 429 AIM word format is comprised of three types of data files: initial word, intermediate word(s), and final word. As the name

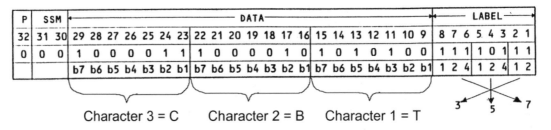

Figure 2-8-20. A typical ARINC 429 data word format

implies the initial word is always the first in the series. The initial word defines the total number of words in the message to be transmitted. A maximum of 126 words can follow the initial word. Initial words are identified by the SSM being set to 01. Intermediate word(s) contain the data to be transmitted (SSM=00). The final word (SSM=10) is used to signal the end of the data words. After the final word a 4-bit null would be transmitted and a new message would begin. The exact formats for the initial, intermediate, and final word all vary slightly and are identified in the ARINC 429 specification.

Discrete Words

The discrete word format is used to transmit the status of several individual components or systems in a 32-bit word. These discrete words are not defined by the ARINC specification. The manufacturer can determine what discrete signals the various bits of the data field will represent. For example, the manufacturer may assign bit 11 as engine 1, fire loop 1 - fail, and bit 12 as engine 1, fire loop 2 - fail. In this example, if bits 11 or 12 were binary 1 it would indicate a failure in that specific fire loop. If bits 11 or 12 are binary 0 it would indicate the respective fire loop has not failed. Discrete data word formats use bits 11 through 29 for the data field. Any discrete word bits that are not used are referred to as pad bits and set to binary 0.

For a typical aircraft using the ARINC 429 data bus system most of the information is transmitted using the binary or binary coded decimal word formats. On the Boeing 767 for example, approximately 65 percent of all 429 data is transmitted in either the BNR or BCD formats.

ARINC 629

ARINC 629 is a digital data bus format that offers more flexibility and greater speed than the 429 system. ARINC 629 permits up to 120 receiver/transmitters to share a bidirectional serial data bus. The bus can be either a twisted wire pair or a fiber optic cable up to 100 meters long. ARINC 629 was introduced to the commercial aviation industry on the Boeing 777. A similar version of 629 can also be found on some military aircraft. ARINC 629 was developed as a potential replacement for ARINC 429, but has found limited use within the aircraft industry. Although it is difficult to predict the future, it is likely that ARINC 629 will be employed on only one transport category aircraft, the Boeing 777. Since the development of this aircraft, a

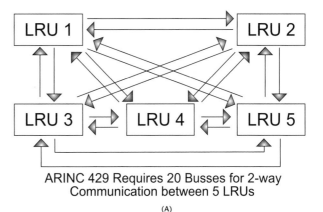

ARINC 429 Requires 20 Busses for 2-way Communication between 5 LRUs

(A)

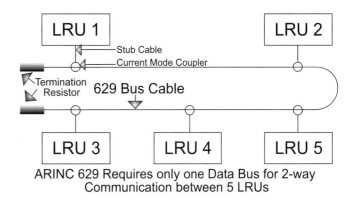

ARINC 629 Requires only one Data Bus for 2-way Communication between 5 LRUs

(B)

Figure 2-8-21. Typical ARINC data bus structures: (A) One-way ARINC, (B) Two-way ARINC 629

more modern data bus standard, ARINC 664, has become popular and employed on newer aircraft. ARINC 664 will be discussed later in this text.

ARINC 629 has several major improvements over other data bus systems. First, there is a substantial weight saving when compared to the ARINC 429 system. The older 429 specifications is a one-way data bus requiring a separate bus for each transmitter; 629 is bidirectional and can accommodate up to 120 receiver/transmitters on each bus. Figure 2-8-21 shows a simplified diagram of the 429 and 629 bus structures. Here it can be seen that the bidirectional 629 system requires much less data cable than the older 429 system. Second, the 629 bus operates at a speed of 2 megabits/second; this is much faster than many older bus formats. ARINC 429 operates at a maximum of 100 kilobits/second. Third, the 629 requires no separate bus control unit since each LRU connected to the bus is designed to independently coordinate its own transmission activities. Fourth, the 629 bus system employs an inductive coupling unit to connect each LRU to the data bus. The inductive coupler permits easy connection to the bus without physical interruption of the bus wires; hence improving bus reliability, (i.e., fewer loose connections).

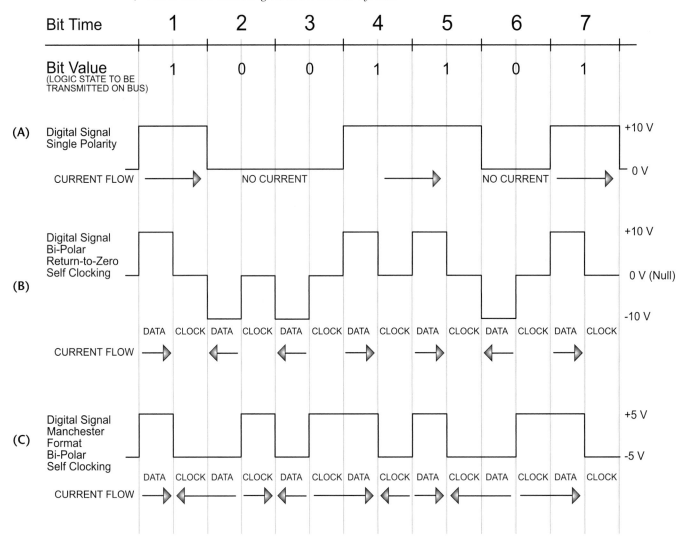

Figure 2-8-22. Digital signal characteristics: (A) Single polarity signal, (B) Bipolar signal typical of ARINC 429 data, (C) Bipolar Manchester signal

Signal Characteristics

The ARINC 629 data signal actually changes format during the transmission process. The data initially produced by the transmitting LRU is a simple digital signal in a Manchester format discussed earlier. This current signal is then converted to a signal known as a *doublet* for transmission on the 629 data bus. A doublet signal emits a short pulse whenever the bit value changes from binary 1 to 0 or vice versa. The doublet signal will also transmit a short pulse for any timing signals transmitted on the bus, i.e., a doublet pulse occurs whenever there is a change in logic state. The receiving LRU converts the doublet signals back into a Manchester format for use by the receiver's processor circuitry.

In order to better understand the ARINC 629 signal characteristics, a short review of digital data signals will be presented here. As seen in Figure 2-8-22A, a simple digital signal requires a +10 V for binary 1 and 0 V for binary 0. Note

that the voltage values may vary for different systems. A very weak current flow occurs for binary 1 and no current flows for binary 0. The voltage never changes polarity since this is a single polarity format. A bipolar return-to-zero- type signal is shown in Figure 2-8-22B. Here the polarity on the bus is such that a +10 V equals binary 1. The bus polarity reverses to –10 V for Binary 0. This is a bipolar format used by ARINC 429 systems. The signal also contains a clock pulse with each data bit. When voltage equals zero, the clock (null) signal is produced. There is no current flow during the clock pulse. Figure 2-8-22C shows a typical Manchester II format. This Manchester signal is bipolar and contains a clock pulse with each data bit. A +5 V produces a binary 1 and –5 V produces binary 0.

A *doublet signal* produces a short positive and negative pulse or spike on the bus whenever the data value changes from binary 1 to binary 0 (Figure 2-8-23). Notice that with a doublet signal, the bus is active only for a short period

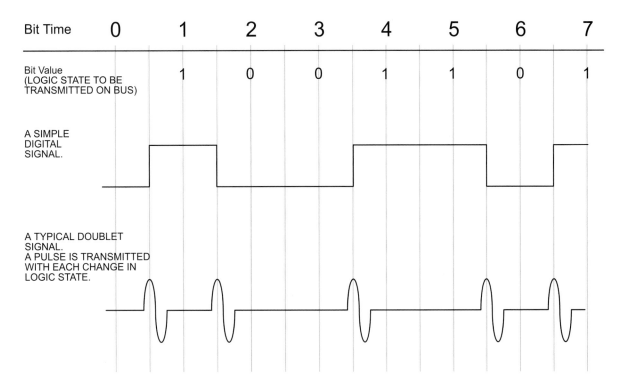

Figure 2-8-23. A typical doublet signal creates a voltage (or current) spike at each data bit change

of time during each change of the binary data. The short spike alerts the receiver that the data bit has changed. From this short spike signal, the processor can conclude the value of the binary data (1 or 0).

A doublet can be either a voltage specific or current specific signal. That is, a voltage doublet will have limits set to the voltage parameters of the signal. A current doublet will have limits set to the current parameters of the signal. Figure 2-8-24 shows an example of typical voltage and current doublets.

Figure 2-8-25 shows the various forms of data required for transmission and reception by an ARINC 629 system. Study this figure during the following discussion. ARINC 629 data goes though several steps between transmitter and receiver.

- Step 1: The Manchester signal is generated in the transmitting LRU.

- Step 2: The Manchester signal is transformed into a voltage doublet signal. The voltage doublet leaving the transmitter will have a maximum and minimum differential voltage. (In this case, the voltage values are +4.5 V and –4.5 V.)

- Step 3: The voltage doublet is sent to the bus via the transmitting stub cable. Stub cables will be discussed in the next section of this chapter.

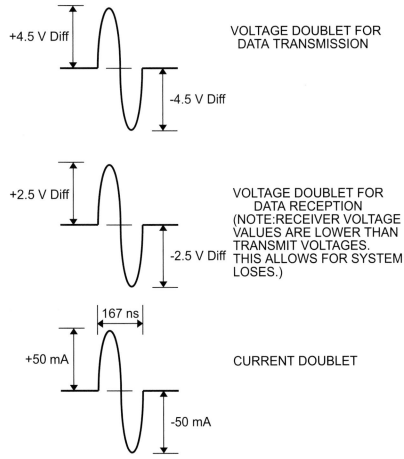

Figure 2-8-24. Voltage and current doublets typical of those used for ARINC 629 data transmission

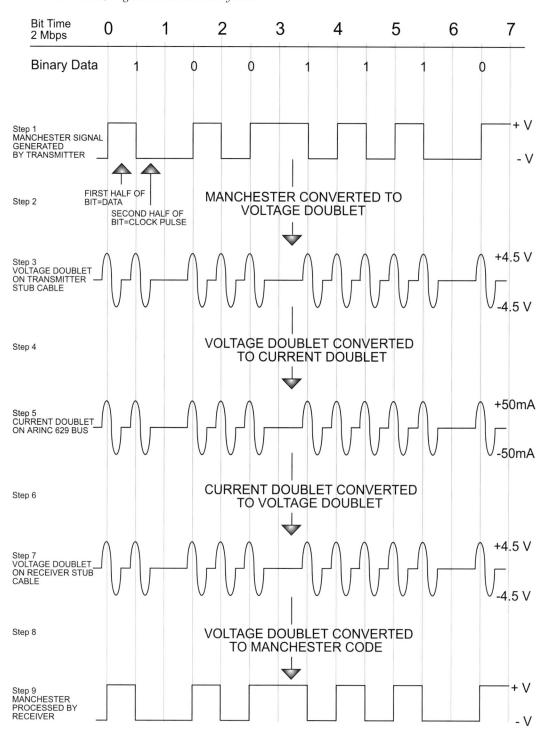

Figure 2-8-25. ARINC 629 data changes format several times between transmitter and receiver

- Step 4: The signal is changed into a current doublet for transmission on the 629 data bus.

- Step 5: The current doublet is transmitted on the 629 data bus. The current doublet has a maximum value of + 50 mA and a minimum value of –50 mA.

- Step 6: The data is changed from a current doublet back into a voltage doublet at the receiving end of the data bus. The receiver's voltage doublet parameters are +2.5 V

and –2.5 V; slightly less than the transmission doublet parameters.

- Step 7: The data is sent to the receiving LRU on the receiver's stub cable.

- Step 8: The data is once again converted to a Manchester format that can be processed by the receiving LRU. It should be noted the voltage and current parameters displayed in Figure 2-8-25 are not fixed values. The ARINC 629 specification allows the manufacture substantial lee-

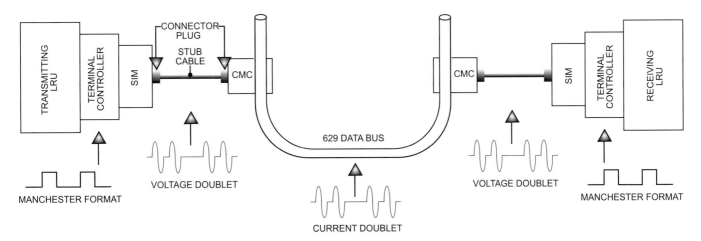

Figure 2-8-26. ARINC 629 bus structure and related signal formats

way in the design of their specific systems. This flexibility allows for new designs to be developed without limiting engineers; hence, different aircraft may have slightly different system parameters. Be sure to check the aircraft specifications for exact details.

Data Bus Physical Properties

ARINC 629 signals may be transmitted on a traditional copper data bus or a fiber optic network. The fiber optic network is used to transmit an optical (light) signal that can be adapted to a variety of formats. Fiber optics have the advantage of being immune to electromagnetic interference and lightning transients. Fiber optics will be discussed in more detail later in this text. The traditional copper wire data bus used for ARINC 664 may be up to 100 meters in length with impedance of 130 Ω ±2%. The data bus consists of two insulated conductors of #20 AWG stranded wire. The conductor pair must be twisted together and may, or may not, be shielded. The insulation consists of a foam core with a Teflon outer shell. During maintenance, be sure not to damage the insulation since water may penetrate the cable and change the impedance of the bus. Special tools are typically required for installing and removing components on the data bus. Be sure to follow approved maintenance procedures. The bus must have a termination resistor of 130 Ω ±4 percent at each end. It is recommended the outer insulation of all ARINC 629 data bus cable be a unique color that is easy to distinguish from other wires in the aircraft.

The information transmitted using ARINC 629 must travel through several subsystems to enter or leave the data bus. Figure 2-8-26 shows the relationship of the various LRUs, and their related subsystems. The terminal controller (TC) is an LRU subsystem that

moves data to and from the LRU memory. The terminal controller is the first interface between the transmitting LRU and the data bus. The TC provides various control functions for data transmission. From the terminal controller the data is sent to the serial interface module (SIM). The SIM connects the TC to the data bus stub cable. The SIM changes the Manchester current signal from the LRU into an analog voltage doublet signal. The SIM also provides a fault monitoring function for the various signals received from the data bus. The SIM examines each data word for correct format and informs the TC in the event of a system failure.

As seen in Figure 2-8-26, the SIM is connected to the actual data bus through a stub cable and the current mode coupler. The *stub cable* is a four-wire cable used to connect the TC to the current mode coupler, and can be a maximum of 75 feet long. The stub cable uses two wires for receive mode and two wires for transmit mode. The stub cable also supplies power from the LRU to the current mode coupler.

The *current mode coupler* (CMC) is an inductive coupling device that connects the LRU to the 629 data bus. The CMC contains a coupler transformer, which receives induced signals from the data bus cable (Figure 2-8-27). The CMC sends and receives electrical signals from the data bus without being physically connected to the bus. This helps to ensure integrity of the main data bus cable. The CMC actually contains two inductive couplers in one unit; one operates in active mode and one in standby. The SIM selects which coupler is operated in the active mode. As the CMC couples to the data bus the data signal changes format. During transmission the CMC changes the voltage signal (doublet) from the stub cable into a current signal for transmission on the data bus. In the receive mode the CMC

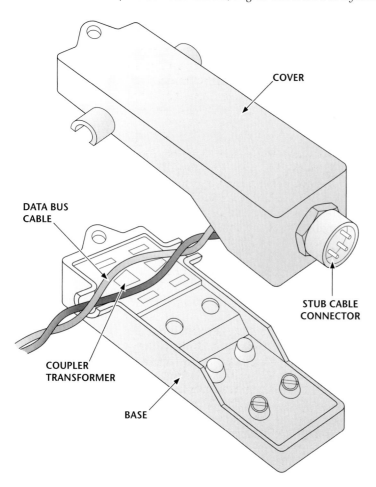

Figure 2-8-27. A typical current mode coupler

pulse (PSSP) and the pre-presync sync pulse (PPSSP) distinguish the start of a label word from the start of a data word. The PSSP and PPSSP are bits that transmit for approximately one-half the time of a standard bit. These special bits are used to inform all receivers on the data bus that a new label word is being transmitted.

Transmission Protocol

Any particular unit can transmit on the bus or "listen" for information. At any given time, only one user can transmit, however one or more units can receive data. The 629 system allows each transmitter to be completely independent of each other, and there is no bus controller used to coordinate transmission activities. The software of each transmitter must determine when that transmitter can talk on the bus. This "open bus" scenario creates some interesting challenges for the 629 system. The transmitter software must ensure that no one transmitter dominates the use of the bus and that the higher priority systems have a chance to talk first.

ARINC 629 is often referred to as a periodic/aperiodic multi-transmitter bus. Multi-transmitter simply means the bus will support more than one transmitter; ARINC 629 allows up to 120 receiver/transmitters. According to the dictionary, *Periodic* is defined as something that occurs at regular intervals, *aperiodic* means not occurring periodically, or something that occurs at irregular intervals. ARINC 629 allows certain systems to transmit at regular intervals, i.e., in the periodic mode. Periodic transmissions would include information that requires updating at regular intervals, such as aircraft attitude or airspeed. Information transmitted aperiodically would be data such as flap position, system failure data, or fire warning information. This type of signal is only transmitted when the event occurs, i.e., not at regular intervals.

changes the current signal received from the data bus into a voltage doublet sent to the stub cable and onto the receiving LRU.

Data Word Format

Reference Figure 2-8-28 during the following discussions on the ARINC 629 data word format. ARINC 629 transmits a series of messages separated by various timing signals. Each message has up to 31 word strings. A 4-bit sync gap identifies the beginning of each word string. Each word string may contain up to 256 data words.

Each message begins with a *label word*. The label words are followed by a sequence of data words. The label word consists of 20 bits, containing a 12-bit label field, a 4-bit label extension field, a single parity bit, and a 3-bit sync pulse. The *data word* also contains 20 bits. The data word is divided into a 16-bit data field, a single parity bit, and a 3-bit sync pulse.

Before the start of each label word a special pulse is added just prior to the 3-bit sync pulse. As seen in Figure 2-8-28 the presync sync

Transmission Priority

ARINC 629 uses a series of timing signals to ensure each LRU transmits in the correct sequence. The most important LRU must have the option to transmit first, and only one transmitter can occupy the bus at any given time. There are three timing signals used to sequence the transmission of 629 data: the terminal gap (TG), the synchronization gap (SG), and the transmit interval (TI). The TG is a unique time period for each transmitter; the SG is always longer than the longest TG and is common to all transmitters. The TI is the longest time interval and is also common to all transmitters on that bus. These three timing signals are stored in the memory of the LRU software. Each LRU is responsible for its own timing. The LRUs monitor the bus for activity and keep track of all transmissions. When the bus becomes idle and the correct timing occurs, the LRU software transmits information onto the data bus.

Terminal gap. The terminal gap (TG) is a time period unique for each transmitter connected to the bus. The terminal gap determines the transmission priority for each LRU. LRUs with a high priority have a short TG. LRUs with a low priority have a long TG. In other words, the most important information will be transmitted from an LRU with a short TG; least critical information will be transmitted from an LRU with a long TG. No two LRUs connected to the same bus can ever have the same terminal gap. The TG priority is flexible and can be determined through software changes in the receiver/transmitters; however, all LRUs on that bus must have a different TG. Terminal gap time periods range from 3.68 to 127.68 microseconds.

Synchronization gap. The synchronization gap (SG) is a time period common to all transmitters on the bus. The synchronization gap can be thought of as the reset signal for the bus.

Once re-set, the bus is open to all transmitters. Each LRU can transmit only once until after the next SG. The LRUs always transmit in order of their priority determined by their respective terminal gaps. Remember, the synchronization gap is longer than any terminal gap, and will only occur on the bus after each user has had a chance to transmit.

The terminal gaps and synchronization gaps are time periods that and measured by the software of each LRU with a timer that starts whenever the bus is idle. Idle meaning there is no data transmission on that bus. If that LRU has information to transmit, the transmission will begin at the end of that LRUs TG. If the LRU does not wish to transmit, the timer continues until another LRU transmits or the SG is reached. Once the SG is complete, all LRUs reset their TG and SG timers.

Transmit interval. The transmit interval (TI) is a common time period for all transmitters on the bus. The TI for a specific LRU begins immediately when that LRU starts transmission. The TI inhibits any transmission from that same user until after the TI time period. Even if the TG and the SG time periods are satisfied, the LRU will not transmit until after the TI time has expired. The TI therefore creates a relatively long waiting period between

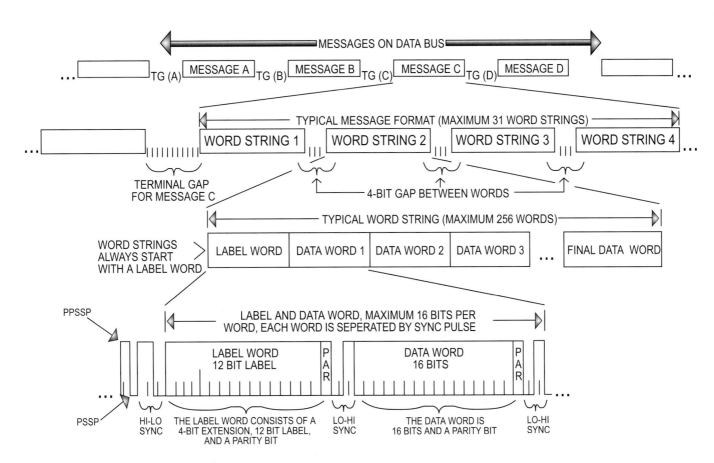

Figure 2-8-28. ARINC 629 message format

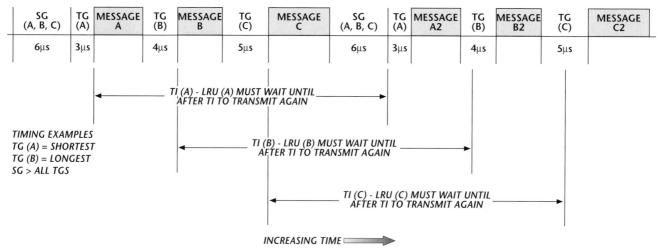

Figure 2-8-29. ARINC 629 typical timing sequence for various LRU transmissions

two transmissions from the same LRU. In effect, the TI keeps any one user from dominating the bus. The transmit interval ranges from 0.5 to 64 milliseconds and are determined by the software engineers developing the system.

Timing description. Each LRU on an ARINC 629 data bus must determine when that LRU can transmit. The TG, SG, and TI are all used to determine the LRU transmission timing. To understand this system, study the example in Figure 2-8-29 during the following discussion on transmission timing. Each LRU can use the bus if it meets a certain set of conditions. First, each transmitter is inactive until the terminal gap time for that transmitter is complete. Second, each LRU can only make one transmission then must wait until after the synchronization gap occurs before a second transmission. Third, the LRU can only make one transmission per each transmit interval.

Think of Figure 2-8-29 as a snapshot of the data being transmitted on a 629 bus. The snapshot begins at the SG in the upper left corner of the diagram, time progresses as we move to the right of the diagram. Remember the SG is the "re-set signal" for all LRUs. At the end of the SG all TGs begin their timers. The first TG to be satisfied is TG(A). TG(A) is the shortest since LRU(A) has the highest transmission priority. At completion of TG(A), LRU(A) begins transmission. At that time the TI timer for LRU(A) begins. LRU(A) cannot transmit again until TI(A) is complete. When message A is complete the bus goes idle and the TG timers start again. LRU(B) has the next highest transmission priority. The next TG to be satisfied is TG(B).

When TG(B) is complete LRU(B) begins transmission. At that time the TI timer for LRU(B) begins. LRU(B) cannot transmit again until TI(B) is complete. When message B is complete the bus goes idle and the TG timers start again. LRU(C) has the next highest transmission priority. The next TG to be satisfied is TG(C). This sequence continues until all LRUs on the bus have had an opportunity to transmit. Once all LRUs have had an opportunity to transmit, the SG will repeat and the bus will once again be open to all users. Each LRU is not required to transmit after its given TG. At that time the LRU may not have information to "share" with other LRUs on the bus. In this case, the LRU makes no transmission.

Periodic/Aperiodic Mode

As mentioned earlier, ARINC 629 operates in a periodic mode when all users complete their desired transmission prior to completion of the TI. Systems designed for operation in the periodic mode will transmit messages that have a consistent length. This ensures the update rate remains relatively constant and all transmitters have a chance to transmit within a stable time period (i.e., periodically). If an LRU transmits a longer than average message, the TI is exceeded and the bus is operating in the aperiodic mode. Systems designed to operate in the aperiodic mode transmit messages that vary in length. These systems include data to control various flight operations, such as flap position, and data for system status. Some systems are designed to operate in both the periodic and aperiodic modes depending on the specific data being transmitted.

Section 9

Avionics Full-Duplex Switched Communication System (ARINC 664)

The latest digital data transfer system to gain popularity in commercial aircraft is commonly referred to as AFDX (avionics full-duplex switched Ethernet) communication system. This data transfer system was developed by Airbus S.A.S. and first employed on the Airbus A-380. Shortly after the A-380 was developed, the Boeing 787 made its first flight using the same data transfer system. Today AFDX is also employed on a handful of other hi-tech aircraft and will most likely gain popularity in the future. AFDX is typically used as the "backbone" data transfer system for a variety of systems, including flight deck avionics, autoflight systems, central maintenance systems, air conditioning, power control/distribution, and others. After the introduction of AFDX, the aviation organization ARINC established a data transfer standard known as ARINC 664, which reflects the characteristics of AFDX. AFDX is an Airbus S.A.S. registered trademark; and ARINC 664 is the specification that defines the AFDX data transfer system. For all practical purposes AFDX and ARINC 664 can be thought of as the same and the terms are often used interchangeably. The ARINC 664 specification was designed similar to the IEEE 802.3 Ethernet standard in an effort to use off-the-shelf hardware and reduce development costs. IEEE (Institute of Electrical and Electronics Engineers) is an international non-profit organization for the advancement of technology related to electricity. One major function of IEEE is to set worldwide standards related to electrical systems. ARINC, being an aviation related organization, has modified the IEEE standard for aircraft.

AFDX allows for asynchronous data transfer at two rates, 10 Mbits/s or 100 Mbits/s. Critical data is transferred at the higher speed. Since the system operates in an asynchronous mode, the transmitters and receivers of the data operate completely independently. This format makes AFDX one thousand times faster than its predecessor the ARINC 429 data transfer system. The asynchronous transfer mode (ATM) of operation is commonly used in the telecommunications industry and has had years of proven reliability. This proven track record made the IEEE 802.3 data transfer system very attractive to the aircraft industry. Through relatively modest modifications ARINC 664 transformed IEEE 802.3 into the new aircraft data transfer standard. The A-380 and B-787, the latest transport category aircraft released worldwide, employ the AFDX data transfer system, which adheres to the ARINC 664 specification.

AFDX data transfer system increases data speed, as well as, decreases aircraft weight when compared to an ARINC 429 system. Using 664 allows for data transfer in both directions between LRUs, therefore, requiring substantially less cable. Figure 2-9-1 shows a com-

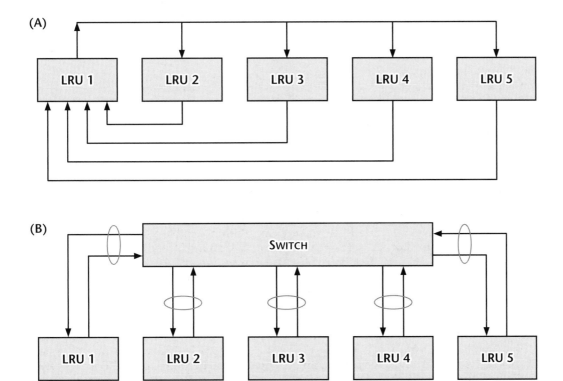

Figure 2-9-1. (A) Comparison of ARINC 429 and AFDX, (B) AFDX (bidirectional directional data bus)

parison of ARINC 429 and 664. Using ARINC 429 requires five separate data busses if LRU number 1 wishes to transmit and receive to/from LRU 2, 3, 4, and 5. Following the 664 standard allows for communication between LRUs 1 through 5 with only one data bus. All modern aircraft rely heavily on computers for system control and contain large numbers of LRUs. The unidirectional 429 data transfer system therefore must employ numerous data busses, adding hundreds of pounds and creating large wire bundles. With AFDX all end users (LRUs) are connected through a bi-directional bus to a distribution switch. The switch controls traffic to and from the LRUs. This allows any given LRU to transmit to any number of receivers and receive from any number of transmitters.

In order to fully understand the AFDX system, one must first become familiar with the terminology. A glossary of common terms related to AFDX appears in Table 2-9-1.

The ADFX Network

AFDX is considered a full-duplex, switched Ethernet interconnect system. It generally consists of a network of switches that forward Ethernet frames to their appropriate destinations through a bidirectional data bus. AFDX employs redundancy in the various components of the systems in order to achieve the reliability needed for aircraft use.

AFDX uses a series of smart switches and end systems to control the flow of data (digital message) through the network. Smart switches, commonly called switches, control the flow of data through electronic circuitry and software

programming. End systems are basically electronic (computer) LRUs which transmit/receive data in order to perform a variety of functions necessary to operate the aircraft. The program contained in each switch and all end systems decide where and when to distribute the data. The software programming used to determine the flow of each message employs complex mathematical equations, which are beyond the scope of this text. However, each smart switch and all end systems can make decisions to control data flow using their individual software. This concept is often referred to as a *virtual link*. Of course the data must travel through some type of interconnect, typically a copper data bus (a twisted pair of wires). The combination of the data bus, the smart switches, and all related connectors are often referred to as the *physical layer*. The physical layer is basically the hardware needed to make all the electrical connections between end system LRUs. The physical layer creates a pathway for the virtual link.

Figure 2-9-2 shows the basic structure of an AFDX data bus system. An avionics computer system receives inputs from a variety of sensors, actuators, and other subsystems. Since AFDX is a bidirectional system, the avionics computer can also send messages to the subsystems. The avionics computer provides a computational environment for the avionics subsystems and routes all incoming and outgoing messages to the appropriate location. The avionics computer contains an embedded end system that connects the avionics subsystems to an AFDX interconnect. The end system sends or receives data to/from the AFDX switch. Remember, these are computer controlled electronic "smart switches." Each switch then determines when to transfer the data packets and where each data packet is to be sent.

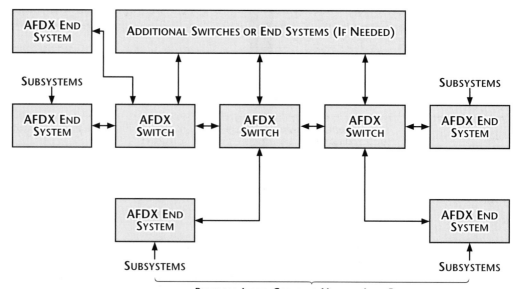

Figure 2-9-2. AFDX Network example

COMMON TERM	DEFINITION
AFDX	Avionics Full-Duplex Switched Ethernet. A bidirectional Ethernet-based digital data transfer system found on modern aircraft.
AFDX Network	A network of switches used to forward Ethernet frames to their appropriate destinations through a bidirectional data bus.
API	Application Program Interface. Enables avionics subsystems to communicate with each other through a simple message interface.
ARINC 664	The data transfer standard that defines the AFDX data transfer system, similar to the IEEE 802.3 Ethernet standard.
BAG	Bandwidth Allocation Gap. Represents the minimum time interval between Ethernet frames transmitted on the same virtual link ranging from 1 to 128 milliseconds.
End System	Electronic LRUs (computers) that transmit and receive data.
Ethernet Header	Digital information used to route a message to the appropriate destination.
FIFO	First In First Out
Full Duplex	Ethernet data transfer systems that can transmit and receive data simultaneously, employing two separate wire pairs, one to transmit and one to receive.
IP	Internet Protocol. Contains the digital information necessary to route the data to the correct end system(s).
Jitter	A possible delay when data packets are stored and waiting within buffers that are waiting for transmission on a data bus.
Physical Layer	The combination of the data bus, the smart switches, and all related connectors. Basically, the 'hardware' needed to create the data bus structure.
Port ID	Defines which port will be used to receive/transmit data.
Protocol Stack	Stacks or layers containing complex set of rules or protocols. A stack can be thought of as a module of the entire set of communication instructions.
Queuing Port	A port that has enough storage space to retain several data packets in the queue.
RX	Receive
Sampling Port	A port that can only contain one data packet in the queue.
Smart Switch	Commonly called switches on electronic units to control the flow of data using electronic circuitry and software programming.
TX	Transmit
UDP	User Datagram Protocol. The actual message being transmitted by an Ethernet-based system.
Virtual Link	The concept that using software, each smart switch, and all end systems can make decisions to control data flow using their individual software.
Virtual Link ID	Digital identifier used to direct all packets within the network.

Table 2-9-1. Glossary of AFDX terms

The AFDX network utilizes traditional subsystems, such as engine monitoring computers, accelerometers, position sensors, flight control computers, navigation equipment, and more. These subsystems can communicate to/from the avionics computer in a variety of analog or digital formats. Of course all analog data must be converted to a digital format prior to transmission on the digital data bus. Many subsystems utilize ARINC 429 to transfer data to the avionics computer. The use of ARINC 429 helps to reduce costs, since 429-based equipment is

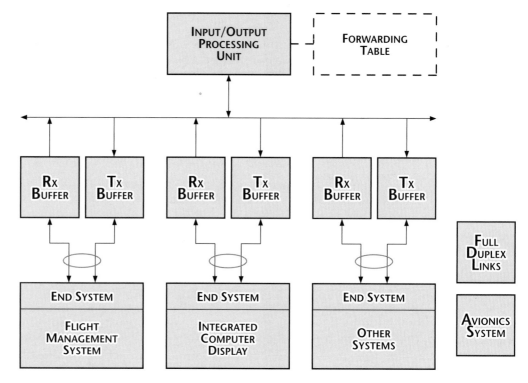

NOTE:
There is one data bus for transmit and one bus for receive.

Figure 2-9-3. Simple AFDX data transfer system

already available, meet aircraft standards, and has a proven record of reliability.

The end system within the avionics' computer provides an interface between the avionics subsystems and the AFDX interconnect. Each avionics subsystem must communicate through an end system interface to guarantee a secure and reliable data interchange with other avionics subsystems. This interface exports an application program interface (API) to the various avionics subsystems, enabling them to communicate with each other through a simple message interface. The message provides the instructions needed by the AFDX switch to create the correct path (connections) for incoming and outgoing data.

Full-duplex Ethernet interconnects are specified by ARINC 664 in order to eliminate the possibility of data collisions. When using half-duplex Ethernet systems data collisions may occur if two or more end systems connected to the same bus transmits simultaneously. In this case, the data would be lost and each end user would repeat the transmission at a random time interval. Theoretically the collision could repeat itself and create a long time period before a successful transmission is complete. As shown in Figure 2-9-3, the full-duplex switched Ethernet employs two separate wire pairs twisted and shielded as needed; one pair to transmit (TX) and one pair to receive (RX). Each end system is connected to the switch through two twisted wire pairs.

Switches

Within each switch are three essential circuit elements: the buffers (RX and TX), the I/O processor, and the forwarding table. The buffer circuit is used to store multiple incoming/outgoing data packets. Each packet is sorted and released on a FIFO (first in/first out) order. The I/O processor is used to direct data packets

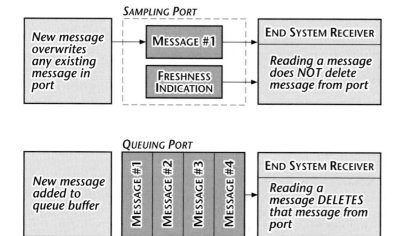

Figure 2-9-4. AFDX ports: (A) Sampling port, (B) Queuing port

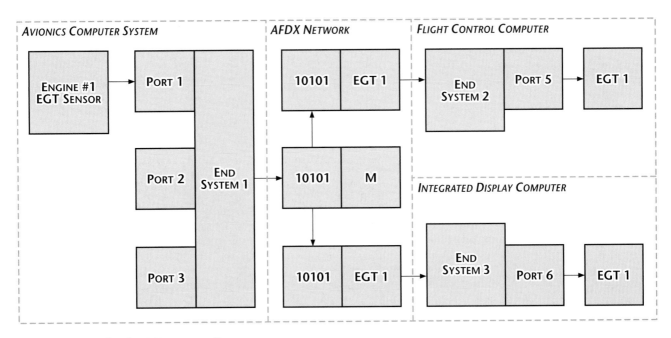

Figure 2-9-5. Example of AFDX message flow

from the incoming (RX) buffers to the appropriate outgoing (TX) buffers. The I/O processor must "read" the destination address (virtual link identifier) and refer to the forwarding table to determine the correct outgoing path. The forwarding table is contained in the software of the I/O processor. The I/O processor then sends the outgoing data packet to the correct TX buffer through the memory bus. The outgoing data is then transmitted to the correct end systems on a FIFO basis.

The buffers within the switch must be designed with the capacity to store any and all data packets awaiting transmission/reception. If the appropriate buffers are used and a full duplex cable is employed to all end systems there will be no delays due to collisions. However, this system may cause a delay, known as jitter, whenever data packets are stored and waiting within the buffers. The extent of jitter is controlled through software and switch design in order to ensure a timely transmission of all data.

Communication Ports

All software for AFDX end systems is designed to contain two types of communication ports: sampling ports and queuing ports. These ports provide the input/output pathway for data to/from the end systems. All AFDX end systems must support both ports according to the ARINC 653 specification. The two ports found in the end systems receive and manipulate data packets in different ways (Figure 2-9-4). A sampling port can only contain one data packet in the buffer. As a new message arrives into a sam-

pling port, the old message is over written. If no new message arrives in the port, the buffer will retain any existing message, therefore, a sampling port must provide information, which defines the freshness of the data in the buffer. A queuing port, on the other hand, has enough storage space to retain several data packets or messages. Messages are then read on a FIFO basis and when a message is read it is removed from the buffer. The data packet sent to each avionics computer and end system must contain a port ID. The port ID defines which port will be used to receive/transmit the data.

Routing Data

In order to ensure all data packets are transmitted and received by the correct end systems (avionics computers), an identification system must be transmitted with all data packets. The smart switches, avionics computers, and receiver ports are all controlled by software and will link each data packet according to its unique digital identifier. This type of connection is known as a virtual link. The digital identifier used to direct all data packets within the network is known as the virtual link ID. It is very important that all end systems within a given network contain a unique virtual link ID. The software of an AFDX switch "reads" the virtual link ID contained in the data packet and routes the message to the correct end system(s). This type of identification is similar to the ARINC 429 eight-bit label; however, there are several layers of AFDX identifiers and they are more than eight bits. Each end system can originate several virtual links and each virtual link can carry messages from

ETHERNET HEADER	IP HEADER	UDP HEADERS	UDP PAYLOAD AVIONICS SUBSYSTEM MESSAGE	PAD	FCS
BYTES: 14	20	8	0 - 1472	4	

Figure 2-9-6. Example of a typical AFDX Ethernet frame

more than one communication port. This complexity requires ADFX to utilize a multilevel identification system.

When an application sends a message to a communication port, the source end system, the AFDX network, the avionics computer, and the destination end systems are configured to deliver the message to the appropriate receive ports. This configuration occurs through the software contained within each LRU. Figure 2-9-5 shows the engine 1 EGT message being sent to port 1 by an avionics subsystem. End system 1 encapsulates the message in an Ethernet frame and sends the Ethernet frame to the AFDX switched network on virtual link 10101 (the Ethernet destination address specifies VLID 10101). The network switch refers to the forwarding tables within the switch software. The forwarding table contains the instructions to deliver the Ethernet frame to the flight control computer and the integrated display computer (LRU). The end systems that receive the data packet use their internal software to determine the correct destination ports for the EGT message. In this example (Figure 2-9-5) the system's software ensures the data packet (EGT 1) is delivered by end system 2, port 5 and to end system 3, port 6. This is a simplified example and actual AFDX messages contain several layers of ID information, each with at least eight bytes.

During transmission each end system receives information from an individual port; it will then encode header information to define the routing and scheduling for that message. There are three distinct headers used by each AFDX virtual link: the Ethernet header, the Internet protocol (IP) header and the user datagram protocol (UDP) header. The software, located in various components of the virtual link, decode the header information and route the message to the appropriate destination. Figure 2-9-6 shows an example of the headers that make up the data packet as transmitted by a typical avionics computer. The UDP payload is the actual message being transmitted; in this case engine number one EGT. A sequencing number known as the frame sequence number (FCS) is added to the data packet to ensure proper sequencing of the message.

The complete Ethernet frame may also include pad bits if the UDP payload is smaller than 18 bytes. In this case, the pad bits plus the UDP payload will total 18 bytes. If the UDP payload is larger than 18 bytes the payloads will be fragmented among multiple IP payloads. An important function of the IP header is to provide fragmentation control for large UDP packets. The IP header contains a destination end system identification and port identifiers or is a multicast address. If the message contains a

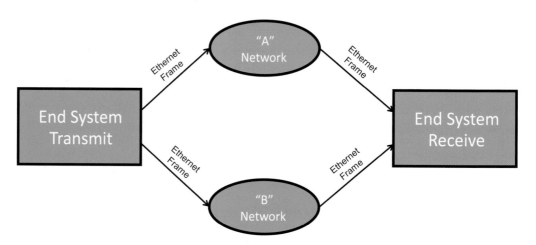

Figure 2-9-7. Two parallel networks employed by AFDX

multicast address, the IP destination address contains the virtual link ID, which provides enough information in the headers for the switches and an end system to determine all destination ports for the message.

In order to achieve the reliability needed for aircraft systems, AFDX must employ two independent networks. Each data packet transmitted by an end system is sent on both networks (network A and B). Therefore, under normal operation, each end system will receive two copies of each data packet. Figure 2-9-7 shows a simplified example of this process. Keep in mind each network is typically made up of several switches.

In order to avoid confusion, each end system must identify duplicate data packets as they arrive on the A and B networks. Therefore, all packets transmitted over an AFDX virtual link are provided with a 1-byte sequence number field. The sequence number field is part of each transmitted Ethernet frame. The receiving end systems can then make a determination that a redundant data field has arrived and discard the second data packet, or compare the first and second data packets for the purpose of a validity check.

Isolating data packets. All messages transmitted by an AFDX end system must travel from an individual port, through the physical layer to eventually reach the receiving end system. Remember, the physical layer is made up of the Ethernet data cable and one or more smart switches. The AFDX system is designed to operate at 100 Mbps. This 100 Mbps bandwidth carried by the physical layer can support several virtual links at any given time. The virtual links must share the 100 Mbps bandwidth. Figure 2-9-8 shows three virtual links being carried by a single 100 Mbps physical link (data bus cable). Also note that each virtual link carries data from more than one end system port. Since several virtual links share the same physical layer and the 100 Mbps of bandwidth, a mechanism is necessary to isolate the data on each virtual link.

This virtual link isolation is necessary to prevent the traffic on one virtual link from interfering with traffic from other virtual links using the same physical link. This is done by limiting the rate at which Ethernet frames can be transmitted on a virtual link and by limiting the size of the Ethernet frames. To do this each virtual link is assigned a bandwidth allocation gap (BAG) and a largest Ethernet frame (Lmax) limiting value. The BAG represents the minimum time interval between Ethernet frames transmitted on the same virtual link. The BAG ranges from 1 to 128 milliseconds. The Lmax for any given virtual link is chosen

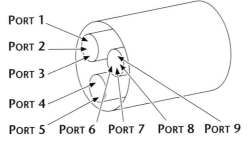

Figure 2-9-8. An AFDX with three virtual links

to accommodate the largest Ethernet frame to be transmitted by all ports using that link.

The AFDX Message

The AFDX message is contained in the UDP payload section of the transmitted data. The structure of the message can be arranged in a variety of formats in order to accommodate design flexibility. Within limits, the aircraft design team is free to choose the desired format for each payload. The encoding and decoding of each message is a matter of the end system software and the AFDX data link becomes a method to transport the message to the correct destination. Since ARINC 429 data is commonly used by many aircraft subsystems, the AFDX payload often carries the message in an ARINC 429 format. Of course an ARINC 429 message contains only 32 bites of data; therefore, each payload can carry several messages. Other common UDP payloads include messages written using Boolean algebra, Signed_32(64) integer, Float_32(64), and Opaque data. This is a simplified explanation of the complex AFDX payload structure, which allows for a variety of data formats.

The ARINC 664 specification defines two types of message structures: explicit and implicit. Explicit message structures include format information that enables the receiver to correctly interpret the data. Implicit message structures do not contain any descriptive information to aid the receiver in interpreting the data. Explicit messages therefore require more bandwidth since they contain the decoding instructions for the receiving end system.

However, an implicit message structure requires a way to identify the message format. This is accomplished by associating implicit message structures with specific AFDX receive ports. In other words, specific implicit messages can only be sent to specific receive ports. The ARINC 664 specification defines all implicit data ports. For example, any message sent in a Boolean algebra format would be assigned a specific AFDX data port number(s). That port would eventually send the data to an

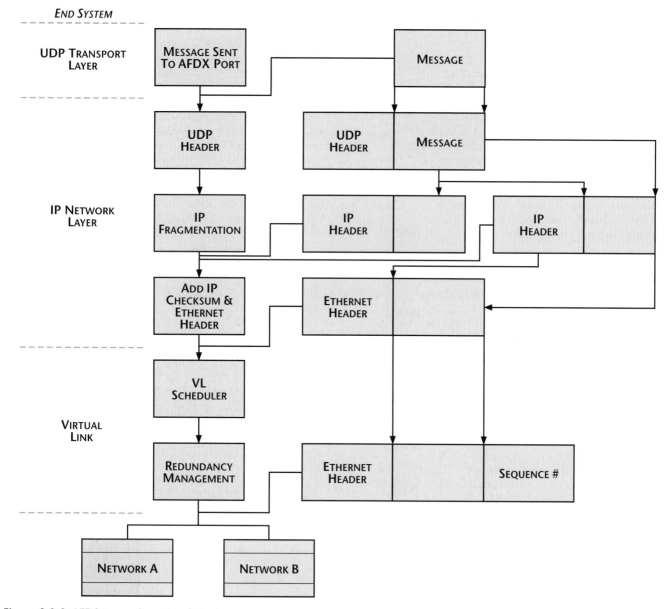

Figure 2-9-9. AFDS transmit protocol stack

AFDX end system that would be programmed to decode Boolean algebra.

AFDX Protocol

In its simplest form, a protocol can be defined as the rules governing communication. An AFDX protocol is a set of rules each component must follow in order to communicate with others across the ADFX network. In other words, the protocol rules control the data transfer (communication) between ADFX end systems (computers). Protocols may be implemented by hardware, software, or a combination of the two. At the lowest level, a protocol defines the behavior of a hardware connection; at the highest level the protocol defines the software needed for the network operation.

Protocols are often made up of a complex set of rules and are therefore divided into layers or stacks. A protocol stack can be thought of as a module of the entire set of communication instructions. This modularization makes design and evaluation easier. Because each protocol module usually communicates with two others, they are commonly imagined as layers in a stack of protocols. The lowest protocol always deals with low-level, physical interaction of the hardware. Every higher layer adds more features and diversity to the software. The entire protocol stack will define all the rules necessary for each end system to communicate on the network.

For AFDX operation the protocol layers are divided into AFDX communications services, UDP transport layer, and virtual link services. The TX protocol shown in Figure 2-9-9 begins with a message being sent to an AFDX port. The UDP transport layer is responsible for adding the UDP header, which includes the appropriate source and destination UDP port numbers. These numbers are, in most cases, determined by the system design and are set values for each AFDX communications port. The IP network layer receives the UDP packet and determines whether it needs to be fragmented. (Remember fragmentation is done for messages that are too large to be sent in one data packet. The Lmax value is used to determine whether fragmentation is necessary.) The IP header is added, and IP checksum is calculated for each fragment. The checksum is used to ensure valid transmission of data. The IP layer also adds the Ethernet header.

The virtual link layer is responsible for scheduling the Ethernet frames for transmission, adding the sequence numbers, and passing the frames to the redundancy management unit (RMU). The RMU replicates frames as needed and the Ethernet source address is updated with the physical port ID on which the frame is transmitted (network A or network B).

In general, the reception protocol is simply the reverse of transmission (Figure 2-9-10). The process starts at the virtual link level. The virtual link receives an Ethernet frame, which is checked for correctness using the frame check Sequence (FCS) byte. If there is no error, the FCS is removed and the AFDX frame is passed through integrity checking and redundancy management circuitry. The resulting IP packet is passed on to the IP network level without the Ethernet header. The IP network level will ensure the data validity using the IP checksum field and any fragmented data will be reassembled. The UDP (user datagram protocol) packet is passed to the UDP transport layer. Here the ADFX message is demultiplexed then delivered to the appropriate end system communication port.

Data transmitted on an AFDX network must travel through a series of avionics computers, ports, queues, and smart switches. This migration allows data packets to travel from end system to end system. The protocols just described are achieved through the design concepts

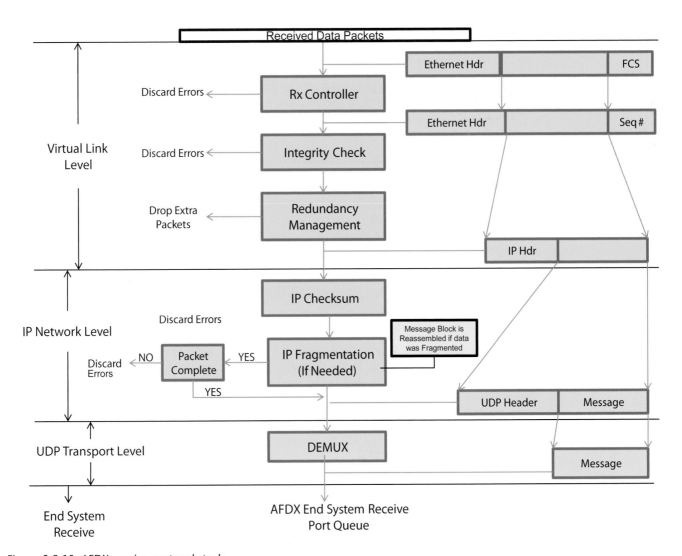

Figure 2-9-10. AFDX receive protocol stack

incorporated in the software of the various network components. The hardware components along with the software design create the network necessary to achieve a reliable AFDX communications system.

Testing and Maintenance of AFDX Systems

The complexity of AFDX creates a challenge for any technician called upon to repair the data transfer system. Similar to troubleshooting other avionics, there are a variety of tools available to test and analyze AFDX data signals. In general, these troubleshooting tools are computer-based and designed to operate using commonly available software like Windows and Linux. A typical AFDX analyzer consists of one or more circuit cards designed to fit a standard personal computer and the appropriate software to operate the system. Of course, the aircraft or system manufacturer may also offer an AFDX analyzer specifically designed for a given aircraft.

During testing and analysis of an ADFX system one must always employ logical troubleshooting techniques. Like all electrical circuits, the data transfer system must output, transmit, and input electrical signals. When testing the system always understand where the electrical information originates (which end system and port) and the pathways (data cable and switches) the data must take to eventually get to the receiving port and end system. Of course, the data may travel to multiple end systems through a maximum of 128 virtual links. A typical analyzer will allow the technician to test virtual links, to visualize raw data, to read MAC, IP, and UPD address, payload data, and more. Virtually all aspects of the AFDX protocols can be brought to the display screen of the technicians PC in order to analyze the system. The technician can use this information to help troubleshoot any defects and replace the defective component accordingly.

One should also keep in mind that the AFDX physical layer is made up of a variety of hardware components, each connected by one or more digital data busses, power cables, and often a variety of analog input/output wires. Each electrical wire must be connected to the hardware component via some type of connector. History has shown that electrical connectors are often a source of system failures. The AFDX system currently has a short history and limited flight time; however, one could anticipate that the electrical connectors would be suspect when looking for a data bus failure. During troubleshooting, carefully inspect all end system and switch electrical connections as well as the related wiring. If a wire or connector looks discolored, worn, or unusually loose, troubleshooting should begin in this area.

Many of the AFDX analyzers also incorporate a data loader, which is designed to upload and download configuration data to end systems and smart switches. All data loading must adhere to the ARINC 664A specification in order to be compatible with the AFDX system. When the configuration of the aircraft is changed for modifications, improvements, or hardware additions the software of any related end system and switch may require an update. A typical ADFX analyzer will present a menu-driven set of instructions for software updates to allow for easy installation of any new software.

Section 10

Digital Bus System Troubleshooting and Repair

In general, the digital data bus is a relatively trouble free system. Many aircraft operate for years without ever experiencing a data bus problem. The actual data bus is nothing more than a pair of twisted wires. Most of the data bus problems occur at the bus connections to the various LRUs. As aircraft gain flight hours, many of the LRUs are periodically replaced. Virtually all connector pins and sockets "wear out" through repetitive LRU replacement or due to the vibrations encountered during normal aircraft operations. Since data bus signals are very low power and rapidly changing, a slight increase in connector impedance can easily create a faulty bus or bus connection.

Corrosion can also create connector problems on digital systems. If moisture enters into the connector, pins and sockets can corrode and eventually create problems. Aircraft operating in unusually moist environments are often prone to connector problems caused by corrosion. This is especially true for aircraft operated in coastal environments. The moist, salty air accelerates the corrosion of all metals including connector pins. When experiencing data bus problems on aircraft operated in harsh environments, always suspect connector corrosion. Many manufacturers recommend the use of an anticorrosion spray on electrical connectors. A common anticorrosion spray is manufactured according to military specification MIL-C-81309. The fluid is simply sprayed

DATA SIGNAL FAULTS

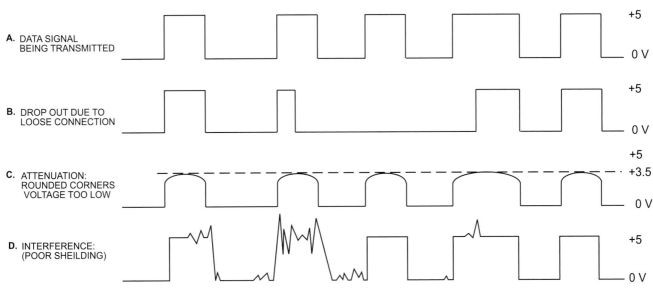

Figure 2-10-1. Digital data transmission faults: (A) Proper data signal, (B) Drop out, (C) Attenuation, (D) Interference

on various electrical components at regular intervals to help prevent corrosion.

All data bus systems have a given limit as to the impedance of the bus and connectors. If a connector becomes worn or even dirty, the transmitted signal will experience loss or even stop completely. A minimal loss is acceptable; however if the signal loss exceeds given parameters, the receiver will not recognize the transmitted data. Adding to troubleshooting difficulties, a change in data bus impedance is difficult to measure. So, in most cases the actual transmitted signal is monitored and if excess loss is detected, the data bus, or most likely a connector, must be repaired.

Testing the Data Bus

There are several ways to measure the transmission characteristics of a digital data bus system (the "system" includes the connectors and/or bus couplers). The simplest test is a visual inspection of the bus and connectors. Obviously, if a bus cable has been physically broken the defect has been found. A visual inspection of the connector pins and sockets may also reveal problems. On the connector plugs, look for worn, bent, or dirty contacts. A magnifying glass may help to identify connector problems. Connectors that are worn beyond limits may be slightly discolored due to a loss of the protective plating on the pins or sockets. Most pins and sockets are made of copper and plated with silver, gold, or other metal to prevent corrosion and enhance continuity. Dirty connectors may be cleaned using approved solvents and nonabrasive swabs. Be sure not to damage the connector during cleaning. If a

connector socket is forced open by a cleaning swab, it will most likely bend out of shape and permanently damage the socket. Also never clean the contacts with an abrasive, such as sand paper. This will damage the contact plating and allow for corrosion.

An ohmmeter can also be used to detect an open in a bus wire or the bus shielding. Be careful, however when using an ohmmeter as the bus cable may have low resistance to a direct current signal. Remember that ohmmeters use direct current to measure resistance, and yet, have high impedance to a high frequency digital signal. An ohmmeter will only detect a catastrophic bus fault. A bus may pass an ohmmeter test and still fail to properly transmit digital data due to high impedance.

Data Signal Faults

There are three basic ways in which transmitted digital data can be affected by the data bus system. As shown in Figure 2-10-1, the digital signal can experience drop out(s), the signal can be attenuated, or the signal can be overridden by interference.

Drop outs. Drop outs occur when a portion or portions of the data signal are completely lost during transmission. Drop outs are typically caused by loose connections at a data bus connector plug. The loose connection causes an intermittent signal transmission and partial loss of the data. Intermittent connections often occur only during certain flight conditions, such as, turbulent air, large bank angle, or temperature extremes. This phenomenon often makes it difficult to find intermittent

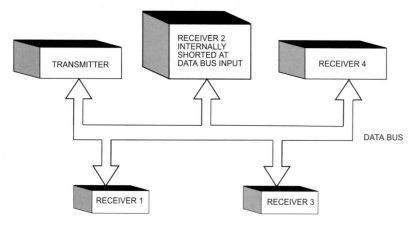

Figure 2-10-2. Simplified data bus structure showing a shorted LRU

connections. In some cases, the technician can simulate turbulent air by wiggling the LRU or wiring harness. In many situations, the circuit must be analyzed under actual flight conditions to detect the fault.

Attenuation. Attenuation occurs when a digital data signal becomes weak or the transmitted voltage becomes too low. The attenuated signal shown in Figure 2-10-1C is weak in voltage and the square waveform has been rounded somewhat. All data transmissions experience some attenuation; however, if excessive signal loss occurs, the digital data receiver will not recognize the signal. Excessive attenuation is caused

by a change in data bus impedance. This can be caused by dirty connection pins, poorly soldered or crimped connections, pinched data bus cable, or excess moisture which has penetrated the data bus insulation.

Attenuation can also be caused by an excessive load placed on the data bus by one of the units connected to the bus. If this occurs, the problem can be isolated by systematically removing items connected to the bus. Keep in mind this type of problem only occurs for LRUs with a direct connection to the data bus. LRUs, which use an inductive coupling device to connect to the data bus, are immune to this problem. Let us assume that receiver number 2 in Figure 2-10-2 has an internal short, which places excess load on the data bus. To isolate the problem, operate the system normally then remove receiver number 1. In this example the problem still remains. Next, disconnect receiver number 2 from the bus. In this case, the bus operations return to normal, therefore a likely cause of the problem is receiver number 2. However, the transmitter on the bus can also cause excess attenuation. To confirm this the output signal of the transmitter must be checked using an oscilloscope or data bus analyzer.

Interference. Data bus interference occurs when an unwanted signal is induced into the bus wiring, inhibiting data transmission. Improper shielding of the data bus cable often causes interference. If a bus cable is not shielded properly and it is exposed to a changing magnetic field, the digital signal will become distorted as in Figure 2-10-1D. An excessively strong electromagnetic field near a data bus termination point (connector plug) can be another common cause of interference. In this case, it is likely that a component independent of the data bus, but located near the bus, is actually creating too much interference. This excessively strong interference signal induces unwanted current onto the bus and degrades the original signal beyond limits. The bus cable may actually have no faults and yet the interference "finds" its way onto the bus through "leaks" in the bus shielding at the connector plug. The source of the interference must be eliminated. Keep in mind, if the interference only occurs during certain flight conditions, such as operation of the flap motor, the technician should suspect either the data bus cable shielding or the flap motor brushes the shielding. One of the best ways to detect interference is to view the digital signal using an oscilloscope. To view the original digital signal and the interference, connect the oscilloscope in parallel with the bus connections.

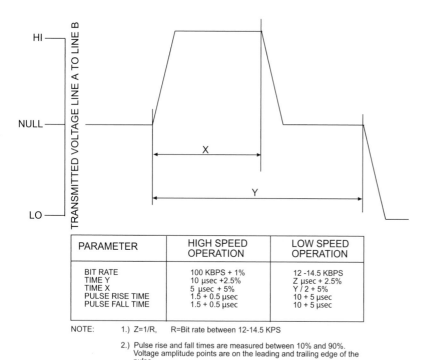

PARAMETER	HIGH SPEED OPERATION	LOW SPEED OPERATION
BIT RATE	100 KBPS + 1%	12 -14.5 KBPS
TIME Y	10 µsec +2.5%	Z µsec + 2.5%
TIME X	5 µsec + 5%	Y / 2 + 5%
PULSE RISE TIME	1.5 + 0.5 µsec	10 + 5 µsec
PULSE FALL TIME	1.5 + 0.5 µsec	10 + 5 µsec

NOTE: 1.) Z=1/R, R=Bit rate between 12-14.5 KPS

2.) Pulse rise and fall times are measured between 10% and 90%. Voltage amplitude points are on the leading and trailing edge of the pulse.

Figure 2-10-3. ARINC 429 digital signal timing tolerance. Notes: (1) Z = 1/R; where R=bit rate selected from 12-14.5 Kbps/s, (2) Pulse rise and fall times are measured between the 10 percent and 90 percent voltage amplitude points on the leading and trailing edges of the pulses

There are several other types of faults that can cause loss of digital data transmission; however, the aforementioned faults are those most likely to occur due to data bus problems. Other faults can be caused by failure of the data transmitter or receiver circuitry. Whenever analyzing the transmission of digital data, it is often wise to check the output signal at the data transmitter. The signal at this point must be within specifications and not subject to drop out, attenuation, or interference. If one of these problems is present at the transmitter with the data bus disconnected, the fault lays within the transmitter not the data bus.

Timing tolerance faults. A timing tolerance fault occurs when the rise or fall of the digital signal responds too slowly to be within specifications of the data bus standard. A defective transmitter typically causes this type of fault. The ARINC 429 output signal timing tolerances are shown in Figure 2-10-3. As can be seen in this diagram, total bit time is designated by the letter Y. For high-speed data transmission $Y=10$ μsec. For low-speed, $Y=Z$, where Z is determined by $Z=1/R$. Remember, the bit rate (R) for low speed data is between 12-14.5 Kbps and 100 Kbps for high-speed data.

Analyzing Digital Data Signals

There are two basic test instruments that can be used to detect loss of digital data: the oscilloscope, and the data bus analyzer. The oscilloscope is basically a sophisticated voltmeter with a two-dimensional graph display that can be used to measure voltage (amplitude) and frequency (time) of an electric signal. This allows the operator to view voltage and/or frequency changes over time; hence, this instrument is ideal for analysis of digital data. An oscilloscope is typically used to analyze signals, such as digital data that change too fast to be monitored by a common multimeter.

The storage scope. A storage scope or storage-type oscilloscope allows the technician to hold a given portion of digital signal on the oscilloscope's CRT display even though the signal may continue to change. This feature is extremely

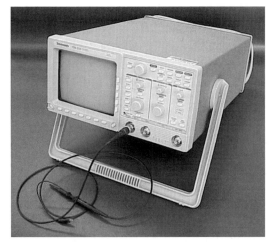

Figure 2-10-4. A typical storage oscilloscope for display of digital data

helpful for data bus troubleshooting since digital data is transmitted in a series of changing voltages. A typical storage-type oscilloscope is shown in Figure 2-10-4. Whenever connecting any oscilloscope to a digital data bus, the scope probe should be connected between the two twisted wires of the data bus as shown in Figure 2-10-5. This connection will typically be made at the plug to an LRU. In most cases, the plug must be disconnected and two extra pins slipped into the correct sockets of the connector. The oscilloscope can then be connected to the extra pins.

The data bus analyzer. The data bus analyzer is a common carry-on piece of test equipment used to troubleshoot digital systems. There are many types of data bus analyzers, but the basic functions are quite similar. Bus analyzers are used to receive and review transmitted data, or transmit data to a bus user. Before using any analyzer, first be sure the bus language is compatible with the bus analyzer. For example, the DATATRAC 400, shown in Figure 2-10-6, can monitor, simulate, and record data transmissions for avionics equipment using ARINC 429 or CSDB standard data buses. Data bus analyzers are also incorporated into the software

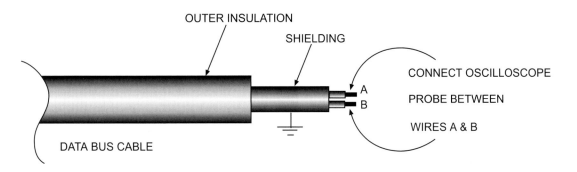

Figure 2-10-5. Connecting an oscilloscope to a data bus cable

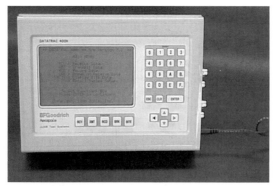

Figure 2-10-6. A typical data bus analyzer

of many modern digital aircraft systems. For example, the Boeing 747-400 central maintenance computer system input monitoring function could be used to detect ARINC 429 bus problems similar to a carry-on data bus analyzer. One major advantage of the built-in test equipment (BITE) systems is that they often record the exact time when the fault occurred. This can be of great benefit when dealing with intermittent problems.

Using a typical data bus analyzer. During system troubleshooting a data bus analyzer is often used to capture a stream of data being transmitted between digital devices. The recorded data can then be displayed for evaluation. If inconsistencies are detected, the transmitter or the data bus system is faulty. Remember, a data bus analyzer should be used to check the input and output signals of the transmitter, the receiver, and the data bus. Use the bus analyzer to determine if the correct signal is being transmitted and/or reaching its destination. Some data bus analyzers are capable of reading several transmission lines, or channels, at one time. This allows for comparisons to be made that might expedite troubleshooting. When using a data bus analyzer for ARINC 429 data, always be sure the word label and equipment identifier are both programmed into the test unit. The label/identifier combination is needed to ensure the bus analyzer monitors the data correctly.

For the DATATRAC 400 (Figure 2-10-6) there are five basic modes of operation:

- Receive: This allows the technician to receive either high or low speed ARINC 429 data. The data received can be displayed in various formats, such as hexadecimal, decimal, or binary. The data may also be down loaded via an RS-232 port.

- Transmit: The data bus analyzer is capable of sending digital data in order to simulate communications between avionics equipment. This mode is often used to

test the operation of a specific LRU when given the correct data bus signal.

- Record: A particular data label and record rate are selected and the avionics bus analyzer collects information sent to that address. All the data can then be displayed in a numerical format.

- Break: This mode allows for "trapping" intermittent data in memory and reviewing the data as time permits.

- BITE: This mode supports the built-in-test equipment (BITE) for various transport category aircraft, such as, the B-747-400, A-320, and MD-11. Up to 240 BITE displays can be saved in the analyzer memory and viewed or downloaded at a later time.

In most cases, the data bus analyzer must be connected to the system at the connector plug of an LRU. For example, if you plan to receive data transmitted to the generator control unit (GCU), you would unplug the GCU from the bus and connect the analyzer to the data bus cable at the GCU connector plug. If the correct connections were made and the test equipment properly adjusted, the analyzer will display all messages sent to the GCU on that bus.

On some aircraft, a specific connector is available to allow for direct connection into the data bus system. Figure 2-10-7 shows a carry-on bus analyzer connected to the data bus. In this case, the technician is monitoring the fuel quantity. Note that the selector is set to the CMC (central maintenance computer) and the analyzer indicates 3,246 lbs.

In the transmit mode, a digital data bus analyzer can be very helpful to simulate different avionics equipment. For example, the analyzer can be used to send a signal to an electronic flight instrument system (EFIS). The operator can then verify that the correct display appears on the CRT. Some analyzers provide for dynamic transmission of data. For example, with dynamic data, an EFIS course display can be made to rotate 360°, or the autoflight system can be tested through simulation of aircraft movement.

Fiber Optic Data Bus

Many aircraft today employ fiber optic cable to distribute data between various computer-based systems throughout the aircraft. Both the ARINC 629 and ARINC 664 (AFDX) data bus standards are designed for data transmission using fiber optics, as well as traditional copper data busses. Fiber optics is a term used to describe the process of transmitting data signals using light. This transmission is based on the principle of total internal reflection discovered in 1841. This concept basically says that light will travel through water and will reflect internally following the contour of the water stream. Total internal reflection also applies to light traveling through glass fibers. Fiber optic cables employ glass fibers to direct the light from one place to another.

Fiber optic cable can be found on many military aircraft, on the B-777 to connect various LRUs, and on the B-787 that employs fiber optics for information sent to the flight deck displays, as well as various other systems. The B-777 employs optical cable using the ARINC 629 standard; the B-787 employs optical cable using the ARINC 664 standard. In the future, fiber optic data transmission will most likely become more widespread on aircraft since it has several advantages over traditional copper cable. These advantages include:

- Greater bandwidth capability
- Lower signal loss (attenuation)
- Signals are immune to electromagnetic interference
- Optical fiber does not radiate any electromagnetic signal that could interfere in other systems
- Fiber cable is lighter than copper (a big advantage on aircraft)
- Optical cable is immune to corrosion
- Less signal loss and therefore fiber requires less power during transmissions

Some disadvantages of fiber cable are the specialized tools and technical training needed to work with fiber optics. Another disadvantage is that an optical signal converter must be used to change all electrical signals into light signals for transmission and then back into electrical signals at the receiver. This requires additional circuitry and adds complexity to equipment.

In order to understand the optical transmission of light using glass fiber, one must first have a basic understanding of the properties of light. Two terms to understand: *reflection*, a change of direction of a light beam; and *refraction*, the bending of light as it passes through two different materials. The glass fiber within a fiber optic cable is surrounded by an outer reflective coating (cladding), which reflects the light beam inward along the fiber. This cladding material must have a very low refractive index in order to keep the light beam consistent. Light travels in waves, similar to waves on the water or electromagnetic radio waves. Each color of the light spectrum is a different wavelength/frequency of the electromagnetic spectrum. Fiber optics typically use frequencies that extend beyond the visible range of humans. Using

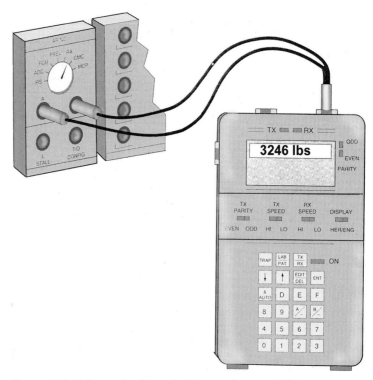

Figure 2-10-7. Example of a typical bus analyzer installed on a transport category aircraft

these infrared wavelengths increases the efficiency of the system as it passes through the glass fiber. Typical wavelengths employed in fiber optic transmissions are 800 or 1,550 nanometers. Light traveling through a medium (like glass fiber) is called propagation. The basic concept of moving data signals with light includes three steps:

1. The transmitting unit converts electrical information into a modulated light signal.

2. The light wave modulated with digital information, connected to a fiber cable, propagates through that cable.

3. The receiving unit converts the incoming light signal from the fiber into an electrical information signal for processing.

Properties of a typical glass fiber used in the construction of an optic data bus can be found in Figure 2-10-8, this glass fiber is often referred to as the optical fiber. Here it can be seen that optical fiber that carries the data signal is made up of three essential layers: the core, the cladding and the coating. The core material is the light conductor and carries the signal. The cladding is an ultra thin layer around the core, similar to a mirror finish, used to reflect the light inward. The cladding prohibits the light from escaping the core while traveling through the fiber optic data bus cable. The coating is a hard outer layer used to protect the glass core and coating of the cable. There

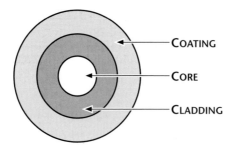

Figure 2-10-8. End view showing layers of glass fiber

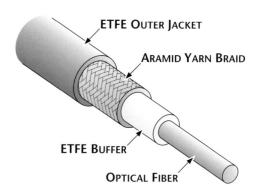

Figure 2-10-9. Layers of a typical fiber optic data bus cable

are three basic types of glass fiber in common use today: single mode, step index multimode, and graded index multimode. Each of these optical fibers has different refraction characteristics and may be found in various applications. The more practical example of an actual fiber optic data bus cable is shown in Figure 2-10-9 where it can be seen that there are typically at least three additional layers surrounding the optical fiber. The inner layer is known as the ETFE buffer. ETFE is polymer known as ethylene-tetrafluoroethylene made by DuPont under the trade name Tefzel. This layer is surrounded by the Aramid Yarn Braid which is a strong, heat-resistant synthetic fiber constructed into a braid to help provide structural support. The external layer is another coating of ETFE, forming an extremely hard, yet flexible, protective layer very similar to modern copper wire found on aircraft.

The light source used for the transmission of data can be one of two types: a laser or a light emitting diode (LED). It is essential that the optical fiber and the light source match in order to ensure signal propagation. For long runs of cable or large quantities of data transfer, the laser source is preferred. Lasers are matched with a single-mode optical fiber typically having a nine-micron diameter core. Single-mode cable has only one data path and requires a laser with a 1,310 or 1,550 nm wavelength. Transmission of data using light from an LED employs a fiber cable having hundreds of light paths and each carries the same data. This is known as multimode cable and uses an LED with an output wavelength of 950 or 1,300 nm.

Safety When Using Fiber Optic Cable

It is very important to remember the following safety precautions whenever working with fiber optics. Fiber optic cable is constructed of extremely fine strands of glass fiber, which can penetrate the skin and are very difficult to remove. These glass fibers are referred to as shards and are a byproduct whenever installing fiber optic cable terminations. It is there-

fore very important to follow all manufactures safety precautions whenever working with fiber optics. Common precautions when working with fiber optics include:

- Always wear safety glasses
- Never touch the end of the fiber cable
- Never eat or drink when glass fibers (shards) might be present
- Never look directly into the fiber unless there is no light source connected
- Wear aprons that can be removed after work is completed
- Always dispose of all glass shards properly

Installation and Testing of Optical Cable

Installation and testing of fiber optics have many similarities to that of copper wire and for the most part fiber cable can be routed through the aircraft along the routes with standard wire bundles or along a separate path for fiber only. In either case, it is important to follow all manufacturers' recommendations. Fiber optic cable installed in most aircraft is typically a unique color in order that it may be identified easily during maintenance activities. Fiber cable should be clamped in specific spacing and often with special clamps. Whenever clamping fiber cable it is important not to apply too much pressure to the cable. Incorrect clamping can create micro-bends that "kink" the fiber. A kinked fiber will distort the signal propagation through the cable as shown in Figure 2-10-10. Standard tie wraps can be used for securing fiber optic cable, but only if they are installed with limited pressure. Fiber cable must also be installed using a bend radius that will not distort the light signal. A tighter bend creates a greater signal loss in the cable. Typically a bend of greater than approximately 50 mm is acceptable. The fiber optic core may also be distorted if too much tension is applied to the cable. Whenever routing cable never exceed a pull tension of 100 pounds. During installation or maintenance it is also very important never step on, kink, or subject the cable to a heavy impact due to a sudden force; this could damage the fiber. To help protect the fiber core, the cable is often routed inside additional protection, such as a conduit.

For the most part, fiber optic cable is very reliable. Many of the problems that occur are due to signal loss created at the cable connector. This area is extremely sensitive to misalignment or contamination from dirt. To understand the problems that can occur in a fiber connector, it is important to understand how connectors function. In general, a fiber connector is simply any location where the end of an optical fiber meets another fiber or optical sensor. Since the fiber core is extremely small, the connectors must be held to an extremely tight tolerance. There are four common connector problems that can typically occur on an aircraft: (Figure 2-10-11):

- Loose connectors, causing an excessive end separation

- Extreme bend radius causing an angular misalignment

- Too tight of a clamp, which might cause lateral misalignment

- Dirt, oil, or other contaminants located at the end of the fiber, which can distort or eliminate the optical signal

From a practical stand point, any time a optical connector is suspected as being a problem area, the technician should clean the optical surface using an approved cleaner (typically alcohol), and reconnect the fiber using proper techniques. Typically connectors are secured with a relatively low torque and may require safety wire.

Whenever troubleshooting a fiber optic data bus system, if the receiving LRU fails to obtain an appropriate light signal, the fiber cable may be at fault. One of the first steps in the process would be to verify that the transmitting LRU is functioning properly and supplying the correct light signal. This test can often be performed using the aircraft's built-in-test equipment, or it may require carry-on test equipment to measure the light signal output of the transmitter. Once the transmitter operation has been verified, the cable should be tested to verify that the fibers and connectors are without damage. There are three essential properties that must meet minimum standards for a functioning fiber cable: continuity, attenuation, and proper cleaving.

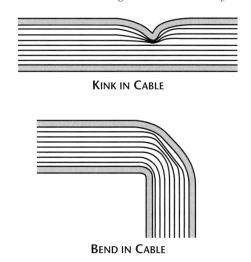

Figure 2-10-10. Signal loss is caused by a kink or bend in the fiber cable

Fiber cables must also have complete continuity between the transmitter and the receiver. This means that the fiber core must create a continuous light circuit (path) from the transmitter to the receiver. It is important to know that fiber optic cable can appear undamaged on the outside and yet the internal glass fiber can be cracked or broken. If the fiber is broken the light circuit is not continuous. A common method used to test fiber cable for continuity is through a visual inspection, where a laser is applied to one end of the fiber and the technician inspects the length of the suspect cable. If the cable is cracked or experiences too much bend loss, the jacket of the cable will glow red from the applied laser.

Attenuation is critical to the proper operation of optical fiber. Attenuation is the signal loss from the transmitter to the receiver. If this loss is too great, the light signals at the receiving LRU will be too weak for the optical sensors

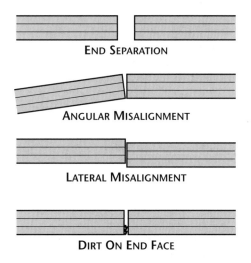

Figure 2-10-11. Common fiber optic connector problems

GOOD CLEAVE

SMOOTH, MIRRORED
SURFACE

BAD CLEAVES

CHIPS AND
SHARDS

CRACKS AND
PITS

DIRT AND
OIL

Figure 2-10-12.
Inspection of the optical fiber end face

to read the digital information. Attenuation is measured in a unit known as the *decibel* (dB). A decibel is a logarithmic unit of measure used to express a power ratio. In the case of fiber optics, attenuation is the ratio of the power (light) transmitted and the power (light) lost through the cable. Typically attenuation of 3.5 dB or less is acceptable for proper light transmission using fiber cable. An optical loss test set (OLTS) is typically used to detect fiber cable attenuation. This test set consists of two units, the light source and the optical power meter. The light source is a calibrated unit connected to one end of the cable; the power meter is connected to the other. Prior to testing, the light source must be set for the correct cable type, multimode or single mode, since they operate at different wavelengths. The light source simply transmits a signal and the power meter records the attenuation. If the attenuation exceeds the specified limits for the aircraft, the fiber cable must be replaced.

Cleaving is the process used to terminate the fiber at each connector. This process is similar to the cutting of copper wire; however, special tools and procedures are required in order to achieve the optical clarity at the end of the glass fiber. It is vital that the very end tip of the fiber be clean and free of any scratches or cracks that may distort the light signal. This is also an area of the cable that is easy to damage during the cleaving process or during installation/removal of fiber connectors. Figure 2-10-12 shows typical problems, which may occur at the end of a fiber cable. For the most part, the naked eye is not sensitive enough to find cleaving defects. A specialized inspection scope designed for fiber is typically used to view the cable end. The inspection tool will typically magnify the fiber 400 times or more in order to identify any possible defects.

Another test instrument often used for detecting faults in fiber optic cable is known as the optical time domain reflectometer (OTDR). The OTDR transmits a light signal into the fiber and measures the backscatter that returns to a sensor on the same end of the cable. Through measurements of the backscatter, the test instrument can calculate the location of the fault within the cable. The OTDR has limitations when used on short cable runs and is typically not suitable for aircraft fiber optic testing.

Terminating Fiber Optics

Fiber optic cable is always run between two LRUs containing an optical transmitter and/or sensor. The transmitting LRU contains the circuitry, which changes an electrical signal into a digital light signal; the receiving LRU contains an optical sensor and the conversion

circuitry needed to reproduce the original electrical signal. Special optical fiber connectors must be used to terminate the cable at both ends. Fiber cable is typically run directly between end users without any splice connections. The termination of any fiber cable requires practice, patience, and special tools (Figure 2-10-13). The function of the optical connector is to accurately position the fiber optic core adjacent to the receiver/transmitter connector. The connector must also offer physical protection and security of the cable. The aircraft maintenance manual will outline all procedures used for the termination of fiber cable. In most cases, this procedure will be similar to those established by the NASA standard 8739.5 for fiber optic terminations, cable assemblies, and installations. Although all fiber connectors are very similar, there may be slight differences. It is therefore imperative to ensure you have the proper instructions and tools necessary for the specific connector and cable for the aircraft. One commonality for all fiber optic termination is cleanliness. It is vital that the work area, connector components, the termination tools, and the optical fiber remain dirt free.

The components of a typical fiber connector assembly are shown in Figure 2-10-14. A connector consists of multiple components designed to hold the extremely small strand(s) of optical fiber in the correct alignment with the mating connector. The following steps illustrate a simplified version of a typical optical fiber connector installation:

1. Install the strain relief boot and remove approximately 2.5 inches of the outer jacket using the appropriate stripping tool.

2. Fold back the supportive sleeve (yarn braid) which surrounds the EFTE inner coating.

3. Using another specialized stripping tool; remove the ETFE buffer coating from the glass fiber. This will expose the extremely fragile glass fiber that is approximately the diameter of a human hair. Be very careful at this point not to damage the fiber.

4. Fold the yarn braid back over the fiber and the first crimp is made onto a support sleeve.

5. The braid is typically folded back over the already installed sleeve and a ferrule instated over the braid and crimped in place.

6. The connector body is now installed over the ferrule and aligned according to the manufacturer. Alignment at this stage becomes critical.

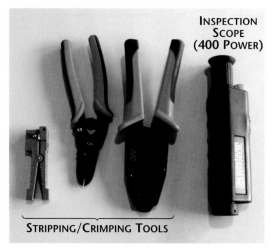

Figure 2-10-13. Special tools required for a typical optical fiber termination

Figure 2-10-14. Typical connector components

7. The final crimp is made to hold the connector body in place.

8. The optical fiber is now cut to the proper length in a process known as cleaving (discussed earlier). Be sure to dispose of the fiber in an appropriate container.

9. The face of the glass fiber is now cleaned with an alcohol pad and the protective cover is placed over the connector or the connector is installed into its final destination.

In order to ensure precise data transmissions, many optical fiber termination procedures require a polishing process on the end of the fiber. Whenever the optical fiber is cut to length (cleaved) the tip of the fiber becomes irregular. To ensure optical clarity, a polishing tool can be used to flatten and clean the end of the fiber. A portable polishing system typically used during field installations is shown in Figure 2-10-15. This unit contains a polishing turntable, which rotates during operation. The cut end of the optical fiber is placed in the machine with the tip against the polishing surface. The equipment automatically moves the fiber along the rotating surface of the turntable. In approximately five minutes, the optical fiber is polished. The technician would then use the inspection scope to determine the quality of the optic fiber and repeat the process if necessary.

Figure 2-10-15. Typical optical fiber polishing tool

Another process used to improve the quality of light transmission through the end of a cut fiber is to encapsulate the fiber in a gel. The gel fills any small chips, cracks, or pits and closes any minute gaps that may exist at the cut end of the fiber. The gel is designed to have a refractive index as close to the glass fiber as possible; therefore, as the light passes from the fiber to the gel there will be little or no distortion. When two terminated fibers are connected the optical gel surrounds the cleaved ends. The gel then creates a transmission medium at the connector ensuring optimum performance.

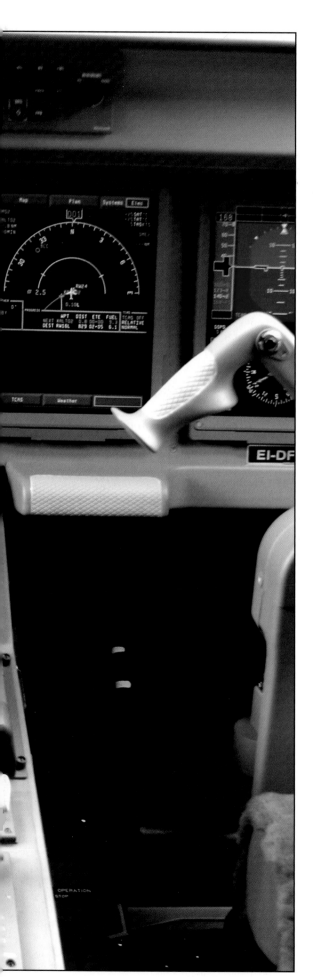

Chapter 3

ELECTRONIC
instrument systems

Section 1

Introduction

Over the past three decades, aircraft instruments have gone through a revolution. Older mechanical-type instruments have been replaced with electronic instruments offering improved accuracy and reliability. electronic flight instruments systems (EFIS) became possible with the development of a sunlight readable cathode ray tube (CRT) display and sophisticated aircraft computer systems. A digital data bus system is used to transfer the majority of information between the various components of EFIS. With large amounts of data to transfer, using digital (not analog) systems saves hundreds of pounds in additional wiring. Today's electronic instruments employ flat-panel displays and integrated system technologies. The modern instrument panel has become more computerized than ever before, offering the pilot a multitude of information including moving map displays, night vision systems, and a paperless flight deck.

Electronic flight instruments began to enter the transport category and corporate-type aircraft market in the early 80s. Due to their flexibility in data displays and proven reliability, EFIS have become commonplace in modern aircraft. However, modern integrated systems resemble the original. Early electronic flight instrument systems were designed to replace high maintenance electromechanical HSIs (horizontal situation indicators) and ADIs (attitude director indicators). Newer electronic instrument systems also provide HSI and ADI data, and are thoroughly integrated with other aircraft systems. This allows for the presentation of a multitude of other display functions. In many cases, state-of-the-art systems referred to as Integrated Display

Learning Objectives

- *First Generation Electronic Instrument Systems*

- *Second Generation Instrument Systems*

- *Third Generation Instrument Systems*

Left. Most modern turbine aircraft feature multiple flat-panel instrument displays offering a wealth of information to the flight crew.

Systems (IDS) or electronic instrument systems (EIS) help distinguish them from the original EFIS. Keep in mind; the *F* in EFIS stands for *flight*. The early version electronic instruments were used to display only flight data, such as attitude, heading, and navigational references. Newer electronic display systems show flight data, engine parameters, airframe systems, as well as warning information.

Electronic instrument systems have been in common use for nearly two decades. During that time, systems have grown from single display units to fully integrated systems. State-of-the-art EIS computers interact with autoflight, autothrottle, and integrated test equipment systems. This type of integration allows for more system flexibility, better redundancy, and a substantial weight savings over earlier stand-alone EIS.

It is the purpose of this chapter to present the basic principles of electronic instrument systems. To do so, three different generations of EIS will be examined. The first system will be an early version stand-alone EFIS. The second is an integrated advanced system. The third is a system, which employs compact modularized electronics and state-of-the-art displays. Although only a few basic systems will be presented in detail, all EIS operate in a similar fashion. The information in this chapter will allow a technician to easily adapt to the specifics of any of the electronic instrument systems in use today.

Section 2

First Generation Electronic Instrument Systems

An electronic flight instrument system typically employs two or more CRT displays to present alphanumeric data and graphical representations of aircraft instruments. Each EFIS display replaces several conventional instruments, and various caution/warning annunciators. For the

most part, EFIS was introduced on the Boeing 757 and 767 in the early 80s. About this time, similar systems began to find a home in corporate type aircraft, such as the Beechcraft King Air, Falcon jet, and Gulfstream G-3. Figure 3-2-1 shows the instrument panel of a Dassault Falcon 900 with a five tube EFIS.

NOTE: *the word "tube" is often used to refer to the CRT display units.*

This system, made by Honeywell, provides two EFIS displays for both the pilot and first officer. The fifth tube installed in the center of the panel is called a multifunction display. The multifunction display (MFD) is used to display weather radar data, course and flight plan information, and system checklist. Most importantly, the MFD provides backup functions in the event of a partial system failure. There are only a limited number of corporations producing electronic aircraft instruments. Common manufacturers include, Honeywell, Inc, Rockwell Collins, Smith Industries, Garmin International, and Avidyne Corporation.

This text will present details on several specific electronic instrument systems. Each of these systems is typical of those found on modern aircraft and should be used as general examples for educational purposes only, and not for specific aircraft maintenance.

Collins EFIS-85 and 86 System Components

The Collins EFIS-85 and 86 are two first generation electronic flight instrument systems, which are very popular in corporate and commuter-type aircraft. The Collins Corporation eventually merged with Rockwell International and is now Rockwell Collins. This text will use the name *Collins* when referring to the EFIS 85/86. The EFIS-85/86 is part of the Collins Pro Line II series of avionics equipment and is designed to operate in conjunction with a Collins autoflight system. The EFIS-85/86 operates virtually the same; however, the CRT display of the 86 is larger. The Collins EFIS equipment will be examined in detail in the following pages; the autoflight system will be presented in Chapter 6.

The basic system components of the EFIS-85/86 systems are shown in Figure 3-2-2. A five-tube system consists of four electronic flight displays (EFD), two display processor units (DPU), one multifunction display (MFD), one multifunction processor unit (MPU), a course heading panel (CHP), and a display control panel (DCP). On some aircraft a single unit display select panel (DSP) is used to replace the CHP and the

Figure 3-2-1. Instrument panel of a Dassault Falcon 900 showing a five-tube EFIS

DCP. Information from the aircraft weather radar feeds all three EFIS processor units, and can be displayed on the CRTs; however, the weather radar system (center of Figure 3-2-2) is not part of the EFIS system.

The displays. The electronic flight displays (EFD) are used to provide primary flight and navigation data on a high-resolution color CRT. The EFIS-85 system employs a CRT of approximately five inches, measured diagonally; the EFIS-86 system employs a CRT of approximately six inches, measured diagonally. The EFDs are located directly in front of the pilot and co-pilot. The EFDs are typically arranged in a vertical fashion with electronic attitude director indicator (EADI) on top and electronic horizontal situation indicator (EHSI) on the bottom. It should be noted that, although uncommon, some single-tube systems are also in use. On a single tube EFIS the display is typically used as an EHSI. As shown in Figure 3-2-3, the EHSI can

provide three different formats: Rose, ARC, and MAP. The *Rose* format will display the complete 360° compass rose; the *ARC* format will display an 80° arc of the compass rose and the *MAP* format will display an 80° arc of the compass along with a map of the current flight plan. In each of these formats the EHSI shows basic navigational data, such as compass and selected heading, navigational source and course deviation, selected course and track, and ground speed. Weather radar information can also be imposed on the EHSI in the approach or en route modes. The EADI operates in one format as shown in Figure 3-2-4. The EADI provides information such as heading, vertical speed, radio altitude, pitch and roll data, aircraft attitudes, and autopilot status.

In general, all EHSI and EADI of different aircraft formats are very similar. However, different EFIS manufacturers will present information in slightly different locations on the

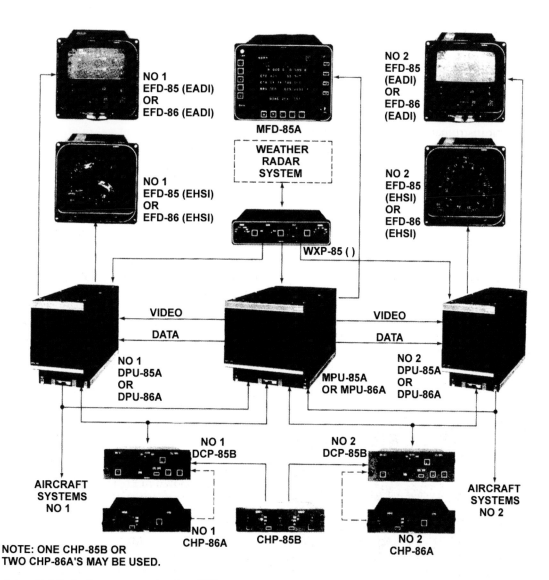

NOTE: ONE CHP-85B OR TWO CHP-86A'S MAY BE USED.

Figure 3-2-2. Basic components of an EFIS

Courtesy of Rockwell International, Collins Avionics Divisions

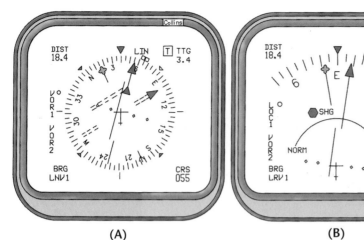
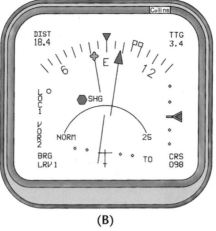
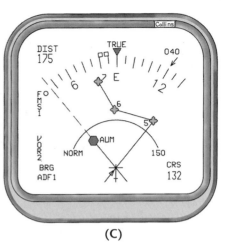

(A)   (B)   (C)

Figure 3-2-3. Typical EHSI display: (A) Rose format, (B) ARC format, (C) MAP format

Courtesy of Rockwell International, Collins Avionics Divisions

display or use different symbols or colors for some basic data. The EADI and EHSI shown in Figure 3-2-5 represent the formats used on the Bendix/King EFIS-10. From this figure one can see differences in the display formats between the Bendix/King EFIS-10 and the Collins EFIS 85/86 shown in Figures 3-2-3 and 3-2-4.

The Collins EFIS-85/86 EFDs consist of a CRT assembly, deflection and video amplifiers, a phosphor protection monitor, and a high-voltage power supply. As shown in Figure 3-2-6, these basic functions are all contained on various circuit cards within the EFD. The inputs to the EFD include:

1. A power monitor from the EFD circuit breaker

2. A video intensity (VINT) signal from the DCP

3. Video information from the DPU

4. Power inputs from the DPU

The processor unit. The DPU 85/86 is a line replaceable unit approximately 12.5 inches (316 mm) by 7.6 inches (194 mm) by 5 inches (128mm) and weighs 13.6 lbs. (6.2kg). The DPU receives inputs from various aircraft systems and other EFIS components. This information is processed within the unit to produce the deflection and video signals needed by the displays. During normal operation of the EFIS-85/86, each DPU is used to produce the signals necessary to drive one EADI and one EHSI. During isolated backup mode, one DPU can power four EFDs. In isolated backup mode both the pilot's and co-pilot's displays will show identical information since the same processor is used to drive all four displays.

As seen in the block diagram of Figure 3-2-7, there are eight separate circuit boards (cards) within the DPU. The cards each have a specific function:

1. Input/output interface (card A7 & A8)

2. Display processor (card A4)

3. Symbol generator (card A5)

4. ADI multiplexer (card A6)

5. Power supply (cards A3 & A9), and 6) the relay card (A10).

The input/output (I/O) cards receive AC and DC analog signals, AC and DC discrete signals, audio identification signals from the DME, and serial data from the cross-side DPU. Table 3-2-1 shows a list of DPU inputs. The I/O cards convert this data into a serial data signal that is sent to the processor card. The DPU also processes a variety of other data that is used by various systems, such as the autopilot (Table 3-2-2).

Figure 3-2-4. Typical EADI display

Courtesy of Rockwell International, Collins Avionics Divisions

(A)

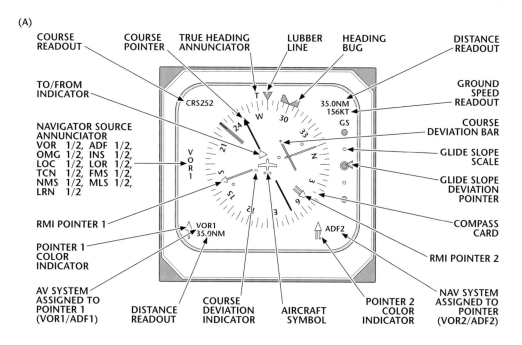

COURSE READOUT
COURSE POINTER
TRUE HEADING ANNUNCIATOR
LUBBER LINE
HEADING BUG
DISTANCE READOUT
TO/FROM INDICATOR
GROUND SPEED READOUT
NAVIGATOR SOURCE ANNUNCIATOR
VOR 1/2, ADF 1/2,
OMG 1/2, INS 1/2,
LOC 1/2, LOR 1/2,
TCN 1/2, FMS 1/2,
NMS 1/2, MLS 1/2,
LRN 1/2
COURSE DEVIATION BAR
GLIDE SLOPE SCALE
GLIDE SLOPE DEVIATION POINTER
RMI POINTER 1
COMPASS CARD
POINTER 1 COLOR INDICATOR
RMI POINTER 2
AV SYSTEM ASSIGNED TO POINTER 1 (VOR1/ADF1)
DISTANCE READOUT
COURSE DEVIATION INDICATOR
AIRCRAFT SYMBOL
POINTER 2 COLOR INDICATOR
NAV SYSTEM ASSIGNED TO POINTER (VOR2/ADF2)

CRS252
T
W 30
35.0NM
156KT
GS
VOR1
VOR1 35.0NM
ADF2

(B)

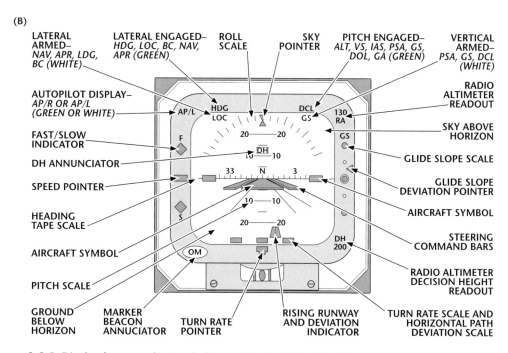

LATERAL ARMED– NAV, APR, LDG, BC (WHITE)
LATERAL ENGAGED– HDG, LOC, BC, NAV, APR (GREEN)
ROLL SCALE
SKY POINTER
PITCH ENGAGED– ALT, VS, IAS, PSA, GS, DOL, GA (GREEN)
VERTICAL ARMED– PSA, GS, DCL (WHITE)
AUTOPILOT DISPLAY– AP/R OR AP/L (GREEN OR WHITE)
RADIO ALTIMETER READOUT
FAST/SLOW INDICATOR
SKY ABOVE HORIZON
DH ANNUNCIATOR
GLIDE SLOPE SCALE
SPEED POINTER
GLIDE SLOPE DEVIATION POINTER
HEADING TAPE SCALE
AIRCRAFT SYMBOL
AIRCRAFT SYMBOL
STEERING COMMAND BARS
PITCH SCALE
RADIO ALTIMETER DECISION HEIGHT READOUT
GROUND BELOW HORIZON
MARKER BEACON ANNUCIATOR
TURN RATE POINTER
RISING RUNWAY AND DEVIATION INDICATOR
TURN RATE SCALE AND HORIZONTAL PATH DEVIATION SCALE

AP/L
HDG LOC
DCL GS
130 RA
F
GS
DH 10
33 N 3
10 10
20 20
OM
DH 200

Figure 3-2-5. Display formats of a Bendix/King EFIS: (A) EHSI, (B) EADI

The display processor card receives inputs from the I/O cards through a multiport RAM. A multiport RAM is a random access memory with multiple input/output connections. The display processor uses discrete 12-bit TTL circuitry to control the operation of the symbol generator and ADI/MUX cards. The symbol generator produces the information needed to *draw* the picture on the EHSI and EADI CRTs. The symbol generator is divided into two separate functions: the character generator and the vector generator. A character generator is used to produce letters, numbers, and symbols. A vector generator is used to draw lines on the CRTs. The ADI/MUX card is used to produce the deflection and video signals needed for the EADI sky and ground background displays. There are three separate power supplies in the DPU contained on two power supply cards. One power supply (card A3) is used for DPU operations. The two power supplies on card A9 are used to power the EFDs. During troubleshooting of the EFIS-85/86, remember the EFDs receive power from the DPU, the EFDs do not contain their own low-voltage power supplies. The A10 relay card provides switching to the EFDs during partial failure of the EFIS.

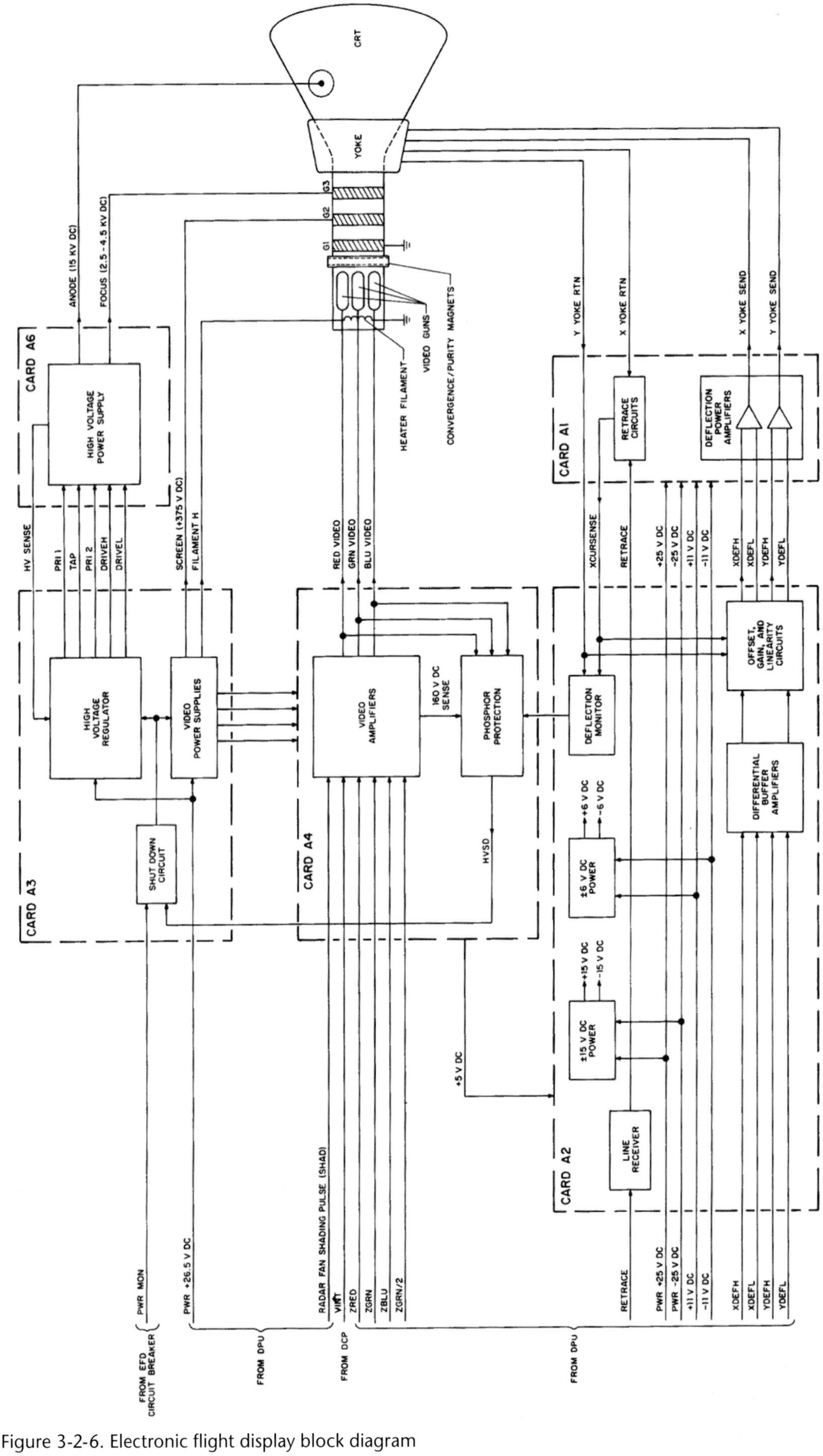

Figure 3-2-6. Electronic flight display block diagram

Courtesy of Rockwell International, Collins Avionics Divisions–for training purposes only

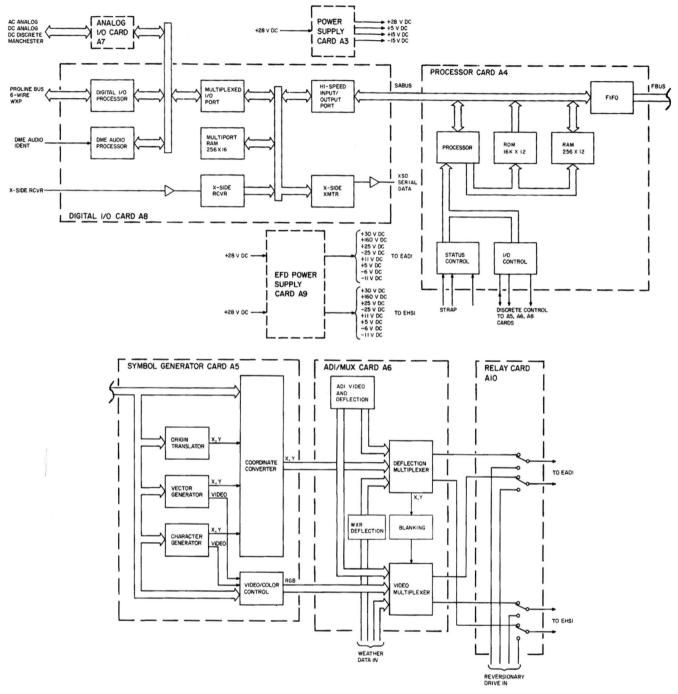

Figure 3-2-7. Block diagram of the display processor unit

The multifunction processor. The multifunction processor unit is used to process system data and provide video signals to the MFD. The MPU also stores checklist information that can be recalled and displayed on the MFD. The MPU has input/output capabilities to allow for redundant monitoring of various systems. The MPU can be thought of as a safety net or backup processor for the DPUs. In the event of a single or dual DPU failure, the MPU can provide the deflection and video signals to the EFDs. The MPU cannot supply operational power to the EFDs.

The multifunction display. As the name implies, the multifunction display unit performs multiple functions. The MFD provides weather radar data, navigational maps, and page data for checklists and maintenance functions. The MFD contains a CRT display surrounded by several control switches, making the MFD slightly larger than the EFDs (top portion of Figure 3-2-2). The mode selector, line select keys, and display function controls for the MFD are installed on the face of the unit. Unlike an EFD, the MFD contains its own power supply circuitry. The MFD is typically

INPUTS		
AC Analog Inputs	*Desired track or track angle error & drift angle *Drift angle/bearing to waypoint *VOR/RNV bearing	*Pitch angle *Heading *ADF bearing
DC Analog Inputs	*Radio altitude *Localizer or MLS azimuth deviation *Glideslope, MLS glide path, or VNAV deviation *Roll steering command	*Pitch steering command *AOA/Fast/slow deviation *ADF bearing *Vertical deviation
DC Discrete Inputs	*To/From *Instrument flags *Instrument modes *EFI and I/O staps *Marker beacon *FCS modes *CAT II request *Localizer mode	*VOR test *DME hold *DG mode *Master warn reset *Comparator reset *Comparator reset *Accelerometer monitor
Serial Digital Inputs	*DME distance *Long-range navigation information *DPU cross-side data *DCP/DSP data *Air data computer information	*Vertical navigation information *FCS modes *Altitude/heading *WXPdata *REF airspeed
Audio Identfication Input	*DME audio identification	
AC Discretes	*Marker beacon (inner, middle, outer)	

Table 3-2-1. Display processor unit inputs   *For training purposes only*

OUTPUTS		
AC Analog	*Heading error *Course datum	
DC Analog	*Crosstrack deviation	
DC Discrete	*Comparator warns *CAT II lights *Master warn *To/From *Back localizer	*Modes/status/flags *Over temperature *Decision height *Decision height *Radio Altimeter test
Serial Digital	*DCP/DSP control *FMS-90 WPT load	*Cross-side DPU data *FCS data

Table 3-2-2. Display processor unit outputs   *For training purposes only*

located in the center section of the instrument panel so it is accessible by both pilots. It should be noted that there is no backup display for the MFD. If this unit should fail, most MFD functions would be lost; however, weather radar and some navigational data can be displayed on the EHSI.

The control panels. There are three types of control panels available for the EFIS-85/86: the CHP, DCP, and DSP (Figure 3-2-8). Any given EFIS will be equipped with one DSP or a combination of one CHP and one DCP for each DPU in the system. The specific configuration is a function of aircraft/EFIS design and will

be shown under the system architecture later in this chapter. The control panels are used to select various EFIS display configurations, navigational data selection, and other similar parameters. The DSP provides selection of various navigational parameters. As the pilot activates the various EFIS controls, the DSP produces a digital signal for transmission to the MPU and DPU(s). The MPU or DPU then changes this information on the appropriate display(s).

If a CHP and DCP are used instead of a single DSP, the control functions are distributed between two separate panels. As seen in Figure

3-2-9, the CHP produces signals that are sent to the DCP. The DCP scans the front panel switches of the DCP and CHP for pilot inputs. The DCP then processes the data and sends a serial data transmission to the DPU and MPU.

Collins EFIS-85/86 System Architecture. The Collins EFIS-85/86 can be installed in several different configurations as specified by the type of aircraft and the owner's request. The major configuration changes are found in the number of display tubes and the type and number of control panels installed. The simplest EFIS-85/86 system is shown in Figure 3-2-10. In this system, there are two EFDs (one EADI and one EHSI) controlled by a display select panel. The weather radar system (shown inside the dashed box) is an optional system separate from the EFIS. This two-tube EFIS could also be controlled using one DCP and one CHP.

Figure 3-2-11 shows the different types of data that are transmitted to and from the various EFIS components. The DPU receives six different types of data from the various aircraft systems and digital data from the DSP. The DPU receives control signals from the DSP through a CSDB digital data link. Weather data is sent to the DPU from the weather radar receiver/transmitter unit via a digital data bus. The DPU processes the data and sends video control signals to the EFDs. The DPU also sends all low voltage DC power signals to the EFDs.

A three-tube EFIS consists of two EFDs and one MFD (Figure 3-2-12). A multifunction processor unit is needed to drive the MFD. This arrangement provides system redundancy since the MPU can be used to supply video data to the EHSI and EADI in the event of a DPU failure.

A four-tube EFIS employs two EADIs, two EHSIs, and two DPUs. As seen in Figure 3-2-13, the two DPUs "cross talk" video information and system data for comparison purposes. This system provides redundancy through the ability of one DPU to drive two EHSIs and two EADIs with identical displays in the event the opposite side DPU fails.

A five-tube EFIS contains four EFDs, one MFD, two DPUs, and one MPU (Figure 3-2-14). This system provides the maximum redundancy of the Collins EFIS-85/86 series. With three processor units (two-DPUs, and one-MPU) the system is extremely unlikely to experience complete EFIS display failures. In a five-tube system each DPU will use the MPU as back up for video information in the event of a failure. As mentioned earlier, the relay cards in the DPUs control the output signals to the EFDs. Figure 3-2-15 shows the ability for the relay card to receive data from the MPU or on-side DPU.

(A)

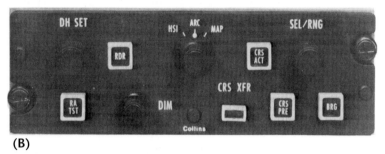

(B)

(C)

Figure 3-2-8. EFIS control panels: (A) CHP, (B) DCP, (C) DSP
Courtesy of Rockwell International, Collins Avionics Divisions

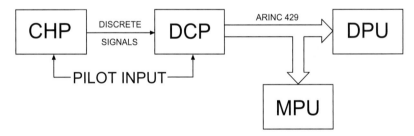

Figure 3-2-9. EFIS control panel block diagrams
Courtesy of Rockwell International, Collins Avionics Divisions

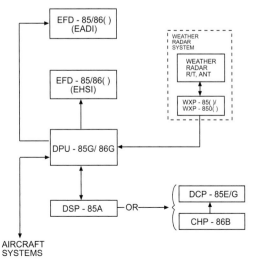

Figure 3-2-10. Block diagram of a two-tube EFIS
Courtesy of Rockwell International, Collins Avionics Divisions

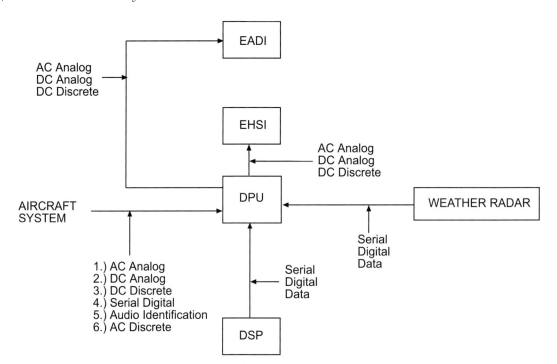

Figure 3-2-11. Data in interface to various EFIS components

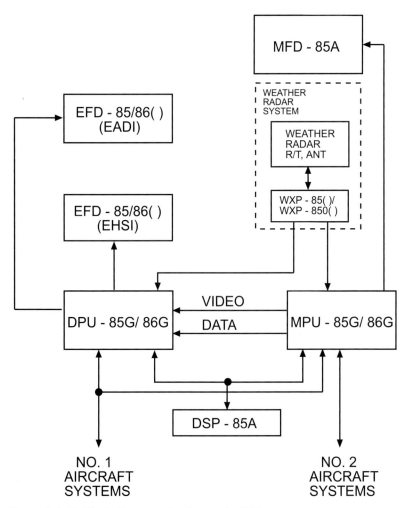

Figure 3-2-12. Block diagram of a three-tube EFIS

Courtesy of Rockwell International, Collins Avionics Divisions

Collins EFIS-85/86 System Operation. The EFIS-85/86 is capable of a multitude of flight related functions that require a thorough understanding of aerial navigation. The materials presented here will provide a brief summary of the flight related controls and concentrate on operation of the system for maintenance purposes. Initial operation of the EFIS is quite simple; the system is active at any time the appropriate electrical busses are powered.

In most installations the EFIS equipment will be connected to the aircraft's avionics bus(ses). This will allow for isolation of EFIS during engine starting and other times when the aircraft's electrical power is considered unstable. The system is capable of withstanding voltage spikes of ± 600 volts at a rate of 50 spikes per minute. However, it is considered that exposure to voltage spikes at any level and rate can compromise system reliability. The voltage required for normal operation of the EFIS-85/86 is between 22.0 VDC and 29.5 VDC. In emergency operations the EFIS will function with a voltage as low as 18.0 VDC. The entire EFIS will require between 13.5 and 22.2 amps at 27.5 VDC for operation.

During operation the DPU, MPU, and EFDs require forced air cooling. If cooling is not adequate, EFDs will blank out and processor units may be damaged. The MPU has the capability to drive an *overheat* annunciator installed on the aircraft's panel. During any EFIS operation, be sure the proper cooling is provided, and discontinue operation if temperature limits are exceeded.

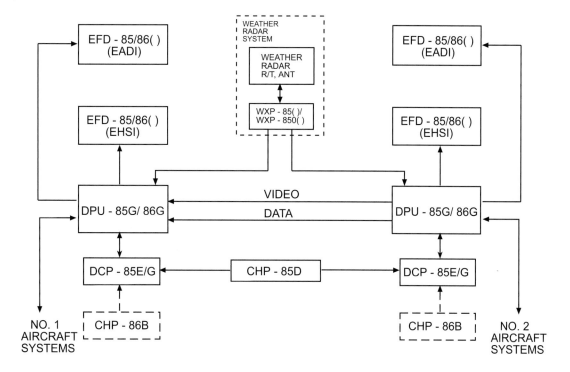

Figure 3-2-13. Block diagram of a four-tube EFIS

Courtesy of Rockwell International, Collins Avionics Divisions

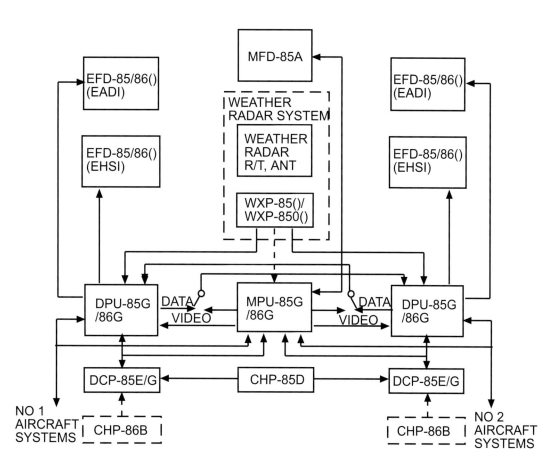

Figure 3-2-14. Block diagram of a five-tube EFIS

Courtesy of Rockwell International, Collins Avionics Divisions

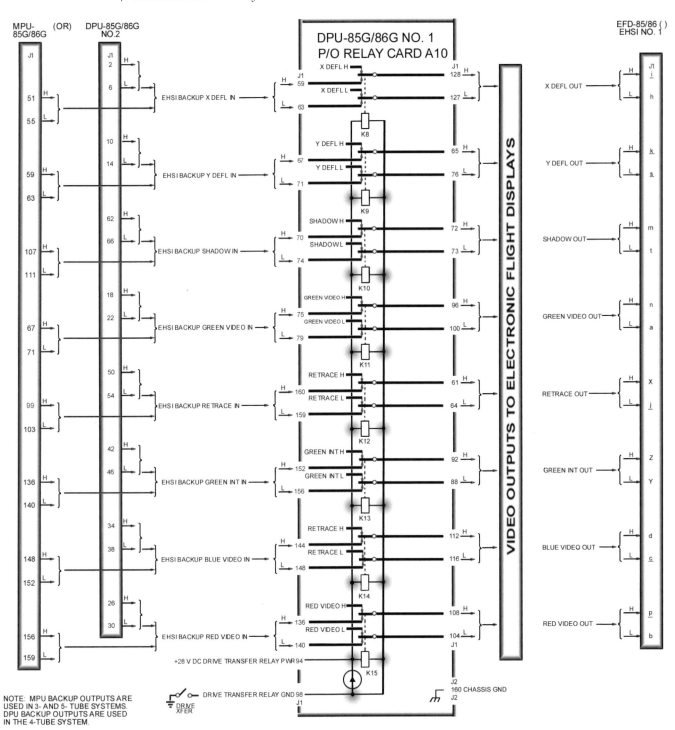

Figure 3-2-15. Display processor unit relay card

Courtesy of Rockwell International, Collins Avionics Divisions

Configuration strapping. Many of the features available for display by EFIS are controlled through strapped inputs. *Strapping* is a term given to an electrical input which is permanently connected to a digital 1 or digital 0 signal. Strapping connections are typically installed any time a system is modified and during initial system installation. EFIS configuration strapping is used to determine various display, input/output signal formats, and other system parameters. A specific system configuration is achieved by connecting a strap connection (or multiple connections) to ground or by leaving it (them) open. On the Collins EFIS- 85/86, strapped inputs open (not grounded) are considered logic 1. Strapped inputs connected to ground are considered logic 0. For example, the DPU may be strapped to allow for operation of global positioning system (GPS) equipment. If DPU plug 2 pin 136 is connected to ground (set to 0), and pin 116 is open (set to 1) it will allow for EFIS display of GPS operation. The specific configuration/operation of any EFIS is always a function of the strapped inputs. It should be noted that some form of configuration strapping is needed on almost all systems. In some cases, configuration set-ups are done using software. On some systems configuration strapping is known as pin programming.

Logic 1 and logic 0 could be defined in various ways for different aircraft or different systems. For example, a logic 1 may be obtained as + 28VDC, open, or connection to ground. The voltage value for strapped inputs is determined during the design of any system and memorized by the operational software.

During maintenance and troubleshooting, strapped inputs should not be overlooked. If a system fails to operate properly, the fault may be caused by a poor strap connection. If a strapped input changes from binary 1 to binary 0, the EFIS processor may not recognize a given input signal, or may not process the signal correctly. As in the previous example, if plug 2 pin 136 (strapped connection for GPS) comes loose, the processor will no longer be configured for GPS display. Even if all systems are operating properly, the EFIS will not display GPS information until the strap is repaired.

Display control panel (DCP) operation. The DCP provides dimming for both the EHSI and EADI along with other controls used to select EHSI format and functions. Refer to the controls as numbered in Figure 3-2-16 for the following discussion on the operation of the DCP. The **DH SET** (1) is used to set the decision height displayed on the EADI. The **RDR**, push-button (2), is used to bring the radar display to the EHSI in the ARC or MAP formats. The **HSI/ARC/MAP** three-position rotary switch (3) allows the pilot to select the EHSI display format. The **HSI** position displays the full compass rose similar to a conventional HSI. The **ARC** position displays approximately 80° of an expanded compass rose on the HSI. The **MAP** position is similar to the **ARC** display except it also includes VOR and waypoint symbols. The three different modes are shown in Figure 3-2-3.

The **CRS ACT** button (4) will allow for selection of the active course. When the **CRS ACT** button is pressed, the **SEL/RNG** knob is used to make the appropriate selections from a menu on the EHSI display. The **SEL/RNG** knob (5) is used to select the range of the ARC or MAP displays. Available ranges are 5, 10, 25, 50, 100, 200, 300, and 600 nautical miles. The **BRG** button (6) allows the operator to set the bearing pointer using the **SEL/RNG** knob. The **CRS PRE** button (7) is used in conjunction with the **SEL/RNG** to select the second (standby) navigational source parameters.

The **CRS XFR** button (8) is used to select the preset second (standby) course. The **DIM** (9) control is used to adjust the intensity of the EHSI and the EADI. There are two concentric controls used for dimming, each changes the brightness of one of the displays. During troubleshooting be sure these controls are in the correct position if the EFDs are not illuminating. The **RA TST** push-button (10) is used to initiate the radio altimeter test sequence. If the system is operating normally the EADI should display a specified altitude (typically 600 feet). The **RA TST** button is also used to control the BITS mode troubleshooting function, as will be discussed later.

Course heading panel (CHP) operation. The flight crew uses the CHP to select course and heading functions. The CHP sends all data to the DPUs and MPU through the DCP. Refer to the controls as numbered in Figure 3-2-17 during the following discussion of the CHP operation. It should be noted that this CHP is used to control a single EHSI; in a four-tube EFIS two CHP-86Bs or one CHP-85D (a dual control version) would be required. The **HDG** knob (1) is used to position the heading cursor on the HSI. The **PUSH/HDG/SYNC** button (2) is used to synchronize the airplane heading and the heading cursor under the lubber line on the EHSI. The **NAV/DTA** button (3) provides the displays of time-to-go, ground speed, elapsed time, and wind information in the upper corner of the EHSI. The **CRS/CTL** button (4) determines whether the course knob (5) is controlling the active or preselect (standby) course. The **CRS** knob (5) is used to control the course arrow on the display. The **PUSH CRS DIRECT** is located inside the CRS knob and is used to zero the VOR course deviation by rotating the course arrow on the EHSI. The **ET** button (7) controls the elapsed timer functions.

Display select panel (DSP) operation. On many EFIS-85/86 systems, the DSP is used to replace a DCP/CHP combination. For the most

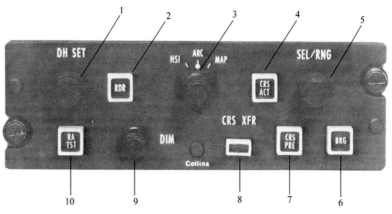

Figure 3-2-16. DCP panel   *Courtesy of Rockwell International, Collins Avionics Divisions*

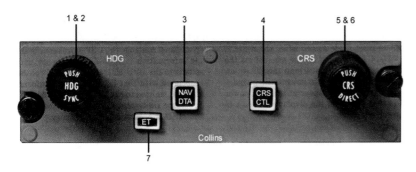

Figure 3-2-17. A course heading panel (CHP)

Courtesy of Rockwell International, Collins Avionics Divisions

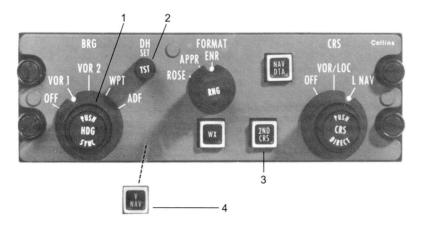

Figure 3-2-18. A display select panel (DSP)

Courtesy of Rockwell International, Collins Avionics Divisions

Figure 3-2-19. A multifunction display (MFD)

Courtesy of Rockwell International, Collins Avionics Divisions

part, the controls found on the DSP are similar to those of the DCP and CHP. Reference Figure 3-2-18 during the following discussion for some of the differences. Switch 1 is used to select a given navigational source. The five possible selections are OFF, VOR 1, VOR 2, waypoint, and ADF. The radio altimeter test switch (2) is located in the center of the decision height knob. The **2ND CRS** push-button (3) replaces the **CRS / CNT** button on the CHP. The **VNAV** push-button (4) is used to allow the presentation of vertical navigation data on the EHSI.

Multifunction display (MFD) operation. The multifunction display is used to access a variety of page data, as well as display weather and navigational maps. The controls located on the perimeter of the MFD are used to control the functions of the MFD (Figure 3-2-19). The **PWR** (power) switch in the upper right corner of the MFD is used to turn on/off the display. The **RDR** (radar) button will allow display of the weather radar data on the MFD (Figure 3-2-20). The **NAV** allows the navigational map data to be displayed (Figure 3-2-21). Both navigational and radar data can be selected simultaneously. The **RMT** (remote) button allows the MFD to display page data from four different remote sources. The **PGE** (page) and **EMG** (emergency) buttons allow for the selection, control, and data entry of up to 100 pages of checklist information (Figure 3-2-22).

In the bottom right corner of the MFD is an electrical jack labeled *DATA*. This jack is used to connect the remote data programmer for page data entry. The data programmer will be discussed later in this chapter. The *INT* (intensity) control is used to adjust the brightness of the MFD display. On the right-hand side of the MFD are four display select keys (DSK). The labels that appear on the CRT during various MFD operations identify the push-button DSKs functions. Pressing the appropriate key will select a given menu function as displayed on the MFD (Figure 3-2-23).

The joystick found in the lower right corner of the MFD is used in the NAV, RMT, PGE and EMG modes. In the NAV mode the joystick is used to aid in the entry of waypoints. In the RMT, PGE, and EMG modes the joystick is used to scroll through pages or chapters of data. The *advance* and *reverse* arrows (located on the bottom left side of the MFD) are used to move a cursor though the list of page data. As the cursor is moved through the data field, each respective line changes from yellow to white to indicate the item on that line has been completed. The *RCL* (recall) switch brings back previously viewed data while operating in the PGE or EMG modes. The *SKP* (skip) mode moves the cursor past a given line on the data page. The *CLR* (clear) key resets all PGE and EMG data page lines to yellow.

Abnormal system operation. The EFIS equipment can experience two basic types of failures: 1) failure causing a comparator warning, or 2) failure of a system element. A comparator warning occurs whenever two signals compared by the processor units disagree. A comparator error will occur only in systems that employ two or more processors (i.e., 1-DPU and 1-MPU, 2-DPUs, or 2-DPUs and 1-MPU). Comparator monitoring is performed by each DPU during normal operation. Both DPUs convert on-side signals received from aircraft systems into digital data. This data is then stored in memory and also transmitted to the opposite-side DPU. Since both processors now have the data from redundant aircraft systems, a comparison can be made by the DPU circuitry. If the processor receives two input signals that disagree, a comparator flag is displayed on the EFDs. A comparator error will show up as a two-, three-, or four-letter system identifier in a small box on the display (Figure 3-2-24). Comparator errors are displayed in yellow. In the event of a single DPU failure, the MPU is capable of comparator monitoring. The MPU would then compare its input signals with the operable DPU. Once again, if the comparison shows a disagreement a fault flag will be displayed.

In most aircraft, an installer supplied (i.e., not part of the EFIS equipment) comparator annunciator is used to alert the flight crew to a comparator disagree fault. A master warning light is also installed on some systems for both the pilot and co-pilot. The master warning light will most likely be a light/switch assembly that illuminates for all EFIS warnings. The switch is used to reset the warning signal activated by the DPU or MPU. Momentarily pressing the reset switch will extinguish the master warning. If the fault still exists, the flashing EFD warning message(s) will change to a steady display and the comparator annunciator will also remain lit. If the fault no longer exists after the reset is pressed, all warning messages will be extinguished.

There are a variety of warnings that can be displayed by EFIS. The master warning is activated and the EFDs will display a specific message related to the failure. If the aircraft is approved for category II landing capabilities a separate CAT II annunciator will also be installed. Each system failure will have a specific message on one or both EFDs.

Failure of an EFIS system element can be caused by a loss of data from an aircraft system which reports to EFIS or it can be caused by a loss of one or more of the EFIS components. In either

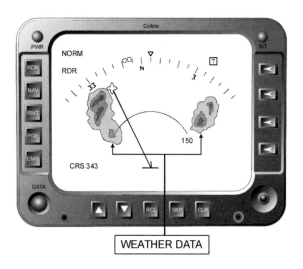

Figure 3-2-20. Weather data shown on the MFD

Courtesy of Rockwell International, Collins Avionics Divisions

Figure 3-2-21. Navigational data shown on a MFD

Courtesy of Rockwell International, Collins Avionics Divisions

Figure 3-2-22. MFD showing PGE (checklist) data

Courtesy of Rockwell International, Collins Avionics Divisions

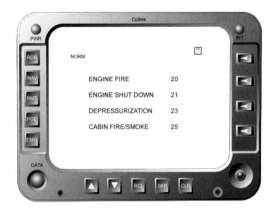

Figure 3-2-23. MFD showing a typical menu function
Courtesy of Rockwell International, Collins Avionics Divisions

case, the warning associated with the fault will typically create a red warning message on the appropriate EFD. An example of the system element fault flags on the EADI is shown in Figure 3-2-24. Fault flags are displayed in red. (Comparator errors are displayed in yellow.) EFIS fault flags are any indication displayed on one or more of the EFDs related to a complete or partial system failure. If a failure occurs in a system that feeds EFIS, the flag will be displayed until the fault is repaired. If one of the processor units fails, the flag *DPU FAIL* or *MPU FAIL* will be displayed on the CRT for five seconds. After five seconds the display will go blank

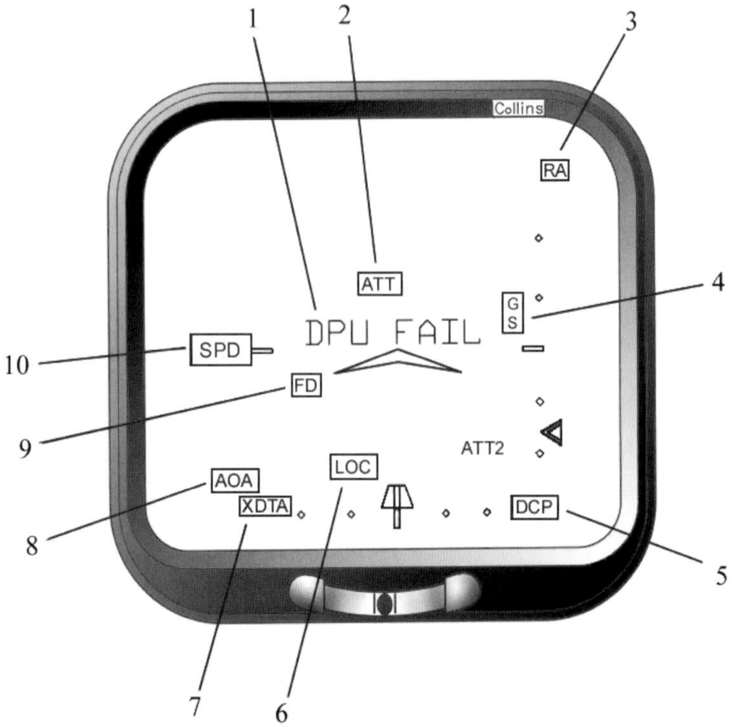

except for the failure message. In this case, the pilot can activate a transfer switch to regain the display driven by another processor.

A reversionary mode is available on EFIS to allow for system operation in the event of a component failure. The reversionary switches are supplier installed (i.e., not part of the EFIS equipment) and should be located in some area accessible to both pilots. The reversionary switches can be installed to bypass failures of the EFDs, DPUs, MPU, DCP, and/or the DSP. Switching of input data, such as attitude and compass information, is also an available option.

In the event of a single EFD failure, a reversionary switch can be used to transfer the EADI to the EHSI or vice versa. In either case, the information typically shown on both displays will be compacted and displayed on the operable EFD. The EFIS compact mode is called the composite format (Figure 3-2-25). Another option allows external switching to move the full format EADI to the EHSI; the EHSI is then transferred to the MFD.

In the event of a DPU failure, the EADI and EHSI driven by that DPU will display the flag *DPU FAIL* in red. At this time, the pilot would activate the reversionary switch to allow the MPU to drive the EADI and EHSI. In this configuration, no degradation of the EADI and EHSI display occurs. The MFD will only be able to display the EHSI or a composite format. The MFD controls (except power and brightness) are inactive. No backup is provided for an MFD failure. As usual, navigation and weather data are available on the EHSI; however, checklist data is lost.

On a four- or five-tube EFIS an external switch can be used to select a cross-side control panel if a DSP or DCP fails. If the cross-side DSP is selected, both DPUs are controlled by the selected DSP. Whenever a cross-side control panel is selected, the appropriate message *XDSP* or *XDCP* will be displayed on the EADI in yellow.

The EFIS-85/86 systems have the capability to allow for reversionary switching of attitude and heading information sent to the MPU or DPUs. The reversionary switches are external to EFIS, installed in an accessible area of the flight deck. Switching is typically done only in the event of an on-side failure of attitude or heading data. If the pilot's attitude data source fails, the pilot can retrieve attitude data from the co-pilot's source, and vice versa. If the pilot's on-side heading data source should fail, the information can be received from the co-pilot's source and vice versa. The pilot's on-side sources are called *ATT1* and *HDG1*. The co-pilot's on-side

Figure 3-2-24. Typical EADI fault flags (all in red): (1)Display processor unit, (2) Attitude, (3) Radio altimeter, (4) Glide slope, (5) Display control panel, (6) Localizer, (7) Cross-side data bus, (8) Angle of attack, (9) Flight director, (10) Speed command

sources are called *ATT2* and *HDG2*. Whenever receiving cross-side attitude data, the appropriate message will be displayed in yellow. For example, if the pilot has selected cross-side attitude data, the message *ATT2* will be displayed on the pilot's EADI.

Programming the MFD checklist data. The MPU can store up to 100 pages of data each containing 12 lines of up to 20 characters each. Remember the MPU stores the data that is displayed on the MFD. This data is used for checklists accessed by the **PGE** (page) or **EMG** (emergency) buttons on the MFD (Figure 3-2-19). Checklist data programmed into the MFD is stored in a nonvolatile memory. The Collins remote data programmer is used to install new checklist data or modify an existing checklist.

As seen in Figure 3-2-26, the remote data programmer (RDP-300) is a calculator-type unit with an alphanumeric keyboard. The data programmer plugs into the jack found in the lower-left corner of the MFD. To program the MPU, apply system power and turn on the MFD. Connect the data programmer, and allow 15 seconds warm up time. Pressing the **PGE** button will access the page checklist; pressing **EMG** will access the emergency checklist. The data loader is then used to enter the appropriate data to each page of the checklist. The EFIS installation manual offers detailed instructions for data entry.

Another method for entering EFIS checklist data is by entering it into a checklist entry unit using a personal computer; then transferring to the MFD. Figure 3-2-27 shows a CEU-85/85A checklist entry unit. The CEU-85 is interfaced with Apple computers through a Collins Pro Line II interface card. The CEU-85A interfaces with a personal computer using the computer's RS-232 port. Using the *Checklist Editor* software the appropriate checklist can be made or an existing list modified. The checklist is then downloaded into the checklist entry unit. The checklist entry unit is then plugged into the data jack of a powered MFD and the information is transferred from the entry unit to the MFD. It requires approximately 11 minutes to complete the data transfer. To avoid losing data, be sure there are no power interruptions to the system during data loading.

EFIS-85/86 maintenance. The Collins EFIS-85/86, like most other EFIS equipment, requires very little routine maintenance. Most of the maintenance is performed on an as needed basis (i.e., after a fault occurs). However, to ensure long life of any EFIS, it is essential that the components receive proper cooling. To ensure proper cooling, all air passages and equipment housings must be clean and free of dust and lint. It is good practice

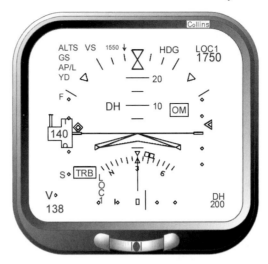

Figure 3-2-25. EFIS composite format combining information from ADI and HSI on one display

Figure 3-2-26. Remote data programmer (RPD-300) used to enter checklist data into the MP memory

Courtesy of Rockwell International, Collins Avionics Divisions

Figure 3-2-27. The CEU-85/58A checklist entry unit

Courtesy of Rockwell International, Collins Avionics Divisions

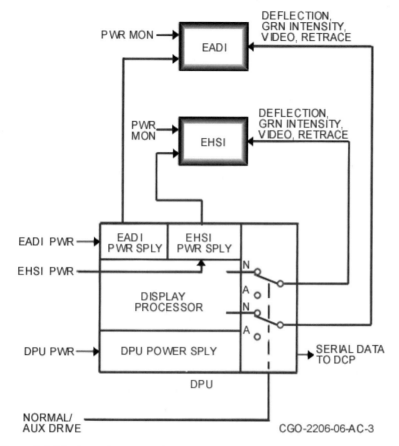

Figure 3-2-28. Power inputs to the DPU
Courtesy of Rockwell International, Collins Avionics Divisions–for training purposes only

to include the cleaning of all cooling fans and equipment housings in the aircraft's scheduled maintenance program. This will improve the reliability and longevity of all EFIS components.

Technicians should always take care when cleaning the EFIS display units. The face of the CRT is coated with a special non-glare film. This film is extremely thin and can easily be damaged by abrasive cleaners or dirty cleaning rags. Proper cleaning should be accomplished with a denatured alcohol cleaner and a soft dirt and lint free rag. Lens cleaning tissues may also be used for cleaning the CRT.

Proper brightness of the EFIS CRTs should be confirmed at regular intervals. The Collins Maintenance Manual suggests the formal test be done at approximately 5,000 hours of EFIS operation; however, a quick visual check can easily be performed during routine inspections. A quick check is done by simply noting the brightness of the display and adjusting the appropriate brightness knob. The display should be easily read in full sunlight with the brightness control set to approximately 3/4 maximum. If the CRT is difficult to read with the control near the maximum position, perform the formal brightness test described below.

The formal brightness test is done using a Minolta LS-100 light meter. The aircraft must be located in an area of low ambient light for this test. The light meter is held in front of the CRT while operating in the appropriate test mode. The proper level of light must be indicated on the meter when the brightness control is turned to maximum. If the proper level is not achieved, the display should be replaced.

Fault monitoring. During normal operation the EFIS-85/86 performs on-line monitoring which is a continuous fault monitoring process designed to detect internal and external faults. Each processor in the system (MPUs and DPUs) contains an on-line monitoring software program. The program tests for a periodic *keep-alive* signal generated by the processor. If this signal is not detected at the proper time intervals, it is assumed that the DPU/MPU is not processing data correctly and a fault flag is displayed.

The data busses, which connect the aircraft subsystems to EFIS, are tested using a series of test words. The processor initiates a test word that is transmitted through a given data bus then returned to the processor. If the test is not successful, the appropriate fault flag is displayed. Activity monitors are used to monitor each EFIS data bus to ensure data transmission at regular intervals. If the activity monitor determines a given subsystem is not transmitting data, the appropriate flag is displayed.

On some systems, the voltage level transmitted to the processor is monitored for proper values. If the voltage is found to be out of limits, a flag is displayed. The I/O section of the processor and the display processors routinely compare their program checksums. If the checksum monitor determines an invalid calculation, the DPU/MPU fault flag is displayed. A RAM monitor circuit is activated during each power up of the EFIS processors to test the I/O and display processors RAM storage. If the test fails, the flag *DPU/MPU FAIL* will be displayed.

Self-test functions. The EFIS DPUs and MFDs contain a self-test software program, which is used to test the health of the system upon request by the pilot or technician. The self-test function is comprised of two modes: confidence test and diagnostics/maintenance routines. The self-test function is activated through a momentary contact switch supplied during EFIS installation. The confidence test is typically used during preflight to ensure reliance in the EFIS equipment, or during maintenance to verify proper system operation after a repair. Maintenance personnel typically use the diagnostics/maintenance routines during in-depth troubleshooting. The diagnostic/maintenance routines will be discussed under the *BITS mode* section of this chapter.

Pressing the test switch will connect a ground signal to the processor unit (DPU and/or MPU) and initiate the confidence test. During the confidence test, the pilot's EADI pitch and roll values will change by an increment of +10°, +20° respectively. The co-pilot's EADI will change -10° and -20° respectively. The word *test* and *pitch and roll miscompare* messages are displayed on the EADI while the confidence test is in progress. If the test switch is pressed again for an additional four seconds the pitch, roll and heading values return to normal and all EADI and EHSI test flags are displayed. This *flag* display will continue until the test switch is released.

Troubleshooting techniques. Whenever troubleshooting any system that is comprised of several LRUs, it is very important to become familiar with the power distribution of the components. The input power to a five-tube EFIS-85/86 system comes from over a dozen circuit breakers; all must be in operable condition before any troubleshooting can begin. As seen in Figure 3-2-28, the power required to operate the EHSI and EADI comes through the on-side DPU. The DPU receives three separate power

inputs, and the EFDs receive a +28 VDC power monitor signal from the bus. Also make sure adequate power is available for proper operation. A weak battery or poor quality ground power unit may cause EFIS problems, which are not system related. The EFIS-85/86 can operate successfully down to approximately 18 VDC; however, CRT displays may dim and images may shrink at low voltage. Remember, 27.5 VDC is the recommended voltage.

Proper equipment cooling is imperative during EFIS operation and troubleshooting. The EFDs, DPUs, MFD, and MPU must be cooled by forced air produced by blower fans located in the appropriate areas. The EFIS-85/86 employs an over temperature monitor circuit in the MPU. If the temperature rises above a 135° C (275° F) the MPU will illuminate an installer-supplied annunciator. If the EFDs overheat, the displays will automatically blank to help cool the unit. A fan motor module is available to monitor the speed of the system cooling fans. If the fan motor module detects a low-speed cooling fan, a supplier-installed annunciator light is illuminated. Keep in mind many aircraft do not have an overheat indicator. If a display goes blank after prolonged operation and functions again when cool, suspect poor cooling. Look for inoperable fans, clogged filters, or discon-

EFIS-85B (2/12) and EFIS-86B (2/12) Input Specifications

STRAP INPUTS	UNIT PIN NUMBER	STRAP NUMBER	SIGNAL CHARACTERISTICS
12.0 DPU/MPU-85G/86G STRAPPING DEFINITIONS			
Autopilot Type	DPU (P2-61,115,121) MPU (P2-1,5,9)1 MPU (P2-4,8,12)2	1 2 3 0 0 0 1 0 0 0 1 0 1 1 0 0 0 1 1 0 1 0 1 1 1 1 1	FCS-80 dual FD Sperry A, (G III) FCS-A dual FD Sperry B, (citation) FCS-B APS-86 (dual FD) APS-65 FCS-80 single FD Sperry A, FCS-A single FD APS-85 (single FD)
Radio Alt Type	DPU (P2-133,134) MPU (P2-61,65)1 MPU (P2-64,68)2	4 5 0 0 1 0 0 1 1 1	ALT-50 (0 to 2000 feet) ALT-55 (0 to 2500 feet) AL-101 (0 to 2500 feet) Undefined
DPU Side	DPU (P2-140)	6 0 1 **Note** 0 = Ground 1 = Open	DPU use only Left DPU (left side I/O annunciated) Right DPU (right side I/O annunciated)

Figure 3-2-29. Sample of strapping definitions

Courtesy of Rockwell International, Collins Avionics Divisions–for training purposes only

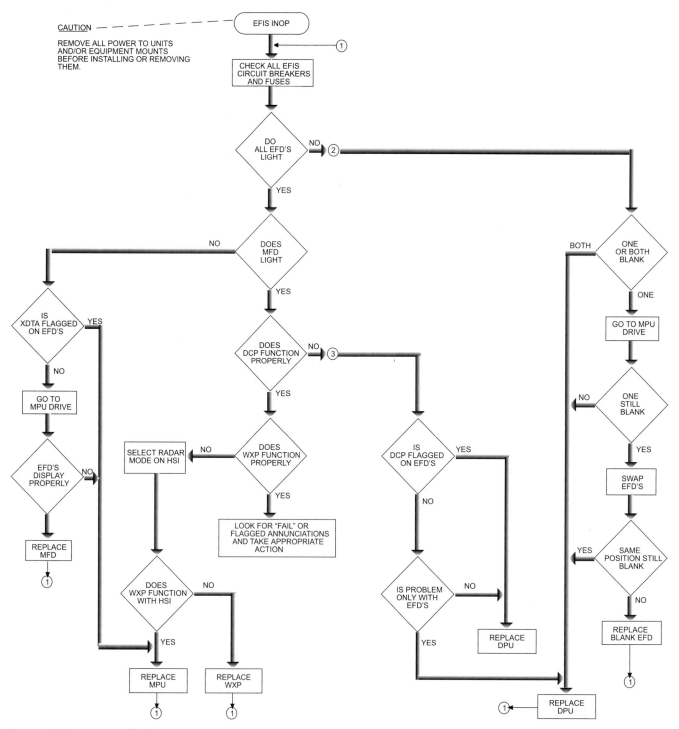

Figure 3-2-30. EFIS troubleshooting flow chart

Courtesy of Rockwell International, Collins Avionics Divisions

nected air ducts. In addition, do not forget that birds and rodents tend to build nests inside air inlet ducts.

The strap configuration of the EFIS equipment is also an important consideration during troubleshooting. A specific strap connection (plug and pin number) is defined by the system interconnect diagrams. Figure 3-2-29 shows a sample of the strapping definitions table for the EFIS-85/86. In this example, it can be seen that the specific autopilot model can be selected using

three strap connections, the radio altimeter setting requires two strap connections, and the DPU side is determined using one strap connection.

An intermittent or disconnected strap connection can create difficult troubleshooting problems. For example, if the straps, which define autopilot type, should become disconnected the DPU will not recognize the data transmitted from the autopilot and a flag will be displayed. The technician may fault the autopilot, the DPU or the associated bus without suspecting the strap configuration. Since the strap is simply a connection to ground (or not) they are typically not a problem; however, the technician should always be aware that strapping controls a variety of EFIS functions and if a strap is incorrect (or has a loose connection) that portion of the system is inoperative.

Isolating faulty LRUs. In most cases, troubleshooting EFIS is simply a matter of finding the defective LRU. On two-tube systems, EFDs can be switched or suspect LRUs can be swapped with the same component "off the shelf" or from another aircraft. Isolating defective LRUs on aircraft with a four- or five-tube system can often be done from the flight deck as follows:

Example 1

IF – the pilot's EHSI has a blank display,

THEN – the pilot's processors can be switched to drive the co-pilot's displays.

IF – the fault shows up on the co-pilot's EHSI also, the processor is most likely defective.

IF – the fault does not show on the co-pilot's EHSI, the fault most likely lays in the pilot's EHSI. (Swapping the pilots EFDs would then verify if the fault was in the pilot's EADI.)

Example 2

IF – the co-pilot's side EADI is blank,

THEN – switch to MPU drive for the co-pilot's displays.

IF – The EADI operates normally, the co-pilot's DPU is most likely defective.

IF – The EADI is still blank, the display unit is most likely defective.

Be careful, in this example it is easy to be fooled. Remember the DPU supplies power to both on-side displays. A power supply problem in the DPU could cause a blank EADI. In this case, the EADI would stay blank when driven by either the MPU or the DPU. (The MPU only supplies video signals to the EADI and EHSI; the MPU does not supply power to these displays.)

Example 3

(assumes the aircraft has reversionary switching of attitude and heading data)

IF – the pilot reports inaccurate attitude data on left side EFDs,

THEN – use the reversionary switching to drive the left side displays with the right side attitude sensors.

IF – the problem is corrected, the left side sensors or sensor processing unit (which normally drive the left side displays) are defective.

IF – the problem still exists, the left side DPU or associated wiring is most likely at fault.

THEN – Swap the left and right DPUs.

IF – the fault moves to the right side displays, the DPU now in the left side (previously in the right side) is defective.

The procedure for reversionary switching of various EFIS LRUs can vary from aircraft to aircraft. It is always best to refer to the aircraft's manuals to verify proper LRU reversionary switching procedures. To aid in troubleshooting, always become familiar with the maintenance manuals for the system in question. The EFIS-85/86 maintenance manual contains a flow chart for isolating defective LRUs (Figure 3-2-30).

If replacement of an LRU is ineffective at repairing a system fault, the system's wiring should be considered a primary suspect. Connector plugs become worn or dirty as the LRUs are removed or replaced. Be sure to check the plugs and sockets carefully. Employ standard techniques for testing continuity of system wiring. Also be sure to test for grounded wiring due to failed insulation or a shorted connector. Oscilloscopes or data analyzers can be used to test digital data signals.

Remember one of the best tools for troubleshooting defective pin connectors is a flashlight, a magnifying glass, and a good pair of eyes. Defective connections are typically discolored, burnt, pitted, bent, or corroded. These traits can often be detected visually. The hardest part for a visual inspection is getting into a position where the connections can be viewed properly. During troubleshooting, if a wiring problem is suspect, remove the LRU and inspect the connections related to the faulty system. First look at the LRU connections; if a defect is found, remember the mating contact in the LRU rack is most likely also defective. For proper repair, the faulty pin and mating socket must be repaired. In many cases, a thorough cleaning will correct the problem. However, when cleaning contacts be extremely careful, it is easy to bend connections and cause greater damage.

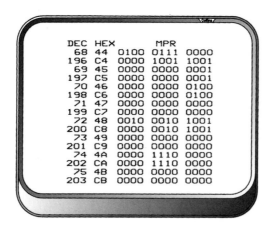

Figure 3-2-31. Electronic flight display BITS model
Courtesy of Rockwell International, Collins Avionics Divisions

BITS mode. The diagnostic/maintenance routines available under the *self-test* function of the Collins EFIS-85/86 are also referred to as the BITS mode. The EFIS BITS mode is used to access real-time data from the RAM memory of the DPUs and/or MPUs. As a processor receives data from the various systems and sensors throughout the aircraft, the information is stored in the processor's multiport RAM (MPR). The MPR stores data from both on-side and cross-side systems, some of which is transmitted due to internal EFIS operations. The MPR stores the data in a digital format and updates the information as new data arrives. This data can be displayed on the EFDs or MFD using the EFIS BITS mode. In other words, BITS mode allows you to *look inside* the processor circuitry at the digital signals being processed. This function is a type of built-in data bus analyzer for the EFIS processors. Keep in mind that BITS mode is an in-depth troubleshooting tool and is typically used by only the most experienced technicians.

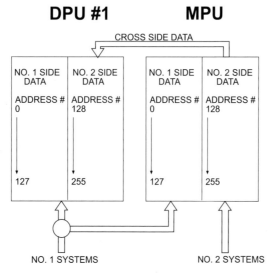

Figure 3-2-32. Diagram showing on-side and off-side data to the DPU and MPU

BITS mode functions. There are five functions available in the EFIS BITS mode:

1. Multiport RAM (MPR)

2. RAM

3. MFD

4. Alignment tests, and

5. Flight control system (FCS) diagnostics

The MPR is most useful for operational verification after system installation and in-depth troubleshooting. MPR is the BITS mode most often used by line technicians for EFIS troubleshooting. The RAM function allows access to the display processor RAM. The MFD function displays data stored in the MFD page data (checklist) RAM. The alignment tests function displays a test pattern generated by the processor to verify symbol generator and EFD operation. The RAM, MFD, and alignment tests are most often used during in-shop repair of an LRU. The FCS diagnostics are available only if the aircraft is equipped with the Collins APS-85/86 flight control system. The MPR function of the BITS mode will be discussed here since it relates specifically to flight-line EFIS maintenance.

The MPR function of the BITS mode can be displayed for either the left or right DPU, or the MPU. To activate the pilot's side DPU BITS mode, press and hold the EFIS test switch and the radio altimeter test (RA TEST) switch on the pilot's DSP or DCP. This will cause the BITS mode to be displayed on the pilot's EFD. The test and RA switches can now be released. To page through the BITS data, press the RA test switch sequentially. The co-pilot's DPU multiport RAM is accessed similarly using the on-side EFIS test and RA switches.

On a five-tube system, the BITS mode for the MPU memory can be accessed by pressing the pilot's EFIS test switches while simultaneously pressing the co-pilot's RA test switch. The BITS mode will be displayed on the MFD. To exit from the BITS mode, system power can be cycled, or repeatedly press the RA test switch for the appropriate processor (DPU or MPU).

Interpreting BITS mode. Figure 3-2-31 shows a typical BITS mode display. BITS mode can monitor 256 EFIS processor's inputs. Each input is given a specific address. Address numbers 000 through 127 are parameters from left-side (pilot's-side) systems. Address numbers 128 through 255 are parameters from right-side (co-pilot's-side) systems. Figure 3-2-32 shows the relationship of the on-side/off-side locations for the pilot's DPU and the MPU. The BITS mode address numbers are simply identifiers for the given systems or sensors which feed the EFIS processor. The address numbers may also be referred to as *locations*. In other words, each system is assigned a given BITS mode number. Figure 3-2-33 shows a partial alphabetized list of systems and their associated addresses. This information would be found in the maintenance manual. In this figure, it is shown that address 097 is IAS (indicated airspeed) from the ADC (air data computer). We will use address 097 for the following example on decoding the data field.

To interpret BITS mode data, reference an EFIS maintenance manual while accessing the BITS mode display. It should be noted that different systems might have slightly different display formats; however, the general concepts are the same. Refer to Figure 3-2-34 during the following discussion. The address of each system is shown in the two left columns of the BITS mode display. The left-most column is a decimal (DEC) address; the next column is the hexadecimal (HEX) address. To the right of each address is the data field for that system. The data field shows the digital data of the given system (1s and 0s). Each data field is made up of 12 bits shown in three groups of four. The least significant data bits are shown in the right-most column (bits 3, 2, 1, and 0). The most significant digits are shown on the left.

Address 097 and 225 are displayed on the top two lines (Figure 3-2-34). Remember the DH knob can be used to scroll through different address numbers. Address 097 defines IAS for the right side and 225 is IAS for the left side. Table 3-2-3 is an example of a typical MPR RAM description table for a Collins EFIS-85/86. This table shows the on-side and off-side decimal addresses and the values of the related binary data field bits. This table must be used to decode the data. In our example, IAS (address 097/225) is shown at the top of the table.

To interpret the meaning of a specific data field, simply determine the value of each bit as given in the table. The data for address 097 is 0100 0011 0100 (Figure 3-2-34). Only bits set to binary 1 will have value. Bits set to binary 0 will have no value. Figure 3-2-35 shows a sample of the decoded data for IAS (address 097). To determine the numerical value of a given data field add all the values of each bit set to binary 1. In this example bits 10, 5, 4, and 2 are set to binary 1. Therefore, the values of 256, 8, 4, and 1 are added to find the total value of the data field (269 knots). The indicated airspeed value calculated by the on-side processor is therefore 269 knots. Decoding information in this manner may be extremely helpful during system troubleshooting. If the pilot reports an indicated airspeed error, the technician can use BITS mode to look *inside* the processor and determine if the correct signal has reached the DPU or MPU. BITS

EFIS-85B MULTIPORT RAM (MPR) ALPHABETICAL INDEX	
On-Side Address (Decimal)	**MPR Description**
124	FD ANNS
075	FD Modes
017	Flag Inputs, Discretes
121	Full-time GS DEVN (DOTS)
120	Full-time LOC DEVN (DOTS)
010	Glideslope DEVN (DOTS)
110	GND SPD (BCD-1000,100,10), LRN
028	GND SPD (Binary), ACT LATL Source
011	GND SPD (Binary), DME
121	GS DEVN (DOTS), Full-time
102	HDG Angle (ACT)
071	HDG Angle (RAW)
031	HDG Error (ACT Filtered)
109	HDG Error (ACT), Selected
067	HDG Error (RAW)
089	HDG, DCP Selected
039	HSI Mode Discretes
123	HSI VNAV DEVN, Scaled
097	IAS (Filtered), ADC
107	IAS (RAW), ADC
119	IAS Display Base Offset in VGUs
113	IDENT, DME
055	Joystick Cursor BRG
056	Joystick Cursor DIST
019	LATL ACT ANNS
116	LATL DEVN (DOTS), ACT
020	LATL Preset ANNS
022	LATL Source TTG (Binary), ACT
043	LATL Source BRG-to-WPT (ACT)
009	LATL Source DEVN (VGU), ACT
046	LATL Source DEVN, Preset
040	LATL Source DIST (Binary), ACT
028	LATL Source GND SPD (Binary), ACT

Indicated Airspeed (IAS) ← (points to 097)

Figure 3-2-33. Alphabetical list of BITS mode address and system titles

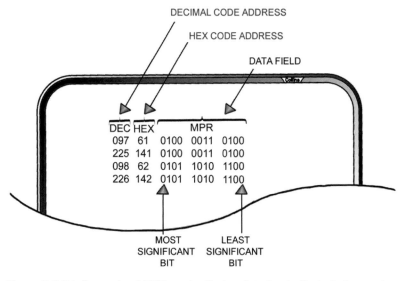

Figure 3-2-34. Example of BITS mode display showing indicated air speed (IAS) data

LOCATION (ADDRESS) ON-SIDE/OFFSIDE	DESCRIPTION
097/225	IAS
	11 - (512) KNOTS
	10 - 256 KNOTS
	9 - 128 KNOTS
	8 - 64 KNOTS
	7 - 32 KNOTS
	6 - 16 KNOTS
	5 - 8 KNOTS
	4 - 4 KNOTS
	3 - 2 KNOTS
	2 - 1 KNOT
	1 - 0.5 KNOT
	0 - 0.25 KNOT
098/226	TAS
	11 - (1024) KNOTS
	10 - 512 KNOTS
	9 - 256 KNOTS
	8 - 128 KNOTS
	7 - 64 KNOTS
	6 - 32 KNOTS
	5 - 16 KNOTS
	4 - 8 KNOTS
	3 - 4 KNOTS
	2 - 2 KNOTS
	1 - 1 KNOT
	0 - 0.5 KNOT

Table 3-2-3. Multiple RAM numerical index   *For training purposes only*

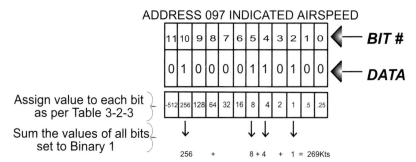

Figure 3-2-35. Method used to decode BITS mode data, in this example IAS equals 269 knots

mode is another tool for technicians to use during flight line troubleshooting.

Other MPR data can be analyzed in a manner similar to the one just described. Table 3-2-4 shows the MPR locations and descriptions for the flight director modes (location 074/202). A binary 1 displayed for a given bit indicates the description is there as labeled. For example, if location 074 bit number 13 is 1 it would indicate an *ELE OUT OF TRIM* (elevator out of trim) condition is detected by the left side flight director. If bit 13 was set to 0 it would indicate an elevator out of trim condition did not exist (i.e., the elevator trim was within tolerance).

Fault isolation using BITS mode. During troubleshooting if the technician suspects a faulty component transmitting to EFIS it can be verified using the BITS mode. Let us assume the pilot reports an indicated airspeed (IAS) disagree message on the EFIS display (remember a disagree flag will be displayed in yellow). The technician could take the following steps:

1. Find the location code for IAS in the MPR alphabetical index (Figure 3-2-33). The IAS data comes from the Air Data Computer (ADC).

2. Enter the MPR test function for the pilot's side DPU. Page through the data until location 097/225 is displayed.

3. Determine if the binary bits for the right and left side IAS agree or disagree. (Keep in mind the data transmitted from various sensors will most likely have some tolerance. To be considered in disagreement, the data must differ by more than the tolerance allows.)

4. If the values agree, configure the aircraft to various airspeeds (using ground test equipment) and determine if the values agree at various airspeeds. If the bit values for IASs consistently agree, the fault must be in one of the EFIS processors.

5. To determine which processor may be faulty, use the switching procedures previously discussed under troubleshooting techniques.

6. If the right- and left-side IASs disagree as measured in step 4, the fault must be in one of the ADCs (or the related wiring).

7. To determine which ADC is faulty, the IAS values on the ground test equipment can be compared to the values presented on the BITS mode display. The faulty LRU is whichever system (left or right ADC) disagrees with the ground equipment.

The BITS mode is an extremely valuable troubleshooting tool, which can be used for in-depth troubleshooting of EFIS faults. In many respects, the BITS mode is similar to a "built-in" data bus analyzer. Whenever using BITS mode, remember that the data is real time and there is no data memory if power to the system is interrupted. If an in-flight fault occurs, the power to EFIS should be left on so a technician can examine the BITS mode or fault flag caused by the fault. If the fault is intermittent the bits data and fault flags will only appear when the fault is active.

Also, keep in mind the BITS mode looks at data stored in the MPR. This data has been manipulated by the I/O section of the processor and is not necessarily identical to what is being transmitted to the processor. In many cases, what is transmitted to the DPU/MPU is an analog signal converted to digital for use in the processor. If you wish to look specifically at what is being transmitted to the I/O section of the DPU/MPU, measure the transmitted signal directly at the DPU/MPU connector plug. Collins breakout box (CTS-9) has all the necessary connections to allow access to the 160 pin connectors found on the back of the EFIS processors. Using a breakout box, the technician can keep EFIS operable and still access the pin connectors on the back of the processors.

Bendix/King EFS-10

The Bendix/King EFS-10 was a first generation system designed for corporate and commuter-type aircraft. The EFS-10 was the first in a series of stand-alone electronic flight instrumenta-tion systems designed by Bendix/King. The EFS-40 and EFS-50 were the next systems to be developed and incorporated several changes to improve performance, cooling, and reliabil-ity. The general architecture of the EFS-40/50 remains similar to the EFS-10. Bendix/King chose the acronym EFS (electronic flight sys-tem) for the designator of their systems. The EFS-10, -40, and -50 models are similar to the Collins EFIS-85/86 discussed previously in this chapter. Due to the overwhelming similarities between the two systems, the Bendix/King EFS discussion will be brief and concentrate on the unique features of the system. It should be noted that Bendix/King has gone through several corporate mergers since the develop-ment of the EFS-10. Originally the company was a sole entity of Bendix/King Avionics; the company was sold to AlliedSignal Aerospace division. In late 1999, AlliedSignal merged with Honeywell, and at the time this text was written Bendix/King remains a light aircraft division of Honeywell Aerospace. Most people refer to the EFS-10 as a Bendix/King system. Under any name, the EFS-10 and the EFS-40/50 are reliable systems, which are extremely pop-ular in a variety of corporate aircraft.

System Description

The EFS-10 can be configured to match the needs of the aircraft, and provide various options selected by the owner. The system is available in a simplified version containing only one processor and two displays (one-EHSI, one-EADI). This system provides very

DPU/MPS MULTIPORT RAM (MPR) LOCATIONS WITH DESCRIPTIONS				
DECIMAL LOCATION NO. 1 SIDE/ NO. 2 SIDE	HEX LOCATION NO. 1 SIDE/NO. 2 SIDE	SOURCE	DESTINATION	DESCRIPTION
074/202	4A/12A	DIGITAL I/O PROCESSOR	ANALOG I/O PROCESSOR	FLIGHT DIRECTOR MODES APS-65/85/86 MODES 15 – RUD OUT OF TRIM 14 – AIL OUT OF TRIM 13 – ELE OUT OF TRIM 12 – YD DISCONNECT 11 – AP DISCONNECT 10 – V-BAR IN VIEWS 9 – STEERING VALID 8 – NOT DEFINED 7 – HEADING SYNC 6 – HEAVY 5 – MEDIUM 4 – LIGHT 3 – YD ENGAGE 2 – AP ENGAGE 1 – DESCENT 0 – NOT DEFINED

**Table 3-2-4. Multiport RAM locations and description

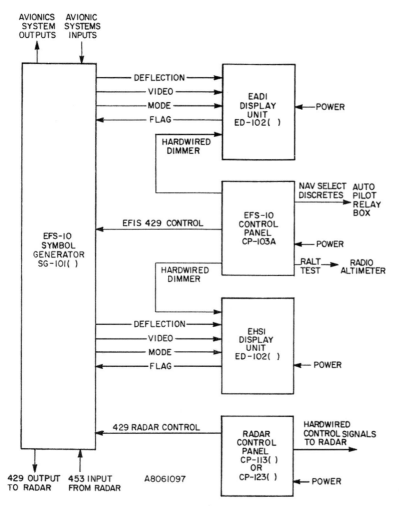

Figure 3-2-36. EFS-10 three-tube block diagram

Courtesy of Allied-Signal Aerospace Company

limited backup; however, it is the least expensive. The processor unit for the EFS-10 is called a symbol generator (SG). The SG receives inputs from various aircraft systems. The SG processes the data and creates the video and deflection signals, which are sent to the display units (Figure 3-2-36). The SG can also process data, and supplies information to the autopilot and other navigational systems. The majority of the data transmitted to and from the SG is digital ARINC 429 data. Radar data is transmitted in the ARINC 453 format. This system can be configured with either one or two control panels.

Another popular version of the EFS-10 is configured with two symbol generators. This is often referred to as a four-tube system since there are four display units. Bendix/King refers to their display units as electronic displays (ED). The EFS-10 is typically equipped with the ED-102 display. This configuration allows both the pilot and co-pilot to have an electronic HSI and ADI. The four-tube system employs two symbol generators (Figure 3-2-37). The installation of a second SG provides for enhanced reliability. The SGs cross talk with information on pertinent data. Comparison checks are made and the system can flag any discrepancies.

The most popular configuration for the EFS-10 is the five-tube system. In this arrangement there are three SGs, two EADIs, two EHSIs, and one MFD. The MFD is used to display weather radar, checklist data, and provide

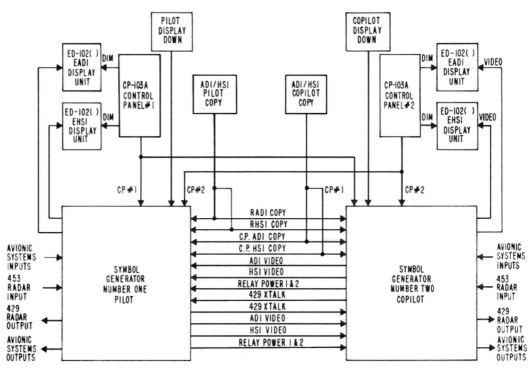

Figure 3-2-37. EFS-10 four-block diagram

Courtesy of Allied-Signal Aerospace Company

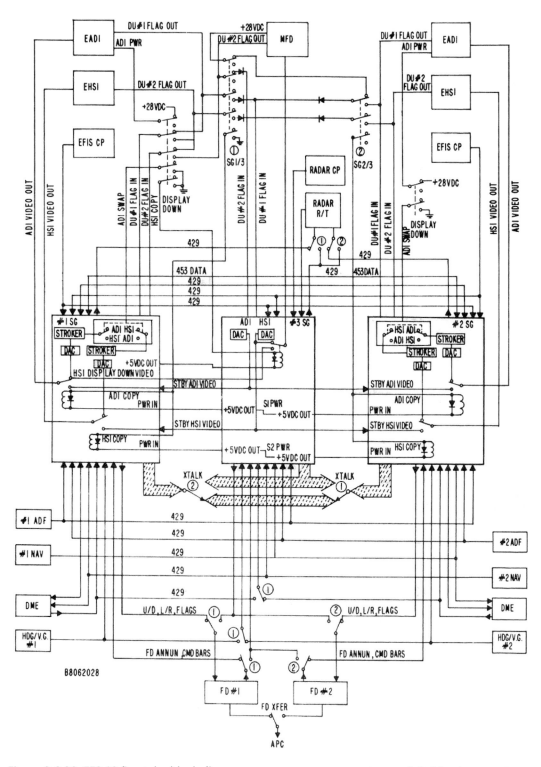

Figure 3-2-38. EFS-10 five-tube block diagram

Courtesy of Allied-Signal Aerospace Company

backup for the EHSI. The MFD receives video information sent from the third (center) SG. As seen in Figure 3-2-38, the five-tube EFS-10 enhances reliability through the use of cross talk busses for all three SGs, and a variety of switches used to control backup operations. These switches are called *standby switches*. Near the middle of the diagram notice the SGs are represented by three rectangles. The

left SG is #1, the center is #3, and #2 is on the right. Look carefully to find the SG number at the top of each rectangle. Notice the cross talk (*XTALK*) bus is located just below the SGs and is a double wide shaded arrow. The cross talk bus is currently switched to connect SG #1 and SG #2. SG #3 is currently being used for MFD operations and provides no comparator monitoring. Below the SGs are the navigational

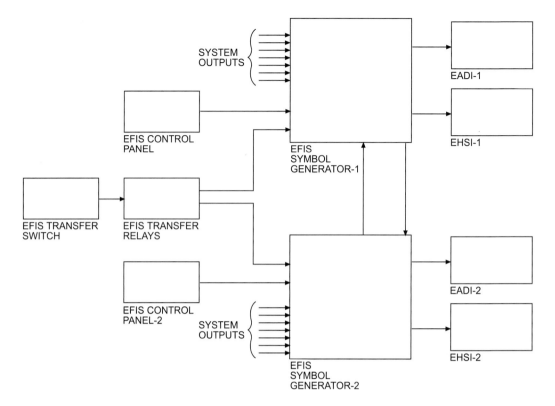

Figure 3-2-39. Boeing 737-300 EFIS block diagram

Courtesy of the Boeing Commercial Airplane Company

radios and the flight directors (FD #1 and #2 at the bottom of the diagram). Notice the flight directors can receive input data from either SG #1 or SG #2. The five displays are located at the top of Figure 3-2-38. On the diagram, the four main reconfiguration switches are located between the displays and the SGs. The two *DISPLAY DOWN* reconfiguration switches are used to move the pilot's or co-pilot's EADI into the EHSI position in the event of a system failure. If you look at the switch carefully, you will notice that 28 VDC power is removed from the EADI when the *display down* switch is activated. This turns off power to the EADI. In that same general area of the diagram are the two symbol generator reconfiguration switches. The switch labeled SG 1/3 enables the pilot to transfer left side EADI and EHSI operations from SG #1 to SG #3. The switch labeled SG 2/3 enables the co-pilot to transfer right side EADI and EHSI operations from SG #2 to SG #3. Reconfiguration switches are typically lighted push-button switches located on the instrument panel near the EFS displays. The reconfiguration switches are *not* part of the EFS control panels.

Comparing EFS-10 and EFIS-85/86

Major differences between the Bendix/King EFS-10 and the Collins EFIS 85/86 are apparent when comparing both companies five-tube systems.

- The EFS-10 contains no push-button controls on the MFD. The EFIS 85/86 MFD has line select keys and a joystick control mounted on the perimeter of the MFD. On the EFS-10 system all MFD controls are installed on a separate EFS control panel; therefore, all five displays are typically interchangeable.

- The Bendix/King (EFS-10) system uses three identical symbol generators that are completely interchangeable. The Collins (EFIS 85/86) employs only two interchangeable processors, the DPUs (display processor unit); the third processor is an MPU (multifunction processor unit). The Collins MPU is not interchangeable with the DPUs.

- The two systems differ in the event of a display (EHSI or EADI) failure. If a display unit fails on the EFS-10, the crew would press the *display down* switch and the EADI would move to the lower display. On the EFIS 85/86 when a display unit fails the crewmember would select the *compact mode*, which shows both the EADI and EHSI information on the operable CRT in a compact format.

EFS-10 Troubleshooting

The EFS-10 has three basic levels of fault isolation.

1. The system performs continuous fault monitoring and shows any appropriate fault flags on the display units.

2. The pilot or technician can activate a pre-flight self-test that runs a sequence of system tests to verify correct operation.

3. EFS-10 has a *maintenance page* format that allows the technician to check system configuration, system strapping, radar straps, software codes, LRU status tests, and activate a test pattern.

The various troubleshooting modes are activated through the EFS control panel by rotating the function select switch to the test (TST) position. The various test options are then displayed on the CRT. Simply follow the prompts on the display to retrieve the appropriate data.

The EFS-10 employs strapped connections to ensure proper system configuration. Like the Collins EFIS 85/86, the EFS-10 can interface with a variety of different systems. During installation or system modification, the correct strap connections must be made to ensure the EFS-10 operates properly. Be sure to consider improper strapping when troubleshooting defects. If a strap wire comes loose, the system is no longer configured properly and may not function. Configuration strapping can be checked using the *maintenance pages* option. Like all electronic systems, cooling is critical for the EFS-10. All displays and all symbol generators contain a cooling fan. During inspection and/or maintenance, be sure these fans are clear from obstruction and operating properly.

Transport Category First Generation EFIS

The first generation EFIS found on transport category aircraft are very similar to the EFIS 85/86 systems described earlier in this chapter. The transport category system studied here will be the EFIS found on the Boeing 737-300. Other transport category aircraft that employ first generation EFIS include the Boeing B-757, B-767, some B-737s, the Airbus A-300, and the McDonnell Douglas MD-80. Keep in mind; general concepts of the B-737 EFIS can be applied to virtually all first generation EFIS found on transport category aircraft. Only a brief description of the systems will be presented to avoid repetition from previous sections of this chapter.

B-737-300 EFIS General Description

The B-737 EFIS employs four display units used for the pilot's and co-pilot's EADI and EHSI. The displays employ a typical color CRT design each weighing approximately 23 pounds and powered by 115 VAC at 400 Hz. Located on the rear of each display is an inlet and outlet port for cooling air. The aircraft's forced-air cooling sends air through each display. The inlet port contains a fine-mesh filter assembly that must remain clean for proper unit cooling. As with the Collins 85/86 systems, only flight data is displayed; the B-737 EFIS cannot display engine parameters or airframe systems information. The 737-300 contains a separate electronic display system for engine and airframe system data. As seen in Figure 3-2-39, the main components of the system include two EADIs, two EHSIs, and two symbol generators. The symbol generators are LRUs, which process data from the various aircraft systems and provide data to the four EFIS displays. The symbol generators perform basically the same functions as the display processors units found on the Collins EFIS 85/86 or the symbol generators of the Bendix/King EFS-10.

The symbol generators receive data from a variety of aircraft systems, such as attitude sensors, air data computers, navigational radios, and the EFIS control panels (Figure 3-2-40). Most of the information sent to the symbol generators is transmitted as an ARINC 429 digital data signal. The symbol generators also receive weather radar (ARINC 453) and discrete signals from various sources. The symbol generators monitor the built-in-test equipment (BITE) of each display to ensure the health of the four CRTs. The symbol generator monitors the cooling function of each display unit. If a display reaches the first over-temperature threshold the symbol generator turns off the raster signal to that display and part of the video is lost. This helps to cool the display unit since the CRT is using less power. If the second over-temperature threshold is reached, the symbol generator will turn off both the raster and the stroke signals and the display goes blank. (The raster generator provides the background video, such as the attitude ball and weather images. The stroke generator provides the signals to generate lines and characters, such as arrows, letters, and numbers.)

Pin programming is used to set various display parameters and to identify each unit as symbol generator number one or number two. Pin programming is identical to configuration strapping mentioned during the discussion of the Collins 85/86 EFIS. For more information on pin programming (configuration strapping), refer to the previous sections of this chapter. The symbol generators also contain BITE circuitry to record EFIS faults. BITE access will be discussed further during system troubleshooting.

Component location and operation. The symbol generators are located in the aircraft's

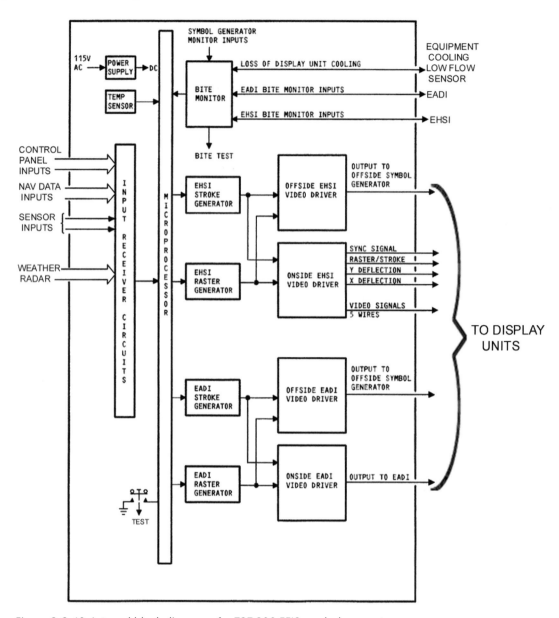

Figure 3-2-40. Internal block diagram of a 737-300 EFIS symbol generator

Courtesy of the Boeing Commercial Airplane Company

electrical equipment compartment, weigh approximately 21 pounds and receive power from an 115V, 400 Hz bus. Symbol generator number one normally drives the left side EADI and EHSI. The number 2 symbol generator normally drives the right EADI and EHSI. In the event of a unit failure, either the right or left symbol generator can drive all four displays. The switching for the symbol generators is controlled by two EFIS transfer relays located in the electronic equipment compartment installed behind the flight management computer. The EFI switch located on the flight deck controls the relays.

During the following discussions refer to Figure 3-2-41 to determine EFIS component locations. The EFI switch, used to select a symbol generator, is located on the pilot's overhead panel. The main EFIS control panels are located on the pedestal between the pilot and co-pilot seats. There are two EFIS control panels; the pilot and co-pilot can each select a different format for their EFIS displays. The EFIS control panel is divided into two sections. On the left are the controls for the ADI. The decision height (DH) selector and display are located on the top left side of the panel. The ADI brightness control is located in the lower left of the EFIS control panel. The EHSI controls are located on the right side of the EFIS control panel. The mode selector allows the flight crew to select the EHSI display mode. The range switch determines the weather radar and TCAS operational range. The WXR switch turns on/off the weather radar. The five map buttons on the lower left of the display determine the map mode of the EHSI.

During normal operation each display operates using a three-color CRT. In the event one of the color guns fails, the system reverts to a monochromatic mode. If a system, which feeds the EFIS, symbol generators should fail, that data will be removed from the display and the appropriate fault flag will be presented. Most fault flags are short abbreviations of the failed system, such as ATT (displayed in red) for loss of attitude data. Both the pilot's and co-pilot's displays are equipped with a remote light sensor (photo diode), which is used to ensure proper brightness of the CRT. Figure 3-2-41 shows the light sensors located on the glare shield of the aircraft facing forward in order to monitor ambient light. Each display unit also contains a local light sensor to monitor cabin light. The symbol generator uses inputs from both the remote and local light sensors and adjusts the CRT brightness accordingly.

System maintenance and troubleshooting. On most aircraft, the display units are identical and can be interchanged. An inclinometer must be installed on any display in the EADI location and removed if the display is used as an EHSI. Pin programming is used to "tell" the display where it is installed. To remove a display, depress the handle firmly against the front of the display and remove the two screws located in the handle (Figure 3-2-42). Lower the handle to unlock the display and pull firmly. To reinstall the display, reverse the procedure. Be sure to depress the handle whenever installing or removing the screws. This will help protect the screws from being stripped. Both symbol generators are also identical and can be swapped for maintenance or troubleshooting purposes.

The 737-300 EFIS is equipped with BITE circuitry, which allows for access to fault data for maintenance purposes. The BITE information is accessed through the right or left control display unit (CDU) located in the center console between the pilot's and co-pilot's seats. The CDU is an alphanumeric keypad and display used primarily for operation of the flight management system (FMS). The CDU will be discussed in greater detail later in this text. The EFIS BITE is part of the symbol generator circuitry; however, it is accessed through the FMS computer and only when the aircraft is on the ground. The BITE circuitry contains four categories:

1. In-flight faults

2. Ground faults

3. Current status, and

4. Discrete status

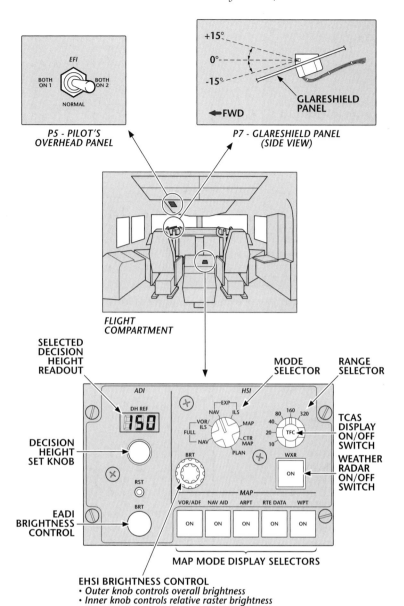

Figure 3-2-41. Locations of various EFIS components on the 737-300

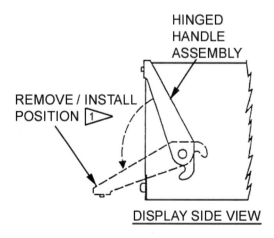

Figure 3-2-42. Removal and installation techniques for a typical Boeing display unit

Courtesy of the Boeing Commercial Airplane Company

Figure 3-3-1. Instrument panel of the A-320

In-flight faults are faults, which occur whenever the aircraft is airborne. The EFIS symbol generators continuously monitor the system and its various input/output signals. If a fault occurs, it is stored in a nonvolatile memory for up to nine consecutive flights. The in-flight faults display will indicate whether the fault was intermittent or continuous. Ground faults can store up to seven pages of faults, which occurred while the aircraft was on the ground. The faults are listed in alphabetical order. *Current status* is used to test the various components of the EFIS and related interfaces. It should be noted that if an in-flight fault shows as intermittent, it might appear to be functioning properly during the current status check. *Discrete status* is used to test any discrete signals that connect to EFIS. Access of the B-737 EFIS BITE information is a menu driven process similar to BITE data from other systems. BITE access will be covered in greater detail in Chapter 5.

If the EFIS symbol generator should fail, the BITE information for that system is inaccessible. During a failure of the symbol generator processor circuitry, the message *SG FAIL* will be displayed on the EADI and EHSI. If symbol generator power should fail, any related EADI and EHSI will go blank.

During initial power-up of the EFIS, a self-test is initiated. If a fault is detected, the appropriate fault flag will be displayed. During this self-test, the EFIS pin programming is also checked for accuracy. A parity pin is used to verify the correct pin programming of the system. If a pin programming error exists, both displays go blank and a message *PARITY ERROR* is displayed. If a pin programming error exists, the BITE for that symbol generator remains operational.

Section 3

Second Generation Electronic Instrument Systems

In the late 1980s a new generation of electronic flight instrument systems was beginning to emerge. These second generation EFIS were designed to allow for better flight deck management, improved reliability, and maintainability. Improved computer software allowed the integration of the aircraft's flight management, engine indicating, central maintenance computer, and the electronic flight instrument systems. Although many of these systems take on different names or features for different aircraft, the end result is still the same: a fully integrated electronic instrument system.

Today, second generation electronic instrument systems (EIS) can be found on a variety of transport category aircraft including the Boeing 747-400 and the Airbus A-320. Many corporate and commuter aircraft also employ advanced EIS, such as, the Gulfstream G-5, the Beechjet 400A and the Cessna Citation X. Obvious changes over early EFIS systems are the larger displays and a less cluttered instrument panel. Figure 3-3-1 shows the instrument panels of the A-320; note the six IDS (instrument display system) displays.

The Beechjet 400A Electronic Instrument System

The electronic instrument system used in the Beechjet 400A is a fully integrated system

Figure 3-3-2. Typical four-display EFIS system in a corporate jet aircraft   *Courtesy of Rockwell International, Collins Avionics Divisions*

which incorporates state-of-the-art CRT or LCD displays. The Beechjet can be configured for a two-, three- or four-display system. In the two-display system, one electronic flight display (EFD) and one multifunction display (MFD) are installed on the pilot's side of the instrument panel. In a three-display system, one EFD is installed for the co-pilot; one EFD and one MFD are installed on the pilot's side. The four-display system can be configured with three EFDs and one MFD or two EFDs and two MFDs. It should be noted this system also employs two monochrome backup displays located outboard of the main displays. The material presented in this portion of the text will discuss the four-display (three EFD, one MFD) configuration as seen in Figure 3-3-2.

The Pro Line 4 systems installed in the Beechjet 400A have many similarities to the earlier version electronic flight instrument systems (Pro Line II). For example, both systems:

1. Utilize CRT displays for navigational and attitude data

2. Employ MFD for display redundancy

3. Interface with various aircraft systems using digital data busses

4. Interface with the aircraft's autoflight system

5. Have a built-in diagnostics

The advanced EIS has improved on many of the early system's shortcomings. Some of the changes made to the advanced EIS include:

1. New CRT displays that are larger and easier to read

2. Control display units (CDUs) using a central control center for EIS and flight guidance data selection

3. Integrated avionics processor providing system integration for improved interfacing capabilities

4. Built-in diagnostics that are easier to interpret and provide improved reliability

Major differences between the first- and second-generation systems are presented in the following section.

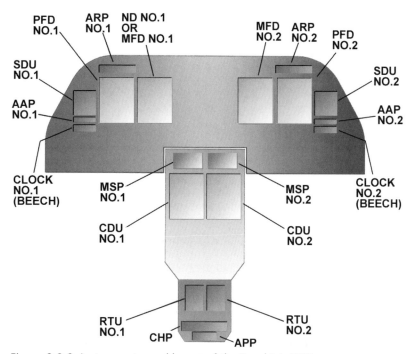

Figure 3-3-3. Instrument panel layout of the Beechjet 400A

Courtesy of Rockwell International, Collins Avionics Divisions

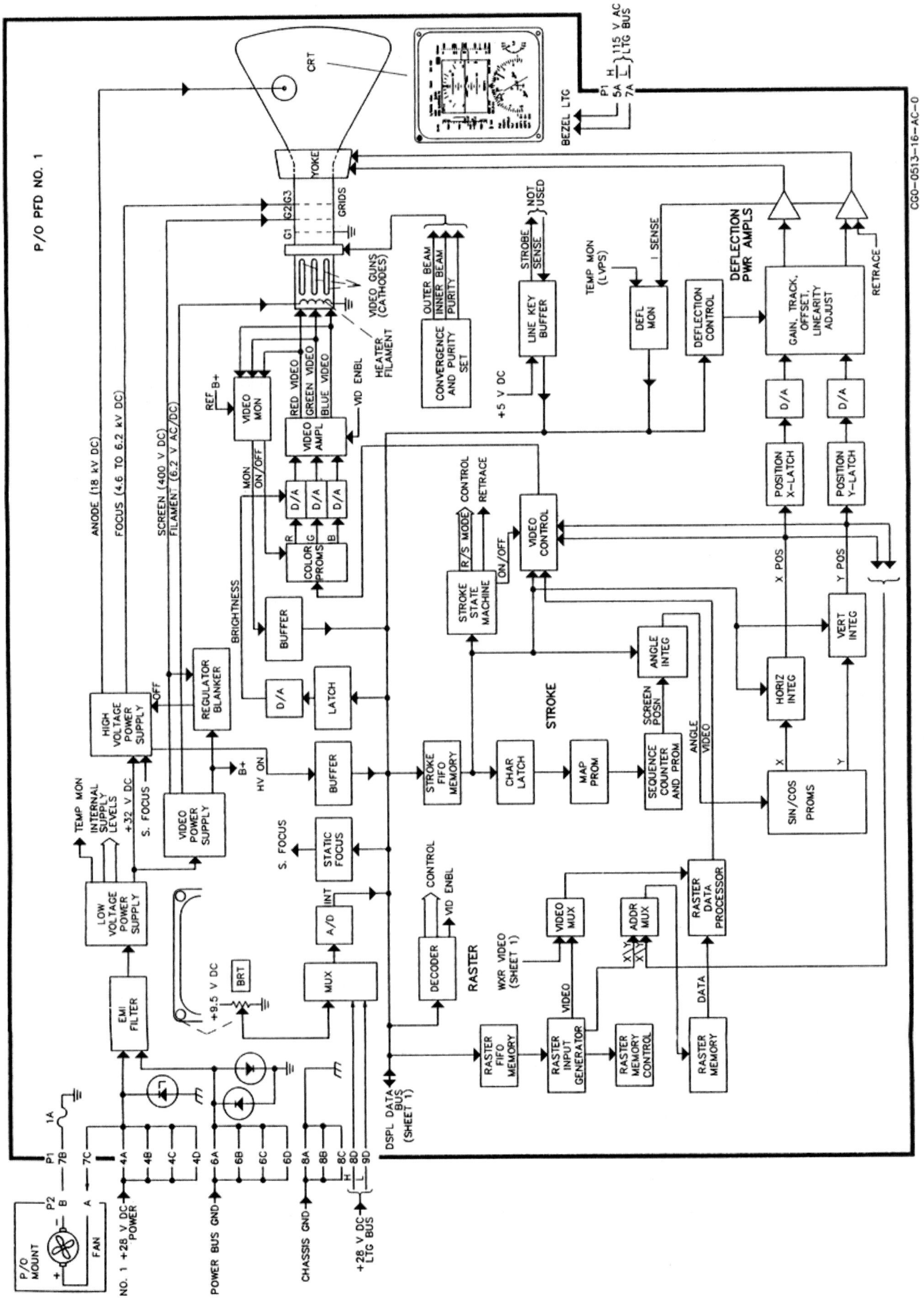

Figure 3-3-4. Internal block diagram of a display unit

Courtesy of Rockwell International, Collins Avionics Divisions–for training purposes only

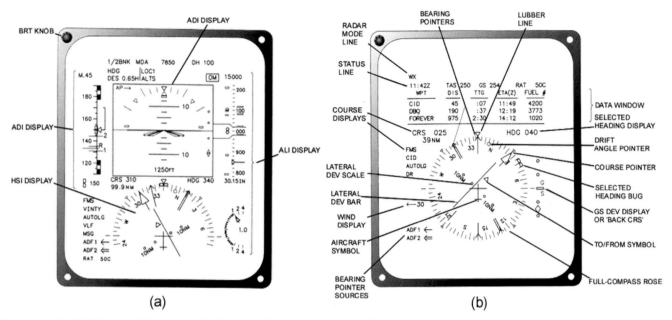

Figure 3-3-5. (A) Primary flight display (PFD), and (B) Navigational display (ND)

System Components

In general, the individual components of the Pro Line 4 system are lighter, more powerful, and use less electrical power than their predecessors use. Some of the components' internal functions have been combined into integrated avionics processor system (IAPS). Many of the redundant LRUs, such as the display units, are still interchangeable which is very helpful when troubleshooting the system.

System displays. The Beechjet 400A system is referred to as the IDS. The IDS contains four main displays as shown in Figure 3-3-3. The PFD is the primary flight display. There is one PFD for both the pilot and co-pilot. The MFD is the multifunction display. This unit is typically installed on the co-pilot's side only; however, the pilot's side may also employ an MFD instead of a navigational display (ND). There is one ND installed on the pilot's side of the instrument panel. The MFD is used as a backup for the PFD in the event of a failure.

Any of the PFDs and NDs is completely interchangeable for maintenance purposes. If two MFDs are installed, they are also interchangeable. The specific installation locations are determined by configuration strapping. The displays on this advanced system differ from the EFIS-85/86 in that they contain their own power supplies as well as video circuitry. (The EFIS-85/86 displays operated in conjunction with a DPU power supply.) Figure 3-3-4 shows an internal block diagram of a display unit. The top left corner of the diagram shows the dedicated fan circuitry. Each display unit has a

cooling fan that operates any time the display receives power.

As can be seen in Figure 3-3-5, the PFD combines the HSI and ADI displays on one CRT. This provides the pilot with complete navigational data in one location. The ND is used to display lateral navigation data, course map displays, and radar information. For both the PFD and ND, several variations of the display formats are pilot selectable.

On the Pro Line 4 system there are two backup displays called sensor display units (SDU).

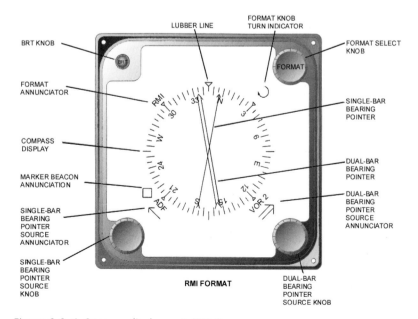

Figure 3-3-6. Sensory display unit (SDU)

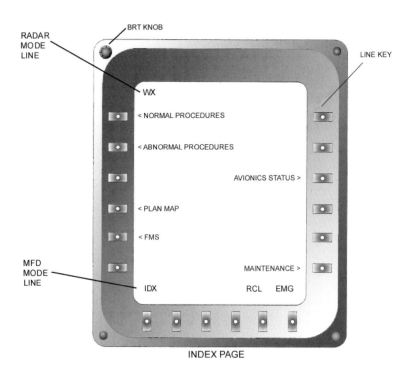

RADAR MODE LINE

BRT KNOB

LINE KEY

WX

< NORMAL PROCEDURES

< ABNORMAL PROCEDURES

AVIONICS STATUS >

< PLAN MAP

< FMS

MFD MODE LINE

MAINTENANCE >

IDX

RCL   EMG

INDEX PAGE

Figure 3-3-7. A multifunction display unit (MFD)

Courtesy of Rockwell International, Collins Avionics Divisions

These displays are located outboard of the pilot's and co-pilot's PFDs (Figure 3-3-3). The SDUs are monochrome displays used to show basic compass and navigational data. The SDUs receive their input data from an independent processor; allowing for operation free of any possible main display system failures. A sensory display unit showing the RMI format is shown in Figure 3-3-6.

The multifunction display units contain 18 line select keys (LSK) located around the perimeter of the CRT. The LSKs are used to select a given function from the MFD display. For example, to select the *AVIONICS STATUS* function, (Figure 3-3-7) press the third LKS from the top on the right side of the display. The LSKs will perform different functions according to the information currently displayed by the MFD. Similar to the PFD or ND, the MFD contains its own processor and power supply circuitry. A dedicated cooling fan is also used for the MFD.

Many of the newest electronic instrument systems are being equipped with flat-panel displays that offer several advantages over conventional CRT displays. For a review of flat-panel displays see Chapter 2.

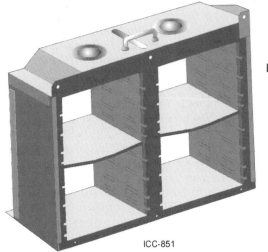

IAPS CARD CAGE

ICC-851

VARIOUS LINE REPLACABLE MODULES

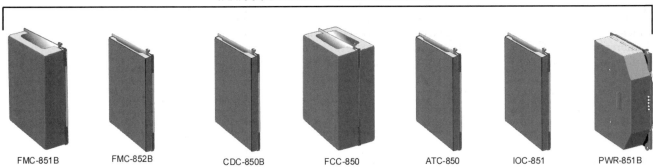

FMC-851B   FMC-852B   CDC-850B   FCC-850   ATC-850   IOC-851   PWR-851B

Figure 3-3-8. Components of the integrated avionics processor assembly

Courtesy of Rockwell International, Collins Avionics Divisions

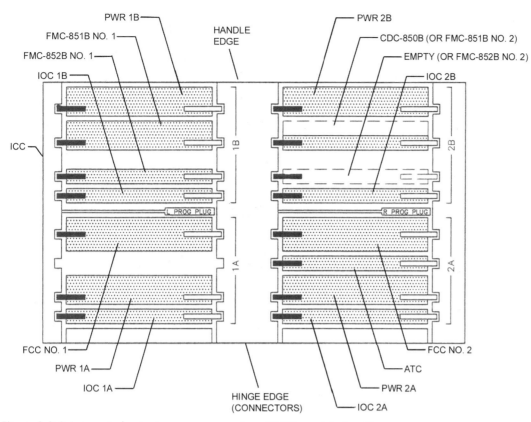

Figure 3-3-9. Integrated avionics processor system (IAPS) module locations

Courtesy of Rockwell International, Collins Avionics Divisions–for training purposes only

Integrated avionics processor assembly. As mentioned earlier, the Beechjet 400A employs an integrated avionics processor system (IAPS) that provides a central location for integration and interfacing of the avionics equipment. This integration design concept is one of the major changes found in second-generation display systems. The IAPS functions similar to a distribution network. Data from the various avionics equipment is sent to IAPS. The input data is monitored, checked for integrity, sequenced in the correct order, and transmitted to the display units and flight management/control system. The integrated avionics processor assembly consists of an IAPS card cage (ICC) and several line replaceable modules (Figure 3-3-8). This configuration allows for replacement of individual modules without affecting the entire processor unit. The modules perform various power supply, computing and data distribution functions. The six types of modules found in the IAPS include:

1. FMC (flight management computers)

2. CDC (control display coupler)

3. FCC (flight control computer)

4. ATC (automatic trim coupler)

5. IOC (input/output concentrator)

6. PWR (power module)

The IAPS is divided into four quadrants: 1A, 1B, 2A, and 2B. Each quadrant contains its own power supply and I/O data concentrator (IOC) (Figure 3-3-9). The IAPS card cage contains slots for its respective modules, which are each held in place by a cam-latch assembly. The connectors for each module are located at the rear of the ICC and engage as the module is installed. The ICC is located in the aircraft's equipment bay. It should be noted that each power supply card contains an ON lamp that can be seen through the ICC cover. Each lamp should be illuminated any time its respective power supply is operating. This is an extremely handy feature for basic troubleshooting.

The ICC contains two environmental control cards that ensure a stable operating temperature for the line replaceable modules. The environmental control card heaters and fan assemblies are integrated into the ICC. If the IAPS temperature drops below limits the heater is turned on and the fan operates on low speed. If the temperature rises above normal operating limits, the fan operates on high speed.

IDS controls. There are seven different panels used to control the complete integrated display/autoflight system. The altitude awareness panel (AAP), the autopilot panel (APP), the air data reference panel (ARP), course heading panel (CHP), and the mode select panel (MSP)

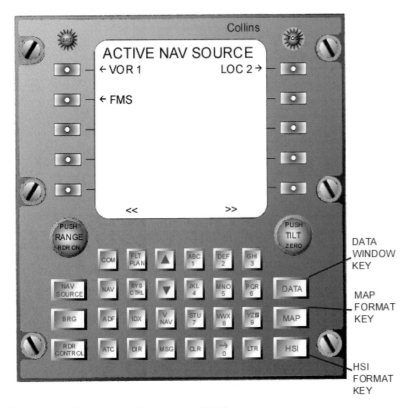

Figure 3-3-10. A control display unit (CDU)
Courtesy of Rockwell International, Collins Avionics Divisions

IDS. The IAPS contains four I/O concentrators (IOC), two on the right side and two on the left side. Each display unit receives digital data from its respective on-side IOCs. All digital data busses to the IDS use ARINC 429 format; except the weather radar, which transmits an ARINC 453 digital signal. As seen in Figure 3-3-12, the IOCs receive inputs from the various aircraft LRUs. The output data is sent directly to the flight deck displays, the flight control system (FCS), and the flight management system (FMS). IOC output data is also transmitted to other avionics systems on three general purpose (GP) busses.

Displays. The PFDs (shown in the upper portion of Figure 3-3-13) each receive critical data directly from the R/L (right and left) attitude heading computers (AHC) and the R/L air data computers (ADC). Since the PFDs also receive data from the IOC (in IAPS), this provides a redundancy for critical data. The ND also receives data from the AHCs, ADCs, and IAPS. The PFDs, ND, and MFD each have an input data bus returned to IAPS to monitor system health. The MFD receives inputs from the IOCs and FMCs within IAPS. The MFD also receives direct data from the AHCs, ADCs, CHP (course heading panel), and WXR (weather radar system).

Two sensor display units (SDU) are installed in the Beechjet 400A IDS to provide backup to navigation and sensor data. Each SDU receives information through a digital data bus from the sensor display driver (SDD). The SDD receives inputs from R/L navigation radios, both AHCs, and an IAPS IOC. The SDD is divided into two channels each channel drives one SDU.

Discrete inputs from various aircraft systems to the IDS are routed through the data acquisition unit (DAU). The DAU is shown in Figure 3-3-11 just to the right and below the right side IAPS. The DAU converts all discrete signals to an ARINC 429 format and transmits that data to IAPS. The DAU also converts digital data received from IAPS for operation of the discrete caution and warning annunciators.

all deal with various navigation and autoflight functions. These panels are used to make changes to the EFD formats according to the appropriate pilot selection. The PFDs and ND are on any time the avionics bus is active. Their brightness is controlled through a knob located in the top left corner of each display unit.

The Beechjet 400A employs a control display unit (CDU) that is the primary controller for the PFDs and ND. There are two CDUs (one pilot's and one co-pilot's) located in the center console of the aircraft; each controls their respective displays. The CDU contains 10 line select keys and 32 dedicated push-buttons (Figure 3-3-10). The *DATA* key is used to select FMS data for display on the MFD and ND. The *MAP* key is used to display a dynamic present position map on the ND and MFD. When the *HSI* button is pressed the conventional "full-compass rose" format is displayed on the MFD and ND. The CDU is also used to control a variety of flight navigation management functions.

System architecture. The Pro Line 4 electronic instrument system is thoroughly integrated with the flight management system as can be seen in Figure 3-3-11. The major integration takes place in the IAPS side 1 and 2. The IAPS receives data from the various aircraft systems through digital data busses, manipulates the data, and transmits the necessary data to the

Reversionary switching. The IDS employs reversionary switching to provide for backup of PFD displays, AHC and ADC inputs, and the CDU. The reversionary switches send discrete signals directly to the IDS as shown in Figure 3-3-13. The reversionary switches allow the flight crew to select which AHRS (discussed in Chapter 6) and AIR DATA source (1 or 2) will be used for inputs to the PFD, ND, and MFD. The PFD backup can be selected, which moves the PFD display to the ND or MFD. The CDU switch allows for selection of the on-side or off-side control display unit.

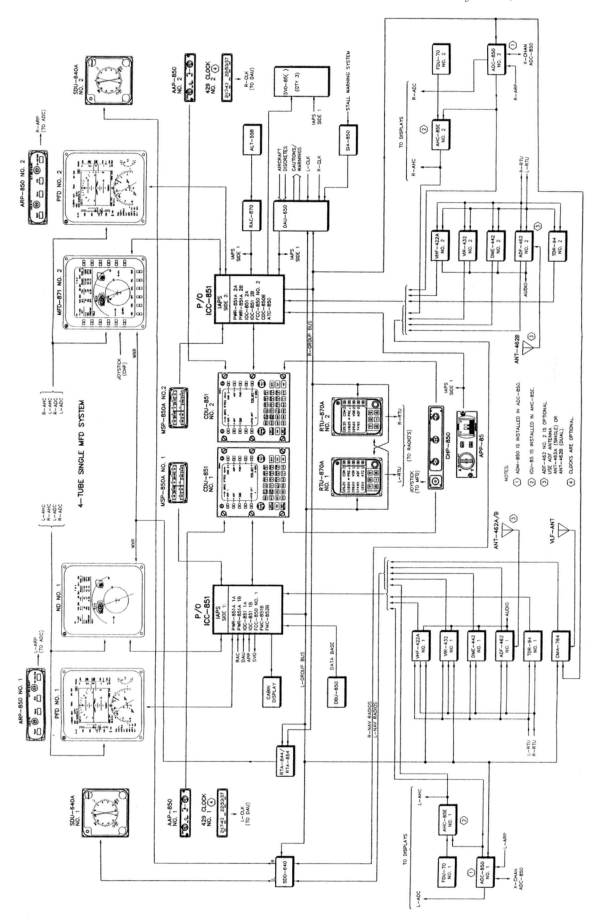

Figure 3-3-11. Beechjet 400A integrated avionics system block diagram

Courtesy of Rockwell International, Collins Avionics Divisions—for training purposes only

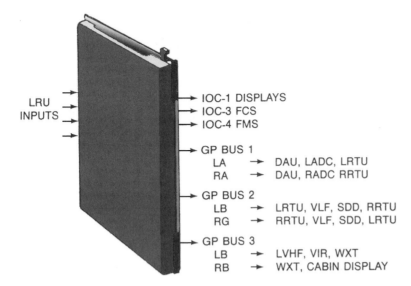

LRU
INPUTS

IOC-1 DISPLAYS
IOC-3 FCS
IOC-4 FMS

GP BUS 1
LA → DAU, LADC, LRTU
RA → DAU, RADC RRTU

GP BUS 2
LB → LRTU, VLF, SDD, RRTU
RG → RRTU, VLF, SDD, LRTU

GP BUS 3
LB → LVHF, VIR, WXT
RB → WXT, CABIN DISPLAY

Figure 3-3-12. Input/Output concentrator (IOC) interface diagram

Courtesy of Rockwell International, Collins Avionics Divisions

Separate display control switches are supplied for both the pilot and co-pilot. Therefore, the pilot and co-pilot's IDS can be operated from independent attitude/heading and air data sources. To select the off-side CDU, press the **REV** button, press **NORM** to select the on-side CDU. To move the PFD to the MFD or ND, press the **REV** button on the PFD reversionary switches.

System troubleshooting. The Collins Pro Line 4 equipment has significant improvements over the Pro Line II series. System diagnostics are still accessed through the MFD as they were on earlier systems; however, a more detailed troubleshooting menu is available. The Pro Line 4 IDS diagnostics resemble the diagnostics systems, such as, EICAS or ECAM, found on transport category aircraft. That is, Pro Line 4 avionics troubleshooting is selected from the diagnostics menu, coordinated through a central location, and displayed on a CRT.

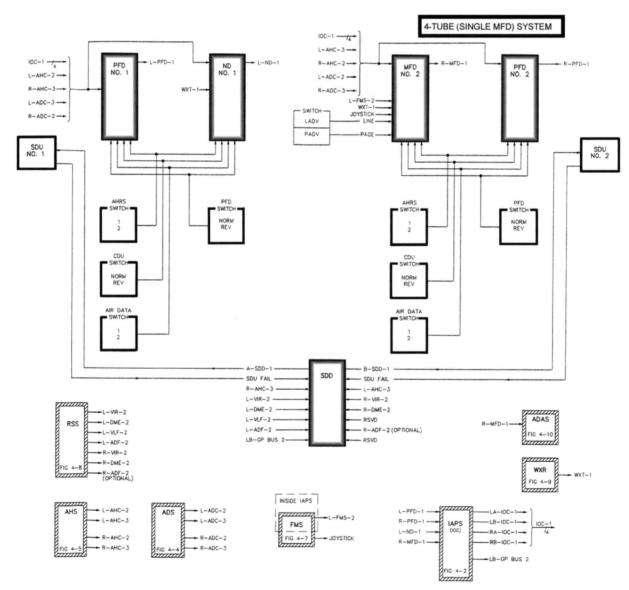

Figure 3-3-13. IDS interface diagram

Courtesy of Rockwell International, Collins Avionics Divisions

Diagnostics architecture. With any built in diagnostics system, the technician should become familiar with capabilities of the system prior to fault isolation. The diagram in Figure 3-3-14 shows the diagnostics architecture for the four-display IDS. The arrows on the diagram indicate the flow of information within the system as a request for diagnostics data is made. Refer to Figure 3-3-14 during the following discussion on the diagnostics architecture.

The flight management computer (FMC), shown in the center right side of the diagram, is the core of the diagnostics system. Each avionics system and subsystem that has internal monitoring reports to the FMC. The FMC contains the central monitoring circuitry. The FMC monitors the IOC data busses for fault messages caused by system failures. The FMC will also detect any system/subsystem that fails to transmit data. If a failure is detected the FMC

stores the information in a nonvolatile memory for retrieval at a later date. If the FMC detects a failure, the yellow message (*MSG*) light on the CDU will flash. The PFD also displays the term *MSG* displayed in yellow.

When a selection is made for diagnostics using the correct multifunction display LSK, the data word is transmitted on bus R-MFD-1 to the 2A-IOC, 2B-IOC, and the DAU (data acquisition unit). The DAU transmits the request on data bus A-DAU-2 to the opposite side IOCs. The IOCs then send the request for diagnostics data to the number 1 FMC. The FCC (flight control computer) also receives the request since the FCC manages diagnostics for the flight control system.

After receiving a request for diagnostic data, the FMC processes the request and transmits the requested data back to the MFD. The diag-

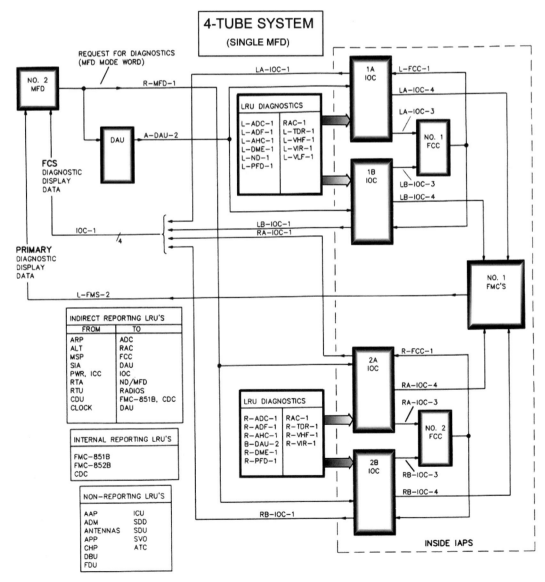

Figure 3-3-14. Diagnostics configuration four-display IDS
Courtesy of Rockwell International, Collins Avionics Divisions—for training purposes only

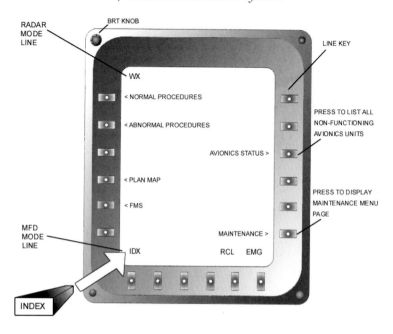

Figure 3-3-15. MFD display used to access troubleshooting of avionics systems

Courtesy of Rockwell International, Collins Avionics Divisions

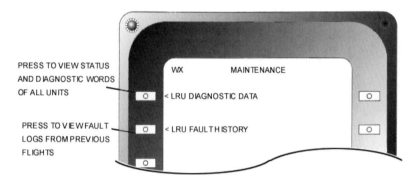

Figure 3-3-16. MFD maintenance menu page

Courtesy of Rockwell International, Collins Avionics Divisions

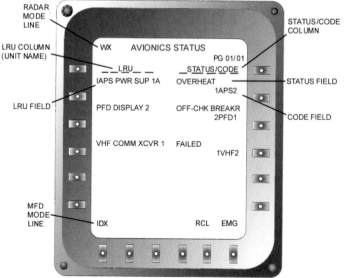

FAULTS DISPLAY ON THE AVIONICS STATUS PAGE 60 SECONDS AFTER THEY OCCUR.

Figure 3-3-17. MFD avionics status page

Courtesy of Rockwell International, Collins Avionics Divisions

nostics data is transmitted on the high-speed bus L-FMS-2 from the FMC to the MFD. If a request is made for additional diagnostic information, the cycle repeats. Recognizing this flow of data becomes important when the diagnostics request fails. For example, if the data bus L-FMS-2 should fail, the diagnostics data would never reach the MFD. In any situation where a request for diagnostics data is not displayed, check the bus structure and LRUs that interact during diagnostic requests.

Troubleshooting levels. The Beechjet 400A avionics troubleshooting is divided into two categories, Level 1 and Level 2. *Level 1* troubleshooting is considered the initial fault isolation procedure used when a problem first arises. Level 1 troubleshooting consists of a preliminary review of diagnostics data to see if the system has detected a faulty LRU. *Level 2* troubleshooting procedures encompass a detailed study of the diagnostics data to find the failed system or component. Level 2 troubleshooting requires the technician to reference the avionics maintenance manual for analysis of fault code data. This analysis will reveal problems created by faults both internal and external of the LRUs.

Whenever a fault occurs in the avionics system, the technician should begin with Level 1 troubleshooting. The first step in the troubleshooting procedure is to access the MFD index page. To access the index page press the line select key adjacent to *IDX* (Index) on the MFD (Figure 3-3-15). The index page can be accessed from any MFD display format. There are three different formats used to display diagnostics data on the Pro Line 4 system: avionics status, LRU diagnostic data, and LRU fault history. The avionics status data is accessed directly from the MDF index page. To access the LRU diagnostics data or fault history, press the *maintenance* LSK on the MFD index page. The LRU diagnostics is then selected from the MFD maintenance menu page as shown in Figure 3-3-16.

Avionics status page. The avionics status page is a real-time display of avionic systems/subsystems that experience a complete or partial failure. The avionics status page is used for Level 1 troubleshooting, and should locate the defective LRU for approximately 90 percent of the possible faults. To access the avionics status page press the LSK adjacent to *AVIONICS STATUS >* on the MFD (Figure 3-3-15). If more than one page exists in the report, access additional pages by using the joystick on the CHP (course/heading panel). To exit from avionics status, press the *IDX* LSK on the MFD.

A typical avionics status page is shown in Figure 3-3-17; refer to the figure for the following discussion. In this example, three faults are

FAULT CODE	DESCRIPTION	REASON
PFD DISPLAY 1 (CONT)		
1PFD5	IOC 2A input failed.	RA-IOC-1 bus from the 2A IOC to the large displays is inactive at the PFD input but active elsewhere.
1PFD6	IOC 2B input failed.	RB-IOC-1 bus from the 2B IOC to the large displays is inactive at the PFD input but active elsewhere.
1PFD7	ADC 1 input failed.	L-ADC-3 bus from the #1 ADC to the left large displays is inactive at the PFD input but active at the ND/MFD input.
PFD DISPLAY 2		
2PFD1	No output data; check circuit breaker	R-PFD-1 bus to the 2A and 2B IOCs inactive at both units.
2PFD2	Various circuits failed	An internal PFD fault is detected.
2PFD3	IOC 1A input failed.	LA-IOC-1 bus from the 1A IOC to the large displays is inactive at the PFD input but active elsewhere.
2PFD4	IOC 1B input failed.	LB-IOC-1 bus from the 1B IOC to the large displays is inactive a the PFD input but active elsewhere
2PFD5	IOC 2A input failed.	RA-IOC-1 bus from the 2A IOC to the large displays is inactive at the PFD input but active elsewhere.
2PFD6	IOC 2B input failed.	RB-IOC-1 bus from the 2B IOC to the large displays is inactive at the PFD input but active elsewhere.
2PFD7	ADC 2 input failed (4-tube system only).	R-ADC-3 bus from the #2 ADC to the right large displays is inactive at the PFD input but active at the MFD.

FAULT CODE 2PFD1 →

Figure 3-3-18. Fault code explanation

Courtesy of Rockwell International, Collins Avionics Divisions

currently active. If there are no active faults, the page will display *"NO FAULTS."* To keep nuisance messages to a minimum, a fault must be active for 60 seconds prior to posting on the Avionics Status page. The top line of the page contains the radar mode. In this example, the weather radar is on. The LRU column is used to list the name of the LRU suspected of failure. The MFD mode line is used to identify the function of the adjacent LSK. *IDX* is used to return to the MFD index page, *RCL* will recall the previous page, and *EMG* is used to display the emergency checklist. The *status/code* column is used to describe the failed condition and list the fault code.

The status field on the avionics status diagnostics page is a plain English message describing the suspected LRU fault. In Figure 3-3-17, the second failure listed is "PFD DISPLAY 2" and has a status field of "OFF-CHK BREAKR." This message tells the technician that the diagnostics system suspects there is no power supplied to PFD 2. The most likely place to start troubleshooting is at the circuit breaker. If the circuit breaker is found to be okay, the next logical step would be to check the power feeding PFD2. If no power is detected at PFD2 the logical conclusion is defective wiring.

Carefully inspect all related wires and connectors between the circuit breaker and PFD2. Also, be sure to verify proper ground connections.

Fault codes. The code field of the avionics status page contains a five-digit fault code, which is used to help isolate the cause of a given fault. The fault code is typically used to solve a problem only if the suggestions given by the status field were ineffective. To determine the specific fault code, the diagnostic circuitry monitors the transmitting busses of various LRUs. From this information, the diagnostics will make a determination as to the problem with the system.

An explanation of the fault codes is given in the avionics maintenance manual. As seen in Figure 3-3-18, the explanation table contains the fault code number, a description of the fault, and the reason the diagnostics reported the failure. In the example, under the subheading PFD DISPLAY 2 the fault code is 2PFD1. According to the explanation table, the reason this fault was reported was "R-PFD-1 bus to the 2A and 2B IOC's inactive at both units." The most likely reason both busses would have no data is the PFD is not trans-

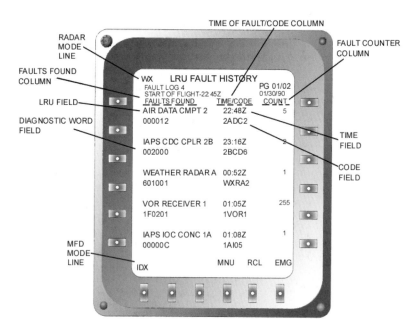

Figure 3-3-19. MFD fault history page

Courtesy of Rockwell International, Collins Avionics Divisions

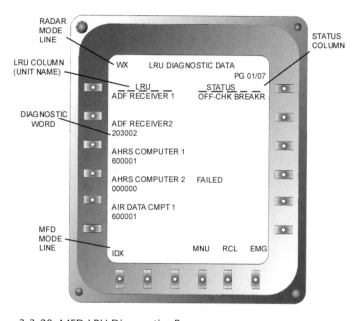

Figure 3-3-20. MFD LRU Diagnostics Page

Courtesy of Rockwell International, Collins Avionics Divisions

mitting. The most likely cause the PFD is not transmitting is a failed PFD or no power to the PFD. Hence the associated status field for this failure is "OFF-CHK BREAKER." Note, fault codes offer more information than the status message, (i.e., the fault code identifies which specific bus has failed). This information can be very helpful if the problem is wiring and not an LRU problem.

For the Beechjet 400A, the three messages that can appear in the status field of the avionics status page are:

1. *OVERHEAT* indicating an IAPS power supply has overheated.

2. *OFF-CHK BREAKR* indicating all transmitting busses from the LRU are inoperative; the diagnostics circuit assumes no power to the unit and the circuit breaker should be checked first.

3. *FAILED* indicates that the LRU is still transmitting but not on all busses, the diagnostics circuitry assumes a failed LRU.

LRU fault history. The fault history page of the IDS diagnostics is used to access a record of faults that occurred during previous flights. The fault history page is helpful when troubleshooting a reoccurring fault or intermittent problem. To view the LRU fault history page, press the following line select keys on the MFD: first press *IDX*, second press the **MAINTENANCE** key, third select **LRU FAULT HISTORY**. The LRU fault history can display up to 40 faults, which are stored in the FMC nonvolatile memory. The faults are organized into individual logs. Each log represents a previous flight. The most recent flight is log number 1.

As seen in Figure 3-3-19, the LRU fault history page includes the:

1. Flight log number

2. Time that flight began

3. Name of the suspect LRU

4. Time at which the fault was first diagnosed

5. Number of times (count) the fault occurred during that flight

6. Diagnostic word

7. Fault code

The fault code is identical to the code given for the same fault on the Avionics Status page. The diagnostic word is a six-digit code used for Level 2 troubleshooting of the Pro Line 4 avionics systems. A thorough examination of the diagnostic word will be presented later in this chapter.

LRU diagnostic data. The Pro Line 4 LRU diagnostic data page provides a real time in-depth analysis of LRU status. Each LRU that reports to the FMC is listed in alphabetical order on the LRU diagnostic data pages. The LRU diagnostics page provides the technician with a list of LRUs along with their status message and diagnostic code (Figure 3-3-20). This diagnostics page is typically accessed only if the avionics status page data does not provide the needed information to repair a fault.

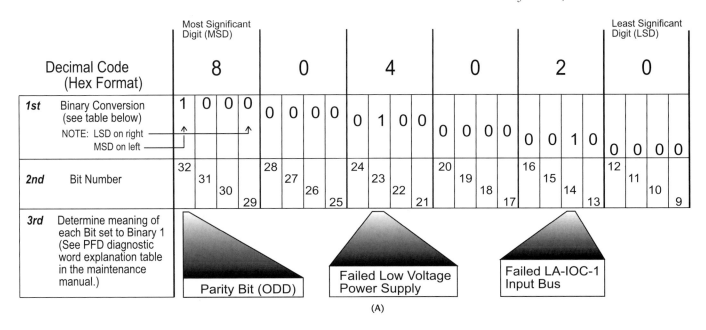

Figure 3-3-21. Diagnostic word interpretation: (A) analysis of code 804020, (B) hex to binary conversion table

The LRU diagnostics page is accessed from the MFD maintenance page. The CHP joystick can be used to scroll through different pages of the LRU diagnostics data. To exit the LRU diagnostics pages, press the index LSK on the MFD.

Diagnostic words. Diagnostic words are six-digit hexadecimal codes, which are displayed on the LRU diagnostics page and LRU fault history page. Each digit of the diagnostic code has specific meaning for the particular LRU in question. The FMC diagnostic circuitry monitors the health of an LRU's internal circuitry and its associated busses to determine a diagnostic code.

In general, the avionics maintenance manual must be used to analyze a given diagnostic code. For normally operating systems the code always contains the number 6 in the left most digit. For example the code 600001 indicates normal operation of the No. 1 unit. The rightmost digit of the code displays the LRU number (1 or 2). For example in Figure 3-3-20 the AHRS computer 1 is reporting a code of 600001. This indicated the number 1 AHRS computer is operating normally. If the most significant digit of the code is 0 the unit has failed. The fault code in Figure 3-3-20 shows that the AHRS computer 2 has failed.

Refer to Figures 3-3-21 and 3-3-22 during the following discussion on diagnostic code analysis. Assume we have a diagnostic code of 804020 for PFD 1. To analyze a diagnostics code, simply determine which bits are set to 1 then identify the description for that bit. To determine which bits are set to 1 or 0, the hexadecimal diagnostics code (804020) must be converted to binary and applied to the diagnostic word explanation table in the maintenance manual.

OCTAL 350 BIT NUMBER	DESCRIPTION
PFD DISPLAY Diagnostic Word	
9	0
10	0
11	Failed R-AHC input bus
12	Failed L-AHC input bus
13	Failed LB-IOC-1 input bus
14	Failed LA-IOC-1 input bus ◀— If bit 14 is set to 1 the diagnostic software has detected a failed LA-IOC-1 input bus.
15	Failed RB-IOC-1 input bus
16	Failed RA-IOC-1 input bus
17	Failed L-ADC input bus
18	Failed R-ADC input bus
19	0
20	0
21	ARINC wraparound failed
22	High-voltage power supply failed
23	Low-voltage power supply failed ◀— If bit 23 is set to 1 the low-voltage power supply has failed.
24	MEM memory module failed
25	I/O processor failed
26	Display processor failed
27	Video amplifier failed
28	Deflection amplifier failed
29	Graphics generator failed
30	**SSM code
31	**SSM code
32	Parity (odd)

** SSM code:	Bit 31	Bit 30	LRU Status
	0	0	Failed
	0	1	No computed data
	1	0	Functional test
	1	1	Normal

Example: Diagnostic word = 600000		
6	Bits 29-32	Bits 30 and 31 are set; SSM = Normal
0	Bits 25-28	No bits are set
0	Bits 21-24	No bits are set
0	Bits 17-20	No bits are set
0	Bits 13-16	No bits are set
0	Bits 9-12	No bits are set

Figure 3-3-22. Diagnostic word explanation

Courtesy of Rockwell International, Collins Avionics Divisions

Example

To convert the code:

First, convert the hexadecimal code to binary (Figure 3-3-21).

Second, assign each binary digit to a specific bit number. The least significant digit (9) represents bits 9 to 12; the most significant digit (8) represents digits 29 to 32. In this case, the hex code (804020) is equivalent to the binary number 100000000100000000100000. Only digits 14, 23, and 32 are binary 1; all other digits are binary 0.

Third, determine the English meaning of each bit set to binary 1. This is done using the diagnostics word explanation table from the avionics manual (Figure 3-3-22). In this example, the diagnostics have detected a failed LA-IOC-1 input bus (bit #14), and a failed low-voltage power supply (bit # 23). The parity (bit # 32) is set to binary 1 to achieve odd parity.

From this example one can see how to decode the PFD diagnostics code 804020. The hex number shows a failure of the low-voltage power supply and an IOC input bus. The technician would then refer to the system manuals to determine the appropriate action. The analysis of a diagnostics code is somewhat time consuming; however, the process provides detailed fault isolation data. This information can be used during troubleshooting, or to verify correct system operation.

Level 2 troubleshooting procedures. As mentioned earlier, Level 2 troubleshooting is an in-depth look at a given portion of the avionics system. In the avionics maintenance manual, there are specific test procedures for 65 LRUs. Each of these one-page tests are considered Level 2 troubleshooting. Figure 3-3-23 shows the troubleshooting procedures for the number 1 MFD. In this example, the procedures direct the technician through a series of tests including reference to the avionics status page, checking data bus outputs, and swapping displays.

General troubleshooting techniques. Most integrated display systems are very symmetrical. The LRUs from the right side can be interchanged with the LRUs from the left side. If a fault occurs in a redundant system, simply swap the LRU with the identical unit on the opposite side of the aircraft. This procedure is simple; but be careful not to damage connector pins; take all necessary ESDS precautions, and be sure power is off to both LRUs before swapping.

In some cases, "swapping" of units for troubleshooting purposes can be done using the reversionary switches. Selecting the PFD reversionary mode blanks the normal PFD, and the ND now displays the PFD format. Reversionary switching can also be used to select different attitude and heading data sources.

If a failed bus is detected through the diagnostics program, the bus structure must be investigated. The transmitting LRU, the receiving LRU, or the bus wiring may cause a failed bus. Remember bus wiring includes all connectors, wiring, and related shielding. Systematically swapping LRUs that transmit and receive on the failed bus may be helpful to isolate a component problem. The bus (both wires) can also be tested for continuity, and isolation from ground. A data bus analyzer or oscilloscope can be used to test voltage levels. An ARINC 429 bus should have ±10 volts from A to B for binary 1 and 0 respectively. Zero volts should be present for null conditions. A RS-422 bus should have ±5 volts from A to B for binary 1 and 0 respectively. Also, be sure the data bus shielding is properly grounded at all termination points. See Chapter 2 for more details on data bus troubleshooting.

Transport Category Aircraft Second Generation Integrated Electronic Instruments

This section of the text will take a brief look at two popular transport category aircraft. Their electronic instruments are advanced systems, which integrate with airframe and engine displays as well as warning systems, flight management systems, the flight data recorder, and the central maintenance computer. The integration of these systems allows for greater redundancy, improved reliability, and greater maintainability. Only a brief discussion of the electronic instruments will be given here since the integrated systems are discussed with more detail in chapters 4 and 5.

The Boeing 747-400 Integrated Display System

The B-747-400 flight instrument displays are part of a complete aircraft monitoring system called the integrated display system (IDS). The integrated display system monitors a variety of aircraft and engine system parameters, as well as various flight parameters, through three electronic interface units (EIUs). As shown in Figure 3-3-24, three EIUs manipulate incoming data and send outputs to five display units. The EIUs are the main processing units for the IDS. The EIUs receive infor-

STEP	PROCEDURE
	Note
	The MFD 1 may be swapped with either of the PFDs to verify aircraft wiring and to isolate a failed unit. Each unit will display (MFD/PFD) information according to its mount; line key operation is enabled only at the MFD mount.
1.0	Check the AVIONICS STATUS page and troubleshoot according to table 5-3.
	If the AVIONICS STATUS page will not display, check the L-MFD-1 bus.
2.0	If the display is blank, check the circuit breaker. Then swap the MFD with an (operational) PFD to isolate a failed MFD or aircraft wiring problem.
3.0	If DISPLAY TEMP annunciates on the MFD, check the cooling fan installed in the MFD mount. The MFD provides fused power and ground for this fan.
4.0	Select the following formats and verify correct MFD display response. On the left CDU select the radar format, the plan map (selected from MFD IDX page), the present position map, and then the HSI format.
5.0	Check that the heading indication is valid and correct (agrees with PFD and SDU headings). If not, check for L-AHC-2 bus activity at P1 pins 3K/3J.
6.0	Check for a TAS readout on the MFD status line. If TAS is dashed or is not displayed, check for L-ADC-3 bus activity at P1 pins 2F/2E.
7.0	Set left AHRS reversion switch to 2, and verify that MAG 2 annunciates on the MFD. Repeat step 5.0. If heading is not correct, check the AHRS switch and the R-AHC-3 bus at P1 pins 11B/11A. Set AHRS switch to 1.
8.0	Set left AIR DATA reversion switch to 2, and verify that ADC 2 annunciates on the PFD 1. Repeat step 6.0. If TAS is not correct, check the AIR DATA switch and the R-ADC-2 bus at P1 pins 12G/13G. Set AIR DATA switch to 1.
9.0	Set left CDU reversion switch to REV, and verify that CDU 2 annunciates on the MFD. Check that the left CDU blanks and that the right CDU now controls the left PFD and MFD displays. Set CDU switch to NORM.
10.0	Set the left PFD reversion switch to REV. Verify that the MFD now displays PFD data and that the left PFD blanks. Set PFD switch to NORM.
11.0	Display the ACTIVE NAV SOURCE page (on CDU 1). Select each NAV source and verify that the MFD course display is white for FMS, green for VOR 1 (or LOC 1), and yellow for VOR 2 (or LOC 2).
12.0	Press the IDX line key to display the MFD index page. Then press the NORMAL PROCEDURES line key and display a checklist. Press the control wheel LINE ADV button to check the cursored line; press the optional PAGE ADV button to exit the checklist and select the next checklist on the NORMAL PROCEDURE menu.
13.0	A checklist should still be displayed on the MFD. Move the CHP joystick up to select the previous checklist page, down to select the next checklist page, right to exit the checklist and select the next one on the menu, and left to exit the checklist and select the previous one on the menu. Then press the IDX line key.
14.0	Select various MFD displays and check all line keys for sticky or improper operation.
15.0	If a problem is suspected with the L-MFD-1 bus to the DAU, display the LRU DIAGNOSTIC DATA page showing the DATA ACQ UNIT B word. Refer to table 5-5. If bit 18 is set, check for bus activity at DAU pins P2-48/47 to isolate a failed DAU input or wiring problem.

Figure 3-3-23. MFD 1 test procedures   *Courtesy of Rockwell International, Collins Avionics Divisions—for training purposes only*

mation from 16 different systems, process the data, and then output signals to the flight deck displays and the other users such as the flight management system and flight data recorder. One major difference between transport category aircraft and light aircraft (such as the Beechjet 400A) is the number of processors used to drive the displays. On transport category aircraft, three processors are required to provide added redundancy; most light aircraft only have two processor units.

The display units on the B-747-400 are called integrated display units (IDU). The six IDUs are located in a *T* configuration as shown in Figure 3-3-25. During normal operation, the primary flight displays (PFD) are used to show flight data such as airspeed, attitude, and altitude. The navigational displays (ND) are used for the display of navigational data, such as the compass rose, course map, and weather radar information. The two engine indicating and crew alerting system (EICAS) displays,

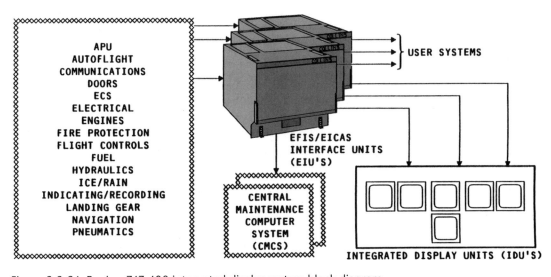

Figure 3-3-24. Boeing 747-400 integrated display system block diagram

Courtesy of the Boeing Commercial Airplane Company

located in the center of the instrument panel, are used to display engine and airframe systems data. EICAS will be discussed in Chapter 4.

Since IDS is an integrated system, it provides greater flexibility in the event of a single or multiple display failure. In other words, there are six displays (not four) driven by the same processors; hence, an integrated system provides more options in the event of a display failure. On integrated systems, the EIUs also talk directly to the central maintenance computers. The central maintenance computer system (CMCS) is used to monitor the health of various aircraft systems. All IDS BITE tests are accessed through the CMCS.

The three EIUs are located in the main equipment center of the aircraft. These computers are static discharge sensitive and should

be handled only when using proper precautions. The electrical leads of the EIUs are made through an 800-pin connector mounted on the rear of the unit. Take special care not to damage the pins or sockets during removal and replacement of the computers. Maintenance of the B-747-400 IDS will be discussed in further detail in chapters 4 and 5.

The Airbus A-320 EFIS

The flight displays found on the Airbus A-320 are part of an integrated display system called the electronic instrument system (EIS). The EIS contains six CRT display units arranged in a *T* configuration similar to the Boeing 747-400. The four outboard displays are used primarily for flight data (EFIS). The remaining two displays are used primarily for the electronic centralized aircraft monitoring sys-

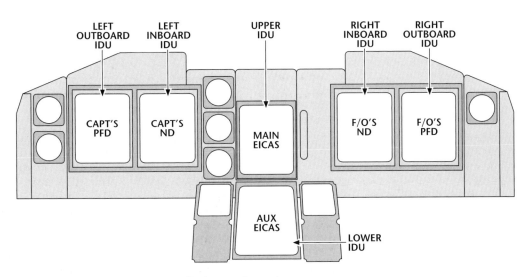

Figure 3-3-25. Boeing 747-400 IDS display configuration

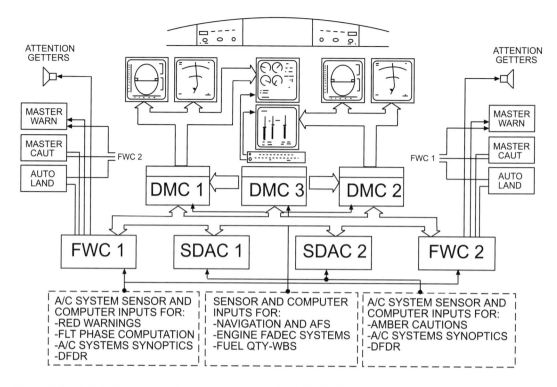

Figure 3-3-26. A-320 electronic instrument system (EIS) block diagram   *Courtesy of Airbus S.A.S.*

tem (ECAM). ECAM is similar to the Boeing EICAS and will be discussed in Chapter 4. On the A-320 the two captain's side EFIS displays are called the primary flight display 1 (PFD 1) and the navigational display 1 (ND 1). The first officer's (co-pilot's) side consists of PFD 2 and ND 2. The two ECAM displays are referred to as the upper and lower ECAM displays (Figure 3-3-26).

As with other integrated systems, the major advantage of the EFIS/ECAM system is the flexibility in the display of data. In the event a display fails, one of the operational displays can be used to show the missing data. The flight crew can switch the PFD and ND data manually. This is accomplished by the PFD/ND XFR switch located just outboard of the captain's and first officer's EFIS displays. The data from the upper ECAM display is automatically transferred to the lower ECAM display in the event of a single ECAM display failure. The upper ECAM data moves to an ND in the event both ECAM displays fail.

Figure 3-3-26 shows the relationship of the EIS processors and the six displays. The main processors for the A-320 EIS are known as the display management computers (DMCs). DMC 1 and 2 drive the displays during normal operation. DMC 3 operates in hot standby and provides backup for DMC 1 and 2. The DMCs receive inputs from four computers, the flight warning computer (FWC) 1 and 2, and the system data acquisition concentrators (SDAC)

1 and 2. The FWCs and the SDACs receive information from the various aircraft systems. Figure 3-3-26 shows the various input/output signals to the DMCs. Here it can be seen the DMC receives ARINC 429, ARINC 453, RS-422 and DSDL inputs. The dedicated serial data line (DSDL) is a digital bus structure unique to Airbus aircraft. Each DMC normally feeds three displays through a DSDL interface. Each display sends a feedback signal (used for health monitoring) to the DMC. The EIS BITE is monitored by the aircraft's centralized fault display system. More details on the various system interfaces and system troubleshooting will be given in chapters 4 and 5.

Section 4

Third Generation Electronic Instruments— Integrated Modular Avionics Systems

In the mid 1990s the concepts of integration for flight deck instruments had begun to mature into a design concept known as Integrated Modular Avionics (IMA). The Boeing 777 was introduced which employed AIMS, the airplane information management system. AIMS could be defined as the first commercial airliner to employ advanced modular electronics fed by a high-speed bi-directional data bus. The B-777 employed the ARINC 629 data bus as the backbone data transfer systems for AIMS. During the next decade (2000-2010) many developments took place, which allowed the integration of aircraft electronic systems to mature even further. Increased processing power, the miniaturization of components and development of large high quality flat-panel displays allowed more systems to be combined into smaller modules. The traditional line replaceable units (LRU) each installed on a separate rack in the aircraft equipment bay were soon replaced with line replaceable modules (LRM) installed in a common mounting rack. In some cases, these modules were small enough to be placed on the flight deck just behind the flat-panel instruments.

Another major advancement, which allowed for the development of IMA, was the use of an Ethernet-type data transfer system. Some manufactures employ their company's own proprietary bus standards although many systems follow industry accepted ARINC 664 data transfer specification. Generally speaking, modern data bus systems support a variety of data formats, operate in full-duplex mode at high speed and can employ fiber cable if desired. The ARINC 664 data transfer system (also referred to as AFDX) was discussed in Chapter 2. The IMA concept defines the Application Programming Interface (API) and not the structure or software of individual modules or end systems. This allows individual manufactures to develop specific software/hardware applications for their needs and simply interface with the complete system using a predefined API. Also, new hardware and applications can be installed with simple software updates. It should be noted that no detailed standards have been defined for components used in IMA architecture; only interface applications such as the API have been standardized. The ARINC 653 standard for example, defines software partitioning for IMA systems. This allows a corporation, like Boeing, to control the large-scale system design and yet offers flexibility to subsystem manufacturers.

Other advancements recently developed include heads up displays, enhanced vision systems, satellite radio weather data, and electronic flight bags. These systems as well as the incorporation of airframe and engine data on virtually all flat-panel displays has allowed the aircraft flight deck to become a high-tech visual sensation with improved systems management and pilot awareness. At the same time, the concepts of integrated modular avionics have allowed the physical size of all necessary components to shrink. This miniaturization has made it possible for even light single-engine aircraft to employ advanced flight deck instrumentation. Both transport category and light aircraft systems that employ the concepts of IMA will be discussed in the following sections of this chapter.

Heads-up Displays

The heads-up display (HUD) is a system which allows pilots to view flight critical instruments while at the same time keeping their "head up" in order to view the area outside the airplane windshield. Traditional, head down, displays require the pilot to divide their attention between inside the aircraft and outside the aircraft. This transition commonly takes a pilot two and one-half seconds to adjust, refocus and create a proper visual awareness. Two and one-half seconds can be critical when the aircraft is in a landing situation. HUD allows pilots immediate access to critical flight information and the environment outside the aircraft without head movement. Tests have shown that HUD significantly increases safety, especially during landing and takeoff.

There are two basic types of HUD systems: helmet mounted displays, used by the military, and systems that employ a transparent display (known as the combiner) mounted in the vision path between the pilot and the aircraft windshield. This text will focus on the combiner-type HUD since these systems are found on corporate, commuter, and transport category aircraft (Figure 3-4-1).

Figure 3-4-1. Typical heads-up display
Courtesy of Boeing Commercial Airplane Company

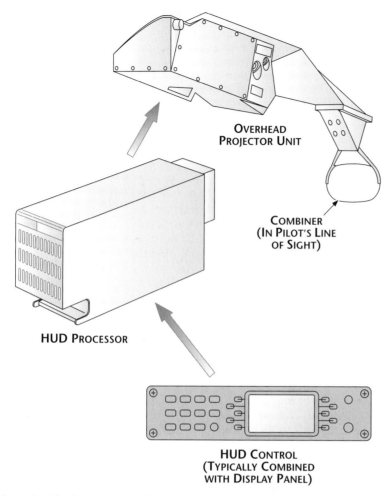

OVERHEAD PROJECTOR UNIT

COMBINER (IN PILOT'S LINE OF SIGHT)

HUD PROCESSOR

HUD CONTROL (TYPICALLY COMBINED WITH DISPLAY PANEL)

Figure 3-4-2. Components of a typical HUD system

phor screen (the combiner). The majority of HUD systems in operation today are of this type. These systems have the disadvantage of being relatively large, have heavy power demands, and the phosphor screen coating can degrade over time.

- Second Generation—use a solid-state light source, typically LEDs, which is modulated to project an image on the display. These systems do not fade, are more compact and require less power than first generation systems. These systems are found on the latest commercial and corporate aircraft.

- Third Generation—employ optical waveguides to produce the image directly in the combiner rather than use a projection system.

- Fourth Generation—Use a scanning laser to display images or even video imagery on a clear transparent screen.

The optical waveguide and laser scanner type HUD systems are not currently used on aircraft; however, as technologies improve, it is likely commercial aircraft will employ third or fourth generation systems.

There are four basic components found in a typical transport category aircraft HUD system. As seen in Figure 3-4-2, the combiner assembly contains the transparent screen mounted in the pilot's line-of-sight. The image displayed on the combiner is projected from the overhead unit, which is typically mounted on the flight deck overhead panel above and slightly behind the pilot's head. The newer, second generation, LCD projector units are much smaller than the CRT-type and are installed on most modern systems. The HUD processor assembly is typically located in the equipment bay or other electronics area. The HUD control panel can be a standalone unit or integrated into the display system controls.

Since their introduction into commercial aircraft in the 1970s, aircraft HUD systems have undergone several changes through technological advancements. There are four distinct generations of HUD:

- First Generation—these systems employ a CRT to project an image on a phos-

Altitude	Guidance Symbols
Airspeed	TO/FROM
Vertical Speed Tape and Digital Indications	Selected Course
Heading	Drift Angle
Autopilot Modes	Distance
Flight Director Commands	Flight Director Modes
Vertical Deviation	Altitude Awareness Cue
Radio Altitude (RA)	Roll Scale
Decision Height (DH)	Slip/Skid Indication
Minimum Descent Altitutde (MDA)	Low-Speed Awareness
Lateral Deviation	Synthetic Airport/Runway

Figure 3-4-3. Possible flight information shown on a typical heads up display

The HUD processor is a digital unit, which typically contains redundant circuits or two channels for comparison and/or backup processing. Inputs to the processor come direct from critical systems, through an input/output concentrator or from the aircraft's integrated display processor. Redundant inputs come from separate sources, to allow for data comparisons, and ensure system validity. These inputs are most commonly ARINC 429 signals, or other digital formats appropriate to the system manufacturer. The unit automatically adjusts system brightness with inputs from ambient light sensors. The processor sends analog video information to the CRT-type overhead projector units. Newer, LCD systems receive digital signals from the HUD processor. The processor unit communicates with the crew alerting and central maintenance systems in the event of a HUD malfunction.

The combiner assembly is a foldaway glass screen approximately eight to ten inches diagonally. The pilot can deploy the combiner for HUD operations or fold the unit upward against the ceiling of the flight deck. The unit is designed with a quick release in the event of a crash and the pilot strikes the screen with his/her head. The combiner glass and projector unit are designed to produce an image, which is in focus as the pilot looks through the windshield while flying the aircraft. The image on the combiner is therefore focused at near infinity. This allows the pilot to easily focus on ground images and the combiner image simultaneously. The glass screen is covered with a synthetic hologram coating, which is frequency selective. In other words, the combiner will only reflect the green color (green frequency) sent by the overhead projector unit, all other colors pass easily through the glass and the pilot's vision is undisturbed.

For the most part, all flight variable information available inside the flight deck that is displayed on conventional instruments is made available by HUD. Most of the HUD display is conformal, meaning that the information matches the outside world degree-for-degree. In order to keep the pilot's orientation correct, as the aircraft rolls to the left 10° the information displayed on HUD also roles 10°. Figure 3-4-3 shows a list of flight variables, which could be displayed by a typical HUD.

As heads-up displays advance it is likely that more aircraft will employ this technology. At the time this text was written, the soon to be released Boeing-787 is scheduled to have HUD installed as standard equipment; the Airbus A-380, the newest airliner in production, employs HUD as an optional system; although, it should be noted that many of the orders for the A-380 include the HUD option. HUD systems of the future will most likely become smaller, require no overhead projector unit, and incorporate synthetic vision for improved safety during bad weather.

Synthetic Vision

The concepts of synthetic vision were developed by NASA in the early 1980s and took several decades to become an affordable reality in today's aircraft. Although still in very limited use, synthetic vision offers promise to improve safety, decrease flight delays, and cancellations, as well as, increase the capacity of the national air space system. Synthetic vision systems create a computer generated flight deck display that represents the visual information, as it would appear in daylight conditions with good visibility. The concept is to provide each pilot with "real-looking" visual information no matter what the weather or daytime conditions. Through advances in technologies available today, several manufacturers offer a synthetic vision option on high performance aircraft. In many ways the imagery presented on a synthetic vision display is very similar to a modern computer, or video game as seen in Figure 3-4-4. The terrain, although somewhat artificial looking, accurately depicts ground terrain and sky currently in front of the aircraft. This display also shows guidance cues for aircraft steerage known as the pathway/tunnel or Highway-

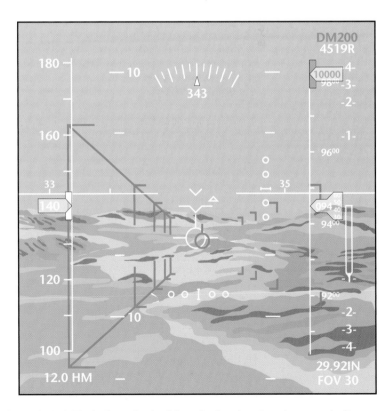

Figure 3-4-4. Typical synthetic vision display showing the tunnel often called the Highway in the Sky

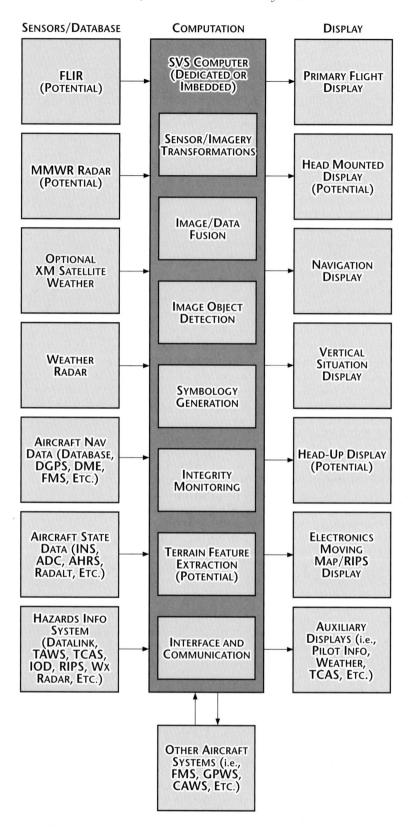

SENSORS/DATABASE COMPUTATION DISPLAY

Figure 3-4-5. General elements of a synthetic vision system

tion alone, studies have shown that over 30 percent of all fatal accidents are categorized as controlled flight into terrain (CFIT). This involves a normally operating aircraft with no mechanical difficulties impacting terrain due to the pilot's lack of outside visual references (most often due to poor visibility). In general aviation, the majority of accidents occur due to non-instrument rated pilots flying into weather that requires an instrument rating. Also, poor visibility due to inclement weather creates airport delays and flight cancellations. Each of these problems can be greatly reduced with the use of synthetic vision systems.

In general, synthetic vision systems utilize airborne equipment to determine the precise position of the aircraft. This is done with modern global positioning systems (GPS), inertial reference systems (INS), and ground proximity systems. The aircraft position is then compared to the information in a large database within the synthetic vision computer(s). The database contains all the necessary terrain data, such as ground levels and contours, roads, railroads and bridges, lakes and rivers, even cities and large buildings. Of course, the database also contains the necessary airspace information and airport specifics such as approach-to-land flight patterns, runway lengths, headings, and taxiway information. The synthetic vision system also monitors current air traffic using TCAS (traffic collision avoidance system). In some cases, specialized high frequency radar and inferred sensors are used to input data to the synthetic vision system. A block diagram of a synthetic vision system is shown in Figure 3-4-5. Here it can be seen there are three general elements to any synthetic vision system: the sensory inputs, the computational element, and the output for pilot interaction.

To understand the synthetic vision systems, think in terms of a virtual reality system that uses an extensive database and real-time data for various inputs. The terrain database is a highly detailed 3-D map of the earth's surface including large human-made objects (like railroads, towers, airports, and cities). This enormous database requires mass quantities of computer memory, which was one of the limiting factors in developing synthetic vision. The various "forward-looking" sensors can detect real time air traffic and other items not in the terrain database. For example, inferred sensors can detect wildlife, like deer, on the runway. The synthetic vision computational elements analyses all this data and creates a virtual display for the pilot that resembles the real world situation regardless of current weather conditions. These systems are also employed for pilotless aircraft such as UAVs (unmanned aerial vehicles).

in-the Sky. The system determines the correct flight path and the pilot simply guides the aircraft into the tunnel shown on the display.

Synthetic vision systems were developed mostly to improve safety. In commercial avia-

Figure 3-4-6. Honeywell PlaneView system installed in Gulfstream G-V

Most synthetic vision systems utilize equipment already familiar to the pilot and already installed on the aircraft for traditional navigational purposes. For example, the synthetic vision displays are incorporated as part of the primary flight displays (PFDs) and navigational displays (NDs) found on modern aircraft. This allows the pilots to seamlessly transition to synthetic vision when visibility decreases. Heads up displays can also show synthetic vision information to further enhance safety.

Honeywell was the first company to certify a synthetic vision system on a FAR Part 25 certified aircraft (corporate-type business jets). The Gulfstream line of jet aircraft can now be equipped with the Honeywell Primus Epic systems know as PlaneView. The PlaneView system employs state-of-the-art avionics in order to present the pilot with an advanced synthetic vision system. The system is based on an open architecture concept, which allows for the installation of upgrades not yet envisioned. To do this the PlaneView system utilizes a flexible data bus structure and system software which is easily modified.

The Honeywell PlaneView system employs four 13 by 10 inch LCD displays arranged horizontally across the instrument panel (Figure 3-4-6) The two outboard displays are PFDs (primary flight displays); the two inboard displays are used as MFDs (multifunction displays). Each display is interchangeable for flexibility and redundancy. The PFDs are used to display necessary flight data. The MFDs support engine and airframe data as well as navigational information, weather information, or synoptic pages. The system incorporates a side-mounted cursor control device, one for each pilot. PlaneView uses a windowed-display for management of the various system features. Similar to a common personal computer, the pilots use the cursor for a point-and-click selection of the options. The synthetic vision system is also displayed on the heads up display, which is available with the PlaneView system.

Synthetic vision systems are also available in a portable electronic tablet-type format. The tablet, similar to a common electronic-book or iPad, designed to be mounted on the control yoke directly in front of the pilot or taken by the pilot for use outside the aircraft. Many of these units can also be used for flight planning and/or other pre-flight operations. The Bendix/King AV80R HORIZON 3D is a table-style unit, which can operate as a multifunction display with synthetic vision (Figure 3-4-7). This system utilizes a touch screen display approximately seven inches, measured diagonally, and weighs only 1.44 pounds. The display assembly is called the mobile computer platform (MCP) and contains the database and processing hardware/software needed for system operation. This unit requires minimal installation and is offered at relatively low cost when compared to other panel mounted systems. The AV80R is actually a portable electronic flight bag (EFB) providing navigational information during flight.

Figure 3-4-7. Bendix/King AV80R with synthetic vision *Courtesy of Bendix/King by Honeywell*

Figure 3-4-8. The Skypad 2; a portable electronic flight bag *Courtesy of Seattle Avionics*

In order for the MCP to create an accurate synthetic vision display during flight operations, the unit must receive vital navigation inputs provided by the AV80R's inertial navigation unit (INU). The INU is a GPS-enabled inertial navigation system which can be mounted almost anywhere in the aircraft. The INU sends aircraft position data to the display unit using a wireless connection, hence, making installation of the systems extremely simple. The INU does require a power input and connection to an external GPS antenna. The INU weighs approximately one pound, is only 63 cubic inches in size, and contains a back up battery that will operate for approximately one and a half hours. The new technology of miniaturization and high-power computer processors has made synthetic vision systems small enough and cost effective even for light aircraft.

Electronic Flight Bags

The electronic flight bag (EFB) is a relatively new feature found on many aircraft designed to perform preflight and/or flight management tasks with greater efficiency and less paper than long-established legacy systems. The term *flight bag* comes from the traditional case in which pilots would carry operational manuals, charts, calculators, plotters, pens/pencils, and the various forms needed for flight management. Electronic flight bags are general purpose computing platforms intended to reduce, or replace, paper-based reference materials and provide automatic functions for operations traditionally performed by hand. For example, in the past pilots would perform take-off performance calculations using graphs and charts, pen and paper, and perhaps a hand held calculator; using the modern EFB pilots can input the data and the EFB automatically determines the correct take-off performance data.

In its most basic form, an EFB is a dedicated computer that contains software designated for a specific aircraft and a series of flight routes. Many modern EFB are completely portable and resemble tablet-type computers similar to the iPad or common e-books (Figure 3-4-8). On larger aircraft, EFBs are permanently mounted on the flight deck for easy interaction with each pilot. These systems continuously interact with aircraft systems, as well as using wireless technologies to communicate with land-based stations while on the ground or during flight. Electronic flight bags found on transport category aircraft are designed to improve safety and reduce pilot workload. Many of the routine calculations traditionally performed by the flight crew, and/or dispatch personnel are now done using EFBs and software applications. While parked at the gate, many of these systems can seamlessly upgrade avionics software through a wireless connection to the airline configuration/revision management center. Some of the benefits associated with EFBs include:

1. Weight savings through reduced paper on the flight deck

2. Reduced flight deck clutter

3. Easy document revisions and cost savings related to automatic updates

4. Efficient transfer of data such as flight and maintenance reports, gate changes, and passenger information

5. Improved safety and decision-making through display of real-time weather information

Electronic flight bags are approved for use with the help of the FAA Advisory Circular AC 120-76A entitled "Guidelines for the Certification, Airworthiness, and Operational Approval of Electronic Flight Bag Computing Devices." As described by the FAA there are three classes of EFB hardware; Class I, Class II, and Class III. Class I EFBs are portable electronic devices (PED), which must be stowed during take-off and landing operations. Class I EFBs do not require an administrative process to remove or install the PED in the aircraft. Class II EFBs are also portable devices, but are typically mounted in a position where they are utilized during various flight phases. Class II EFDs

require an administrative process to install/remove the unit in the aircraft. Both Class I and II EFBs are considered portable electronic devices and do not require certification related to design approval. However, any fixed equipment such as computer mounts or docking stations that remain in the aircraft require an FAA design approval such as a supplemental type certificate (STC). Class III EFBs are fixed units typically permanently installed on the flight deck accessible to each pilot. Class III units typically incorporate a flat-panel display, often with touch screen capabilities, a keyboard and/or track ball device for pilot inputs. Class III units could utilize the aircraft's PFD or ND for the flight bag displays; of course, critical flight data takes priority. All Class III EFBs require FAA design certification and in many cases are certified as part of a new aircraft design as in the case of the Airbus A-380 and the Boeing 787 (Figure 3-4-9). On the A-380 the EFB is integrated into the onboard information terminal (OIT) system and each pilot has a dedicated keyboard similar to a laptop computer for operating the system.

EFB Architecture

In general, the more complex the aircraft the more complex the EFB system. Light aircraft that employ Class I EFBs can be completely self-contained. Class II and Class III units must communicate with various aircraft systems for pertinent data during operation. An

Figure 3-4-9. A Class III electronic flight bag installed on the B-787

Courtesy of Boeing Commercial Airplane Company

ARINC 429 or other data bus system provides this communication link. For aircraft certified under FAR 91 (turbine & large aircraft), navigational charts must be present on the aircraft at all times. This means if the aircraft's sole source for charts is the EFB the operator must demonstrate the EFB will continue to operate throughout a decompression event, and thereafter, regardless of altitude. The only way to achieve this capability is by using a solid-state disk drive or a standard rotating mass drive in a sealed enclosure. Most modern EFBs employ solid-state memory.

As seen in Figure 3-4-10, a Class III electronic flight bag is a complex system that is designed to communicate with a variety of airborne and ground based systems. The airborne components of a transport category aircraft system (Class III) typically contain two dedicated monitors located in easy view of each pilot. These are often mounted slightly in front of and just to the right/left of each pilot on the side panel of the aircraft (Figure 3-4-9). Each display unit is connected to some means of pilot control (such as a keyboard or cursor control device) and a processor unit containing the system database on a solid-state memory. The processors must be equipped with crossover capabilities. Crossover capabilities ensures that the loss of one processor will not adversely affect the EFB display, keyboard, or cursor control device; hence providing redundancy for both pilots.

The modern EFB processors will typically cross-communicate through an Ethernet based data link (ARINC 664) using a conventional copper data bus or fiber optic cable. The processors must also be connected to some form of information management computer in order to receive/transmit to the necessary aircraft systems. The airborne data loader can be used to update system information using carry-aboard DVD or CD data discs or a solid-state memory. Updates can also be made through a wireless connection and a local area network (LAN) similar to those found in a home or office. When the aircraft is parked at the gate or in a maintenance facility, the LAN communicates with the aircraft equipment and upgrades are made wirelessly. Since the information is critical to flight safety, the LAN connection uses encrypting techniques and is done through a secure channel. Ground stations can also communicate with an airborne EFB processor using cellular links or satellite data feeds.

Type A, B, and C Software

The FAA approves three variations of EFB software; each provides different levels of sophistication, safety, and reliability. The software is categorized by type, listed from simple to most complex; Type A, Type B, and Type C. Each type requires a different level of certification and typically different levels of hardware complexity.

Type A software can be used for static display applications such as a document viewer using a PDF or HTML format; electronic check lists; flight operations manuals, and other printed documents like notice to airman (NOTAM) reports. These documents must be static and require change only at the next software update. Typically the user will subscribe to a private service for software updates. Type A systems can be used for flight performance calculations and post flight reports, such as, fault reporting (squawk sheets). Type A systems require FAA operational approval and may be hosted on any class EFB (Class I, II or III).

Type B software can perform all the applications available to Type A software systems as well and non-interactive electronic approach charts or approach charts that require panning, zooming, or scrolling according to FAA AC120-76A, App B. Type B software can be used to create a head-down display for an enhanced vision system, synthetic vision system, or video cameras, as well as a real-time weather data display. All Type B systems require both FAA operational approval and regulatory agency approval since these systems are dynamic in nature and are allowed to use interactive data during flight, such as weather.

Type C software systems are by far the most sophisticated of the three levels and found mostly on corporate, commuter, and transport category aircraft using Class III EFBs. Type C systems employ dynamic interactive displays and processor units. The displays for this software can be part of the aircraft main instrument system and used as a multi-function display (MFD) during various flight operations. Type C applications are subject to airworthiness requirements, such as software certification. Type C software systems can only be run on Class III EFB hardware. Since Class III hardware is permanently installed in the aircraft, all Type C, Class III systems require airworthiness certification (Type Certificate or Supplemental Type Certificate) operational approval, and regulatory agency approval.

EFB Regulations

According to the FAA, Class I, Class II, and Class III EFBs may act as a substitute for the paper manuals that pilots are otherwise required to carry with them. While FAR Part

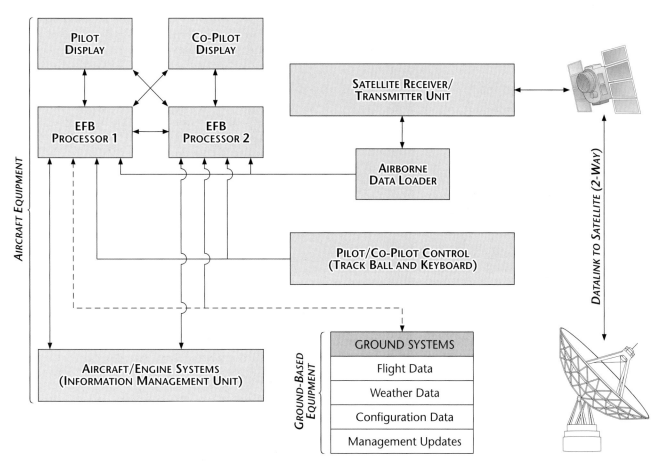

Figure 3-4-10. Architecture of Class III electronic flight bag

91 operators (those not flying for hire, including private and corporate operators) can use their pilot in command (PIC) authority to approve the use of Class I and Class II EFBs as long as they are portable electronic devices (PED). However, these operators must seek operational approval through the FAA process described by the Advisory Circular AC 120-76A.

Any data connectivity of a Class I or II PED to various aircraft systems shall be performed in accordance with a Supplemental Type Certificate, Type Certificate, or Amended Type Certificate. Any PED mounting or attachment device installed in the aircraft shall also be performed in accordance with a Supplemental Type Certificate, Type Certificate, or Amended Type Certificate.

FAA operational approval is only necessary for Part 135 and 121 operators. The operational approval process is individual to each flight operation and involves a detailed process with the local FAA Flight Standards District Office (FSDO). Regardless of whether an EFB has been approved for use in one aircraft, application for operational approval for these operator types has to be done for each aircraft and for each operation.

Light Aircraft Electronic Instruments

Over the past decade several advancements have made it possible to design electronic instruments into lightweight, compact, easy-to-install components. These improvements have made it possible to develop a whole new generation of instrument systems for light aircraft. Typically, the displays are extremely integrated with various airframe and powerplant systems as well as the traditional flight systems in order to process and display necessary information. Most aircraft contain two or more flat-panel displays that replace the traditional electro-mechanical instruments therefore simplifying and improving cockpit management.

There are currently several corporations manufacturing light aircraft electronic instrument systems. Many of the traditional manufactures, such as, Honeywell and Bendix/King produce systems for corporate and some general aviation aircraft. There are also relatively new companies such as Anodyne and Garmin producing electronic instrument systems for light aircraft. In general, these systems can be found as retrofit packages for existing aircraft or as factory installed systems sold as an inte-

Figure 3-4-11. A light aircraft electronic display system produced by Garmin

is typical of many of the light aircraft systems and will be discussed in detail here.

The Cirrus SR-22 was designed with an integrated electronic display system as a core element to the flight deck. The Garmin system chosen for the aircraft utilizes two integrated avionics units (IAU) as the main processors for the two large LCD flat-panel displays. The IAU may also be referred to as a GIAU (Garmin integrated avionics unit). This unit receives inputs from various airframe and engine sensors, flight information, and signals from various control panels. The IAUs then processes the data and sends all necessary signals to the video display units.

As seen in Figure 3-4-12, the number 1 and the number 2 IAUs receive data from four additional processors; the air data computer (ADC), the attitude heading reference system (AHRS), the transponder, and the engine/airframe unit. Heading information from the aircraft's magnetometer is sent indirectly to the IAUs through the AHRS unit. Most information is sent to/from the IAUs on high-speed data busses in one of four formats:

gral part of a new aircraft. Figure 3-4-11 shows an instrument panel from a Cirrus SR-22 four-place light aircraft. This airplane is commonly used as a personal or training aircraft and contains an electronic display system produced by Garmin International. The Garmin system

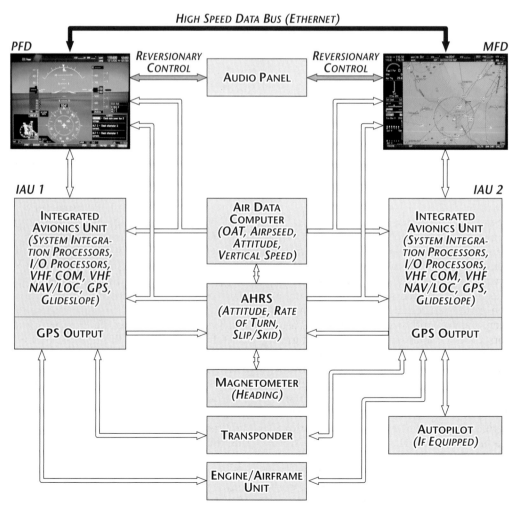

Figure 3-4-12. Garmin G1000 System overviews for a Mooney M20M and M20R

Figure 3-4-13. Integrated avionics processor unit: (A) front view, (B) back view

1. A high-speed Ethernet data bus
2. An RS-232 data bus
3. An RS-485 data bus
4. An ARINC 429 data bus

The IAUs also receive a few discrete analog signals from various sources. Both display units communicate using a high-speed Ethernet-type bi-directional data bus in order to share systems information and monitor for system errors.

There are literally hundreds of sensors located throughout the aircraft that report to the various processors; these processors send information to the integrated avionics units, which eventually feed the PFD and ND. The various sensors are used to monitor variables such as engine temperatures, airframe components, air temperature, and aircraft attitudes. These sensors produce analog, discrete, and digital signals that are sent to the ADC, AHRS, magnetometer, transponder, or engine/airframe unit (Figure 3-4-12). These units then combine, analyze, and process the data and send a modified version of the data to the IAUs. The IAUs then perform various reliability and validity checks and converts the data one more time for display on the PFD and ND. Figure 3-4-13 shows an IAU removed from the aircraft; (A) the front view with the locking handle partially extended, and (B) the rear view showing the D-sub connectors. The unit is approxi-mately 4 inches wide by 6 inches tall by and 10 inches deep.

On many light aircraft, such as the Cirrus SR-22, these processors are installed in the aircraft cabin just behind the two flat-panel display units. Figure 3-4-14A shows the instrument panel of a Cirrus SR-22 with the PFD and ND removed, exposing the various processor units and mounting structure. The processors are installed in an aluminum rack assembly surrounded by a perforated aluminum cage. The cage is used to provide protection from high energy radio frequencies (HERF) that may create electromagnetic interference and adversely affect the system. Figure 3-4-14B shows a typical HERF cage as seen through the aircraft windscreen and the instrument panel glair shield (instrument panel cover) removed.

This system employs two 12-inch (or 10.4 inch) LCD displays with the PFD located on the left and an MFD installed on the right side of the panel. The displays are full color with a resolution of 1024 x 768. The backlighting for the displays automatically adjusts for ambient light conditions. Photocells monitor the flight deck light conditions and the system software adjusts the PFD, MFD, and bezel/key brightness. All brightness levels can also be controlled manually if the pilot so desires. Both units contain a series of controls located around the perimeter of the displays as seen in Figure 3-4-15. These surrounding controls and keys are known as

Figure 3-4-14. Installation rack and Garmin processor units: (A) Cabin view facing forward with the flat-panel displays removed, (B) View of HERF cage

each soft key is listed on the display just above the key (button). If the function of the soft key changes, the key identification located on the display just above the button will also change. Using soft keys allows the system designer to incorporate a multitude of controls using only 12 push-button switches (soft keys). These two Garmin display units contain the necessary processing circuitry needed to activate all soft keys and discrete controls mounted on the displays as well as create the video information as seen by the pilot. The two displays also cross talk to provide redundancy in the event of a system failure.

During normal operation one IAU drives only one display with IAU 1 feeding the PFD and IAU 2 feeding the MFD. The two displays then share data through the Ethernet cross-talk data bus; hence, providing true redundancy. If one IAU fails, both displays can still operate normally. For example if the IAU 1 should fail, the PFD would receive information from IAU 2. The signal would come from IAU 2 through the MFD and the Ethernet data bus to the PFD. If one display should fail, the flight information has priority and the system will operate in reversionary mode. The reversionary mode combines most of the information previously available on two displays in a smaller combined format on one display. If one display fails the cross-talk Ethernet connection is disabled. If a sub-system, such as air data, should fail, the IAU will remove the information from the PFD or ND and provide a red *X* where the information was previously displayed.

Configuration and troubleshooting. Modern electronic display systems are extremely reliant on software for troubleshooting, operational updates, and system configuration. To allow for flexibility in design, the system has the ability to change configuration through modifications to the system software. For example, if one aircraft employs only one GPS receiver and another aircraft uses two, the display system software is modified to configure the systems accordingly. Each aircraft display system could have the exact same hardware; only the software would be different. With this philosophy of software configurations, the manufacturer can produce limited variations of the hardware and still offer a variety of options to an aircraft owner at a reasonable price. For the Garmin IAU discussed here, inputs and outputs on the D-sub connectors are configured using software. In other words, each electrical connection of the RS-232, ARINC 429, and the Garmin proprietary high-speed data busses are defined by the system configuration software and subsequent software updates.

the bezel controls. The right side controls are dedicated to one or more specific functions; the bottom row buttons are called soft keys. A soft key will have a variety of functions described by the system software. The specific function of

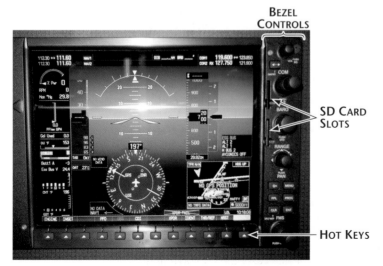

Figure 3-4-15. A PFD with various controls located on the right, soft keys on the bottom, and two SD card slots

If a system needs to be reconfigured due to a change in aircraft equipment or a periodic software modification, the software update would most likely be downloaded from the manufacture's website and loaded into the aircraft system. In the case of the Garmin system found on the Cirrus SR-22, the software update is downloaded from the appropriate source onto a solid-state memory card known as a secure digital (SD) card. The SD card is then installed into a slot on the bezel of the PFD or ND (Figure 3-4-16). The SC card contains all necessary files to configure the system LRUs. It should be noted that on many systems, configuration updates might be necessary when replacing defective LRUs. Even if the system hardware configuration is not changed, any time an LRU is swapped for a new unit (or repaired unit) a software update may be required. It is also very important to ensure the SD cards are installed in the correct aircraft and correct location; they are part number specific and *are not* interchangeable between systems. When installing the SD card always consult the manual, following the specific procedures as the installation of the card is critical. In general, the card is pressed into the slot until one feels an internal latch secure the card. Removal of the card is completed by pressing the card, which releases the latch, and the card should eject.

Figure 3-4-16. Installation of the SD card on the bezel of a typical display

Some configuration of the Garmin system is done through the use of conventional copper wires to a specific pin connection on one of the connectors. This is often referred to as pin programming or configuration strapping as previously discussed in the section on the EFIS-85/86 system. An example would be if an IAU is installed in the number one IAU slot, a specific pin may be set to a digital 0 (or open); if the IAU is installed into slot number two, the pin may be set to digital 1 (connected to ground). In another example, if the aircraft uses identical displays their installation location must be identified; the PFD pin programming may be set to 00 and the ND may be set to 11. These discrete programming pins can fail due to loose or poor connections. Remember, when troubleshooting difficult repairs in any electronic display system, verify appropriate program pins.

Whenever troubleshooting any systems of this type it is important to identify the exact depth of the fault. The technician must understand the system architecture completely and how data is transmitted and received (i.e., through which sensors or probes, through which bus or busses, and processed by which LRUs). For example, if the display shows an EGT (exhaust gas temperature) failure, a red X will be placed over the EGT indication. The EGT information starts at the engine exhaust temperature probe and travels through the engine/airframe unit to the IAU. If all data from the engine/airframe unit is faulty, the technician should suspect the LRU or defective bus (connector) to/from the

engine/airframe unit. If only the EGT information is faulty, suspect the system (EGT probe and related wiring) before the signal enters the engine/airframe unit. Since the Garmin Perspective TM system found on the Cirrus SR-22 is extremely software reliant, the system can perform a variety of troubleshooting and diagnostic functions. This becomes very important since most of the communication between LRUs in the electronic display system is completed through various data busses and troubleshooting can become extremely complex. The Perspective TM software continually monitors each data bus for activity and ensures data validity through a series of cross checks with parallel systems. The maintenance manual for this system contains a troubleshooting section detailing the various steps to help isolate defective components. In the event of an inoperative system, the PFD or ND will display a red *X* in place of the lost data. The technician can then refer to the system troubleshooting tables of the maintenance manual. The troubleshooting section of the manual will take logical steps to guide the technician to the failed component.

In the normal operation mode the systems will display messages, alerts, and/or annunciations on the PFD. To view messages, press the bezel key located below the flashing annunciation. A typical yellow caution message could be *NO GPS DATA*; in this case, the message may indicate that the aircraft is simply not receiving GPS signals (if the aircraft is in the hanger

Figure 3-4-17. The Bendix/King KFD 840 panel layout

Courtesy of Bendix/King by Honeywell

for example). If the GPS message is shown in red, some portion of the GPS is inoperative and further analysis is necessary. The ND is used to identify auxiliary systems status. A list of LRUs is shown on the status page of the ND along with a simple red X or green check mark. Of course the green check indicates normal operation, the red X a failed system. Data bus pathways can also be checked through the ND. In the configuration mode displayed on the ND, a failed path message indicates lack of digital data. The software uses a timing program to determine if a bus is still actively sending data. If the bus times out (that is, allows too much time before sending data) a message is latched into memory until the next power cycle. This allows the maintenance crew to observe any potential faults. Once again, a simple color indication is used to identify data path status; green is OK, red is failed, and black indicates an unknown status. Once the status of a given data path is known the maintenance manual can be consulted for recommended repairs.

Honeywell Systems

Honeywell Aerospace has developed a variety of systems employing the concept of integrated modular avionics for large aircraft used by the major air carriers, corporate, and commuter type aircraft, as well as, for light general aviation aircraft.

The Bendix/King division of Honeywell manufactures a variety of low cost, light aircraft systems installed as original equipment straight from the factory or as retrofit avionics for existing aircraft. These systems designed for a wide range of aircraft types are available in a variety of configurations. Small turboprop or jet aircraft may install the KDU-1080 flat-panel display (measuring 10 by 8 inches); or the KDU-1500 (measuring 15 inches diagonally). These aircraft have the option of installing single or multiple PFDs and/or NDs as well as more advanced systems such as synthetic vision or automatic radio tuning. Two common Honeywell systems are presented here, the Primus avionics package installed on corporate and commuter aircraft, and the Bendix/King Apex system typically installed on smaller aircraft.

Bendix/King Apex. The Bendix/King Apex system found on personal light aircraft employ the KFD 840 flat-panel display. This is an 8.4-inch primary flight display unit, which incorporates virtually all flight data on one backlit LCD display (Figure 3-4-17). The KFD 840 is mostly a self-sufficient unit containing all circuitry needed for processing information received from various systems, such as, navigation sensors. The KFD 840 also contains a built-in air data and attitude heading reference system. This all-inclusive concept makes the KFD 840 ideal for new light aircraft and the retrofit of older aircraft replacing traditional electromechanical instruments. The unit will also interface with various analog, RS-232, or ARINC 429 digital navigation inputs. As seen in Figure 3-4-17, the left side bezel incorporates an SD card slot. The SD card is used for system software updates and any reconfiguration needed when new equipment is installed. The system is certified for aircraft that fly below an airspeed of 230 knots, a maximum altitude of 35,000 feet, and operate in the FAA designated normal or utility categories.

Architecture. The KFD 840 unit (considered a PFD) is designed for panel installation and measures only 7 inches high by 8.5 wide by 5.72 inches deep. Although mostly self contained, the PFD requires inputs from various sensors and systems as seen in Figure 3-4-18. The minimum inputs needed to make the system operational come from an outside air temperature probe, the magnetometer, and connection to the system configuration module. The configuration module is used to store the system's configuration data and stays with the aircraft in the event the PFD is exchanged for another unit or sent out for repaired. The configuration module is a long thin cylindrical unit, approximately 0.05 inches by 17 inches and simply strapped to the wiring harness during installation. The complete system also includes a magnetometer (the CRM 500), which supplies magnetic compass information to the processor. The magnetometer is typically installed in the aircraft wing away from any ferrous metal or electrical cables, which might interfere with the earth's magnetic field.

The rear of the KFD 840 contains two air data pneumatic connections and three D-sub connectors for system power and input/output signals. The air data inputs are for pitot and static pressure. The KFD 840 is designed to receive inputs from both digital and/or analog GPS and VOR systems. The exact number of inputs is dependent on which type of signal (analog or digital) is used. The PFD processor outputs data in two formats, ARINC 429 and RS-232. This data can be used as input signals to a traffic collision avoidance system, ground proximity warning system, weather radar, a multifunction display, or various other systems. The KFD 840 will also interface with an autopilot system if installed on the aircraft.

Installation. The basic installation of any aircraft electrical equipment must comply with the guidelines in the FAA Advisory Circular AC 43.13-1B and -2B. In most cases, the installation would begin with a new instrument panel, which conforms to the aircraft's original specifications for material and dimensions. The PFD is installed directly in front of the pilot along with the traditional airspeed, attitude, and altimeter indicators that are used as standby instruments. Figure 3-4-19 shows the installation layout for a typical light aircraft. This type of installation would typically be considered a major alteration to the aircraft and would require the submission of the appropriate paperwork, most likely a *Form 337*.

After the physical installation is complete, the system-configured database must be set correctly and the attitude and navigation sensors must be adjusted. Each of these procedures requires the technician to access the KFD 840's maintenance menu on the display. The magnetometer alignment involves placing the aircraft on a series of known compass headings (typically using a compass rose) and following the steps outlined on the display. To set the attitude sensors, the aircraft must be placed on jacks and

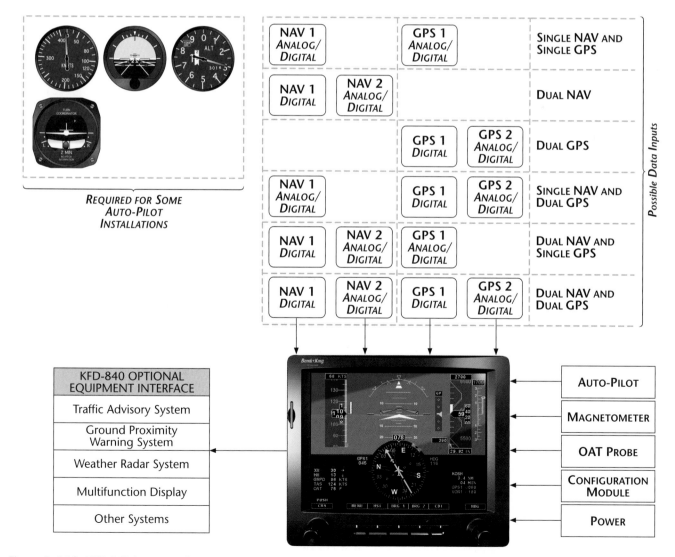

Figure 3-4-18. KFD 840 inputs and outputs

Screen Image courtsey of Bendix/King by Honeywell

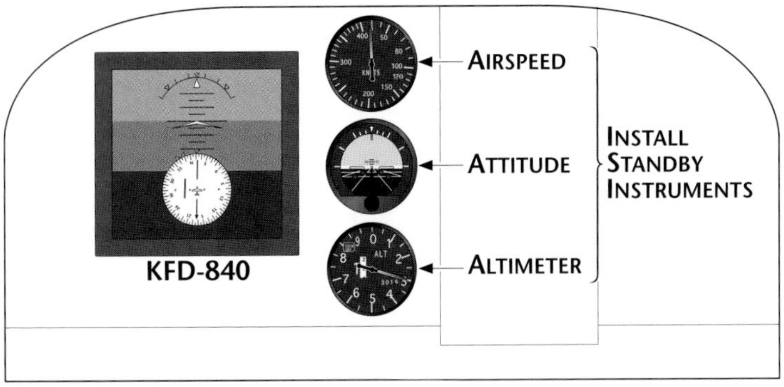

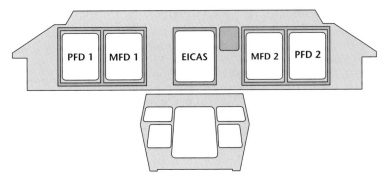

Figure 3-4-19. Typical panel layout of a Bendix King AFD-840

leveled in all directions. The technician would then make the correct selection from the PFD maintenance menu and follow the procedures. In order for the KFD 840 to properly understand all input signals and output all necessary data, the system must be "told" how the aircraft is configured. The configuration process guides the technician through a series of questions, such as, "The GPS Number 1 input signal is analog or digital?". After all configuration data has been entered, the information is saved on the system's configuration module, and the installation is complete.

Honeywell PRIMUS. The Honeywell Primus 2000 is an integrated avionics system designed for a variety of corporate and commuter type aircraft. Primus incorporates an electronics display systems, which will be discussed in this chapter, as well as various navigation systems that will be discussed later in this text. The Primus 2000 also employs an integrated radio management system designed to reduce pilot workload related to radio tuning and signal acquisition. This text will present a typical Primus 2000 system, although keep in mind all aircraft models are slightly different and the current maintenance data for the aircraft should always be consulted prior to any maintenance activities.

The Primus 2000 system incorporates a high level of integration made possible with the use

of two Integrated Avionics Computers (IAC). A typical Primus system would employ five LCD (or CRT) displays for both flight and systems data. As seen in Figure 3-4-20, the two outer units are used as primary flight displays (PFD) the two inboard units are multifunction displays (MFD), and the center most unit is an EICAS display for the engine indicating and crew alerting system. In most cases, the MFD and EICAS displays each have specific rotary knobs and pushbutton function keys located at the bottom of the display. These keys are used to select various display functions for flight activities, as well as, used during maintenance operations for access to the integrated maintenance tests. The system also employs a separate control panel for many other display functions.

The IAC contains a variety of cards with high-level integrated circuits, surface mount technologies, and application specific integrated circuits. The IAC contains the operational circuitry for several systems, which in the past were contained in several stand-alone line replicable units (LRU). As can be seen by the system diagram (Figure 3-4-21), the electronic display system is only one portion of the complete integrated package offered by the Honeywell Primus. Other major systems shown on the diagram are used for flight management, autoflight systems, radar altimeter, and weather radar systems.

The Primus electronic display system (EDS) is made up of six major components; the two integrated avionics computers (IAC), the five electronic display units (EDU), a guidance panel controller, two data acquisition units (DAU), two radio management units (RMU), and one system configuration module. The five display units receive information from the IACs as shown in Figure 3-4-22. The following modules within the IAC are necessary for operation of the electronic display system: the IAC power supplies, the electronic display interface (EDI), the input/output interface, the fault warning computer, and the associated data busses. Other circuits within the IAC not directly related to the display system are the flight management system and the automatic flight control system.

The IAC receives the majority of the input data from the data acquisition units (DAU). The data acquisition units provide the interface between various engine and aircraft systems. The DAUs acquire digital, analog, and discrete signals; processes the information, and concentrates the data into an ASCB data format transmitted through a data bus cable to the IAC. The IAC also receives inputs from the various system controls and eventually sends the information to the DUs for display to the pilots. In order to

Figure 3-4-20. Primus 2000 display system

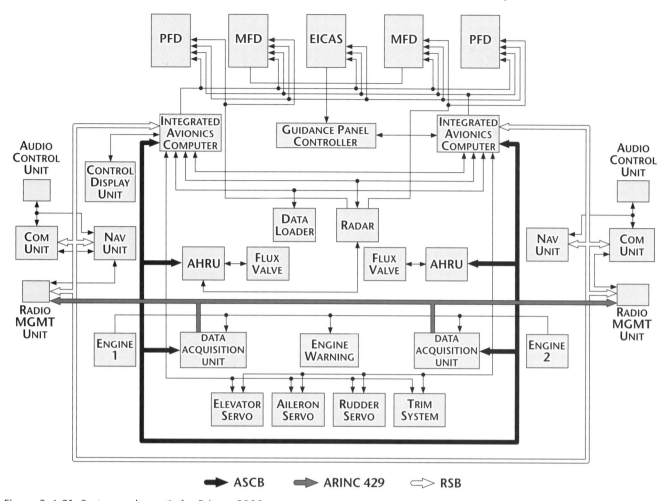

Figure 3-4-21. Systems schematic for Primus 2000

offer various options to aircraft owner/operators, each Primus system incorporates a small (approximately 1 inch by 4.5 inch by 7 inch) configuration module (CM) which stores aircraft specific systems information in a nonvolatile solid-state memory. The CM is programmed through software as well as several configuration straps that "tell" the IAC what systems are found on that aircraft. The CM is mounted in the aircraft and stays in place even when one or more IACs are removed or changed. This arrangement allows newly installed equipment to instantly recognize aircraft configuration.

The two IACs (left and right) are located in the aircraft's equipment bay, operate simultaneously, and cross talk through an ASCB data bus and the IAC I/O cards. The cross-communications between IACs provides for redundancy checks of systems data, which improves reliability. The IACs send information to the five display units through a dual data bus configuration, once again, in order to provide redundancy (Figure 3-4-23). The EFIS/FWC card in the IAC is responsible for signals to the displays and the systems warning unit. The warning unit provides discrete audio and visual indications to the pilot in the event of certain system failures.

It should be noted that most of the integrated electronic displays, flight management, and autoflight systems found on turbine-powered aircraft are very similar. For example, system architecture of the Primus system, and the IAC, share many features found in the Rockwell Collins Proline 4 family of systems (See Proline 4 section for review.)

Boeing 777 Integrated Displays

The Boeing 777, released in the mid 1990s, was developed as a highly electronic aircraft, which employed an integrated display system for flight, airframe, and engine data. Advancements in software, microelectronics, data bus systems, and digital avionics has made it possible to integrate more systems on this aircraft than any previous model. Integration not only reduces components, and saves weight; integration also improves reliability and maintainability therefore reducing operating costs. The main flight deck contains six identical 8 inch by 8 inch LCD flat-panel displays, known as the primary display system (PDS) (Figure 3-4-24). These displays are installed in the standard T fashion with the two outboard displays used

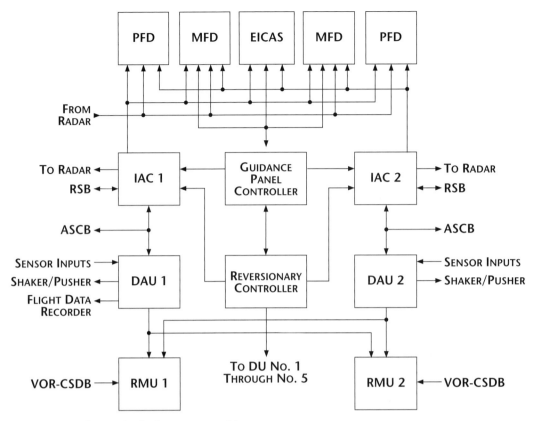

Figure 3-4-22. Electronic display system architecture

as primary flight displays (PFD); the right/left inboard displays are used as navigational displays (ND). The upper center display is used for engine indicating and crew alerting system (EICAS) information and the lower center display is a multifunctional display (MFD).

The six flight deck displays receive their video information from the Airplane Information Management System (AIMS). The AIMS integrates the various digital inputs, any discrete and analog signals, and all control information for operation of the flight deck displays. A brief description of the B-777 AIMS follows.

Airplane Information Management Systems

The airplane information management system (AIMS) found on the Boeing 777 employs line replaceable modules (LRM) to integrate a multitude of electronic systems. The concept behind AIMS is simple; integrate various systems by sharing common functions and components. The AIMS actually integrates data collection, computing functions, power supplies, and output functions for several subsystems. By sharing both hardware and software components AIMS enhances reliability, improves

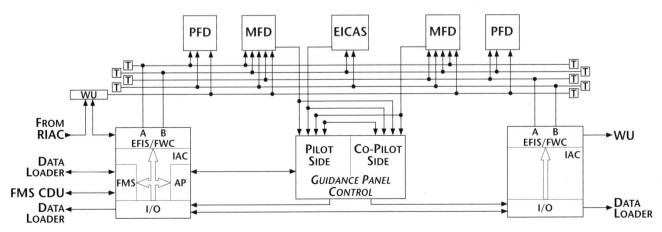

Figure 3-4-23. Honeywell Primus system showing five displays and two IACs

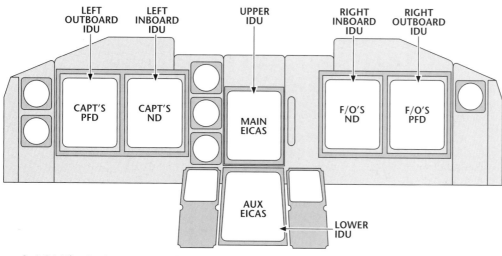

Figure 3-4-24. The Boeing 777 LCD flight deck display

redundancy, and creates a substantial weight savings as compared to conventional systems.

Boeing and Honeywell designed AIMS to incorporate a fault tolerant software design. Fault-tolerance along with proper redundancy of critical systems hardware allows the AIMS software to detect a fault and reconfigure the system for uninterrupted operations. In many cases, the flight crew would not even be aware of the system failure. In theory, this type of design allows the B-777 to continue flying with the failed system until the next convenient maintenance opportunity. Typically, the repair would be performed during scheduled night maintenance, not during the short time period available between flights.

On the B-777, there are two AIMS cabinets, each with eight line replaceable modules (LRM). Each cabinet contains four input-output modules (IOM) and four core processing modules (CPM). The IOMs transfer data to and from the AIMS cabinet and within the four CPMs. The four CPMs perform the calculations for the various avionics systems serviced by AIMS. Each LRM is designed for quick removal and installation. Without the AIMS concept, several LRUs would be required to perform the same tasks. The LRMs are smaller than the conventional LRUs since they share several functions with other LRMs within AIMS. Figure 3-4-25 shows an AIMS cabinet and LRMs on a Boeing 777.

The AIMS has the ability to transfer data in a variety of communication formats (Figure 3-4-26). To help simplify wiring and increase the speed of data transfer, AIMS uses the ARINC 629 data bus to communicate between many of the aircraft systems. A fiber optic data bus is also employed for the onboard local area network (OLAN). This system follows the ARINC 629 fiber optic specifications. As seen in Figure 3-4-27, the OLAN is composed of two major components: an avionics LAN and a cabin

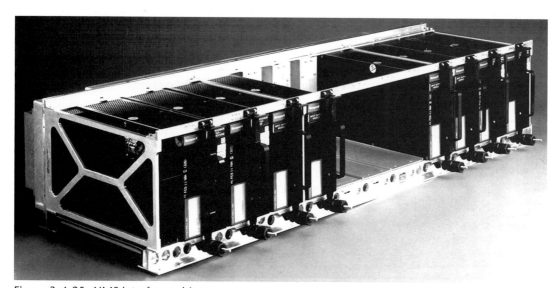

Figure 3-4-25. AIMS interface cabinet

Courtesy of Honeywell, Inc.

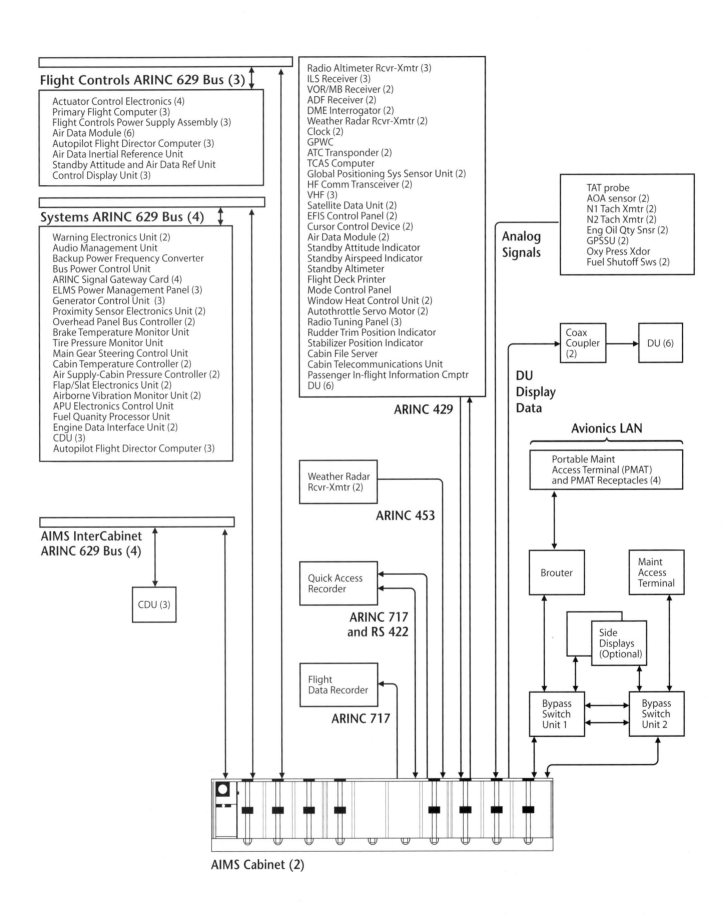

Figure 3-4-26. Boeing 777 airplane information management system (AIMS) interface diagram

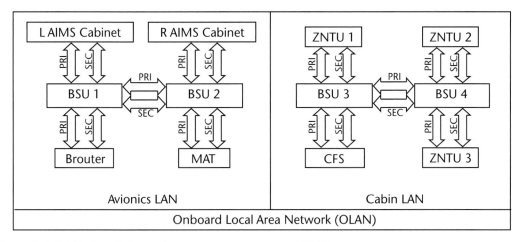

Figure 3-4-27. Airplane information management system (AIMS)

LAN. The avionics LAN connects the left/right AIMS cabinets, the maintenance access terminal (MAT) and the brouter (bridge router). The brouter connects the portable maintenance access terminal (PMAT) to the network. The MAT & PMAT are used to access maintenance data and will be discussed in Chapter 5.

There are three ARINC 629 data busses used for flight control, four ARINC 629 data busses for systems communications, and four ARINC 629 data busses used to connect the LRMs within the AIMS cabinets. As seen in Figure 3-4-26 other data busses include:

1. ARINC 429 for communications with a variety of airframe, engine, and avionics systems

2. ARINC 453 for weather radar information

3. ARINC 717 and RS 422 used for communications with the flight data recorder, and the quick access recorder

As discussed in Chapter 2, the ARINC 629 data bus has the ability to transfer data in a variety of formats, therefore allowing AIMS to communicate with a multitude of systems. AIMS also monitors a variety of analog signals to various components. AIMS coordinates data input and outputs, and performs data processing calculations for a variety of systems. These systems include:

- Primary display system (PDS)

- Central maintenance computing system (CMCS)

- Airplane condition monitoring system (ACMS)

- Digital flight data recorder system (DFDRS)

- Data communication management system (DCMS)

- Flight management computing system (FMCS)

- Thrust management computing system (TMCS)

The display units (PFD, ND, EICAS, and MFD) are connected to AIMS through display unit video busses. The primary display system (PSD) video busses employ coaxial cables that transfer high-speed digitized and compressed video signals to the six flat-panel displays. The B-777 contains two PFDs, two NDs, one main EICAS display, and a multifunction display (MFD). The PDS component locations are shown in Figure 3-4-28. The display units receive high-speed digital data from AIMS through a coaxial cable (and coax couplers) in order to eliminate signal interference and potential display distortion. There are two remote light sensors as well as control panel brightness and contrast controls for the various display units. These controls and sensors input data to AMIS, which in turn controls the displays.

A new feature found on the B-777 are the two cursor control devices (CCD) located on the center console between the two pilots. There is one CCD for the pilot and one for the first officer. The CCD is similar to a common touch pad found on a typical laptop computer. The CCD has differences from a laptop touch pad, as seen in Figure 3-4-29 making it more user friendly for aircraft operations. Obviously, aircraft can encounter turbulent conditions, which would make it difficult to operate a simple touch pad. The CCD incorporates a hand/palm support in order to stabilize the fingers as they operate the CCD. The touch pad moves the cursor position on the active display and the pilot must then press the cursor select switch to make a selection. As can be seen on the CCD in Figure 3-4-

Figure 3-4-28. Primary display system component location   *Courtesy of Boeing Commercial Airplane Company*

29, the press switch is located on the side of the CCD, separate from the touch pad in order to minimize the possibilities of incorrect selections. The pedestal mounted CCDs control only the left/right inboard displays and the lower center display. The PFDs (outboard displays) and EICAS (upper center display) are operated using traditional control panels. There is also a trackball-type CCD located at the maintenance access terminal, which is used for control of the maintenance display only.

In order to provide reliability, the PFDs receive inputs from four separate channels. Each channel has its own graphics generator, a coaxial coupler, and six coaxial cables connected to each PFD. The four graphics generators are contained in the two AIMS cabinets. During normal operation three graphics generators each send signals to two of the display units. In the event of a system failure, each graphics generator has the capability to drive three displays, a maximum of two critical formats and

one essential format. PFDs and the EICAS are critical formats; NDs and the MFD are essential formats. If multiple video channels should fail, one channel can drive three displays and then echo (or duplicate) the images on the other three displays. Of course, since both the AIMS and PDS are considered critical to flight safety, there are multiple power sources available to each. Both the first officers' and captains' 28 VDC busses send power through over a dozen circuit breakers to provide redundancy. The hot battery bus can also power AIMS if other power sources fail.

Troubleshooting the PDS is provided by the central maintenance computing system (CMCS). The major computing functions for this system will be contained in the AIMS processors. On the B-777 technicians will access the central maintenance system data using the maintenance access terminal (MAT). The maintenance access terminal is located on the flight deck just behind the first officer's station.

The technicians can also access maintenance data supplied by AIMS through the local area network and a portable access terminal. More details on assessing B-777 maintenance information is presented in Chapter 5.

The Boeing 787 Display System

The Boeing 787 Dreamliner, which at the time this text was written was soon to be released, will incorporate all aspects of the integrated modular avionics (IMA) design concepts. The Dreamliner will utilize a common core computing system, which Boeing calls the common core system (CCS). The concept is to integrate over 100 different LRUs into a central computing system that will share many of the functions and components of the individual systems found on previous generations of aircraft. The main backbone data bus for this system employs the Ethernet-based ARINC 664 (AFDX) technologies. The AFDX and CCS combination create an integrated system that reduces wire and saves approximately 2,000 pounds compared to earlier-generation aircraft.

The 787 contains five 15 inch (measured diagonally) displays arranged in a landscape format with four displays on the instrument panel and one display on the center console (Figure 3-4-30). The large-scale displays provide more flexibility for display configuration options, such as, interactive airport maps used during taxi. The PFDs are located on the two outermost displays. The two inboard displays are NDs, and the console display is used for a variety of functions including flight planning and redundancy in the event of a display failure. The center console also contains two alphanumeric control panels and a cursor control device, one for each pilot. The cursor control device (often called a scratch pad) contains a touch pad, a palm rest, and validation buttons that must be pressed to activate a selection. The display systems incorporate two heads up displays; this is the first transport category aircraft with HUD as standard equipment. Located on the side panel just to the left/right of each pilot is the electronic flight bag (EFB) display.

System Architecture

The CCS consists of three major elements:

1. Processing circuitry to execute applications software

2. The common data network (CDN) that provides communications between all segments of the CCS

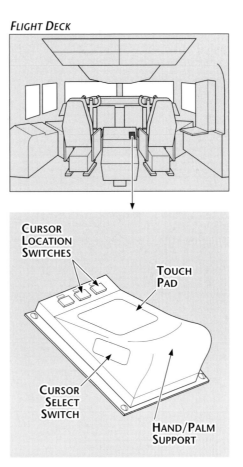

FLIGHT DECK

Figure 3-4-29. Cursor control device (two each, one pilot and one co-pilot)

3. Remote data concentrators that provide gateways to analog, discrete, and ARINC 429-based equipment.

The CCS also hosts specialized functionality; for example, graphics generation used for the instrument display system. In order to accomplish this level of integration in a common system, it is imperative that robust partitioning of all circuitry and software be incorporated into the CCS design. Partitioning is vital to the reliability of the CCS and the ability of systems to fail active. The process of partitioning involves the separation of both hardware and software into individual sections; this allows each section to operate as an individual computer. Partitioning will permit one section of the CCS, hardware/software, to fail without adversely affecting another.

The CSS performs most of the processing for the B-787's display system. The majority of the display applications run using the general processing modules within the CCS. The display information is then sent to the graphics generation module (GGM), also housed in the CCS cabinets. The GGM then converts the information and sends the data to the individual dis-

Figure 3-4-30. Boeing 787 instrument panel

Courtesy Boeing Commercial Airplane Company

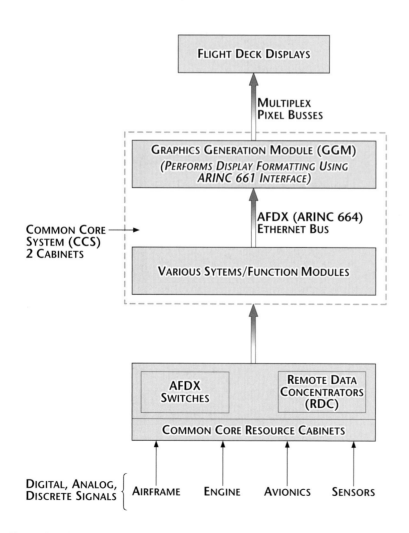

Figure 3-4-31 Boeing 787 common core system input/output

plays using multiple pixel busses. In order to ensure consistency between components, the B-787 display data adheres to the ARINC 661. ARINC 661 contains standards for the flight deck display interface and ensures consistency between all system components.

A key element of CCS is the common data network (CDN), which employs the Ethernet-type (AFDX) network and follows the ARINC 664 protocols and standards. The AFDX/ARINC 664 data bus systems were discussed in Chapter 2 of this text. The CDN connects the processors, the remote data concentrators and specialized LRUs, using fiber or copper connections. Network switches are located in the common computing resources (CCR) cabinets, which are mounted throughout the aircraft. These switched support the distribution process of the CDN.

The design concepts for the B-787 integrated display system and the common core system are outlined in Figure 3-4-31. Follow this diagram during the following discussions. This diagram depicts the information flow from the bottom upward, eventually reaching the flight deck displays at the top of the diagram. Various systems are monitored using independent airframe, engine, and avionics sensors. The sensors connect to one or more remote data concentrators (RDC), which are distributed throughout the aircraft and located in common computing resource cabinets. The RDCs provide a digital gateway for analog, discrete, or digital (ARINC 429) signals. The RDC converts all inputs into an AFDX format in order to transmit the information on the Ethernet data bus. The use of RDCs helps to reduce the long runs of wire that would be otherwise required to reach the main equipment bay.

The RDCs send ARINC 664 (AFDX) data to the two CCS cabinets using traditional copper data bus or fiber optic cable. Inside the CCS cabinets, the information is sent to one or more systems and/or function modules. These modules process the data for a variety of inputs and outputs, perform validity checks, and send the data to the graphics generation module (GGM). The GGM is used to format the data into a display interface format using the ARINC 661 standard. The GGM connects to the various flight deck displays on a dedicated pixel bus, with multiple busses used in order to provide system redundancy.

Airbus A-380 Display System

The Airbus A-380 is a double-deck, wide-body, four-engine transport category aircraft employing state-of-the-art avionics and flight deck display systems. The main instrument panel includes eight identical 15 by 20 cm liquid crystal displays as seen in Figure 3-4-32.

Figure 3-4-32. A-380 flight deck layout

Courtesy of Airbus S.A.S..

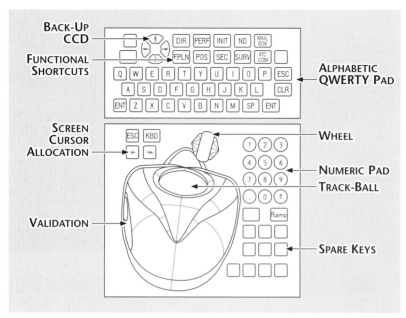

Figure 3-4-33. A-380 keyboard and cursor control device

Airbus calls this the control and display system (CDS). As with other aircraft, the two outboard displays are used for PFDs, moving inboard the next two displays are NDs, and the center most display is the engine/warning display (E/WD). The center console includes three displays; two used as MFDs and one system display (SD). Placed at a slight angle facing the pilots and located between the front and side panels are the onboard information terminals (OITs). Access to flight operations manuals,

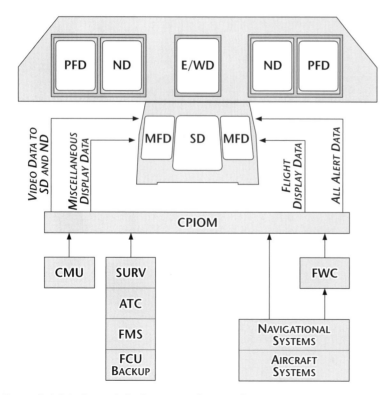

Figure 3-4-34. Control display system basic architecture

takeoff/flight/landing performance, minimum equipment lists, and navigation charts are all provided on the OIT. This display is basically a Class 3 electronics flight bag linked to the aircraft network for processing and data storage. The CDS provides for an optional HUD which uses PFD data for a pilot and co-pilot heads up display. An integrated standby instrument system (ISIS) is located just left of the center (E/WD) display. The two independent ISIS displays provide emergency back up in the event both PFDs and NDs fail.

Controls for the A-380 CDS include a large alphanumeric keyboard placed on a fold-away table directly in front of each pilot. This arrangement is made possible since Airbus employs side stick controllers for pilot/co-pilot aircraft control. (Boeing uses the conventional control yoke/wheel arrangement.) The A-380 has two keyboard and cursor control devices (KCCDs) located on the center console between the pilots. As seen in Figure 3-4-33, the KCCD employs a QWERTY alphabetic keypad, a track-ball with palm rest, validation keys, and a wheel control device. The KCCD allows the pilot/co-pilot to interact directly with the on-side ND, MFD, or sections of the SD. ("On-side" means the pilot's KCCD controls only the pilot side displays; the co-pilot controls only co-pilot's displays.)

Architecture

The control display system uses the star topology Ethernet data bus system, described in Chapter 2 for connection to the various components used to supply data to the CDS. The design concepts of integrated modular avionics (IMA) are also employed on the A-380 to reduce the number of LRUs necessary for processing display data. Similar to the Boeing 787, this combination of IMA concepts and star topology decreases the amount of wire required and allows for integration of systems. In general, these design concepts also reduce aircraft weight while at the same time improving system performance and reliability.

The CDS receives input information from aircraft and navigation systems through an integration of LRUs input/output modules (IOM) and the Core Processing Input Output Module (CPIOM) as seen in Figure 3-4-34. The CPIOM is a cabinet type mounting rack configuration containing 30 different modules which hosts 21 different avionics functions; the CDS is one of those functions. The systems that feed information to the CDS include the Flight Warning System (FWS), navigation systems, Concentrator and Multiplexer for Video (CMV), Surveillance (SURV), Flight

Management System (FMS), Air Traffic Control (ATC); Full Authority Digital Engine Controls (FADEC), and the Primary Flight Computers (PRIM).

The CMV provides multiplexed video signals from several sources located throughout the aircraft. The video can be displayed on the pilot/co-pilot navigational display (ND) as well as the system display (SD). Video cameras are used for surveillance of areas such as the cabin door entrance, external cameras for taxi, and internal cameras for cabin monitoring.

The FWS is a dedicated system to present warning and caution indications to the pilots in the event of a system malfunction. To provide system validity, the FWS operates independently of other display system functions. FWS software is designed with robust partitioning to ensure the operation of all warning displays. As seen in Figure 3-4-35, the FWS controls all electrical signals to the master warning and caution annunciators, the PFDs, the E/WD, and the SD, as well as the flight deck loud speaker for all audio warnings.

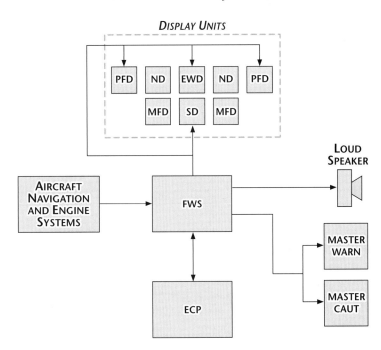

Figure 3-4-35. The A-380 flight warning system

16   FMS1

33

N

KSHD   12.5

TCAS OFF
--.- NM

ETA----Z

SAT	37
TAT	37
TAS	0
GSPD	1

Honeywell

1.53   1.55
[1.00]   [1.00]

T/O
EPR

0.0   0.0

N1

165   179

ITT

	N2	
0.0	N2	0.0
0	FF (PPH)	0
85	OIL TEMP	87
0	OIL PRESS	0

TOTAL FUEL (LBS) [12500]
6200   0   6300
        0

OUTBD BRK LO PRESS
PASSENGER DOOR
SET LDG ELEV
BRAKE TEMP
L-R PACK HIGH FLOW
NOSE STEER OFF
HYD PUMP 3A OFF
L-R ENG SHUTDOWN
L-R VSHLD HEAT OFF
END

GEAR

(DN) (DN) (DN)
IN

0

-TRIMS-

NU

AIL

7.9

LWD   RWD

ND
STAB   NL RUDDER NR

STATIC GROUND OPERATION
PROHIBITED BETWEEN 66-80% N1

INDICATED SPEEDS)

VMO	(BELOW 8,000 FT).	300
VMO	(8,000 TO 30,267 FT).	340
MMO	(30,267 FT TO 35,000 FT).	0.89
MMO	(AT 41,400 FT).	0.88
MMO	(AT 47,000 FT).	0.858
MMO	(AT 51,000 FT).	0.842
VLO	(EXT) (L/G EXTENSION).	200
VLO	(RET) (L/G RETRACTION).	200
VLE	(L/G EXTENDED).	250

SELECT

Chapter 4

INTEGRATED
monitoring and
warning systems

Section 1

Introduction

Early in the history of aviation it became clear that certain systems should be monitored during flight. Even the most basic aircraft had a means to determine engine r.p.m., engine oil pressure, and fuel quantity. As aircraft grew in complexity, many systems were added requiring constant or periodic observations. Airspeed indicators, altimeters, and manifold pressure gauges soon cluttered the instrument panel. Twin engine and turbine engine aircraft brought more engine instruments. Faster, higher, and longer distance flights required more systems and more instruments.

The aircraft grew from the simple machine first launched at Kitty Hawk to a complexity that required multiple flight crewmembers. The pilot, the co-pilot, a navigator, and the radio operator were all standard crew on large aircraft built in the 40s and 50s. Then, in the late 50s and early 60s, aircraft systems began to simplify. With the help of modern electronics, modern for the time period that is, the radio operator's job was combined with the pilot's and navigator's responsibilities. The Boeing 727 required a three-person crew, and the newer B-737 required only a two-person crew. Today aircraft flight is possible using only computers for control and no pilot at all. Of course, these unmanned aerial vehicles (UAV) are in limited use today, mostly within the military; however, whether they will be used in public aviation transportation is yet to be seen.

At the same time when crew size seemed to decrease, more systems were added to the air-

Left: The center screen in most glass cockpits provides engine and aircraft related information. This display shows recently shutdown engines, gear status, fuel load, trim settings, as well as a number of configuration messages.

Figure 4-2-1. Two EICAS displays from a Boeing 757

craft to improve flight safety. Landings with reduced visibility became commonplace. Radar was onboard almost every commercial airliner. Hydraulics, pneumatics, and electrical systems grew in size and complexity. During this time period, aircraft designers and engineers realized that pilot workload was becoming overwhelming and systems monitoring had to be reduced. The first step was to reduce the number of instruments, gauges, and lights that required constant attention.

The invention of the horizontal situation indicator (HSI) and attitude director indicator (ADI) was a major step toward reducing pilot workload. The information on these indicators combined a multitude of instruments to reduce instrument panel clutter. Much of the information contained on these indicators was out of view until needed. This helped to reduce confusion and pilot workload. Today, information is presented to the flight crew on a need to know basis. Only necessary information that is critical at that time is displayed, once again reducing confusion and workload.

It was soon possible to combine many navigation and communication radio systems into a common control panel, and many of the systems that required manual operation could be operated automatically. But still, automation required more equipment. Size and weight were limiting factors as to how far automation could go. More electronics equipment also generated more heat, which presented cooling problems.

Then it happened, with the advent of the Boeing 757 and 767 aircraft, many of the older

analog systems were replaced with lighter, faster, and more reliable digital technologies. Conventional gauges and instruments, with a multitude of moving parts, were replaced with CRT displays. Now for the first time, a real reduction could be made in the flight crew workload. Automatic engine and system monitoring computers coupled to CRT display systems were introduced on the B-757/767. These aircraft contained the first generation of *engine indicating and crew alerting systems* (EICAS) available on transport category aircraft.

From the late 1970s to the present, improvements in computer technologies have allowed aircraft manufacturers to improve safety and reduce pilot workload. Modern aircraft such as the A-380 or the B-777 monitor virtually every system that uses electricity. On modern aircraft, engine and systems monitoring is accomplished through two basic systems: generally speaking, EICAS and *electronic centralized aircraft monitoring* (ECAM). EICAS is used on Boeing, while ECAM is used on the Airbus Industrie aircraft.

The EICAS and ECAM systems have also evolved over their life span. For example, early EICAS were more or less dedicated systems. They used computers dedicated to EICAS and did little else. Today, both EICAS and ECAM are integrated with other aircraft systems, such as the central maintenance system and electronic flight instrument system. The Boeing 777 incorporates an *airplane information management system* (AIMS), which further integrates electronic systems. AIMS actually share some of the computer software between various systems of the aircraft. The newest transport category aircraft designs (B-787 and A-380) incorporate the concepts of *integrated modularized avionics* (IMA). IMA further integrates systems and improves reliability and the capabilities of modern indicating and warning systems. On many modern aircraft, system monitoring can even download systems data to ground facilities. This allows maintenance crews to determine fault activity while the aircraft is still in flight, which reduces repair time and unscheduled delays.

Aircraft monitoring systems have come a long way. Today's computers allow for more control with less pilot interaction. Corporate-types, such as the Challenger, personal light aircraft, and helicopters, including Sikorsky's S-92, now employ automated monitoring systems.

This chapter will examine the automated monitoring systems currently used on transport category aircraft focusing on Boeing's EICAS and Airbus' ECAM systems.

Section 2

Engine Indicating and Crew Alerting Systems (EICAS)

The automated engine and systems monitoring process used on modern Boeing aircraft is called EICAS (Engine Indicating and Crew Alerting System). This system incorporates two LCDs or CRTs used to display engine data, airframe systems data, and warning messages to the flight crew. Early systems such as those found on the Boeing 757 and 767 operate more or less independent of other major aircraft systems. The EICAS found on the Boeing 747-400, 777, 737-800, and the 787, are fully integrated systems that share components and information with the *electronic instrument system, the central maintenance computer system* (CMCS), and the *flight management system* (FMS).

General Description

EICAS has many similarities to the electronic flight instrument system discussed in chapter three. Like EFIS, EICAS employs digitally controlled displays; however, EICAS is used to display various system parameters, such as engine pressure ratio, r.p.m, and exhaust gas temperature. Other systems, such as hydraulic or pneumatic pressures, and electrical system parameters can also be displayed, or in some cases, removed from the screen at the discretion of the pilot. Another vital function of EICAS is to monitor the various aircraft systems and to display caution and warning information in the event of a system failure.

The EICAS displays certain aircraft system and engine parameters on a need to know basis. That is, not all the systems data are displayed continuously. In the event of a system malfunction, any vital information automatically appears on a CRT and the appropriate caution or warning signals are activated. During normal operation, only a minimum of engine data is displayed; additional system data may be displayed if the pilot activates the appropriate EICAS control.

EICAS Displays

EICAS contains two LCD or CRT displays typically placed vertically as shown in Figure 4-2-1. During normal flight configurations, it is typically blank. The lower display is used to display status of any malfunctioning system. The lower display is also a backup display in the event the top display fails. The upper EICAS display is called *the main EICAS display* and the lower display is called the *auxiliary EICAS display*. During normal operation, the upper display shows engine data, such as exhaust gas temperature (EGT), rotor r.p.m. (N_1), and (N_2).

On any given aircraft, the EICAS displays are interchangeable and commonly referred to as line replaceable units (LRUs). On most aircraft, the EICAS CRT/LCD and the EFIS CRT/LCD are also completely interchangeable. Figure 4-2-2 shows a typical CRT, the removal handle and the location of the installation screws. To remove the CRT:

1. Turn system power off

2. Hold the handle flat against the CRT and remove the retaining screws located at the top of the display

3. Pull the handle outward and the CRT will slide from its rack. In general early systems employed cathode ray tubes (CRT) displays and newer aircraft use liquid crystal displays (LCD)

In this text the term *display* can be either a CRT or LCD.

CAUTION:
(1) Most display units are electrostatic discharge sensitive. Be sure to take the necessary precautions to prevent damage to the unit.

(2) Some displays have an antiglare film on the face of the CRT/LCD. These displays are easily scratched and require special cleaning procedures. Always be careful when using tools around the face of the display. Protect the display from rough handling and never set the unit face down.

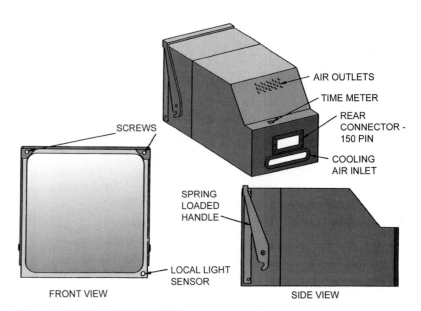

Figure 4-2-2. Typical EICAS CRT

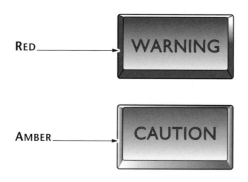

Figure 4-2-3. EICAS master warning/caution light assembly

In general, installation is preformed in reverse order of removal. During installation be sure the handle is in the full-out position and gently push the display into position. Then press the handle slowly toward the CRT. This will seat the display and engage the electrical connector pins at the rear of the unit. After a proper installation, the handle should lay flat against the display held by the retaining screws. In some units the installation screws and handle are located on the side of the CRT not on top. Remember to always use the maintenance manual for removal and installation of equipment.

EICAS Formats

There are several display formats used by the different versions of EICAS. These formats vary with aircraft model. Some common formats include primary, secondary, and compact modes. The primary format shows on the upper CRT during normal operation. Four different colors are used to display information. A change in color indicates a change in system status. Six colors are used by most EICAS displays. White is used for display of various scales, pointers, and digital readouts when the system is in the

normal operating range. Red is used to indicate that a system has exceeded a predetermined value (Redline). If a system exceeds its limits, the entire scale, pointer, and/or digital readout will turn red. Red is also used to show certain warning messages. Green is used to indicate normal operation of a system. Amber is used for certain warning messages and some scale markings. In some cases, a system display will change to amber if that system enters the caution range. Magenta (pink) is used for certain messages and display parameters. Cyan (light blue) is used for various labels and messages.

Alert Messages

During normal operation, EICAS is used to alert the flight crew as to any abnormal powerplant or airframe system operation. There are four types of alert messages used with EICAS:

- Level A are *warning* messages shown in red. These are the most important messages.

- Level B are *caution* messages displayed in amber.

- Level C are *advisories* displayed in amber or cyan depending on the specific model EICAS.

- Level D messages, called *memos*, are displayed in white. Status and maintenance messages are also shown on some systems, and are typically displayed in white.

Warnings are known as Level A messages. Warnings require immediate attention and immediate action by the flight crew. Level A faults include very serious failures, such as engine fire or cabin depressurization. Warnings will also activate additional aural and visual annunciators on the flight deck. For example, a fire bell sounds in the event of an engine fire. There are very few faults that

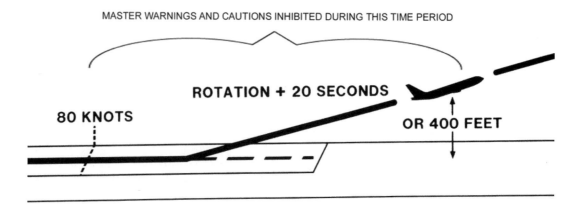

Figure 4-2-4. EICAS take off caution inhibit criteria

require Level A warnings. Caution messages, known as Level B alerts appear on the EICAS display in amber just below any Level A message. Level B alerts also create a distinct audio tone and illuminate a discrete annunciator. Cautions require immediate crew awareness and future crew action. Advisory messages (Level C) require immediate crew awareness and possible future action. Memos (Level D) are used for crew reminders. Typically there is no aural tone or master caution/warning associated with advisories or memos. For all levels of alerts, the most recent message appears at the top of its category. Level A messages appear at the top of the display, Level B is below level A, Level C is next and Level D is shown on the bottom of the list. Status and maintenance messages are used to aid the flight crew and maintenance technicians in determining the aircraft's status prior to dispatch. Maintenance messages used in conjunction with the minimum equipment list for the aircraft will determine what repairs, if any, are required prior to the next flight.

All EICAS equipped aircraft contain two *master warning/caution annunciators*. These annunciators consist of an illuminated switch assembly located on the instrument glare shield directly in front of the pilot and co-pilot. The red warning is located in the top half of the assembly and the amber caution is located on the lower portion of the assembly (Figure 4-2-3). The warning and caution lights illuminate whenever EICAS presents a Level A and B message. If the assembly is pressed, the lamps are extinguished and any aural warning is canceled; however, the related EICAS message remains on the CRT display.

At various times during the flight it is best not to distract the pilots with alert messages. Therefore, EICAS incorporates inhibit software to keep alert messages from being displayed during crucial flight periods. Most inhibits are initiated during takeoff and landing. In Figure 4-2-4 the master warning and caution lights and related audio tones are inhibited from operation between the time the aircraft reaches 80 knots until the aircraft climbs to 400 feet radio altitude or 20 seconds after liftoff, which ever comes first. If the problem still exists after the inhibit period, the appropriate EICAS message, discrete annunciator, and aural tone will be activated.

Compact Mode

The EICAS incorporates a compact mode, which is used when one or more displays are inactive, or being used to display other information. In the compact mode the data is typically displayed in digital format only. In other words, vertical or round dial instrument representations are eliminated. The compact mode is used during flight if one display becomes inactive. Compact information can be shown on either the main or auxiliary displays or in some aircraft on the MFD or ND. While in the compact mode, EICAS is said to have "degraded operation."

Event Recording

Engine indicating and crew alerting systems have the capability to record system data in a nonvolatile memory. Maintenance personnel can access the memory to verify system malfunctions and exceedances. There are two types of data stored in the EICAS nonvolatile memory: manual events, and automatic events.

If a component fails causing a system to exceed limits or parameters, the EICAS will automatically record the event in the *auto event* mode. The EICAS *manual event* mode is activated whenever the **EVENT RECORD** button on the EICAS control panel is pressed. On early systems this nonvolatile memory was quite limited; later versions increased the storage capacity for recording fault information.

Maintenance Pages

For maintenance purposes, EICAS may employ a series of displays called *maintenance pages* to show data related to the various systems monitored by EICAS. For most aircraft, these pages can be accessed on the ground and/or during flight. Maintenance pages are displayed on one of the system's CRTs. Access to a specific page is usually accomplished to verify the system performance during a previous flight. The maintenance technician will typically receive a maintenance discrepancy write-up from the flight crew, indicating a system malfunction. The faulty system is then accessed through the EICAS maintenance pages, which provides more data on the system parameters.

Section 3

First Generation EICAS

The first practical use of EICAS on transport category aircraft came with the introduction of the Boeing 757 and 767 aircraft. First generation systems are also found on some B-737s. The Boeing 757 engine indicating and crew alerting system will be discussed here as an example of a *first generation* EICAS.

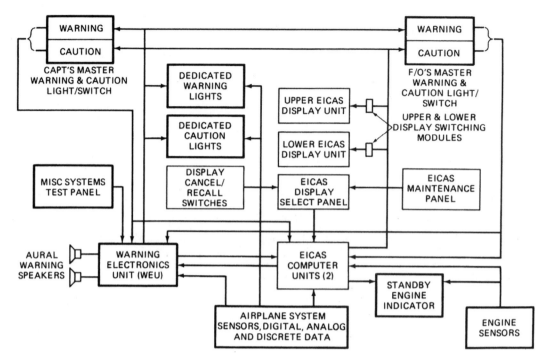

Figure 4-3-1. Block diagram of a B-757 EICAS

Courtesy of Boeing Commercial Airplane Company

System Architecture

A block diagram of the B-757 EICAS is shown in Figure 4-3-1. In this type system, two EICAS computers monitor inputs from engines and airframe systems to display system data and alert the crew in the event of a malfunction. The EICAS computer uses both discrete analog and digital data to communicate with the various components of the system. The ARINC 429 data bus system is used for most EICAS digital data transmission and reception. But, be aware, other data bus formats may be used on certain systems.

The major EICAS components are two computers, two CRT displays, one display select panel, one maintenance panel, and two display switching module. EICAS is also made up of the several discrete components shown surrounded by a bold rectangle in Figure 4-3-1. As shown in Figure 4-3-2, the displays and switching panels are located on the instrument panel. The discrete annunciators are located on the pilot's and co-pilot's glare shield. The EICAS computers and switching modules are located in the main equipment rack.

First generation systems contain several discrete annunciators and a standby engine indicator as shown in Figure 4-3-3. The discrete annunciators are used to get the attention of the flight crew to warn them of system failures, such as a ground proximity warning or autopilot disconnect. The standby engine indicator is comprised of two independent

liquid crystal displays (LCD). This system will be discussed in more detail later in this chapter.

A more detailed block diagram for the B-757 EICAS is shown in Figure 4-3-4. All digital data connections are shown by a wide, double line arrow; analog data is shown by narrow, single line arrows. Reference this diagram during the remainder of the discussion on system architecture.

EICAS receives power from both 115 VAC and 28 VDC sources running through various independent circuit breakers. This system allows for redundant power inputs to EICAS. The right main AC bus powers the right side computer, lower CRT display, and display select panel. The left computer and upper CRT display receive power from the left main AC bus. The right 28 VDC bus powers the upper and lower display switching modules. To offer the least likely possibility of failure, the standby engine indicators receive 28 VDC from the hot battery bus. The master warning/caution lights are powered by the master dim and test circuits, receiving 28 VDC.

The two EICAS computers communicate with each other via a high-speed ARINC 429 digital data bus for the purpose of comparing input data. This bus is called a *cross talk* (XTLK) bus since it allows both EICAS computers to share data or talk to each other. This feature is essential to provide redundancy for the system. The EICAS computers receive over 400

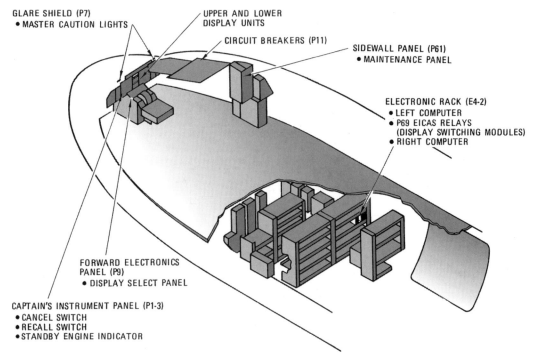

Figure 4-3-2. Location of EICAS system components

Courtesy of Boeing Commercial Airplane Company

analog inputs from various systems and specific switches throughout the aircraft. Discrete signals include information, such as ground-to-air/air-to-ground, and resets for the master warning/caution annunciators. Pin programming for each EICAS computer is also accomplished through discrete input signals. Pin programming will be presented in more detail later in this chapter.

As shown in the upper left corner of Figure 4-3-4, the discrete signals from the ground-to-air/air-to-ground relay, maintenance control panel, and the cancel/recall switches are sent to the computers through the display select panel. In the lower portion of the diagram it can be seen that the discrete right and left engine data is paralleled to both the computers and the standby engine indicator (bottom right).

Digital inputs for the EICAS computers include engine data from the left and right electronic engine controls (EECs). Due to the high priority of engine data, the EECs transmit on a dual channel directly to both EICAS computers. Some items, such as the display select panel (DSP) and the fuel quantity indicating system (FQIS) transmit on a single digital bus to both EICAS computers. A third type of digital input is transmitted to only one computer and that data is shared through the computer cross talk bus. For example, the left flight management computer (FMC) transmits only to the left computer, the right FMC transmits to the right computer.

Switching Modules

On the B-757, two EICAS *switching modules* are used to control the output data signals from the EICAS computers to the upper and lower CRT displays and to the flight data acquisition unit. These modules are made up of circuit cards located in the equipment rack between the left and right EICAS computers. The switching relays are located on the circuit cards. The

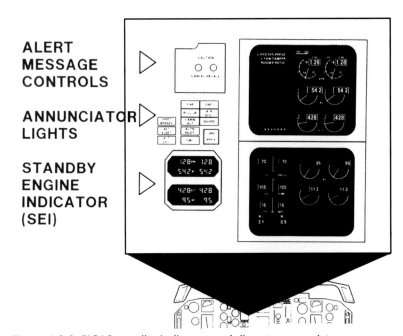

Figure 4-3-3. EICAS standby indicators and discrete annunciators

Courtesy of Boeing Commercial Airplane Company

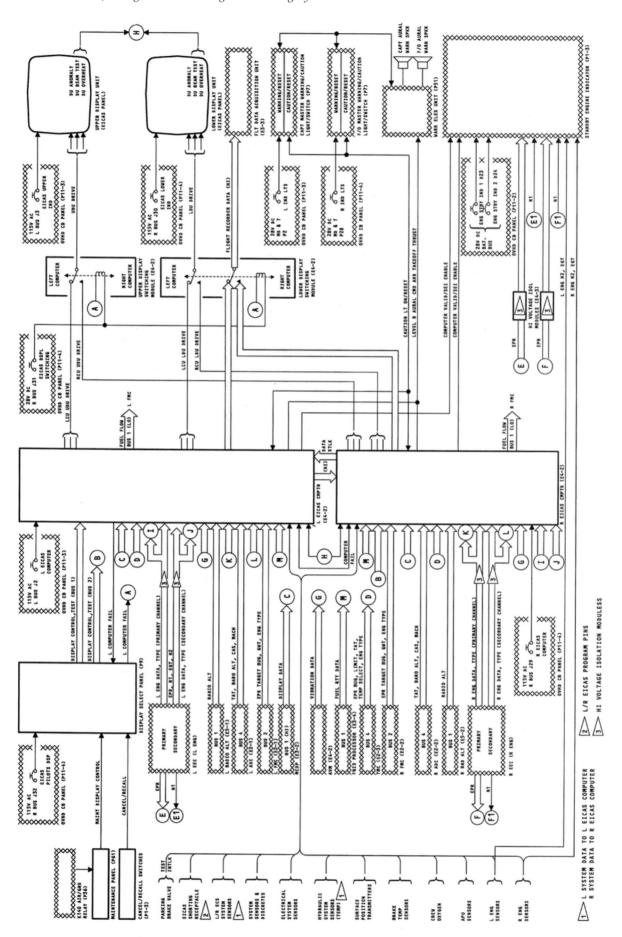

Figure 4-3-4. Detailed block diagram of the B-757 EICAS

Courtesy of Boeing Commercial Airplane Company

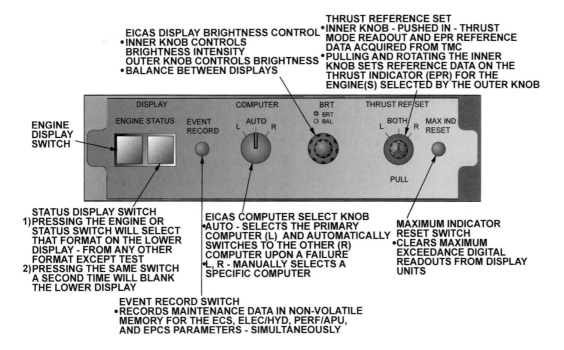

EICAS DISPLAY BRIGHTNESS CONTROL
• INNER KNOB CONTROLS
  BRIGHTNESS INTENSITY
  OUTER KNOB CONTROLS BRIGHTNESS
• BALANCE BETWEEN DISPLAYS

THRUST REFERENCE SET
• INNER KNOB - PUSHED IN - THRUST
  MODE READOUT AND EPR REFERENCE
  DATA ACQUIRED FROM TMC
• PULLING AND ROTATING THE INNER
  KNOB SETS REFERENCE DATA ON THE
  THRUST INDICATOR (EPR) FOR THE
  ENGINE(S) SELECTED BY THE OUTER KNOB

ENGINE
DISPLAY
SWITCH

STATUS DISPLAY SWITCH
1) PRESSING THE ENGINE OR
   STATUS SWITCH WILL SELECT
   THAT FORMAT ON THE LOWER
   DISPLAY - FROM ANY OTHER
   FORMAT EXCEPT TEST
2) PRESSING THE SAME SWITCH
   A SECOND TIME WILL BLANK
   THE LOWER DISPLAY

EICAS COMPUTER SELECT KNOB
• AUTO - SELECTS THE PRIMARY
  COMPUTER (L) AND AUTOMATICALLY
  SWITCHES TO THE OTHER (R)
  COMPUTER UPON A FAILURE
• L, R - MANUALLY SELECTS A
  SPECIFIC COMPUTER

MAXIMUM INDICATOR
RESET SWITCH
• CLEARS MAXIMUM
  EXCEEDANCE DIGITAL
  READOUTS FROM DISPLAY
  UNITS

EVENT RECORD SWITCH
• RECORDS MAINTENANCE DATA IN NON-VOLATILE
  MEMORY FOR THE ECS, ELEC/HYD, PERF/APU,
  AND EPCS PARAMETERS - SIMULTANEOUSLY

Figure 4-3-5. EICAS display select panel

Courtesy of Boeing Commercial Airplane Company

switching relays (upper right portion of Figure 4-3-4) receive input signals from the display select panel.

The EICAS computer outputs are comprised of both digital and analog signals. The digital data is sent to the flight data recorder or flight data acquisition unit, through the lower display switching module. The left and right FMC receive digital data from the left and right computers respectively.

Analog outputs include signals that are supplied directly to specific systems and those transmitted through the upper and lower display switching modules. Typically the left computer output is transmitted to both EICAS displays through the normally closed contacts in the two switching modules. The normally open contacts connect the right computer to the upper and lower displays. The switching modules connect the right computer to the displays only when the display select panel computer switch is in the *AUTO* position and the left computer is inoperative, or when the *R* (right) computer is selected.

System Monitoring

The diagram in Figure 4-3-4 shows some of the monitoring circuitry of the EICAS. For example, the output from the upper and lower display units labeled *H* is used to input a validity signal to the EICAS computers. The circuitry within the displays will monitor the health of the incoming data and the display itself. Then, the display

health is reported back to the computers. This type of self-monitoring system allows EICAS to react correctly when a problem occurs.

EICAS Controls

There are two major control panels used for the engine indicating and crew alerting system on the B-757, the display select panel (DSP) and the maintenance control panel (MCP).

Display Select Panel

The DSP, shown in Figure 4-3-5, is used during normal flight to set various system parameters. The two display push buttons **ENGINE** and **STATUS** are used to select what information will be displayed on the lower EICAS CRT. The **EVENT RECORD** button is used to store system data in the EICAS nonvolatile memory. The *COMPUTER* knob is used to select which EICAS computer will be used to power the displays. In the AUTO mode, the system will automatically switch computers in the event the primary computer fails. The *BRT* control is used to adjust the EICAS display brightness. It should be noted that brightness is also automatically adjusted to compensate for changes in flight deck light levels. The *THRUST REF SET* is used to select which thrust management computer communicates to EICAS. The engine reference data can also be set by this control. The push-button labeled **MAX IND RESET** is pressed to remove any exceedance values on the EICAS displays.

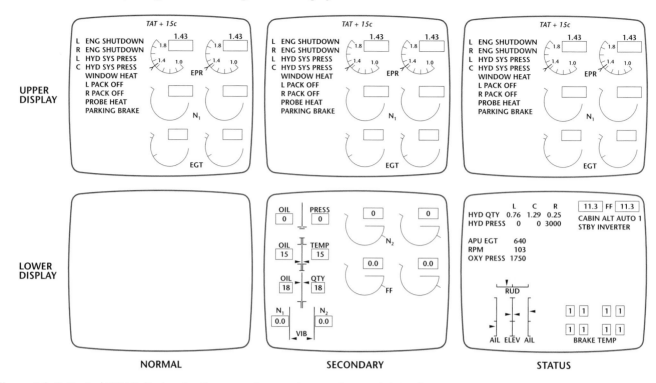

Figure 4-3-6. Typical EICAS displays for the normal, secondary engine, and status formats

Maintenance Control Panel

The controls of the MCP are located on the P61 panel installed aft of the first officer's seat. This panel is used during ground operations and is inhibited from certain functions unless the parking brake is set. The six **DISPLAY SELECT** push buttons located on the left of the panel are used to select which maintenance data will be displayed by EICAS. The label on each button represents the following systems: *ECS/MSG*, environmental control system/messages

(miscellaneous); *ELEC/HYD*, electrical/hydraulic; *PERF/APU*, performance/auxiliary power unit; *CONF/MCDP*, configuration/maintenance control display panel; *ENG EXCD*, engine exceedances; *EPCS*, electronic propulsion control system (engine parameters).

A third EICAS control panel containing the cancel and recall switches is located on the pilot's instrument panel just to the right of the upper EICAS display. The **CANCEL** push-button is used to cancel any caution or advisory message cur-

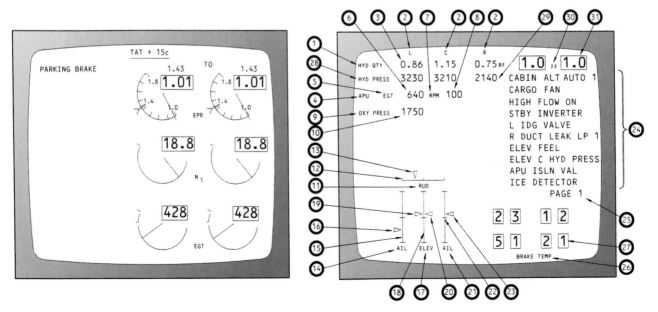

Figure 4-3-7A. EICAS status page format

Courtesy of Boeing Commercial Airplane Company

	STATUS PAGE DISPLAYS			
No.	Display	Range	Source	Color
	SUBSYSTEM DISPLAYS			
1	Subsystem Name – HYD QTY	N/A	EICAS Computer	Cyan
2	Subsystem Identification – HYD QTY	N/A	EICAS Computer	Cyan
3	Actual HYD QTY Readout	0.00 to 1.50 (2)	HYD QTY Sensors	(1)
4	Subsystem Name – APU	N/A	EICAS Computer	Cyan
5	Subsystem Parameter – EGT	N/A	EICAS Computer	Cyan
6	Actual EGT Readout	0 to 900° C	APU Sensor	White
7	Subsystem Parameter – r.p.m.	N/A	EICAS Computer	Cyan
8	Actual R.P.M. Readout	0; 3-120% r.p.m.	APU Sensor	White
9	Subsystem Name – Oxygen Pressure	N/A	EICAS Computer	Cyan
10	Actual Oxygen Pressure Readout	0 to 2500 p.s.i.	Crew OXY Press Sensor	White
	CONTROL SURFACE POSITION DISPLAY			
11	Control Surface Name – Rudder	N/A	EICAS Computer	Cyan
12	Horizontal Scale	-36 to +36°	EICAS Computer	White
13	Actual Rudder Position Pointer	-36 to +36°	Position Transmitter	White
14	Control Surface Name – Aileron	N/A	EICAS Computer	Cyan
15	Vertical Scale	-22 to +22°	EICAS Computer	White
16	Left Outboard All Position Pointer	-22 to +22°	Position Transmitter	White
17	Control Surface Name – Elevator	N/A	EICAS Computer	Cyan
18	Vertical Scale	-24 to +34°	EICAS Computer	White
19	Actual Left ELEV Position Pointer	-24 to +34°	Position Transmitter	White
20	Actual Right ELEV Position Pointer	-24 to +34°	Position Transmitter	White
21	Control Surface Name – Aileron	N/A	EICAS Computer	Cyan
22	Vertical Scale	-22 to +22°	EICAS Computer	White
23	Right Outboard AIL Position Pointer	-22 to +22°	Position Transmitter	White
	STATUS MESSAGE DISPLAY			
24	Status Messages	N/A	Refer to Text	White
25	Overflow Indicator	N/A	EICAS Computer	White
	BRAKE TEMPERATURE DISPLAY			
26	Subsystem Name – Brake Temp	N/A	EICAS Computer	Cyan
27	Actual Brake Temp Readout	0 to 9 units	Brake Temp Sensors (8)	(3)
	HYDRAULIC PRESSURE			
28	Subsystem Name – HYD Press	N/A	EICAS Computer	Cyan
29	Actual Brake Temp Readout	0 to 4000 p.s.i.	Fuel Flow Sensors	White
	FUEL FLOW DISPLAY			
30	Subsystem Name – Fuel Flow	N/A		Cyan
31	Actual Fuel Flow Readout	180 to 12.2K kg/hr		White

(1)	Readout	White
	Refill	Magenta

(2)	HYD QTY <0.75,
	RF Message Appears

(3)	Normal	0 to 2 Units	Box/Numbers Cyan
	Threshold	3 to 4 Units	Box White Numbers Cyan
	Abnormal	5 to 9 Units	Box/Numbers White

Figure 4-3-7B. EICAS status page format

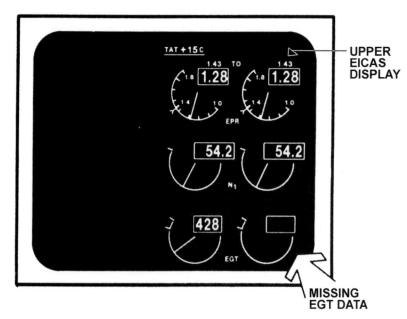

Figure 4-3-8. Example of invalid data on the upper EICAS display

Courtesy of Boeing Commercial Airplane Company

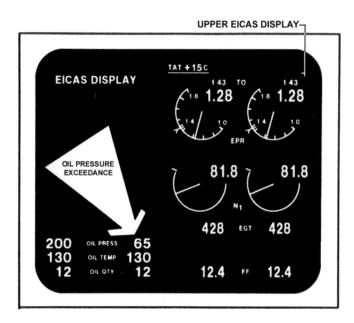

Figure 4-3-9. Example of EICAS compact mode

Courtesy of Boeing Commercial Airplane Company

rently displayed by EICAS. The *recall* switch is used to retrieve previously canceled messages.

Normal Operation of the B-757 EICAS

During regular operation the early versions of the engine indicating and crew alerting systems can display three basic formats: normal, secondary, and status. Figure 4-3-6 shows the typical EICAS displays for each of the three basic formats. Upon starting the first engine, the secondary format automatically appears

on the auxiliary EICAS display. The secondary format displays engine parameters, such as engine pressure ratio (EPR), N_1 and N_2 rotor speeds, fuel flow, oil pressure, oil temperature, oil quantity, and engine vibration. This information is displayed in both a graphic and digital format.

Status Page Format

The status page format is displayed on the lower EICAS CRT when the status switch on the display select panel is momentarily pressed. The crew is alerted to the presence of a new status message by the word STATUS presented in cyan on the lower EICAS display. As shown in Figure 4-3-7 the status format can display up to 11 status messages in the upper right of the display. If more than 11 messages are present, ten status messages appear followed by the term *PAGE 1*. This indicates that a second status page is available by pressing the status button again. The system and control position information (items numbered 1-23, and 26-31 in Figure 4-3-7) will appear on the status page whether status messages are present or not. System information includes hydraulic fluid quantity and pressure, APU data, oxygen system pressure, fuel flow, and brake temperature. The control surface positions are shown in the lower left corner of the status page.

Normal Format

The normal format is displayed any time the status page and secondary engine displays are not active. In the normal format, the lower EICAS display is blank and the upper display shows primary engine data (EPR, N_1, and EGT). This information also remains on the upper CRT during display of secondary engine and status information. To access secondary engine or status page data while in the normal format; depress the corresponding button on the EICAS display select panel.

The EICAS computers use pin programming to determine status page format. PIN programming was discussed in Chapter 3. For example, if pin number 27 is grounded, the hydraulic pressure for all three systems is displayed. If pin number 27 is open, hydraulic pressures will not be displayed. The pins being referred to are found in the electrical connector on the rear of the EICAS computer. Fuel flow is displayed only if pin 10 is grounded; and the readout is in kilograms/hour if pin 13 is grounded, pounds/hour if pin 13 is open. There are literally hundreds of items set through pin programming. Program pins can be used to set display formats or used to configure input/output signals. The use of pin programming allows the EICAS computer to be configured to a variety of different aircraft. Pin programming is typically done during initial system installation; however, technicians should always consider program pins during troubleshooting. If a system fails, it may be due to a loose program pin connection.

Abnormal Operation of the B-757 EICAS

In general, there are four basic ways EICAS can fail and still remain operational. These failures include the loss of:

1. One or more parameters
2. CRT display
3. Both EICAS control panels
4. System computer

In each of these four situations, the EICAS will be operational in some form. The first priority for EICAS is to display primary engine data; the second priority is for the system to display secondary engine data. In any failure mode, the system will automatically try to adhere to these priorities.

The loss of one parameter will result in an EICAS display with a blank in place of the missing parameter(s). Figure 4-3-8 shows an example where the right engine (EGT) data is invalid; that information is therefore not displayed.

A problem with the display unit, a wiring defect, or a problem with the EICAS computer can cause the loss of one CRT display. In any case, all EICAS information except the status page is available on the operable CRT. As discussed earlier during system architecture, a feedback circuit is used to monitor the CRT status. The system software automatically diverts all display data to the operable CRT if the EICAS computer senses a CRT failure or failure of the data to reach the CRT. With only one EICAS CRT operable the compact mode will be used to display systems data. Figure 4-3-9 shows an example of the EICAS compact mode; notice there is an oil pressure exceedance on the right engine. This exceedance valve (65) is displayed in red. The oil system parameters for both engines are displayed. In the compact mode the EGT data is displayed in digital format only, and fuel flow (FF) is displayed below EGT.

In the event that the EICAS maintenance control panel or it's related wiring fails, the systems accessed through that panel are inoperable and all other functions will operate normally. If the display select panel (DSP) fails, the following will occur:

1. The maintenance control panel and cancel/recall switches will become inoperative since their signal feeds through the DSP to the EICAS computers (Figure 4-3-4, upper left corner).

2. The display select panel functions become inoperative except for the brightness/balance controls.

3. The computers automatically display primary engine data on the upper CRT and secondary engine data on the lower CRT.

4. The message *EICAS CONT PNL* will appear on the upper CRT to inform the flight crew of the DSP failure.

Standby Engine Indicator

On the B-757 if both EICAS displays are inoperable, the *standby engine indicator* is used to display primary engine parameters, but all other EICAS information is not available. This type of failure occurs when both EICAS computers fail, both EICAS displays fail, or the associated wiring incurs a catastrophic failure. The standby engine indicator displays can be activated manually by placing the selector to the *ON* position (Figure 4-3-10), or the LCDs will turn on automatically in the event both EICAS computers or displays fail. The standby engine displays are also automatically activated whenever the EICAS test function is in operation.

The standby engine indicator is a LRU, which operates with two completely independent power supplies. Each power supply is fed through its own circuit breaker and related wiring. The indicator is fully functional with only one power supply operable. The indicator contains microprocessor circuitry to analyze both analog and ARINC 429 inputs. As seen in the lower-right corner of Figure 4-3-4, the standby indicators receive all vital engine data directly from the appropriate sensors. Even if both EICAS computers fail, the standby indicators will still function properly.

B-757 Maintenance Pages

The *maintenance pages* of the B-757 are used to verify malfunctions or exceedances, and analyze the systems monitored by EICAS. The maintenance pages are displayed on the lower EICAS display while the aircraft is on the ground. The maintenance control panel (MCP) is used to control the display of maintenance pages. Technicians use the MCP extensively during system troubleshooting and analysis.

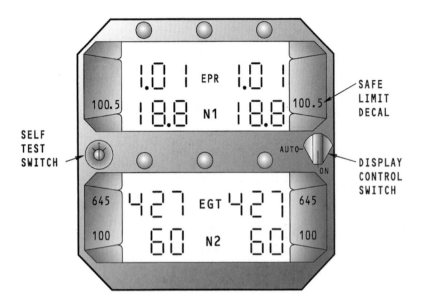

Figure 4-3-10. Standby engine indicator

Courtesy of Boeing Commercial Airplane Company

MCP Controls

The six display select switches on the MCP are used to display real time data for their associated systems (Figure 4-3-11). The **AUTO** and **MAN** push buttons on the MCP are used to access data that was stored in nonvolatile memory during recording of an auto or manual event. As discussed earlier, auto events are recorded by EICAS during system malfunctions or exceedances; manual events are recorded whenever the record (**REC**) button

on the DSP or MCP is pressed. When recording from the DSP, all systems are recorded simultaneously. When recording from the MCP, only the system currently shown on the lower EICAS display is recorded. The *ERASE* switch is used to clear the memory for the auto and manual events. The *TEST* switch is used to initiate the EICAS built-in-test circuitry.

The manual record button is a handy tool, which can be used to document system data during aircraft maintenance. The manual record function is similar to a picture being taken; that is, the EICAS manual record function will store the system parameters at the instant you press the **REC** button. For example, if the technician wishes to record the operating parameters of the left AC generator, he/she would select the electric/hydraulic display; press *REC* in the MCP and EICAS will store the data. The information can be retrieved at a more convenient time and recorded in the maintenance records.

Access to Maintenance Pages

Whenever a specific maintenance page is selected, the engine data page goes into the compact format and the maintenance page is shown on the lower EICAS display. The various maintenance pages and their associated control buttons are shown in Figure 4-3-12. To access the current status of a system using the EICAS maintenance pages, the following sequence should be followed:

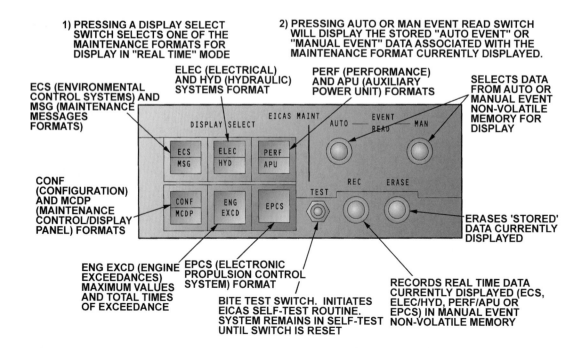

Figure 4-3-11. Maintenance control panel switch layout   *Courtesy of Boeing Commercial Airplane Company*

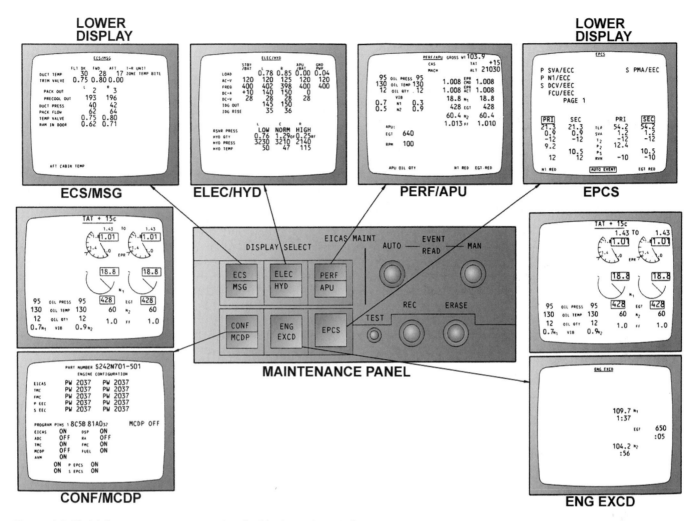

Figure 4-3-12. Maintenance pages associated with the MCP switches

Courtesy of Boeing Commercial Airplane Company

1. Aircraft power must be available through ground power or onboard generators.

2. Any system that is to be monitored must be set into an operating condition.

3. Select the proper system using the display select switches on the EICAS maintenance panel.

The lower EICAS display will now show real time data for the system selected. If an auto event is stored in memory the white display *AUTO EVENT* will appear at the bottom of the page. To access the auto event information, press the **AUTO** button on the MCP. The current system display will change to the data stored during the auto event. During all auto event presentations, the term *AUTO EVENT* will appear in cyan at the lower portion of the page to indicate that the data currently presented is from memory and is not real time data. To access any manually recorded event, the **MAN** push button is pressed. To return back to the primary engine display, exit the event mode by pressing the *MAN* or *AUTO* buttons, and press the system page button.

To erase an auto or manual event, press and hold the **ERASE** button on the MCP. This must be done while the event to be erased is presented on the EICAS display. Once the data presented on the display is removed, the memory has been erased. Only the data, manual or auto event, for the system currently on display has been erased. The recorded events for each system must be erased individually. After viewing and making records as needed of any auto event, it is important to erase the data. A new auto event can only be stored in memory after the previous event has been erased. The recording of any new manual events will automatically erase any manual event in memory.

System Displays

Taking a closer look at a given system display shows the type of data presented. Figure 4-3-13 shows a typical electrical/hydraulic system display. Notice the electrical system data is displayed at the top of the page and hydraulic data is presented in the lower half. At the very bot-

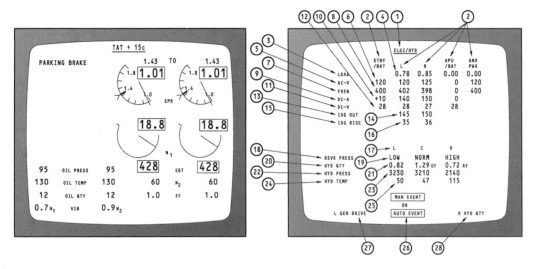

Figure 4-3-13A. Typical EICAS electrical/hydraulic system display

Courtesy of the Boeing Commercial Airplane Company

tom of the display, the terms *MAN EVENT* or *AUTO EVENT* are shown when data displayed is from event memory. Also at the bottom of the display are the terms *L GEN DRIVE* (27) and *R HYD QTY* (28); these are the specific events, which triggered an auto event to be recorded. Each system has a limited number of items that can trigger the recording of auto event data. For example, there are five conditions which can cause the recording of an auto event for the electrical system; they are, *R GEN DRIVE, L GEN DRIVE, R IDG OIL TEMP, L IDG OIL TEMP,* and *IDG RISE TEMP.*

Maintenance Messages

The environmental control system/maintenance message display is used to access any maintenance messages recorded for all systems monitored by EICAS. Maintenance messages are the lowest priority advisory and are not accessible during flight. Maintenance messages would typically be accessed after the maintenance discrepancy write-up of a system malfunction or during routine scheduled maintenance. Figure 4-3-14 shows a typical ECS/MSG display.

In the top right portion of the display are the maintenance messages (item 20). Up to eleven messages can be presented per page. If more than 11 messages are in memory the term PAGE 1 (item 21) is displayed. To view subsequent pages press the *ECS/MSG* switch. Once the maintenance messages have been retrieved, the information can be referenced to the aircraft's maintenance manual for details on system repairs.

System Troubleshooting Using EICAS

In most cases, the aircraft technician will have a copy of the pilot's maintenance discrepancy recorded in the aircraft's log, or the pilot may send a fault code to the technician while the aircraft is still in flight. This will give information as to what system(s) require attention. The pilot's log entry may state that a manual event record was made, or an auto event may have been recorded which relates to the maintenance discrepancy. In either case, the technician should access the EICAS memory for that system and study the parameters that occurred at the time the maintenance discrepancy was initiated. The system should also be operated and the current status observed using EICAS. If the fault has repaired itself, the technician should consider the defect may be intermittent.

It may be wise to access the data contained in the ECS/MSG page. If a maintenance message related to the problem is recorded in the nonvolatile memory, the solution may be found using the maintenance manuals. If no related maintenance messages are available, the next step would be to consult the troubleshooting or maintenance manuals for the system in question. In most cases, the manuals can point you in the proper direction for the repair.

Once the repair has been completed, the system should be operated in the same configuration as when the problem was first noticed. Access the real time information for the system through EICAS and look for any new auto events that may have been recorded. Be sure previous auto events were already erased. If the system operates within specified parameters, the repair

No.	DISPLAY	RANGE	SOURCE	COLOR
	ELECTRICAL SYSTEM/HYDRAULIC SYSTEM DISPLAY			
1	Page Format	N/A	EICAS Computer	Cyan
	Electrical System Display			
2	Power Source	N/A	EICAS Computer	Cyan
3	Parameter Name AC-Load	N/A	EICAS Computer	Cyan
4	Load Readout	0 to 1.50******* (1.00 = 90KVA)	L, R, APU GCU, BPCU	White
5	Parameter Name AC-Volts	N/A	EICAS Computer	Cyan
6	AC-Volts Readout	0/100 to 130 V AC	*	White
7	Parameter Name Frequency	N/A	EICAS Computer	Cyan
8	Frequency Readout	0/380 to 420 Hz	*	White
9	Parameter Name DC-amps	N/A	EICAS Computer	Cyan
10	DC-amperes Readout	0/2 to 150 amps	**	White
11	Parameter Name DC-Volts	N/A	EICAS Computer	Cyan
12	DC-Volts Readout	0 to 40 V DC	***	White
13	Parameter Name IDG Out	N/A	EICAS Computer	Cyan
14	IDG Out Readout	0 to 180°C	L, R, GCU	White
15	Parameter Name IDG Rise	N/A	EICAS Computer	Cyan
16	IDG Rise Readout	0 to 180°C	L, R, GCU	White
	Hydraulic System Display			
17	Hydraulic System Identifier	N/A	EICAS Computer	Cyan
18	Hydraulic Reservoir Pressure	N/A	EICAS Computer	Cyan
19	Hydraulic RSVR Press. Readout	Low, Norm, High	HYD RSVR PRESS Sensors	****
20	Hydraulic Quantity	N/A	EICAS Computer	Cyan
21	Hydraulic Quantity Readout	0.00 to 1.50 (1.00=100%)	HYD QTY Sensors	***
22	Hydraulic Pressure	N/A	EICAS Computer	Cyan
23	Hydraulic Pressure Readout	0 to 4000 p.s.i.	HYD PRESS Sensors	White
24	Hydraulic Reservoir Temp	N/A	EICAS Computer	Cyan
25	Hydraulic Temp Readout	-60 to +200°C	HYD Temp Sensor	White
	Auto/Manual Event Display			
26	Auto/Man Event Mode Display	N/A	EICAS Computer	Cyan
27	Elect Auto Event Message	******	Refer to Text	White
28	Hydraulic Auto Event Message	******	Refer to Text	White

* *Standby Inverter, L & R Generator, APU Generator, Ground Power*
** *Main Battery, L & R TRU, APU Battery*
*** *Main Battery, L & R DC Bus, APU Battery*

****
Reservoir Pressure < 17 p.s.i.	*Low*	*Magenta*
17 p.s.i. < Reservoir Pressure >55 p.s.i.	*Norm*	*White*
Reservoir Pressure > 55 p.s.i.	*High*	*Magenta*

*****
Readout	*White*
Overfill (OF) QTY ≥ 1.22	*Magenta*
Refill (RF) QTY ≤ 0.75	*Magenta*

****** *Auto event messages are not displayed in MAN EVENT read mode*

******* *L.R. APU readout less than 0.05 set to 0.00* <u>*GND PWR*</u> *readout less than 0.10 set to 0.00*

Figure 4-3-13B. Typical EICAS electrical/hydraulic system display

Courtesy of the Boeing Commercial Airplane Company

was successful. If the system is still defective, continue the troubleshooting process.

Troubleshooting EICAS

Both EICAS computers continuously perform fault monitoring of all circuits that can affect the integrity of EICAS data. The fault monitoring system consists of software within the EICAS computers. Incoming digital data is monitored for validity using the ARINC 429 parity bit,

and all input data, both analog and digital, is compared through the EICAS computers cross talk data bus. The health of each computer is monitored, and if one computer fails any necessary switching automatically occurs.

In the event of an EICAS malfunction, the status page will contain a message associated with that failure. For example the message *L EICAS CMPTR* will appear if a left computer fault or computer off is detected. If the EICAS monitoring system detects a fault, the maintenance

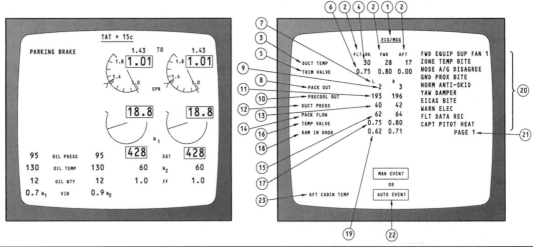

ENVIRONMENTAL CONTROL SYSTEM/MAINTENANCE MESSAGE DISPLAY				
No.	Display	Range	Source	Color
	Environmental Control System Display			
1	Page Format Name	N/A	EICAS Computer	Cyan
2	Function Location	N/A	EICAS Computer	Cyan
3	Function Name – Duct Temp	N/A	EICAS Computer	Cyan
4	Duct Temperature Readout	-60 to +200°C	Duct Temp Sensors	White
5	Function Name – Trim Valve	N/A	EICAS Computer	Cyan
6	Trim Valve Position Readout	0.00 to 1.00 (Open)	Valve Posn Sensors	White
7	L, R Pack Identifier	N/A	EICAS Computer	Cyan
8	Function Name – Pack Out	N/A	EICAS Computer	Cyan
9	Pack Out Temp Readout	-60 to +200°C	Pack Temp Sensors	White
10	Function Name – Precool Out	N/A	EICAS Computer	Cyan
11	Precooler Outlet Temp Readout	-60 to +300°C	Precool Out Temp Sensors	White
12	Function Name – Duct Press	N/A	EICAS Computer	Cyan
13	Duct Pressure Readout	0 to 100 p.s.i.	Duct Press Sensors	White
14	Function Name – Pack Flow	N/A	EICAS Computer	Cyan
15	Pack Air Flow Readout	0 to 60 M³/Min	Pack Flow Sensors	White
16	Function Name – Temp Valve	N/A	EICAS Computer	Cyan
17	Temp Valve Position Readout	0.00 to 1.00 (Open)	Valve Posn. Sensors	White
18	Function Name – RAM in Door	N/A	EICAS Computer	Cyan
19	RAM Inlet Door Position Readout	0.00 to 1.00 (Close)	Door Posn Sensors	White
	Maintenance Message Display			
20	Maintenance Message Readout	*	Refer to Text	White
21	Overflow Indicator	*	EICAS Computer	White
	Automatic/Manual Event Display			
22	Auto/Man Event Read Mode Display	N/A	EICAS Computer	
23	ECS Auto Event Message	*	Refer to Text	White

* Auto Event Message and Maintenance Messages are not displayed in manual event read mode.

Figure 4-3-14. Typical EICAS environmental control system/maintenance message display

Courtesy of the Boeing Commercial Airplane Company

message *EICAS BITE* will appear on the ECS/MSG maintenance page. The ECS/MSG page is accessed as previously described. If this message is present, the EICAS test should be run.

EICAS BITE

On the B-757, the *EICAS BITE* is used to perform testing of the EICAS computers, upper and lower display units, the master caution and warning displays, and various EICAS interfaces. The BITE test is initiated by pressing the TEST switch on the MCP (Figure 4-3-11). The aircraft must be on the ground and the parking brake set for the test to begin. Each EICAS computer is tested individually. The EICAS test can also be initiated automatically due to a pin programming error. In this case, the EICAS test will be initiated at start up of the system.

At the beginning of each test the EICAS message *L EICAS TEST* and *TEST IN PROGRESS* will appear on the upper display. The message *R EICAS TEST* will be presented during the test of the right computer. The message *PROG PIN ERROR* will appear if the test was initiated automatically due to program pin error. During the EICAS test, the master warning and caution lights are activated, the related audio tones will sound, and the standby engine indicators are energized.

After the test is complete, the EICAS will present a display similar to Figure 4-3-15 on both the upper and lower EICAS CRT. The following items describe the test display format.

1. States why the test was initiated (automatically due to program pin error, or manually for L/R EICAS test).

2. Test results; *TEST FAIL* or *TEST OK* are displayed.

3. Any computer failures are displayed.

4. Any display unit failures are displayed.

5. The *KEY* message allows for the operator to test EICAS inputs from the DSP, MCP, and cancel/recall switches. After the test is complete the operator can press any of the associated switches. The switch name will appear next to the *KEY* message. In this example the **ERASE** button has been pressed.

6. This display indicates the current program pins status, in a hexadecimal format, for the left and right computers.

7. The L/R computer test results are displayed in a hexadecimal format. All zeros indicate no computer system failures. NOTE: If the test shows *L/R COMPTR*

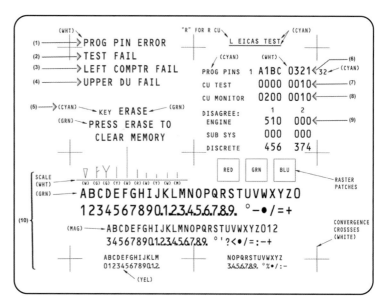

Figure 4-3-15. EICAS test display

Courtesy of the Boeing Commercial Airplane Company

FAIL, be sure to check the CU TEST code shown in the upper right quadrant of the test display (Figure 4-3-15). The *CU TEST code* is used to identify the specific problem with the L/R computer. The code displayed can be found in ATA Chapter 31 of the Fault Isolation Manual (FIM). The FIM will provide further instructions for system repair using the various CU TEST codes. Remember do not merely replace the EICAS computer if a failure is listed on the test display; verify the specific fault using the CU TEST code.

8. The results of the L/R computer input tests are displayed adjacent to CU monitor. The results are displayed in a hexadecimal format, where a display of 0000 indicates no failures. If a fault is detected, use the FIM to determine the specific cause of the failure. This code will direct the technician to a specific EICAS computer input that has failed.

9. Any disagreements between engine, subsystem, and discrete data transmitted to EICAS will be displayed here. A readout displaying all zeros indicates no disagreements. Once again, the aircraft manuals will provide specific information on decoding any engine disagree message.

10. Any special symbols, which are typically displayed by EICAS, are shown on the lower portion of the display during an EICAS test.

To clear the CU test, CU monitor, and parameter disagree codes, the **ERASE** key must be pressed twice. With the first press of the erase key, the message *PRESS ERASE TO CLEAR MEMORY* will appear. Press the **ERASE** key a second time and the codes will revert to all zeroes.

As mentioned above, if the EICAS test results show a CU test, CU monitor, or parameters disagreement, the associated codes can be used to further troubleshooting efforts using the appropriate section of the aircraft manuals. The code(s) will direct the technician to a specific system or LRU suspected of creating the problem. These codes are the key to accurate troubleshooting using the EICAS test function; be sure they are not ignored. After the repair has been completed, the EICAS test should be run again to verify correct system operation.

Configuration/Maintenance Page

The configuration/maintenance page is also a valuable EICAS troubleshooting aid. As discussed earlier, this page is accessed through the maintenance control panel. A typical configuration/maintenance page is shown in Figure 4-3-16. From this example it can be seen that a variety of EICAS computer configuration information is listed on this page. During troubleshooting, items 13 through 34 can be used to verify input bus activity for the EICAS computer currently being displayed. For each bus monitored, one of the five messages will appear: *ON, OFF, TEST, NCD, or FAIL.* No activity on the bus will generate the *OFF* message. Defective bus wiring or no data being transmitted will generate the *OFF* message. An *ON* message indicates that the bus is operating normally. If *TEST* is displayed, the transmitting system or LRU is currently undergoing an internal BITE test. The message *NCD* (no computed data) means the transmitter is operating correctly but there is no data to be transmitted. The message *FAIL* means that the system or LRU in question has failed its own internal BITE test.

Reversionary Switching

Another quick EICAS test often used by line technicians is done through a simple switch of the EICAS computers. This is done if normal operation of EICAS shows missing data for one of the engine, or other system, parameters. In this scenario the missing data is most likely the result of a defective input to one of the computers. To verify which computer is receiving the invalid or lost data, designate a specific computer using the display select panel. This technique is sometimes referred to as *reversionary switching*. For example, for an EGT fault the technician would select the *L* (left) to operate the display directly from the left computer. If the display still has missing EGT data, the EGT system feeding the left computer or its

associated wiring is defective. If moving to the left computer solves the missing EGT data problem, the problem lies in the right side system. The same test could be run for the right side computer by turning the computer select switch to *R*.

Once a technician has become familiar with the aircraft's systems, this type of troubleshooting becomes second nature. If the EGT probe feeding the left computer was suspected as defective, the appropriate manuals would be reviewed and the repair initiated. The test executed in the previous paragraph would be performed and the appropriate paperwork completed. The aircraft would once again be equipped with a fully operational EICAS.

It was noted earlier that for some systems pin programming is used to determine what information is displayed by EICAS. On the B-757, if pin 27 is grounded, hydraulic system pressure is displayed. If pin 27 is open, hydraulic pressure is not presented on the display. This type of pin programming can sometimes cause difficulty when troubleshooting the system. For example, if pin number 27 should become loose on the left EICAS computer; the connection to ground would be eliminated or intermittent. Whenever operating on the left EICAS computer, the pilot would see the hydraulic pressure blink on and off the display as pin 27 connected/disconnected to ground (pin 27 intermittent). Or the hydraulic pressure information would be lost completely (pin 27 open). The technician would most likely suspect a faulty hydraulic pressure input to the left computer. A quick troubleshooting technique would be to manually select the opposite (right) EICAS computer to drive the displays. This switch would indeed fix the problem, leading the technician to assume the left EICAS computer input was faulty, or the left computer itself had a problem. Replacing the left computer would not solve the problem. A quick test of the hydraulic system feeding the left computer would show that system was operational. The cause of the defect is an intermittent or an open connection to ground on pin 27. In this case, an auto EICAS BITE should be initiated. The results of this test would show *PROG PIN ERROR* (program pin error). This should lead the technician to verify all program pins connections related to the hydraulic system.

A Typical Troubleshooting Sequence

The following example illustrates the basic series of events that would take place when an EICAS system failure occurs. In this example the upper EICAS display unit has failed during flight. The following would occur:

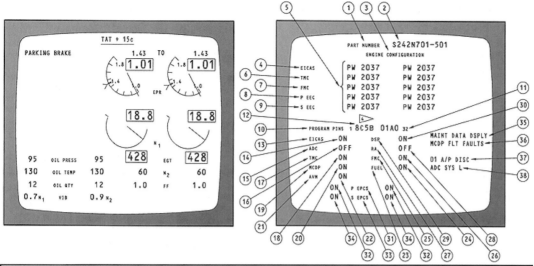

No.	Display	Range	Source	Color
	Engine Configuration Display			
1	Parameter Name – Part Number	N/A	EICAS Computer	Cyan
2	EICAS Computer Part Number	N/A	EICAS Computer	White
3	Parameter Name – ENG Configuration	N/A	EICAS Computer	Cyan
4	Computer Name – EICAS	N/A	L, R, APU GCU, BPCU	Cyan
5	Engine Type/Model	N/A	*	White
6	Computer Name – TMC	N/A	EICAS Computer	Cyan
7	Computer Name – FMC	N/A	EICAS Computer	Cyan
8	Computer Name – P EEC	N/A	EICAS Computer	Cyan
9	Parameter Name – S EEC	N/A	EICAS Computer	Cyan
	Program Pin Configuration			
10	Parameter Name – Program Pin	N/A	EICAS Computer	Cyan
11	Byte Start/Stop Reference	1,32	EICAS Computer	Cyan
12	Program Pin Readout (Hexidecimal)	0000 to FFFF	EICAS Computer	White
	Input Bus Activity			
13	Input Bus Name – EICAS	N/A	EICAS Computer	Cyan
14	Activity/No Activity	**	EICAS CU (Inter-connect)	White
15	Input Bus Name – ADC	N/A	EICAS Computer	Cyan
16	Activity/No Activity	**	ADC (TAT)	White
17	Input Bus Name – TMC	N/A	EICAS Computer	Cyan
18	Activity/No Activity	**	TMC (TAT)	White
19	Input Bus Name – MCDP	N/A	EICAS Computer	Cyan
20	Activity/No Activity	**	MCDP (DSPLY Data Label 357)	White
21	Input Bus Name – AVM	N/A	EICAS Computer	Cyan
22	Activity/No Activity	**	AVM (L ENG Vibration)	White
23	Input Bus Name – Fuel	N/A	EICAS Computer	Cyan
24	Activity/No Activity	**	FQIS (L Tank Fuel Qty)	White
25	Input Bus Name – AVM	N/A	EICAS Computer	Cyan
26	Activity/No Activity	**	FMC (Gross Wt.)	White
27	Input Bus Name – RA	N/A	EICAS Computer	Cyan
28	Activity/No Activity	**	RA (Radio Altitude)	White
29	Input Bus Name – P EPCS	N/A	EICAS Computer	Cyan
30	Activity/No Activity	**	DSP (Discrete Word #1)	White
31	Input Bus Name – P EPCS	N/A	EICAS Computer	Cyan
32	Activity/No Activity	**	L/R EEC (L/R P2)	White
33	Input Bus Name – S EPCS	N/A	EICAS Computer	Cyan
34	Activity/No Activity	**	L/R EEC (L,R Ps)	White
	Maintenance Control & Display Panel			
35	Annunciate Message ***	N/A	MCDP Label 357	Yel/White
36	Mode Message	N/A	MCDP Label 357	Yellow
37	MCDP Top Line Data Message	N/A	MCDP Label 357	White
38	MCDP Bottom Line Data Message	N/A	MCDP Label 357	White

*	Source	Label	Source	Label
	EICAS	N/A	FMC	271
	TMC	270	EEC	271

** On, Test, HCD, Fail/Off
*** If MCDP is off, white "MCDP Off" message replaces all other MCDP messages.
**** SIA – Left computer shown right computer $845B\ D1A1_{32}$

Figure 4-3-16. EICAS configuration/maintenance display format

Courtesy of the Boeing Commercial Airplane Company

BOEING 757
FAULT ISOLATION/
MAINTENANCE MANUAL

FAULT CODE	1. LOG BOOK REPORT 2. FAULT ISOLATION REFERENCE
31 41 04 00	1. Master caution lights did not illluminate when level B caution condition existed. (State condition existing). 2. Examine and repair the circuit from left (right) EICAS computer connector D319A pin F8 (connector D321A pin F8) to captain's (F/O's) master caution lighted switch (pin 16) (WM31-41-14, -24).
31 41 06 00	1. EICAS (ENGINE, STATUS) select switch does not (select, deselect) (secondary engine, status) format. 2. Replace the EICAS select panel, M10195 (MM 31-41-03).
31 41 07 --	1. (03=upper, 04=lower) EICAS display is (blank, out of focus, distorted, wrong color: describe fault). Operation norm on alternate computer. 2. 31-41-00 Fig. 104 Block 1
31 41 08 --	1. (03=upper, 04=lower) EICAS display is (blank, out of focus, distorted, wrong color: describe fault). Fault remains on alternate computer. 2. 31-41-00 Fig. 105 Block 1
31 41 14 00	1. EICAS msg EICAS CONT PNL displayed. 2. 31-41-00 Fig. 106 Block 1
31 41 15 00	1. Max ind reset switch will not reset overlimit readout. 2. Replace the EICAS select panel, M10195 (MM 31-41-03).

EFFECTIVITY

31 FAULT CODE INDEX

ALL

04   Page 5
Jun 20/92

Figure 4-3-17. EICAS Fault Isolation Manual/Fault Code Index

Courtesy of the Boeing Commercial Airplane Company

1. The EICAS would automatically revert to a compact mode and display all necessary information.

2. As standard procedures, the flight crew would switch to the alternate EICAS computer to potentially solve the problem. In this example, the display operates normally on the alternate computer.

3. The first officer would reference a fault codes manual to determine the correct fault code for this failure. Fault codes adhere to ATA chapter/section sequencing; this fault is 31-41-07-04.

4. The first officer would record the fault in the appropriate logbook either by code number or by description. Exactly what information gets recorded is a function of specific airline operating procedures. The maintenance code(s) could also be sent to the ground facility using ACARS. A further discussion of ACARS can be found in Chapter 7.

5. Once the aircraft has landed, the maintenance crew would check the flight logs during aircraft turnaround. The technician would find a log entry and/or code 31-41-07-04 recorded along with the time at which the system failed. The flight crew might also discuss the fault with the technician.

6. The technician would find the fault code in the fault isolation manual. Figure 4-3-17 shows 31-41-07 and states that the "EICAS display is (blank, out of focus, distorted, or wrong color). The 04 defines the upper display." The fault isolation reference tells the technician where to begin troubleshooting. In this case, the fault isolation manual chapter/section 31-41-00, Figure 104, block 1 is referenced.

7. The technician would follow the troubleshooting procedures listed in 31-41-00, Figure 104, block 1 (Figure 4-3-18). Note the prerequisites for the tests are listed at the top of the page. In this case, electrical power must be supplied to the system, and the listed circuit breakers must be in and operable. The maintenance manual (MM) 24-22-00 would be used as a reference for the electrical power procedures.

8. For our test, lets assume the technician followed the fault tree and each time a question was asked the answer was *NO*.

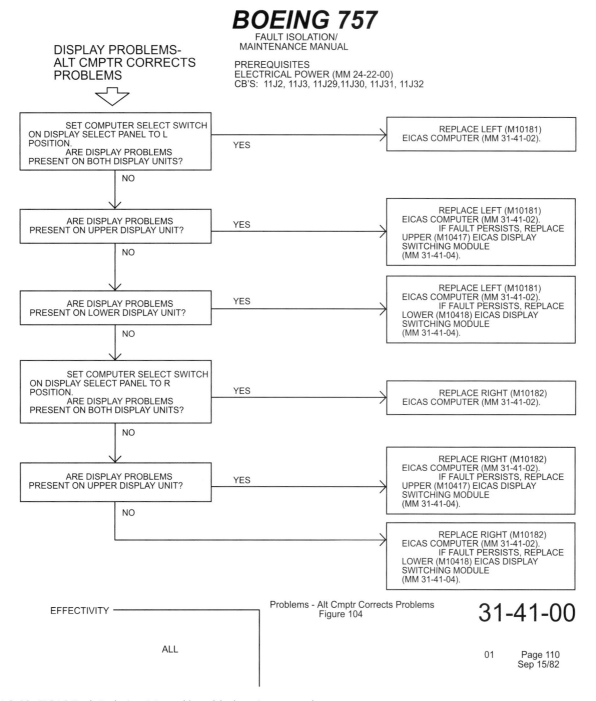

BOEING 757
FAULT ISOLATION/
MAINTENANCE MANUAL

DISPLAY PROBLEMS-
ALT CMPTR CORRECTS
PROBLEMS

PREREQUISITES
ELECTRICAL POWER (MM 24-22-00)
CB'S: 11J2, 11J3, 11J29,11J30, 11J31, 11J32

SET COMPUTER SELECT SWITCH ON DISPLAY SELECT PANEL TO L POSITION. ARE DISPLAY PROBLEMS PRESENT ON BOTH DISPLAY UNITS? — YES → REPLACE LEFT (M10181) EICAS COMPUTER (MM 31-41-02).

ARE DISPLAY PROBLEMS PRESENT ON UPPER DISPLAY UNIT? — YES → REPLACE LEFT (M10181) EICAS COMPUTER (MM 31-41-02). IF FAULT PERSISTS, REPLACE UPPER (M10417) EICAS DISPLAY SWITCHING MODULE (MM 31-41-04).

ARE DISPLAY PROBLEMS PRESENT ON LOWER DISPLAY UNIT? — YES → REPLACE LEFT (M10181) EICAS COMPUTER (MM 31-41-02). IF FAULT PERSISTS, REPLACE LOWER (M10418) EICAS DISPLAY SWITCHING MODULE (MM 31-41-04).

SET COMPUTER SELECT SWITCH ON DISPLAY SELECT PANEL TO R POSITION. ARE DISPLAY PROBLEMS PRESENT ON BOTH DISPLAY UNITS? — YES → REPLACE RIGHT (M10182) EICAS COMPUTER (MM 31-41-02).

ARE DISPLAY PROBLEMS PRESENT ON UPPER DISPLAY UNIT? — YES → REPLACE RIGHT (M10182) EICAS COMPUTER (MM 31-41-02). IF FAULT PERSISTS, REPLACE UPPER (M10417) EICAS DISPLAY SWITCHING MODULE (MM 31-41-04).

REPLACE RIGHT (M10182) EICAS COMPUTER (MM 31-41-02). IF FAULT PERSISTS, REPLACE LOWER (M10418) EICAS DISPLAY SWITCHING MODULE (MM 31-41-04).

EFFECTIVITY ——————

ALL

Problems - Alt Cmptr Corrects Problems
Figure 104

31-41-00

01   Page 110
Sep 15/82

Figure 4-3-18. EICAS Fault Isolation Manual/troubleshooting procedures

Courtesy of the Boeing Commercial Airplane Company

This would bring the solution to block 25, "Replace right EICAS computer (MM 31-41-02). If fault persists, replace upper EICAS display switching module (MM 31-41-04)." The maintenance manual reference is given for each suggested repair.

9. The suggested repair would be accomplished in accordance with the maintenance manual procedures. Once the fault has been repaired, the appropriate operational test(s) would be completed and the aircraft would be returned to service.

Testing the Standby Engine Indicator

A built-in-test equipment (BITE) circuit is contained in the standby engine indicator and accessed using the self-test switch located on the face of the LRU (Figure 4-3-19). Turning the switch clockwise runs the self-test using the number 2 power supply. If the switch is turned counterclockwise the self-test is run using number 1 power supply. This enables the BITE to test the operation of each independent power supply. The test switch is accessed using a screwdriver. During each test the displays should be viewed and compared to the appro-

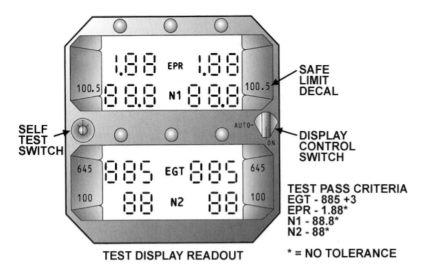

Figure 4-3-19. EICAS standby engine indicators

Courtesy of the Boeing Commercial Airplane Company

priate test values (EGT, 885; EPR, 1.88; N_1, 88.8; N_2, 88).

Section 4

Second Generation EICAS

For the most part, the first generation EICAS was designed in the 1970s and put in production in the 1980s. During that time improvements in microprocessor and computer technologies made great strides. Also, with the first generation systems in production it was easy to see what needed improvement and what portions of the system should be carried forward. Design of the second-generation system began about the time the first generation started flying. In the late 1980s, the second-generation systems were introduced on the Boeing 747-400.

There were two significant changes made for the second generation engine indicating and crew alerting systems: 1) the new EICAS was part of an integrated system, and 2) the systems were better at self diagnostics. Integration of the B-747-400 EICAS made the system more compact through the use of shared components. Integrating the system with other flight deck displays made the EICAS more reliable by providing extended backup capabilities. Improved diagnostics also resulted from integration, in this case, integration with the aircraft's central maintenance system. The B-747-400 EICAS is capable of more accurate troubleshooting and can present information in a more easily understood format compared to previous systems.

The Boeing 747-400 is typical of aircraft that contain a second generation EICAS. The Boeing 747-400 EICAS will be presented during the following discussion.

System Architecture

The B-747-400 EICAS is part of a complete aircraft monitoring system called the *integrated display system* (IDS). The integrated display system monitors a variety of aircraft and engine systems, as well as various flight parameters using the three *electronic interface units* (EIU). As shown in Figure 4-4-1, the three EIUs manipulate the incoming data and send outputs to five display units. The display units on the B-747-400 are called integrated display units (IDU) since the data on each display is interchange-

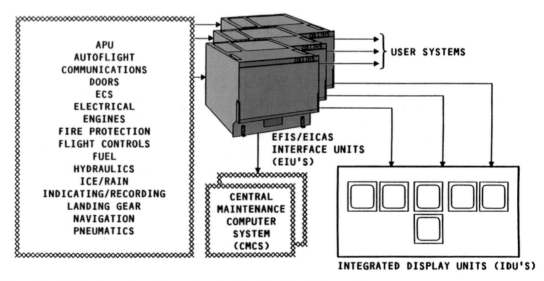

Figure 4-4-1. The three EIUs manipulate the incoming data and send outputs to five display units

Courtesy of the Boeing Commercial Airplane Company

able during operation. This system allows for a greater flexibility of display formats in the event of a single or multiple display failure. The EIUs also talk directly to the central maintenance computers. The central maintenance computer system (CMCS) is used to monitor the health of various aircraft systems.

The three EIUs are located in the main equipment center as shown in Figure 4-4-2. These computers are static discharge sensitive and should be handled only when using proper precautions. The electrical leads of the EIUs are made through an 800-pin connector mounted on the rear of the unit. Take special care not to damage the pins or sockets during removal and replacement of the computers.

The six IDS display units are located as shown in Figure 4-4-3. Shown in this figure, are the normal display configurations. The *primary flight displays* (PFD) are used to display flight data such as airspeed, pitch attitude, and altitude. The *navigational displays* (NDs) are used for the display of navigational data, such as the compass rose, course map, and weather radar information. The *main EICAS* display will usually show primary engine data, flight crew alert messages, and total fuel flow (Figure 4-4-4). The *auxiliary EICAS* display is located beneath the main EICAS display and is used to show one of the following formats: secondary engine data, status, synoptic, or maintenance pages.

Figure 4-4-5 shows a closer look at the extremely complex EIU input/output structure. The EIUs receive ARINC 429 digital data in low- and high-speed formats, digital discrete signals, analog information, and analog discrete signals. Up to 108 digital input busses can transmit to each EIU. Of these 108 busses, a maximum of eight are high-speed and the remainder are

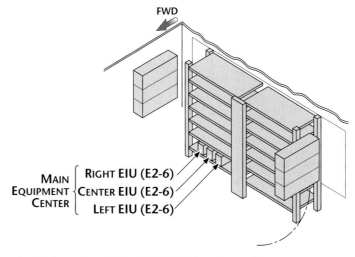

Figure 4-4-2. Location of the EFIS/EICAS interface units

low-speed busses. Lightning protection is provided for up to 22 sensitive input busses. Each EIU transmits to nine digital bus outputs, one low-speed, and eight high-speed formats.

Each EIU receives up to 450 discrete analog inputs from various aircraft systems, LRUs, and control switches. There are also 19 DC analog, 24 AC analog, 4 pulsed, 4 frequency modulated, and 12 tachometer signals transmitted to each EIU.

Each EIU transmits nine discrete analog outputs and 24 AC signals. On the left side of Figure 4-4-5 are the EICAS control panels. The EICAS control panel is connected to the EIUs using a discrete analog connection. A close-up view of the display select panel bus structure is shown in Figure 4-4-6. The EICAS display select panel (DSP) talks directly to the EIUs for cancel/recall functions, all other DSP functions are sent to the EIUs through the L/R EFIS, not

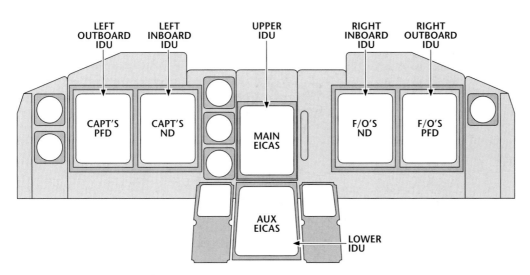

Figure 4-4-3. IDS display locations, normal mode

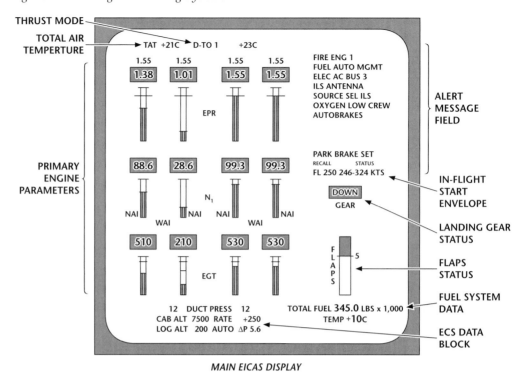

MAIN EICAS DISPLAY

Figure 4-4-4. Main EICAS display, primary format

EICAS, control panels. The EFIS control panels communicate to the EIUs on an ARINC 429 bus through the L/R control display units (CDUs). The CDUs can act as a backup control panel as discussed later in this chapter.

The integrated display units receive inputs from the EIUs on a high-speed ARINC 429 bus. This bus carries the majority of all EICAS information. The IDUs used for electronic flight instruments also receive digital data

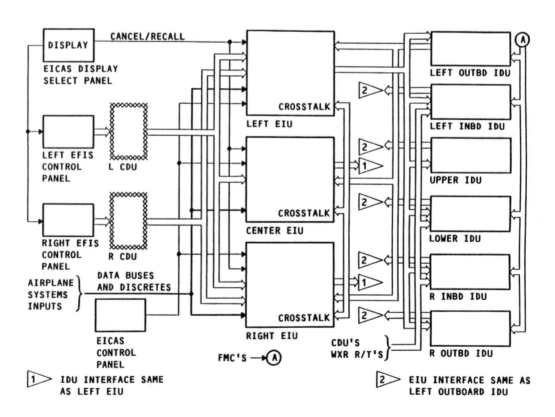

Figure 4-4-5. EUI input/output interface

Courtesy of Northwest Airlines, Inc.

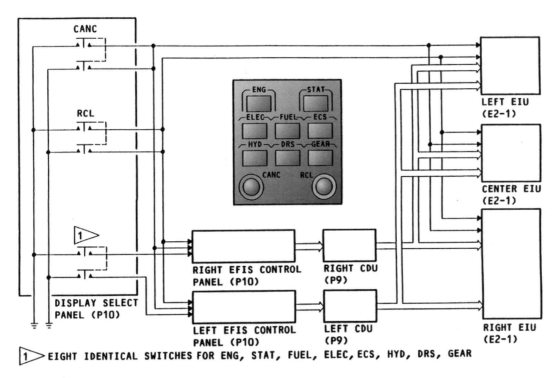

Figure 4-4-6. EICAS display select panel interface

Courtesy of Northwest Airlines, Inc

from the weather radar receiver transmitters (WXR R/T's) on an ARINC 453 data bus and ARINC 429 data from the flight management computers (FMC).

Each EIU has three functional sections: the input/output (I/O) interface, the processing section, and the power supply (Figure 4-4-7). The power supply receives 115 VAC and converts that signal into various voltages for oper-

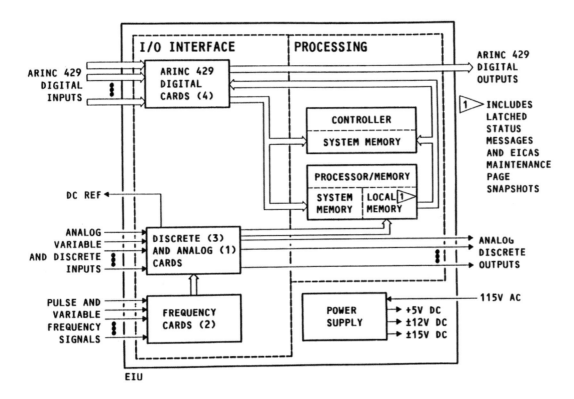

Figure 4-4-7. EIU 3 subsections, I/O interface, processing, and power supply

Courtesy of Northwest Airlines, Inc.

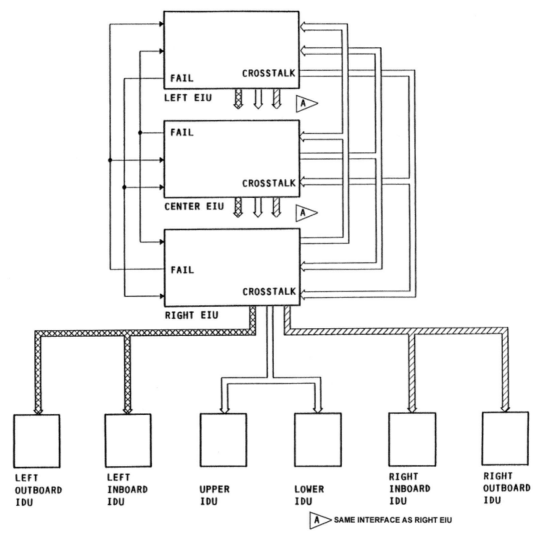

Figure 4-4-8. EIU crosstalk and interface diagram

Courtesy of Northwest Airlines, Inc.

ation of the EIU. The input and output signals are interfaced through four-ARINC 429, three-discrete, one-analog, and two-frequency cards. The signal processing takes place with help of the controller and processor/memory cards. The memory is used to store any latched messages or system data, which may be accessed later.

EIU Interface

Each EIU has the capability to "crosstalk" using a dedicated ARINC 429 bus to receive and transmit between each EIU (Figure 4-4-8). The following digital information is transmitted between EIUs:

1. BITE data consisting of internal fault information and LRU fault data received by the EIU

2. EIU status data

3. EICAS message data that is transmitted from one EIU to the other; hence, causing each EIU to generate the same message

Each EIU also has a discrete signal transmitted to the other two EIUs, which is used to communicate a complete EIU failure. In the event of a total EIU failure or loss of input power, the discrete signal opens. The remaining operable EIU(s) will respond accordingly.

Each EIU uses three separate busses to transmit data to six CRT displays (Figure 4-4-8). The left bus sends data to the left outboard and left inboard IDUs. The center data bus feeds the upper and lower center IDUs. The right bus transmits to the right inboard and outboard IDUs. The display formats on these displays are interchangeable to provide system redundancy. Normal and backup formats will be discussed later in the chapter.

Each IDU transmits ARINC 429 low-speed data to the three EIUs as a feedback signal.

This feedback circuit is used to monitor health of the six IDUs. As shown in Figure 4-4-9, one bus leaves each IDU and transmits the health status of all three EIUs.

IDS power is supplied through three different AC busses and 11 different circuit breakers. In order to provide input power redundancy, the captain's flight instrument transfer bus (115 VAC), the first officer's flight instrument transfer bus (115 VAC), and the standby AC bus (115 VAC) each power one EIU and one or more of the IDUs. The IDS power input diagram is shown in Figure 4-4-10.

Pin Programming

The electronic interface units of the B-747-400 utilize hard-wired discrete analog inputs for *pin programming*. Pin programming is used to configure EICAS specifically for a given aircraft and its related systems. Each EIU has 57 connections, on the 800-pin connector, dedicated to pin programming. One of those 57 pins is used as a parity bit. During aircraft manufacturing or during a system modification, the program pins should be connected/disconnected (ground or open) as specified by the manufacture's data. A ground represents logic 1 and an open equals logic 0. Program pins are also used by the display units to identify what position (inboard/outboard, left/right, upper/lower, or center) of the IDU is installed.

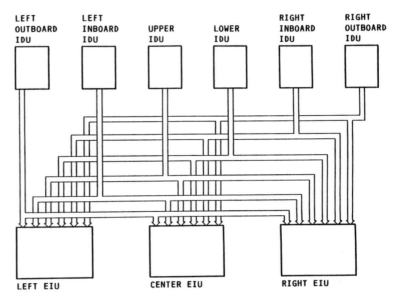

NOTE: ALL INTERCONNECTIONS SHOWN ARE ARINC 429 BUSES

Figure 4-4-9. IDU/EIU data interfaces   *Courtesy of Northwest Airlines, Inc.*

The 57 pin programming connections are assigned the following categories:

1. Pins 1 through 22 are EICAS program pins. These pins determine some of the items to be displayed, such as, tire pressure and fuel quantity units.

2. Pins 23 through 32 are unique option pins. These pins are dedicated for use by any software specifically requested by an airline.

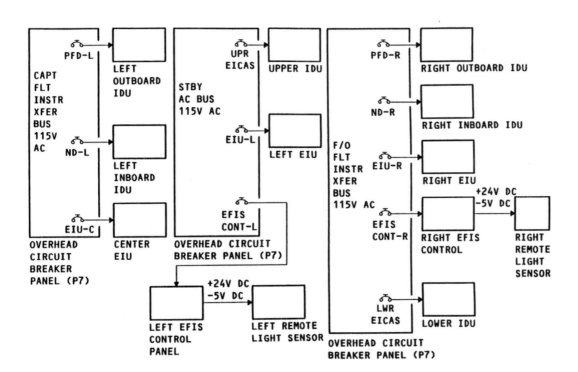

Figure 4-4-10. Integrated display system power inputs   *Courtesy of Northwest Airlines, Inc.*

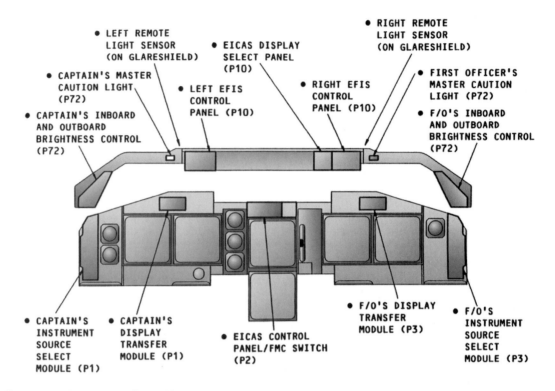

Figure 4-4-11. IDS control panel locations

Courtesy of Northwest Airlines, Inc.

3. Pins 33 through 56 are *displayed* pins; used to determine what information or format is displayed by the PFDs and NDs.

4. Pin number 57 is the parity bit. Note that this parity pin is specifically set accord-

ing to the number of program pins set to logic 1. Odd parity is always maintained. For more information on odd parity see Chapter 2.

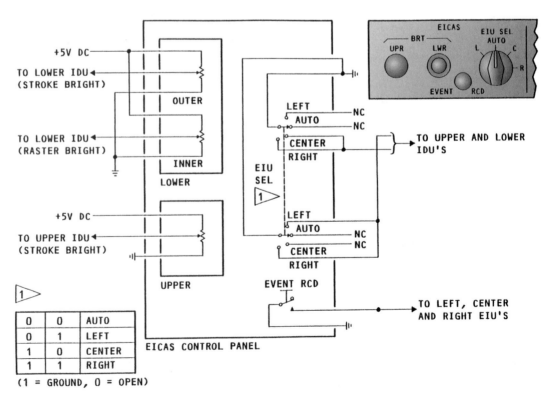

Figure 4-4-12. EICAS control panel and related circuitry

Courtesy of Northwest Airlines, Inc.

EICAS Controls

The diagram of Figure 4-4-11 shows the locations of various IDS control panels. The four panels used to control EICAS functions are the:

1. EICAS control panel, located above the upper EICAS display

2. EICAS display select panel, located on the glare shield just right of center

3. Captain's and first officer's display transfer modules, located above their respective right and left displays

The EICAS Control Panel

The EICAS control panel has three controls: brightness, event record, and EIU selection. As seen in Figure 4-4-12 there are two brightness (*BRT*) controls. The *UPR* knob is used to adjust the brightness of the upper EICAS display. The *LWR* control has an inner and outer knob. The outer knob adjusts the brightness of the lower display. These controls consist of a potentiometer with +5 VDC and ground references (Figure 4-4-12). The *EVENT RCD* (event record) switch is used to record manual events in the EICAS nonvolatile memory. Pressing this button will record the parameters of all EICAS systems simultaneously.

The source selection switch is used to determine which EIU will drive the EICAS displays. The pilot can manually select *L*, *C*, or *R* to choose a specific computer. The *AUTO* position selects the left, center, and right computers respectively. If one EIU fails, the system will automatically select the next priority EIU. The EIU selection switch sends a two bit binary code to each EICAS display; the display's software then selects the correct EIU. Data from all three EIUs is sent to each display unit and the DU "listens" to the selected EIU. As shown in the lower left portion of Figure 4-4-12 the binary code 00 is an AUTO selection, 01 means L, 10 means C, and 11 means R.

The Display Select Panel

The B-747-400 *display select panel* (DSP) combines the engine (*ENG*) and status (*STAT*) switches with the synoptic page switches (*ELEC*, *FUEL*, *ECS*, *HYD*, *DRS*, and *GEAR*). The cancel and recall switches are also part of the DSP. As presented earlier on the B-757, the synoptic pages are found on the EICAS maintenance panel. Figure 4-4-13 shows the layout of a B-747 DSP. The **ENG** (engine) push-button is used to call the secondary engine data to the lower EICAS display. The *STAT* (status) switch, when pressed, will present the status display on the lower EICAS CRT.

Six synoptic pages are available with this EICAS: ELEC (electrical), FUEL (fuel), ECS (environmental control systems), HYD (hydraulic), DRS (doors), and GEAR (landing gear). The *CANC* (cancel) switch is used to cancel any Level B or C EICAS message. The *RCL* (recall) switch is used to recall any messages that are still active. Engine exceedance data is also accessed using the *CANC/RCL* switches.

The B-747-400 allows for EICAS control functions of the DSP to be transferred to the *control display unit* (CDU). The 747's three CDUs, located in the center console of the flight deck, are used to input flight data, access the central maintenance computers, and as backup controls for the IDS. Figure 4-4-14 shows the layout of a typical CDU. They are enabled by failure of the EICAS DSP, or through failure of the EFIS, not EICAS, control panels. To operate the EICAS using the CDU, press the **MENU** key on the CDU (Figure 4-4-15).

Display Transfer Modules

The main and auxiliary EICAS displays can be transferred to either the captain's or first officer's inboard IDU. Likewise, the captain's or first officer's PFD or ND can be transferred to the lower EICAS display. This transfer is controlled through the *display transfer modules* (DTM) and is typically done in the event of a display failure. Figure 4-4-16 shows the pilot's DTM. The first officer's DTM operates identically; however, the DTMs are not interchangeable due to switch position.

Each DTM has two switches, the inboard CRT (*INBD CRT*) and the lower CRT (*LWR CRT*). The term CRT is interchangeable with IDU. The

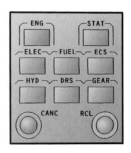

Figure 4-4-13. EICAS display select panel.

Figure 4-4-14. Typical control display unit layout

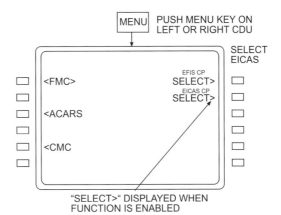

Figure 4-4-15. Using the CDU to access EICAS functions   *Courtesy of Northwest Airlines, Inc.*

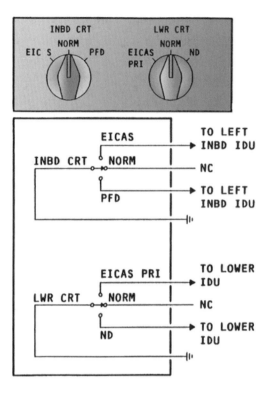

Figure 4-4-16. Pilot's display transfer module
Courtesy of Northwest Airlines, Inc.

inboard CRT control is used to choose what information will be displayed on the captain's or first officer's inboard CRT. The inboard CRT can display EICAS or PFD data if manually selected. The normal position will allow the IDU circuitry to switch display formats in the event of an IDU failure. The lower CRT switch can be set in the EICAS primary or the navigational display position. The normal position for the lower CRT control will also allow the IDU to perform automatic switching. The DTM sends discrete signals of ground or open to the IDU for controlling switch selection.

With all DTM switches in the normal position, if the main EICAS display fails, the main EICAS data is automatically transferred to the lower CRT. EICAS is now operated in a degraded mode. As shown in Figure 4-4-17 full EICAS capacity can be regained by selecting EICAS for display on the inboard CRT. Likewise, if the lower CRT switch is moved to the ND position, navigational data will be substituted for secondary EICAS information and EICAS will be operational in a degraded condition.

Normal Operation of the B-747-400 EICAS

During normal operations the B-747-400 EICAS utilizes two IDUs, the *main EICAS* and *auxiliary EICAS* displays. The main EICAS can display the primary engine, compacted, compacted-partial, and mini-synoptic formats. The auxiliary EICAS can display secondary engine, secondary-partial, status, synoptic, and maintenance pages formats. The compacted or partial formats will be discussed under abnormal operation.

Display Formats

When the AC busses receive power, the EICAS automatically displays the primary and secondary formats (Figure 4-4-18). The primary format consists of the primary engine data, fuel data, landing gear position (when down), flap position, environmental control system (ECS) data, and the requested thrust modes, which are displayed on the upper EICAS display. Any active alert message would also be displayed. The lower EICAS display would show the secondary engine parameters including fuel flow, oil system data, and engine vibration. Pressing the *ENG* switch on the DSP can blank the lower

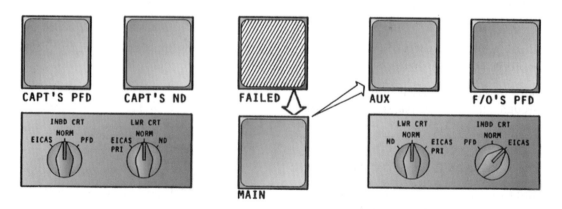

Figure 4-4-17. In EICAS switching, automatic switching moves data from the upper display to the lower display, and Manual switching moves data from lower display to co-pilot's ND display position.
Courtesy of Northwest Airlines, Inc.

display. During normal flight, the crew would most likely blank the lower display.

The display format for the second generation EICAS is similar to the first generation systems. Both digital data and analog displays are used to provide engine parameter information. The B-747-400 analog representations are accomplished using a vertical-scale type display. The B-757 utilizes a round-dial style of analog display.

In the event of a primary engine parameter exceedance, the displayed data turns red and the appropriate advisory message is displayed. With the B-747-400 EICAS in the event of a secondary engine parameter exceedance, the particular system affected will be displayed in the *partial format* (Figure 4-4-19). The parameters for all four engines will always be displayed.

Alerting And Memo Messages

The B-747-400 EICAS utilizes four levels of alerting and memo messages: warnings (Level A), cautions (Level B), advisories (Level C), and memos (Level D). Figure 4-4-20 shows the various messages, as they would appear on the upper EICAS display. Their associated discrete audio and visual annunciators are also shown. All messages except Level A can be canceled to remove them to reduce visual clutter from the EICAS display using the cancel button located on the DSP. If more than one message page is available, the **CANCEL** button is used to scroll through the pages. The recall button will bring back any previously canceled messages.

Status Messages

Pressing the **STAT** button on the DSP will access the EICAS status page. The status page containing APU, hydraulic, and oxygen system data will appear on the auxiliary EICAS display. Status messages and a dynamic flight control surface display will also be shown in the lower portion of the status page (Figure 4-4-21). If more than one status page is available, pressing the **STAT** button will cycle through all available pages and eventually remove the status page data. Technicians and the flight crew typically use the status page data to determine the status of the aircraft prior to dispatch. If the status page shows a given system as inoperative, the technician would consult the minimum equipment list (MEL) to determine the airworthiness of the aircraft. If approved by the MEL, the repair would most likely be deferred until the next maintenance opportunity.

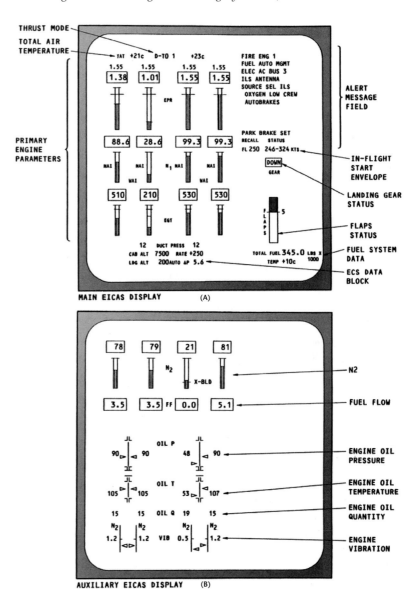

Figure 4-4-18. EICAS displays: (A) primary format, (B) secondary format

Courtesy of Northwest Airlines, Inc.

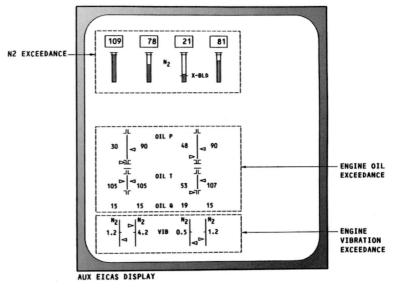

Figure 4-4-19. EICAS secondary partial format   *Courtesy of Northwest Airlines, Inc.*

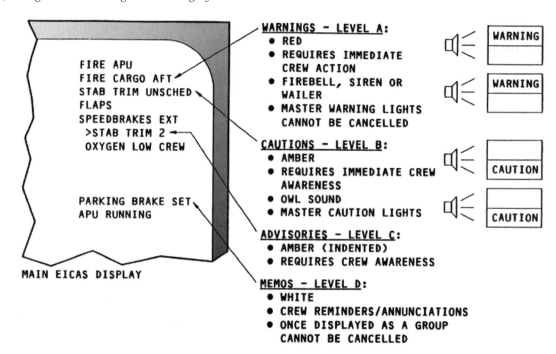

Figure 4-4-20. Four levels of EICAS messages

Courtesy of Northwest Airlines, Inc.

Status pages are initiated automatically by EICAS whenever a malfunction occurs in a system monitored by the EIUs. The EIUs can generate two types of status messages, *latched* and *non-latched*. On the B-747-400 there are three types of latched status messages: *ground only, air only,* and *unconditional*. Their names imply when the messages can be stored (latched) in the EICAS nonvolatile memory. Unconditional messages can be stored with the aircraft in flight or on the ground. The latched messages

are often handy for maintenance of intermittent problems. Non-latched messages are real-time information; if the problem corrects itself, the non-latched message will automatically remove itself.

Synoptic Pages

The synoptic pages for the B-747-400 EICAS are accessed the same as the synoptic pages for the B-757. The major differences are that on the B-747-400, only one system is shown per page and each page has a more graphic format (Figure 4-4-22).

Abnormal Operation of the B-747-400 EICAS

The B-747-400 EICAS can fail in four basic ways and will still operate with the loss of:

1. One or more parameters

2. The main or auxiliary displays

3. One or both EICAS control panels

4. Up to two system computers

In each of these four situations, the EICAS will be operational in some form. The redundancy of these systems makes it extremely unlikely that a complete EICAS failure will occur.

The loss of one parameter to EICAS will cause that particular portion of the display to go

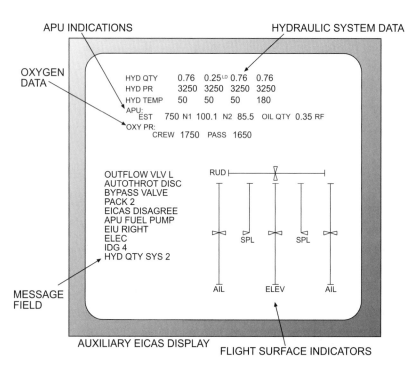

Figure 4-4-21. EICAS status page

Courtesy of Northwest Airlines, Inc.

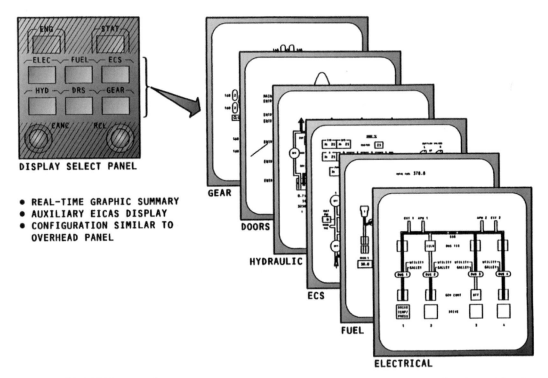

Figure 4-4-22. EICAS synoptic pages

Courtesy of Northwest Airlines, Inc.

blank and the associated message to be displayed. The loss of the main or auxiliary displays will cause the EICAS to go into a compact format. The EICAS *compact-full* format is used when both primary and secondary engine data is requested and only one display is operable (Figure 4-4-23). In compact-full format only EPR is shown in both digital and analog forms. Other information is displayed in digital form only.

A synoptic *mini-format* is available on the second generation EICAS to display certain status data when only one EICAS display is operational. The fuel and gear status pages are the only systems available within the mini-format option and are displayed under the compacted primary engine data. The display is very similar to the compact-full format except the synoptic page replaces the secondary engine data.

In the event of a control panel failure, the EIUs will revert to the fail-safe mode and display necessary information. If the DSP fails the DCU can be used as a backup as previously discussed. If one or two EIUs fail, the third computer can be used as the sole provider of EICAS information. If all three EIUs fail, all EICAS functions are lost.

Maintenance Pages

The second-generation EICAS maintenance pages are accessed through the central maintenance computer system. There are 11 differ-

ent maintenance pages available on the 747-400 system. These pages are each displayed on the auxiliary EICAS display during ground operations or during flight. In general, there are three types of information available for each maintenance page: real time, manual snapshot, and auto snapshot. The real time data is dynamic information and presents information as it currently exists. The snapshot data

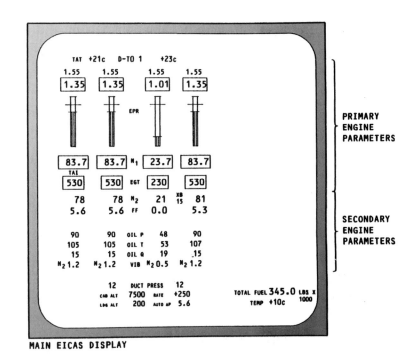

MAIN EICAS DISPLAY

Figure 4-4-23. EICAS compacted mode

Courtesy of Northwest Airlines, Inc.

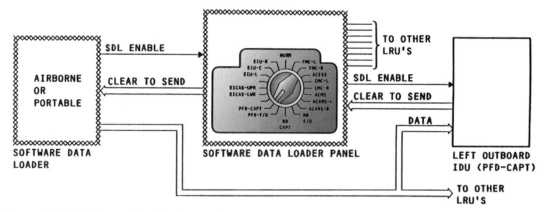

Figure 4-4-24. Software data loader interface diagram

Courtesy of Northwest Airlines, Inc.

is information recalled from memory. EICAS maintenance pages will be studied in further detail in Chapter 5.

Troubleshooting and Maintaining EICAS

Many of the techniques discussed earlier in this chapter concerning troubleshooting the B-757 EICAS also apply to the B-747-400 EICAS. Some of the techniques, which are unique to the second-generation systems, will be presented here.

Software Data Loader

One major change for the second generation EICAS is the ability to change portions of the programming through the use of a software data loader (SDL). The data loader can alter the software in all three EIUs, the upper and lower EICAS displays, and several other LRUs. The updated data and the program needed to access the LRU were originally contained on a 3.5-inch floppy diskette. Many aircraft have been modified to use a solid-state memory stick/flash drive. Two types of data loaders are available for the B-747-400: a portable unit; and a permanently installed data loader located on the observer's console of the flight deck. It should be noted that many of the aircraft data loaders, both permanent and portable have, or will soon be, upgraded to replace the 3.5-inch diskette with a more modern memory device.

According to the Figure 4-4-24 the data loader sends a SDL enable signal to the SDL panel and the selected LRU, in this example the left outboard IDU. The LRU then returns a signal to the SDL when it is ready to accept the updated software. This system has made it possible to make modifications to the LRUs in a relatively easy fashion. In older systems, the same software change would require a LRU change and factory update, or a card replacement.

CRT Displays

The CRT displays on the second generation EICAS are slightly different than those found in the earlier systems. The display units are larger to improve readability, and incorporate an hour meter to track time in use. Cooling is still a critical item for the CRTs and these displays will automatically shut off in two stages when over heated. When the display temperature reaches 110°C, the display will shut down all graphic displays. If the display reaches 125°C the display will blank out all together. When the unit cools, the display will automatically return. Be sure the inlet and outlet air passages are kept clean during display replacement (Figure 4-4-25).

The B-747-400 provides a display unit protective cover located just below the lower EICAS IDU. This cover should be used during all IDU maintenance to protect the CRT screen. The screen of the B-747-400 CRT displays *can* be cleaned with standard glass cleaner. Since

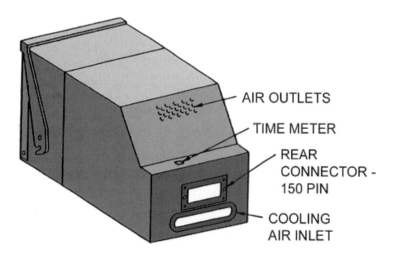

Figure 4-4-25. Rear view of a typical EICAS display unit

Courtesy of Northwest Airlines, Inc.

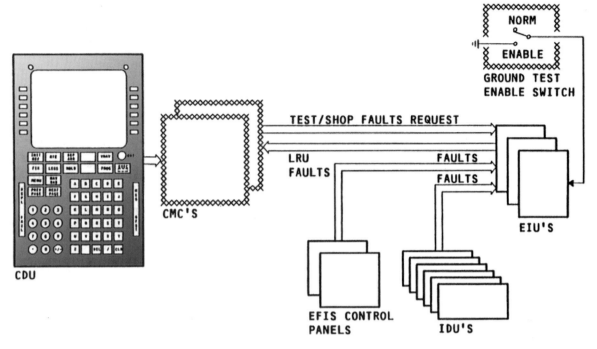

Figure 4-4-26. Block diagram of the integrated display system BITE

Courtesy of Northwest Airlines, Inc.

these displays are always on when their respective bus is energized, it is wise to turn down the brightness if the aircraft is to be powered for long periods during maintenance. This practice will help to lengthen the life of the display. The display circuit breaker could also be opened to "turn off" the IDU. If this is done, take all necessary precautions to ensure that you reactivate the system.

BITE Tests

The BITE test for the B-747-400 is activated through the central maintenance computers. The actual BITE tests and fault memory still occur in the individual LRU of the system. The CMC is used as an access point for all of the aircraft's functional tests. A block diagram of the IDS BITE system is shown in Figure 4-4-26. The CDU is used to access the EICAS EIUs and IDUs. The individual EIUs and IDUs monitor their respective health and input signals, and then transmit that data to the CMCs. The CMCs display all BITE data through the CDU.

To initiate an EICAS BITE test, the ground test *ENABLE* switch must be placed in the *enable* position. The ground test enable switch is located on the overhead panel. The next step would be to access the CMC menu using the CDU. From the menu, the ground tests menu would be selected, and then the appropriate test (IDS) would be selected. During the test, the terms *IDS* and *IN PROGRESS* will be displayed on the CDU. The test takes approximately eight seconds; during that time the test format will

be shown on all operable IDUs. The master warning and caution lights will illuminate and the audio tones will sound during the test. Any detected failures will be displayed on the CDU. There are four different fault messages associated with the EICAS: *IDU failed, EIU failed, EIU disagree,* and *EIU no test response.*

During the IDU BITE test, the IDUs will display a test format similar to Figure 4-4-27. As shown in this figure, the system program pins are displayed in the upper left of the test format. If a program pin error exists the term *ERROR* will be seen adjacent to an error code. The error code can then be found in the aircraft's maintenance manuals.

In general, second generation EICAS offers improved troubleshooting over earlier systems. Use of the CMC has helped to simplify the troubleshooting process. The CMC will display fault codes and fault messages. Using the aircraft's fault isolation manual (FIM), these codes will direct the technician to the proper repair or further troubleshooting sequence.

The CMC can also be used to monitor input data to many of the EICAS LRUs. The *input monitoring* function of the CMC is used to determine the status of ARINC 429 data to the EICAS electronic interface units (EIU). In general, input monitoring is done through the CDU and compares system data received from the EIUs and system data sent directly to the CMCs (Figure 4-4-28). Complete details of the central maintenance computer system will be discussed in Chapter 5.

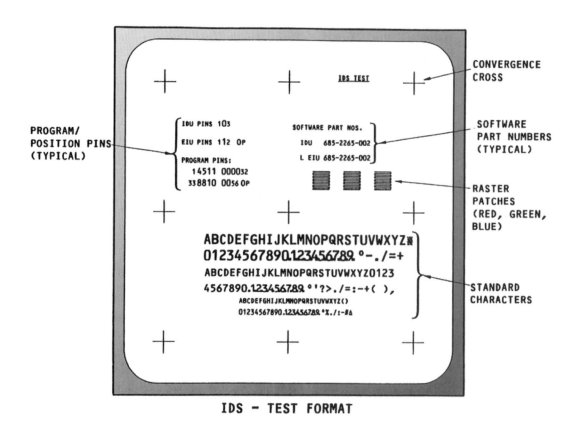

Figure 4-4-27. IDS test format

Courtesy of Northwest Airlines, Inc.

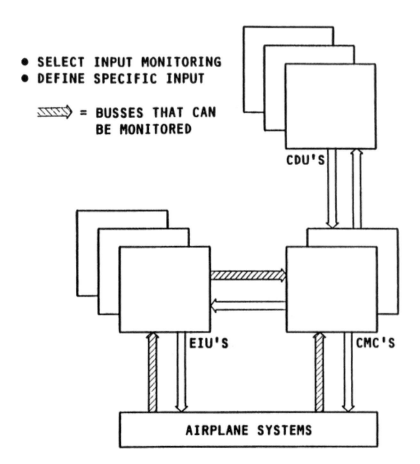

- SELECT INPUT MONITORING
- DEFINE SPECIFIC INPUT

= BUSSES THAT CAN BE MONITORED

Figure 4-4-28. CMC input monitoring diagram

Courtesy of Northwest Airlines, Inc.

Section 5

Electronic Centralized Aircraft Monitoring (ECAM)

Airbus transport category aircraft use a system called *electronic centralized aircraft monitor*, or ECAM, to monitor and display aircraft system and engine parameters. The system utilizes two CRT or LCD displays typically located in the center of the instrument panel. The newest ECAM systems are integrated systems, which operate in conjunction with the aircraft's electronic flight instrument system as well as many other systems. The integrated EFIS and ECAM system found on the A-320 is called the *electronic instrument system* (EIS). The following discussion on ECAM will be based on the A-320 aircraft.

System Description

The EIS consists of six CRT display units as seen in Figure 4-5-1. Four displays are used primarily for flight data (EFIS). The remaining two displays are used primarily for engine and aircraft system data (ECAM). On the

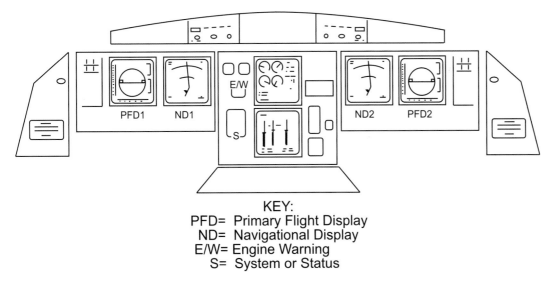

Figure 4-5-1. A-320 Electronic instrument system display locations

Courtesy of Airbus S.A.S.

A-320, the two captain's side EFIS displays are called the *primary flight display* (PFD 1) and the *navigational display* (ND 1). The first officer's side consists of PFD 2 and ND 2. The two ECAM displays are referred to as the upper and lower ECAM displays. The upper ECAM display is used for *engine and warning* (E/W) information, the lower display shows *system and status pages*.

The major advantage of the integrated ECAM/EFIS system is the flexibility in the display of data. In the event of a display failure, one of the remaining displays can be used to show the missing data. Figure 4-5-2 shows the three means by which ECAM data is interchanged between displays. As indicated by the white arrows, the flight crew can switch the PFD and ND data manually. This is accomplished by the *PFD/ND XFR* switch located just outboard of the captain's and first officer's EFIS displays. The shaded arrows show that the data from the upper ECAM display is automatically transferred to the lower ECAM display in the event of a single ECAM display failure (Figure 4-5-2). The black arrows show that the upper ECAM data moves to the EFIS displays in the event that both ECAM displays fail.

Display Formats

There are four basic types of information displayed by ECAM: engine data, messages, system synoptic pages, and system status data. *Engine data* is always displayed any time ECAM is oper-

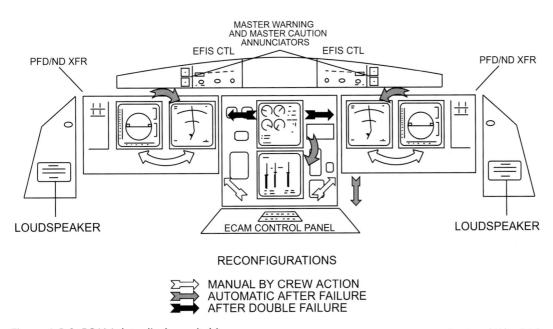

RECONFIGURATIONS

MANUAL BY CREW ACTION
AUTOMATIC AFTER FAILURE
AFTER DOUBLE FAILURE

Figure 4-5-2. ECAM data display switching

Courtesy of Airbus S.A.S.

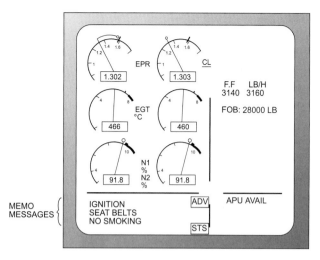

Figure 4-5-3. Typical upper ECAM display, normal operation

Courtesy of Airbus S.A.S.

Similar to the Boeing EICAS, all information displayed by ECAM is color coded to help the flight crew identify a system's status. The different colors also provide a quick reference to the technician during troubleshooting. In general, information in green indicates that all systems are operating normal. Magenta (pink) indicates a critical system failure. Cyan (light blue) indicates a system failure that is not critical, and white indicates general information.

Warning/Caution Annunciators

The ECAM system incorporates two discrete visual annunciators: the *master warning* and *master caution* lights. There is one each of these light assemblies on the pilot's and co-pilot's glare shield (Figure 4-5-2). To ensure visual warnings are displayed in the event of lamp failure, each annunciator contains four light bulbs. ECAM also uses audio tones to gain the pilot's attention during display of critical system failure. Two loudspeakers installed on the instrument panel are used for ECAM audio tones.

System Architecture

The electronic centralized aircraft monitoring system contains seven computers, which receive data from various engine and aircraft systems. The A-320 ECAM computers are classified into three general categories: *display management computers* (DMC), *flight warning computers* (FWC), *and the system data acquisition concentrators* (SDAC).

ating. During normal operation, various memo messages are displayed on the upper ECAM display along with engine data as shown in Figure 4-5-3. In the event of a system failure, a warning or caution message would be displayed on the upper ECAM along with the engine data.

The *system synoptic pages* and *system status data* is displayed on the lower ECAM display. The synoptic display gives a brief general description of the systems operation in a graphic format. The system data displayed is a function of aircraft configuration, pilot selection, or automatically displayed after a system failure. After a failure, status information is automatically recorded on the status page and can be displayed on the lower ECAM CRT when requested. The door/oxygen synoptic page and a typical status page are shown in Figure 4-5-4.

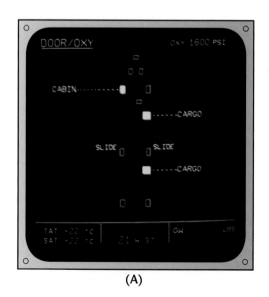

(A)

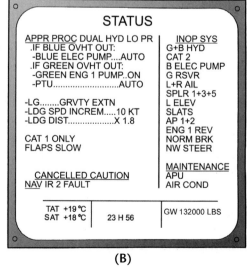

(B)

Figure 4-5-4. Lower ECAM display: (A) System Synoptic Door/Oxygen Page, (B) System status page

Courtesy of Airbus S.A.S.

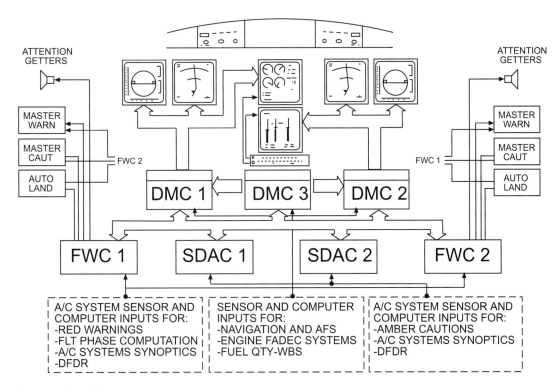

Figure 4-5-5. EIS interface diagram   *Courtesy of Airbus S.A.S.*

Figure 4-5-5 shows a block diagram of the computers and their relationship within the integrated EFIS/ECAM system. It can be seen from this diagram the system computers monitor data for both the EFIS and ECAM displays.

Display Management Computer

The EIS contains three *display management computers* (DMC) that process data from the SDACs, FWCs, and certain discrete inputs. The discrete inputs include digital and analog signals for navigation and autoflight, digital engine control, fuel quantity, and weight and balance data (Figure 4-5-5). The data sent from the SDACs and FWCs is in digital format. A dual ARINC 429 data bus format is used for most systems to provide redundancy of transmitted data to the DMCs. If the data on one bus is invalid, the DMC will take the information from the other bus. If the data is invalid on both busses, the DMC will form a crosshatched pattern on the display where the data would normally appear. The DMCs receive weather data through ARINC 453 busses and the FWC warning messages through RS 422 data busses (Figure 4-5-6).

The DMC sends a digital signal to a maximum of three display units. The output data contains the video information needed to produce the correct image on the EIS displays. The signals from the DMC to the display units are sent on a *dedicated serial data link* (DSDL) Figure 4-5-6. A feedback signal is returned from the display units to the DMC. The feedback signal provides information on the health of the display unit. The status of the display is monitored for automatic display transfer, built-in-test equipment (BITE) data, and various parameters used by the FWC. If any of the data returned by the DSDL does not match the data sent to the display, a system inoperative flag appears on the display.

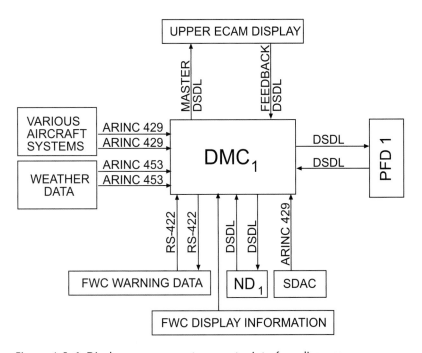

Figure 4-5-6. Display management computer interface diagram

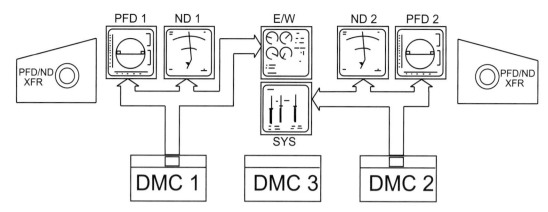

Figure 4-5-7. Normal DMC output configuration

Courtesy of Airbus S.A.S.

During normal operation, DMC 1 supplies video information for PFD 1, ND 1, and the upper ECAM displays; DMC 2 supplies the video information for the PFD 2, ND 2, and the lower ECAM displays. The normal DMC output configuration is shown in Figure 4-5-7. In the event DMC 1 or 2 fails, DMC 3 will supply data to the inoperative displays. During all operations the display management computers compares information to ensure the input and output data is valid. If a discrepancy is detected, the erroneous information is ignored and/or the system reconfigures itself to compensate. If any system fails, the appropriate message will be displayed by ECAM to alert the flight crew.

Flight Warning Computers

The two *flight warning computers* (FWC) monitor the various systems necessary to generate all warnings on the ECAM displays. During normal operation, FWC 1 generates all output signals to the DMCs. The DMCs then create the video information for the display units. FWC 2 is always operational and will automatically take control of the output data if FWC 1 fails.

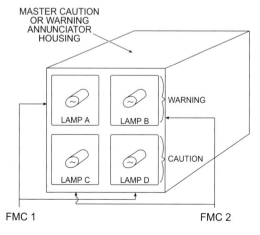

MASTER CAUTION
OR WARNING
ANNUNCIATOR
HOUSING

WARNING

LAMP A   LAMP B

CAUTION

LAMP C   LAMP D

FMC 1   FMC 2

Figure 4-5-8. Master warning and caution annunciator

The software of the FWCs performs the essential computations to generate the warning messages on the ECAM displays via the DMCs. The software also controls the discrete aural and visual warning signals. The master caution and warning annunciators are each powered by both FWCs (Figure 4-5-5). There are four light bulbs in each annunciator as shown in Figure 4-5-8. In this example, the FMC 1 controls bulbs A and D; FWC 2 controls bulbs B and C. This configuration ensures the master annunciators are operational in the event of a single FWC or multiple lamp failure.

Figure 4-5-9 shows the input/output diagram for the flight warning computers. The FWC receives six synchro, 154 discrete, and 40 ARINC 429 data bus inputs for various aircraft and engine systems. The FWC output data is comprised of ARINC 429 data, RS 422 data, and discrete data. The discrete data is sent directly to both the captain's and first officer's caution and warning displays. Discrete data is also sent to both loudspeakers. The RS 422 data outputs are used to transmit warning messages to the DMCs. The ARINC 429 data busses transmit information to three DMCs, the opposite FWC, the centralized fault display interface unit (CFDIU), and the flight data interface unit (FDIU). The CFDIU interfaces with the aircraft's built-in diagnostic system. The FDIU interfaces with the aircraft's flight data recorder.

System Data Acquisition Concentrators

The *system data acquisition concentrators* (SDAC) monitor aircraft systems, perform the necessary computations, and transmit output data to the FWCs and DMCs. The systems monitored by the SDACs are less critical than those monitored by the FWCs. That is, the SDACs control the master caution annunciators; the FWC controls the warning annunciators. Caution messages are considered a lower priority than warnings messages. The SDAC output data is

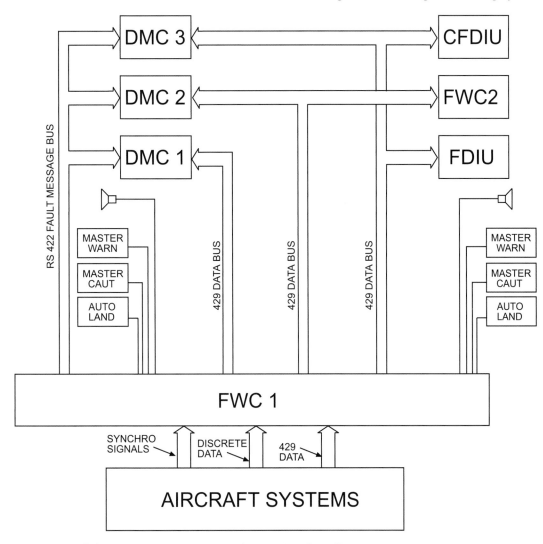

Figure 4-5-9. Flight warning computer input/output interface diagram

sent to the master caution annunciators via FWC 1 and 2. The synoptic page data is sent to the ECAM displays via the CMCs. Figure 4-5-10 shows the input/output diagram for the SDACs.

In general, the LRUs of the ECAM system are interchangeable. That is, the DMC 1 can be installed in the DMC 2 or DMC 3 mountingracks and vice versa. The two FWCs are identical, as well as the two SDACs. All six of the display units are also interchangeable. This becomes very handy when performing troubleshooting. For example, if it is suspected that the FWC 1 is defective; simply swap FWC 1 for FWC 2. If the fault now appears in FWC 2, the LRU, which was moved from the number 1 to number 2 FWC slot, must be defective. *Pin programming* of the FWC's wiring harnesses is used to identify the FWC 1 and FWC 2 installation racks. As the computer is installed into the equipment rack, a given pin is connected to ground (digital 0) to tell the LRU that it is installed in the FWC 2 rack. The same pin would be open (connected to digital 1) to iden-

tify the FWC 1 rack. Pin programming is used to identify the location of all interchangeable LRUs.

EIS Controls

The ECAM and EFIS control panels are shown on the component location diagram (Figure 4-5-11). The ECAM displays are controlled by the ECAM LRU and the 8VU panel located just below the lower ECAM display. The brightness and transfer of the PFDs and NDs are controlled through the 301VU and 302VU panels located to the outboard side of the captain's and first officer's EFIS displays. Using these controls, the PFD and ND can be turned on/off separately as needed for maintenance. Located between the on/off/brightness controls is the **PFD/ND XFR** (transfer) control. This push-button switch is used to transfer the data shown on the PFD to the ND, and vice versa. The captain's and first officer's displays are controlled independently; therefore, identical control panels are accessible to both pilots.

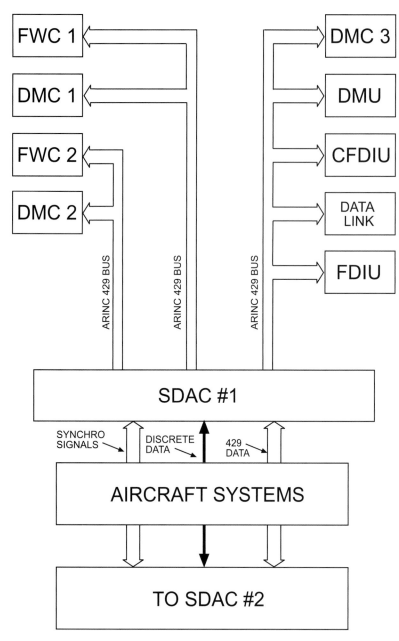

Figure 4-5-10. System data acquisition concentrators (SDACs) input/output interface diagram

ECAM Switching

The ECAM switching controls are used to select the source of information to be presented on the EIS displays. As shown earlier in Figure 4-5-7, during normal operation of the system, DMC 1 supplies data for the captain's EFIS and the upper ECAM displays. In the normal configuration, DMC 2 supplies the first officer's EFIS and lower ECAM displays.

The *ATT HDG* (attitude heading) switch controls the source of attitude heading information. If the ATT HDG switch is placed in the *CAPT 3* position, the captain's EFIS will receive attitude data from DMC 3. If switched to the *F/O 3* position, the first officer's attitude data

will come from DMC 3. The air data switch operates in a like manner.

The *EIS DMC* switch determines which display management computer will drive the captain's and first officer's PFD, ND and the associated ECAM displays. Figure 4-5-12 shows the configuration if *CAPT 3* is selected on the EIS DMC switch. In this situation DMC 3 supplies data to PFD 1, ND 1, and the upper ECAM display. If *F/O 3* is selected, DMC 3 is used to drive PFD 2, ND 2, and the lower ECAM displays.

A simplified diagram of the EIS switching is shown in Figure 4-5-13. In this diagram it can be seen that switching is done through software within the LRUs and through dedicated relays mounted in the equipment bay. Each display unit can be driven by its normal (N) or alternate (A) input. The switching for the PFD and ND displays is controlled via a 28 VDC relay. See the relay box *CAPT*, and relay box *F/O* in Figure 4-5-13. The pilot selects an EIS configuration using the PFD/ND control panel. The control panel commands a relay to send a given signal to the display unit software. The display unit then selects the requested input (alternate or normal). In like manner, the ECAM engine warning (E/W) and status (S) displays are also controlled through a 28 VDC relay and display unit software.

Page Control

The *page control section* of the ECAM control panel contains 18 push-button switches used to access different systems for display on the lower ECAM CRT (Figure 4-5-11). It is important for the electronics/avionics technician to become familiar with this control panel. The various ECAM pages show real time status of all major aircraft systems. During troubleshooting, this information is extremely valuable. The following paragraphs contain a description of the ECAM page control switches and their associated ECAM displays.

The flight crew uses the **TO CONFIG** (takeoff configuration) switch to determine if the aircraft is ready for takeoff. When this button is pressed, the message *TO CONFIG NORMAL* is displayed on the E/W ECAM display. If the system is not in the correct configuration for takeoff, the appropriate warning will be initiated by ECAM.

The *EMER CANC* (emergency cancel) switch will delete any ECAM aural warnings currently active; however, the master warning and associated ECAM message will remain on display. The EMER CANC switch will also cancel any caution messages and extinguish

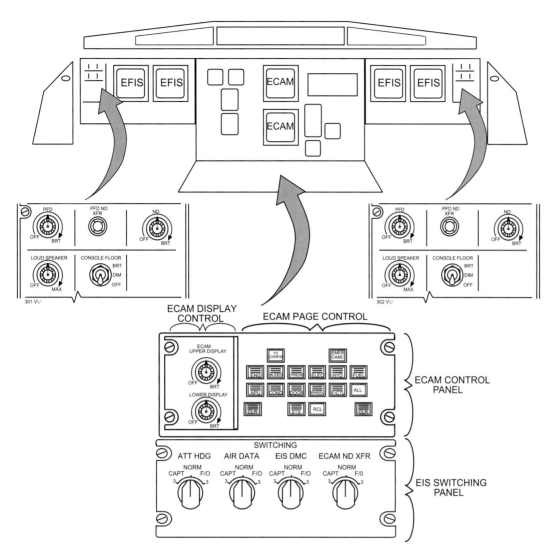

Figure 4-5-11. EIS control panel locations

Courtesy of Airbus S.A.S.

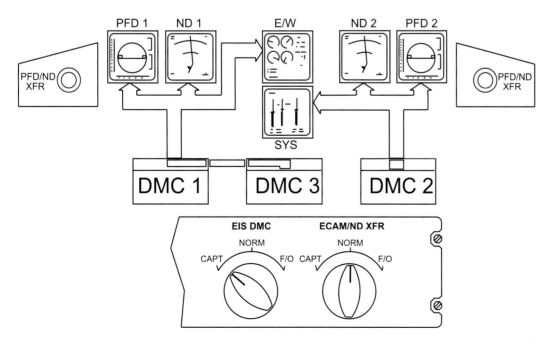

Figure 4-5-12. EIS interface when pilot's DMC 3 is selected

Courtesy of Airbus S.A.S.

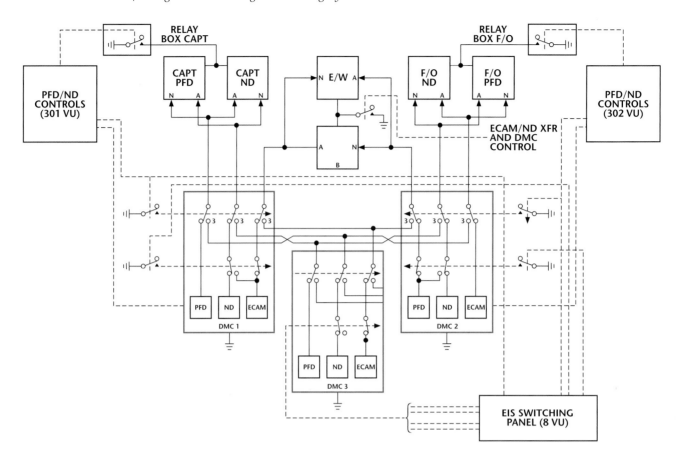

Figure 4-5-13. A320 EIS switching diagram

the master caution light. System cautions and warnings will be discussed later in this chapter.

There are 11 switches each labeled with a given system designator, such as *ENG* (engine), *BLEED* (bleed air), *PRESS* (pressurization), *ELEC* (electrical), etc. Pushing these switches will call the related system information to the lower ECAM display. On an ECAM system page, it is easy to determine the real-time status of the system displayed. During troubleshooting, choosing a given system page will allow the technician to verify system operation.

ECAM Normal Operation

The ECAM system can operate in two basic configurations: *normal mode* and *mono mode*. In normal mode, the top ECAM display shows engine data and warning messages. This display is known as the E/W (engine/warning) display unit. The lower ECAM unit is used to indicate the configuration of systems and aircraft status data. The data indicated on these displays is generally consistent between aircraft. However, some displays show slight differences due to specific configurations requested by an airline.

The Upper Display

The upper ECAM display contains the E/W page, which is divided in to four quadrants (Figure 4-5-14). The top left quadrant contains engine data, such as engine pressure ratio (EPR), exhaust gas temperature (EGT), turbine spool speeds (N_1), and fuel flow (FF). This data is displayed in both a digital and analog format. The analog display provides easy analysis of changing values, while the digital data offers precise reading of stable numbers.

The top right quadrant of the upper ECAM display typically shows *fuel on board* (FOB) in a digital format. A graphic representation of the slat and flap positions is shown below the FOB. The bottom two quadrants of the E/W display contain all warning, caution, and memo messages recognized by ECAM. Primary and independent failure messages are shown in the lower left quadrant. Secondary failures are displayed in the lower right, and memo messages are displayed on either side. The various message types will be discussed in greater detail later in this chapter. The label STS will be displayed if ECAM status messages are available. The label ADV is shown in the mono mode to inform the flight crew of advisory messages not displayed.

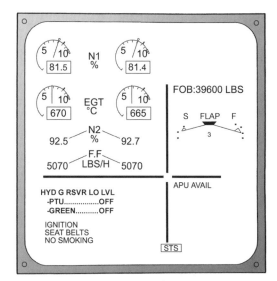

Figure 4-5-14. ECAM Engine/Warning page
Courtesy of Airbus S.A.S.

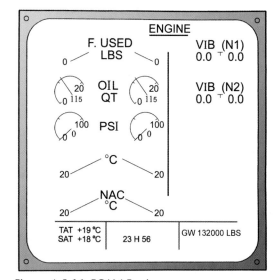

Figure 4-5-16. ECAM Engine page
Courtesy of Airbus S.A.S.

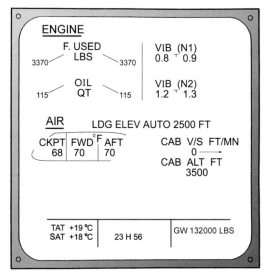

Figure 4-5-15. ECAM Cruise page
Courtesy of Airbus S.A.S.

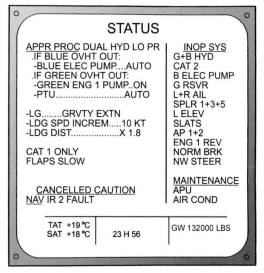

Figure 4-5-17. ECAM Status page
Courtesy of Airbus S.A.S.

The Lower Display

The lower ECAM display will show one of three types of information: the cruise page, system pages, or status pages during normal operation. As the name implies, the *cruise page* is displayed during normal cruise operations. As seen in Figure 4-5-15, this page displays a combination of both engine and cabin data. There are 11 *system pages*, which can be display by ECAM, the engine, bleed air, pressure, electrical, hydraulic, fuel, auxiliary power unit (APU), door, wheel, and flight controls. These pages are displayed under certain aircraft configurations or upon request through the ECAM control panel. Upon failure of any given system, ECAM will automatically call the related system page to the lower CRT. The engine page is shown in Figure 4-5-16.

In the event of a system failure, ECAM will automatically display the message STS (status) on the E/W display (Figure 4-5-14). This indicates that aircraft status information is available on the lower display (*status page*). The status page contains information regarding the aircraft's condition after a system failure. This information tells the flight crew exactly how any given failure has affected the flight characteristics of the aircraft. As seen in Figure 4-5-17, the status page contains procedures and limitations on the left side of the display. The flight crew uses procedures and limitations to determine the aircraft landing capabilities; hence, this page is automatically displayed during approach to land.

It should be noted that all ECAM messages are real time displays. In other words, all the data presented on the ECAM displays is a function

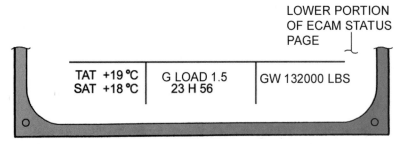

Figure 4-5-18. ECAM status page permanent data   *Courtesy of Airbus S.A.S.*

of the aircraft's configuration at the time of the display. If a system's status changes, the ECAM display will respond immediately. Any warnings, cautions, or memos are presented only if a system has failed and remains inoperative. The status page will also change if systems fail, or repair themselves, during flight. These real time ECAM presentations can be very helpful when troubleshooting various aircraft systems. ECAM can also be used for verification of a system's operation after a repair. In either case, operating the system while monitoring the aircraft's status on ECAM can provide a wealth of information.

ECAM Permanent Data

On the lower portion of the ECAM system/status page, five items known as *permanent data* are always displayed. Figure 4-5-18 shows permanent data displayed on the ECAM status page configuration. The permanent data items include:

1. True air temperature (TAT)

2. Static air temperature (SAT)

3. Gross weight (GW)

4. Greenwich Mean Time (GMT) in hours (h) and minutes

5. Load factor (G load) or pre-selected altitude.

The load factor is displayed in amber if the aircraft exceeds limits of greater than 1.4 g or below 0.7 g. If the load factor is within limits, the altitude selected for the flight guidance system is displayed above GMT.

ECAM Mono Operation

The ECAM system automatically reverts to mono mode in the event of an ECAM display failure. The mono mode is also active whenever only one ECAM display is turned on. In the mono mode, E/W data has priority and will be displayed in its normal configuration. If the E/W display should fail, that information is automatically displayed on the lower ECAM display as seen in Figure 4-5-19. If both ECAM displays fail, ECAM data is temporarily lost. The flight crew will then select an ECAM/ND transfer and the E/W page will appear on the captain's and/or first officer's navigational display (ND).

Display Priority

There are four basic modes of operation that determine the type of data displayed by ECAM. The flight crew selects one of these modes manually and three are displayed automatically. The automatic displays are a function of the current aircraft condition. For example, if a system failure had just occurred, the ECAM system would display a given system page as needed to inform the flight crew of the failure.

The three automatic modes have a distinct priority to ensure the most important data is displayed first. Table 4-5-1 shows that the *automatic mode related to a failure* is the most important and will take priority over all other automatic display modes. After a critical system failure, ECAM will display any associated warnings or cautions on the E/W page and, in most cases, will automatically display an associated system page on the lower display. The *automatic advisory mode* is displayed any time a system parameter nears an out of limits (failure) condition. For example, if the engine oil temperature is higher than normal, but not too hot to trigger a warning, the automatic advisory mode will be active. In this mode the questionable parameter and title of the system page appear blinking on the display. In the advisory mode, the associated system page is also automatically displayed. The third priority is the *manual mode*. The lowest priority automatic mode is the *automatic mode related to a flight phase*. In this mode, certain data is displayed during different phases of the flight. Flight phases will be discussed later in this chapter.

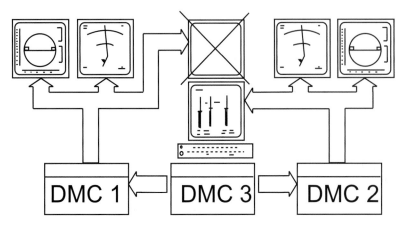

Figure 4-5-19. Automatic transfer of Engine/Warning data to lower ECAM display   *Courtesy of Airbus S.A.S.*

The manual mode is activated by the flight crew to display a given system page. This mode can be activated at any time. However, if the manual mode is selected and a failure or abnormal system parameter occurs, the related automatic mode will take priority. As previously discussed, manual selection of system pages is done through the page control section of the ECAM control panel.

Failure Categories

The failure of any system monitored by ECAM will have a distinct effect on the performance and/or capabilities of the aircraft. Since several hundred different failures can cause an ECAM message, a failure classification system was developed to help the flight crew and maintenance personnel identify the severity of each failure. Each system failure displayed by ECAM can be identified as one of three major categories: independent failures, primary failures, or secondary failures. Independent and primary failures are always displayed on the lower left portion of the E/W page. An *independent failure* will affect only that component and will not have adverse effects on other components or systems on the aircraft. A *primary failure* will cause a given component to fail as well as one or more other components or systems to fail.

Corrective action messages are displayed directly below primary and/or independent failure messages shown on the ECAM display. Corrective action messages are listed in blue and should be performed by the flight crew as soon as practical. Figure 4-5-20 shows a primary failure (*HYD G RSVR LO LVL*) and the corrective actions of (*PTU...OFF*) and (*GREEN ENG 1 PUMP...OFF*). In this situation, one of the flight crewmembers would turn off the PTU (power transfer unit) and the number 1 green engine driven hydraulic pump. The A-320 has three separate hydraulic systems: green, blue, and yellow.

Secondary failures are defined as a loss of a component or system as a result of a primary failure. Secondary failures are always displayed on the lower right side of the E/W page (Figure 4-5-14). To learn more about the displayed secondary failure the flight crew would manually select the appropriate system page.

ECAM Warning Classes and Levels

For most system or component failures monitored by ECAM, a warning message is displayed on the E/W page. These warnings are divided into three *classes* and three *levels*. The highest priority class is *Class 1*. Class 1 warn-

PRIORITIES: ECAM SYSTEM PAGES	
1 – Automatic Mode Related to a Failure	An ECAM system page call is associated with most of the warnings.
	The page is automatically called and has priority over other display modes.
2 – Automatic Advisory Mode	The advisory mode indicates a parameter which drifts out of its normal range before triggering a warning.
	The parameter and the title of the system page are displayed pulsing.
	In advisory mode, the ECAM system page is automatically displayed and the corresponding pushbutton comes on.
3 – Manual Mode	When pressed in, an ECAM control panel push button will come on and display the corresponding system page in normal mode.
	Pressing it again will return to the previous page.
4 – Automatic Mode Related to the Flight Phase	If no other mode is selected, the ECAM system pages are automatically displayed according to the flight phase.

Table 4-5-1. ECAM mode priorities

ings are always displayed on ECAM and may also present a discrete aural and visual annunciator. Class 1 messages are divided into three levels (levels 3 through 1). Level 3 messages are displayed for the most significant failures. *Class 1, Level 3* warnings occur only for a handful of serious failures, which require immediate crew action. For example, Class 1, Level 3 warning messages occur during cabin depressurization, engine fire, inappropriate slat or flap position during take off, low pressure in all three hydraulic systems, and dual engine failure.

Class 1, Level 3 warnings always appear in red on the ECAM display, create an aural warning, and lights the red flashing master warning annunciators on the captain's and first officer's glare shield. The audio signal is a continuous, repetitive chime broadcast by two flight deck speakers. The master warning annunciators are actually momentary contact switches illuminated with the words *caution* or *warning*, when appropriate. Depressing either of these annunciators will cancel the audio warning and turn off the annunciator lights. The ECAM message will remain on the CRT display as long as the fault exists. ECAM Class 1, Level 3 messages are similar to the

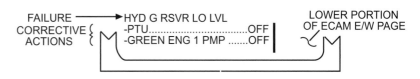

Figure 4-5-20. Example of corrective actions during a system failure

Courtesy of Airbus S.A.S.

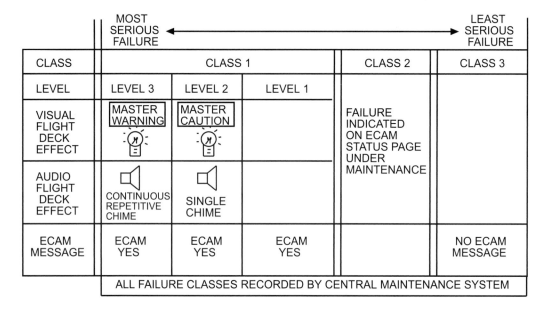

Figure 4-5-21. Various classes, levels, and related flight deck effects reported by ECAM

Level A (warning) messages displayed by the Boeing EICAS, which was discussed earlier in this chapter.

Class 1, Level 2 ECAM warnings are generated by failures less serious than level 3 faults. Most of the ECAM warning messages are class 1, Level 2. Level 2 messages occur from a fault that requires immediate attention and possible future action by the flight crew. The loss of number 1 engine-driven generator causes no threat to flight safety but typically requires the crew to start the auxiliary power unit (APU). For this situation ECAM would display the amber message *GEN 1 FAULT*. ECAM Class 1, Level 2 messages are similar to the Level B (caution) messages displayed by the Boeing EICAS.

Class 1, Level 2 messages appear in amber on the ECAM display and illuminate the amber master caution annunciators. A single chime will sound for all Class 1, Level 2 faults. Pressing either the pilot's or co-pilot's annunciator switch turns off the master caution annunciator. The ECAM message will remain as long as the fault exists.

Class 1, Level 1 messages are the least critical Class 1 message that occur mainly on component failures which are used for redundancy or emergency backups. For example, failure of the static inverter produces a Level 1 message. In this case, the ECAM would show the amber message *STATIC INV FAULT*. There is no related aural or discrete annunciator for Level 1 messages. Class 1, Level 1 faults are similar to the advisory messages found on the Boeing EICAS.

Class 2 messages are caused by faults, which have no flight deck effect, but require flight crew attention prior to leaving the aircraft. These messages appear on the status page of ECAM. These messages are present mainly to allow the flight crew the opportunity to record the failure in the aircraft log.

Class 3 failure messages are recorded only by the aircraft's central maintenance system. The flight crew is totally unaware of these failures. Repair of class 3 faults becomes mandatory at the next heavy maintenance check performed approximately every 400 flight hours.

Figure 4-5-21 shows a graphic representation of the various classes, levels, and their related flight deck effects. All ECAM messages, regardless of class, are reported to the aircraft's central maintenance system. All ECAM messages are real time and are lost from memory after engine shut down. Actual fault data is available throughout the aircraft's central maintenance system.

Local Warnings

A *local warning* is considered any fault message presented on the specific control panel for a given system. For example, if the number 1 AC generator should fail, the number 1 generator control switch will illuminate with the word *FAULT*. This switch is located on the electrical control panel on the overhead panel of the flight deck. The ECAM upper display will also show the message *GEN 1 FAULT* and the lower ECAM display will call the electrical status page. There are numerous local warnings associated with different ECAM messages.

Memo Messages

Items that the flight crew should be aware of, but that are not related to a failure are displayed in the form of a *memo message*. These messages are displayed in the lower portion of the ECAM engine/warning page. Typical memo messages are items such as: *SEAT BELTS, NO SMOKING,* and *APU AVAIL*. These messages inform the flight crew that the seat belt and no smoking signs are illuminated, and the APU electrical power is currently available for use.

Flight Phases

The various segments of a typical flight are broken down into ten segments known as *flight phases*. A breakdown of the ten flight phases is shown in Figure 4-5-22. The flight phases are computed by the FWC and used by ECAM to determine system page presentations and inhibit warning messages. During each flight phase, the ECAM system determines which system page will be presented on the lower ECAM display. For example, during flight phase one, the door/oxygen page is displayed. It is important to remember that the flight phase system pages are the lowest priority. If a failure occurs, or if a manual page selection is made, the flight phase system page is replaced.

Under critical flight phases, such as landing and take off, it is best not to distract the flight crew with ECAM fault messages. Most of the ECAM messages do not require immediate attention and may divert the flight crew from more crucial tasks. The FWC uses the ten flight phases to determine which messages will be displayed and which will be inhibited. Figure 4-5-22 shows some of the major inhibits which occur during the various flight phases.

ECAM System Failures

Let's look at a typical system failure, which is monitored by ECAM. Refer to Figure 4-5-23 (ECAM displays labeled 1 through 6) during the following discussion of the ECAM presentations. For this example, we will start with the aircraft in flight and the E/W page on the upper ECAM display; the lower ECAM will be showing the cruise page (see ECAM display a of Figure 4-5-23). If the green hydraulic system suffers a severe leak, the following scenario should occur:

1. Prior to the leak, the ECAM displays show all systems normal.

2. When the hydraulic fluid reaches a critical level, an amber caution message will appear on the E/W display. Beneath the caution message, the corrective actions will be displayed. At the same time, the lower display will be changed to the hydraulic page. A single chime will sound and the master caution annunciators will illuminate.

Current Aircraft Events which begin and end a flight phase	Flight Phase	ECAM System Page Displayed	ECAM Inhibits
Electrical Power applied to the Aircraft Busses	1	Door/Oxygen	None
1st Engine to Takeoff power	2	Wheel or Flight Control Page for 20 sec. when either side stick is moved or rudder deflected more than 22°	None
Aircraft reaches 80 Knots Airspeed	3	Engine Page	Most Warnings Inhibited
Lift off	4		
Slats are retracted and Engine thrust is reduced from takeoff power	5		
Aircraft Descends to 800 Ft.	6	Cruise Page (portion of Engine and Air Conditioning Page) or Wheel Page Displayed if landing gear are lowered	None
Touchdown	7	Wheel Page	Most Warnings Inhibited
Aircraft slows to 80 Knots Airspeed	8		
2nd engine to shut down	9		None
5 Minutes after 2nd engine shut down	10	Door Page	None

Figure 4-5-22. Ten flight phases, aircraft events, ECAM displays, and ECAM inhibits

Courtesy of Airbus S.A.S.

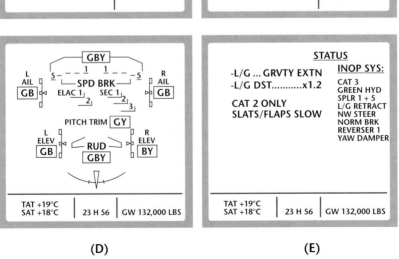

Figure 4-5-23. Typical ECAM displays: (A) All systems normal, (B) Amber caution message, (C) Primary and secondary failures, (D) Memo messages returned and failed system displayed, (E) Status page displayed and engine/warning page returns to it's pre-failure condition

3. The flight crew will then depress the master caution annunciator to turn it off. Next, according to the corrective actions on the E/W page the PTU and green engine pump #1 will be turned off.

4. Once the corrective actions are completed, the primary and secondary failures will be shown on the bottom of the E/W display. The "*" next to the secondary failure indicates a system page available to display that failure.

5. The flight crew will then press the **CLEAR** button on the ECAM control panel. This will replace the primary failure messages with the memo messages. The flight control page will appear on the lower CRT. If the clear button is pressed again the wheel page will appear on the lower ECAM display

6. The status page will be displayed when the **CLEAR** button is pressed again. The E/W page also returns to its original pre-failure condition. After the flight crew reviews the status page, the clear button is pressed again. The lower ECAM reverts back to its flight phase mode (the cruise page), and *STS* will appear on the upper ECAM display to indicate the status page contains information regarding system failures. The recall button can be used to retrieve the status page.

Troubleshooting ECAM

The majority of ECAM troubleshooting is done through the aircraft's central maintenance system. The onboard diagnostics will pinpoint defective LRUs and many wiring problems. The defect should then be corrected and the system tested for proper operation. This troubleshooting sequence is relatively simple if the defect is a continuous fault. However, in cases where intermittent faults exist the troubleshooting process requires additional attention.

During flight, the various temperature extremes and vibrations often cause intermittent problems. Many of these faults repair themselves prior to landing, which makes accurate troubleshooting difficult. In this type of situation, the central maintenance system will most likely recommend a LRU replacement; however, it is always wise to review the aircraft's history for this same defect.

If the suggested repair has already been done, and the problem still exists, performing that same repair will not fix the system. Consider something new. Intermittent faults are often caused by loose connections. Perform a thorough inspection of the system's wiring and all related connector pins.

Whenever repairing ECAM or replacing components, be sure to remove power from the system and use electrostatic discharge sensitive (ESDS) precautions. ECAM computers can be damaged if removed during operation, and just as easily damaged by static electricity. Be sure to read the manual for all precautions.

Section 6
Third Generation Indicating and Warning Systems

The past decade has brought tremendous change to the design of aircraft electrical/electronic systems. The introduction of the integrated modular avionics (IMA) design concepts revolutionized the airplane's avionics and information distribution systems. The concepts of IMA have also allowed for the integration of the aircraft displays; what was formally two dedicated EICAS or ECAM displays will now share functions with other systems. Many third generation EICAS or ECAM employ one dedicated display and share a multifunction display (MFD) or navigational display (ND). This sharing concept provides flexibility and simplifies the information displayed to the pilots. On some aircraft the EICAS or ECAM information is presented on only a given section (top/bottom/left/right) of a display. The concepts of IMA were discussed in Chapter 3 and will only be presented here as they relate to indication and warning systems.

Third generation indication systems benefit from the increased memory, improved processing power, and greater data transfer speeds found in modern electronics. These improvements have made it possible for aircraft to increase the abilities of the indicating and warning systems, as well as, improve reliability and maintainability. Additional capabilities translate to improved safety. Additional safety systems to be discussed in this section include enhanced ground proximity warning, traffic collision avoidance, and satellite weather systems. As integration of systems advance, aircraft seem to become more similar. For example, light aircraft now contain a basic form of EICAS and virtually all modern corporate and commuter aircraft contain most of the indicating and warning features found on the newest transport category planes.

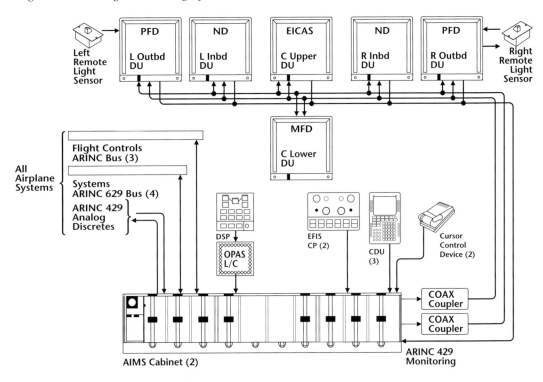

Figure 4-6-1. Boeing 777 Primary display system

Courtesy of Boeing Commercial Airplane Company

For the most part, third generation indicating and warning systems are very similar to previous versions of EICAS and ECAM. For example, first, second, and third generations all continually display critical engine data, all systems have master warning and caution annunciators, they each use the color red to display warnings, and green for normal operations. Major advancements found in the third generation systems include miniaturization of components, use of a more efficient data bus, integration of functions and systems, and advanced software design. This allows modern indicating and warning systems to do more using a small modular design, and improve display flexibility, which enhances flight safety. The use of advanced software allows for easy configuration changes and system upgrades without changing hardware or system LRUs. The following paragraphs will provide a brief description of some changes found in third generation EICAS and ECAM systems. More information about data transfer systems and integrated displays can be found in Chapters 2 and 3 of this text.

Transport Category Aircraft Systems

Boeing 777

It could easily be said that the Boeing 777 was one of the first commercial aircraft to employ

the modern concepts of IMA. This aircraft uses modular avionics to integrate many functions including EICAS, flight displays, flight management, and more. The EICAS display units receive high-speed digital data from the airplane information management system (AIMS) through multiple coax couplers and coaxial cables. The AIMS is presented in greater detail in Chapter 3. As seen in Figure 4-6-1, all information for the EICAS display is sent to AIMS, which performs validity checks, ensures data reliability, and processes the information before sending it on to the display(s). The B-777 employs only one dedicated EICAS display (the upper center display). The secondary EICAS information, when displayed, is shown on the MFD located on the pedestal between the pilots. The MFD is also used to display flight management data, weather information, and other systems as needed. The integration provided by the AIMS allows the software to decide what information should be presented on the MFD. Of course, basic flight instruments and any engine displays necessary for safe flight always take priority.

The Boeing 777 was the first airliner to break away from the dedicated-type data bus system employed on earlier aircraft. The newer ARINC 629 data bus replaced the legacy ARINC 429 data transfer system. Although many subsystems still utilized 429, the backbone of B-777 data transfer utilizes ARINC 629, which improves integration.

Another innovation found on the B-777 is the use of two cursor control devices (CCD) located on the center console between the two pilots. There is one CCD for the pilot and one for the first officer. The pedestal mounted CCDs can be used to control the left/right inboard displays and MFD. Any EICAS information displayed on the MFD can be controlled using the CCD; the main EICAS (upper center display) operates using traditional control panels.

Airbus A-380

The Airbus A-380, which began commercial operations in 2007, is the first aircraft to employ the latest data transfer system known as avionics full-duplex switched Ethernet communication system (AFDX). This system follows the ARINC 664 standard, and as the name implies, is an open architecture Ethernet-type data bus. The flexibility provided by this data bus system enhances the integration concepts of IMA. AFDX/ARINC 664 is discussed in Chapter 2. As with all modern aircraft, the indication and warning system used on the A-380 is deeply integrated with other A-380 avionics systems. Airbus uses the indicating systems know as ECAM that was discussed earlier in this chapter.

As with earlier Airbus aircraft, the A-380 ECAM provides the necessary information to assist the flight crew with aircraft operations and systems monitoring. System synoptic pages are displayed automatically in accordance with flight phase, or can be requested manually. Memos and checklist data can also be displayed to help manage the flight. In the new third generation systems, ECAM is also responsible for the automated aural altitude callouts and decision-height announcements during approach to land. During abnormal operations ECAM also helps the crew manage system failures. ECAM provides the standard warnings and cautions as well as procedures and limitations related to a failure. All information for ECAM is displayed in different colors to help pilots easily recognize system status (Figure 4-6-2).

The ECAM is part of the A-380 control display system (CDS) that contains two dedicated ECAM LCDs: the engine warning display (E/WD) and the systems display (SD). ECAM also displays information during various flight phases (such as flap position) on the PFD. The ECAM control panel and displays are shown in Figure 4-6-3. The CDS is a highly integrated system providing flexibility in display formats, and allows the system to set priority for the display of critical data. Controls for the A-380 CDS include a large alphanumeric keyboard located directly in front of each pilot and cur-

RED	For configurations or failures requiring immediate action
AMBER	For configurations or failures requiring awareness but not immediate action
GREEN	• Normal Operations: memo • Information in procedure or on STATUS page • Items checked in a normal checklist
WHITE	• Actions completed in a procedure • Conditional items • Title of a menu
CYAN	• Actions to be done in procedure; item to be checked in checklist • Limitations • Title of a not completed normal checklist
MAGENTA	For a specific memo (e.g. TO or LDG inhibition)
GRAY	• Items that are not valid/active (e.g. actions subsequent to a condition that is not detected, and not validated) • Completed checklist items; title of a completed normal checklist

Figure 4-6-2. ECAM information color codes

sor control devices (CCD) located on the center console between the pilots.

Aircraft and engine systems send information to the ECAM displays through the core processing input output module (CPIOM) as seen in Figure 4-6-4. The CPIOM performs all necessary processing functions for ECAM. The FWS is a dedicated system used to present warning and caution indications to the pilots in the event of a system malfunction. In order to provide system validity, the FWS operates independent of other display functions. Any software used for the FWS is designed with robust partitioning to ensure the operation of all warning displays. As seen in Figure 4-6-5, the FWS controls signals to the master warning and caution annunciators, the PFDs, the E/WD, and the SD, as well as the flight deck loud speaker for all audio warnings.

The Boeing 787 Display System

The Boeing 787 Dreamliner was also designed under the umbrella of integrated modular avionics (IMA) concepts. The Dreamliner utilizes a common core computing system that Boeing calls the common core system (CCS). The backbone data bus for this system employs the Ethernet-based ARINC 664 (AFDX) technologies. The flight deck displays found on the 787 are the largest of any aircraft and are arranged horizontally, four across, in a landscape format (Figure 4-6-6). This aircraft has no dedicated EICAS display. All EICAS information is shown on a portion of a given display. For

Figure 4-6-3. Location of ECAM displays and controls in an A-380   *Courtesy of Airbus S.A.S.*

example, during normal flight operations the engine data (N_1, EGT, and EPR) are shown on the inboard half of one navigational display.

The design concepts for the B-787 integrated display system are shown in Figure 4-6-7. At the bottom of the diagram are the independent airframe, engine, and avionics sensors used to monitor the various systems. The sensors connect to one or more remote data concentrators (RDC) distributed throughout the aircraft and located in common computing resource cabinets. The RDCs provide a digital gateway for analog, discrete, or digital (ARINC 429) signals. The RDC converts all inputs into an AFDX format in order to transmit the information on the Ethernet data bus to the two CCS cabinets. Inside the CCS cabinets, information is sent to one or more systems and/or function modules. These modules process the data, perform validity checks, and send the information to the graphics generation module (GGM). The GGM is used to format the data into a display interface format using the ARINC 661 standard. The GGM connects to the various flight deck displays on a dedicated pixel bus; with the information sent to the displays includes all EICAS data.

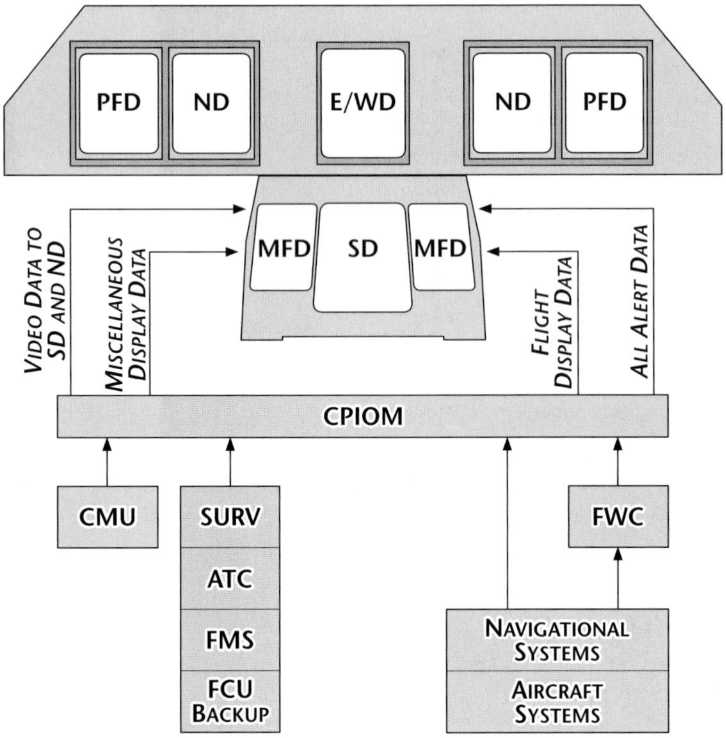

Light Aircraft Systems

The newest aircraft for personal use, business travel, and flight training have incorporated many of the design concepts formally used only on large aircraft. As electronic systems became more reliable, and at the same time, shrank in size, it became practical for light aircraft to employ electronic indicating and warning systems. Liquid crystal displays are now common in light aircraft and most of the traditional engine instruments and standalone panel indicators are displayed on LCDs using integrated system processing circuitry and software. Two aircraft will be presented

Figure 4-6-4. Control display system basic architecture

here, the Cirrus SR-20 and the Cessna Mustang light business jet. Both aircraft employ a digital integrated indication and monition system produced by Garmin International.

The Cirrus SR-20 is designed with an integrated electronic display system as a core element to the flight deck. The two processors, known as the integrated avionics units (IAU), feed data to the two large LCD flat panel displays. The integrated processors are basically small computers, which receive inputs from a variety of airframe and engine sensors, process the information, and send video data to the LCDs. Of course the IAU process flight data also. As can be seen in Figure 4-6-8, most information to and from the IAUs is sent on high-speed data busses in one of four formats:

1. High-speed Ethernet data bus

2. RS-232 data bus

3. RS-485 data bus

4. ARINC 429 data bus

The IAUs also receive a few discrete analog signals from various sources. The engine/airframe units are responsible for all data gathered by engine and airframe sensors.

The Garmin system found on the Cirrus aircraft is similar to that found on the Cessna Mustang that will be discussed in the following paragraphs. The Cirrus is a single piston-

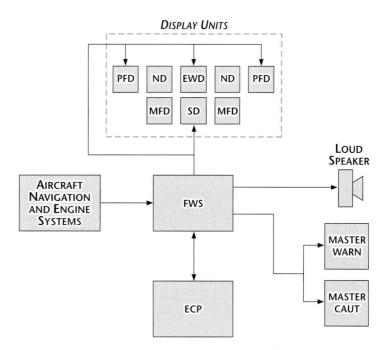

Figure 4-6-5. Electronic centralized aircraft monitoring (ECAM)

engine trainer and the indication and warning system is relatively simple compared to the Mustang; yet the displays, operation, and troubleshooting have much in common. In general, the concepts presented next on the Garmin G1000 can be applied to the entire Garmin line of integrated display/aircraft management systems. Please remember the materials presented in this text are for training purposes only and all maintenance should

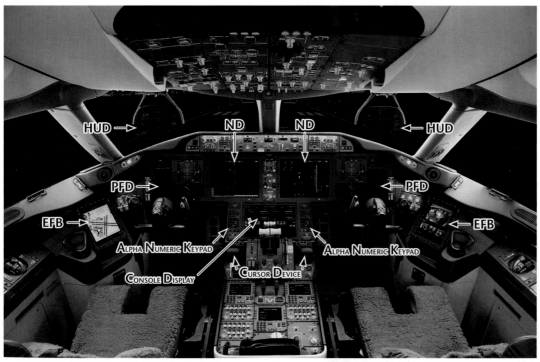

Figure 4-6-6. Boeing 777 flight deck showing large flat panel displays

Courtesy Of Boeing Commercial Airplane Company

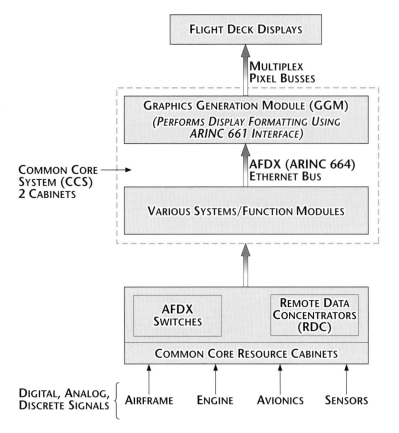

Figure 4-6-7. B-787 integrated display system functional block diagram

N_2, and ITT), electrical power, cabin pressure, and fuel system data, as well as, control surface positions (flaps, ailerons, and rudder). In the event of a parameter exceedance, that portion of the display will turn red, if the system fails a red X will be placed over that portion of the display. In some cases an aural warning is also broadcast over the cabin speaker and/or pilot headset.

System architecture. The Mustang's indication and warning system gathers basic systems data using a variety of engine and airframe sensors (Figure 4-6-10). These sensors gather analog, discrete, and digital information such as engine RPMs, oil temperature and pressure, battery voltage and amperage, bus voltage and amperage, and sends that information to one or both of the GEAs (Garmin engine airframe units). The GEAs then process the information, perform health monitoring checks, and send the data to the aircraft integrated avionics unit using an RS-485 data bus. The Garmin Integrated Avionics unit (GIA) performs additional processing in order to establish the display and any audio related to the information and warning system. The GEA and GIA are both line replaceable units that, on most aircraft, reside in the avionics rack behind the instrument panel.

System troubleshooting. A Garmin flight deck display system incorporates two, three or, in some cases, four flat panel displays. The G1000 system shown in Figure 4-6-9 employs two PFDs, one located in front of each pilot. The PFDs generally display a variety of flight information, such as, aircraft attitude. However if certain systems fail, an alert message related to the failure may be displayed on the PFD. According to Garmin this message is generated by the crew alert system (CAS). The CAS functions similar to EICAS or ECAM, which are found on large transport category aircraft. The message displayed on the Garmin PFD is used as an attention getter that draws the pilot's awareness to the problem. At that time the pilot would most likely view the inoperative system on the MFD and take corrective actions as necessary. It should be noted that not all system failures create a CAS message on the PFD, and the exact type and location of display data changes with the specific aircraft and/or system.

An oil temperature fault will be used during the following discussion, yet keep in mind most Garmin systems will respond in a similar fashion in the event of a failure. If oil temperature rises above a predetermined limit, the oil temperature gauge displayed on the MFD will blink yellow and the audio system will sound a double chime and stop. Depending on how the PFD/MFD is setup this may be slightly differ-

only be performed using airplane specific maintenance data.

Cessna Mustang. The Cessna Mustang employs the Garmin G1000 integrated display system and is one of a new generation of business/personal jets known as VLJs (very light jets). These aircraft typically seat four passengers and are designed to be flown with only one pilot. They offer the high performance and reliability of twin turbine engines and the redundancy of most critical avionics systems. The G1000 integrates a variety of display functions similar to the Garmin system discussed in Chapter 3. The flight management portions of the G1000 will be discussed later in this text.

As can be seen in Figure 4-6-9, the Mustang uses a 15-inch MFD in the center of the instrument panel. During normal operation the MFD contains engine and airframe data on the left quadrant of the display. The format and location of engine/airframe data changes during partial system failures or when the pilot selects other display options. The MFD is known as the GDU 1500 and presents a full color image with a 1024 x 768 resolution. The standard engine/airframe instrumentation for a twin-turbine aircraft would be similar to the display shown in Figure 4-6-8. The indications include engine data (oil pressure and temperature, N_1,

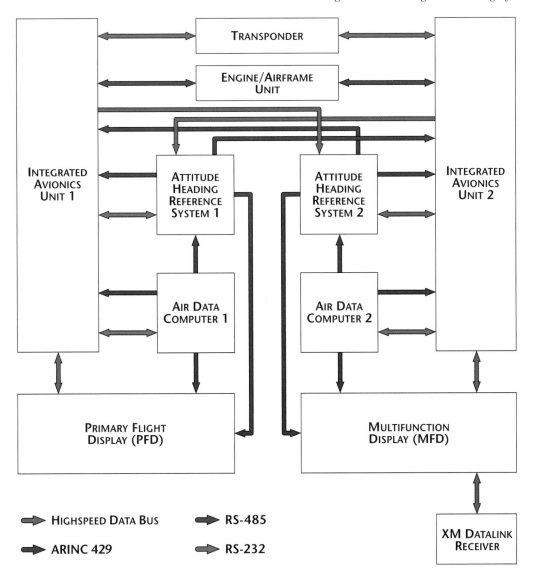

Figure 4-6-8. Block diagram of a typical two-display Garmin system

ent. If the oil temperature continues to progress to a higher limit (the oil temperature redline) that specific oil temperature display will turn red, blink, and the CAS will sound a continuous double chime until the oil temperature falls back within limits. A CAS alert message will also appear on the PFD. Once again the exact type/location of this alert message will depend on how the specific system is set up.

Another type of failure displayed by the CAS would be caused by lack of valid data from a system. Lack of data is different than the system going out of limits, such as high oil temperature. This problem occurs when one or more processors are not receiving any information from the failed system. In this case, the data is typically removed and a red *X* will be placed on the display where the invalid data would normally appear (Figure 4-6-11). Keep in mind, system data must not only arrive at a given processor within a set time, it must also be valid. Validity of digital data was discussed

Figure 4-6-9. Cessna Mustang Instrument Panel

Courtesy of the Cessna Aircraft Company

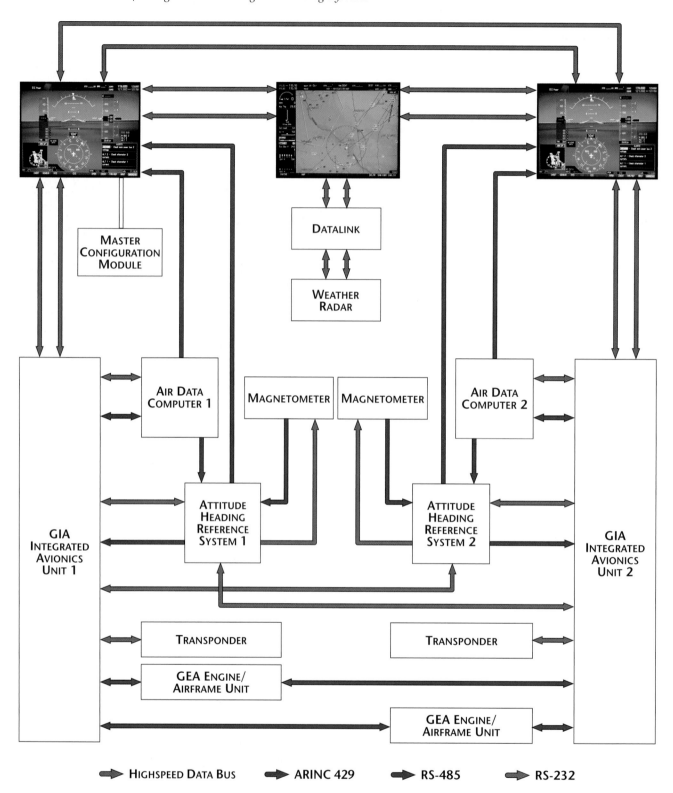

Figure 4-6-10. A G-1000 simplified block diagram

in Chapter 2 of this text. The lack of valid data can come from only a few sources as described in items 1 through 3 below. Once again let's use oil temperature as an example and refer to Figure 4-6-9 during this discussion.

1. If the oil temperature sensor is defective a red X will be placed over the display

for oil temperature only; other systems will display normally. The fault is most likely in the oil temperature sensor, or the related wiring to the engine/airframe unit, or software configuration error.

2. If the problem is caused by lack of data from the engine/airframe unit, the

Garmin display will place a red *X* over all engine/airframe systems. A defective engine/airframe unit, associated wiring, or a software configuration error would typically cause this.

3. If the Garmin display system experiences software errors related to equipment configuration a red *X* will be placed over the faulty system or group of systems. In this case, the software cannot recognize the information sent to the integrated avionics unit or other processor. To correct this problem the software is typically reloaded into the system. Keep in mind any time an LRU is removed and/or replaced, even with the exact same unit, software may need to be updated.

Much of the troubleshooting and repair of any problem with the Garmin system should start on the flight deck. The technician should view all alert messages related to the failure and activate the status checks on the PFD or MFD. The exact procedure used to view system status will be outlined in the maintenance data for your specific aircraft. Typically, during troubleshooting the technician can access system messages presented on the PFD by pressing the MSG soft key on the bezel of the display. This message will provide first level troubleshooting information to the technician. For example, an alert message may direct the technician to a software configuration error, or system power failure.

The Garmin software performs a variety of system tests; the results of these tests are displayed on the system status page. The system status page will provide troubleshooting guidelines, which direct the technician to the suspected fault. For example, when troubleshooting a problem with an engine or airframe system, the technician might find a status message *software configuration error – reload software*. In this case, the configuration software has been corrupted and must be reloaded. Another status message for the same problem could be *verify internal, external, and reference voltage*. In this case, the technician should access an additional status page, which will show the voltage values for various sections of the engine/airframe unit. Generally speaking, the Garmin status pages are simple to understand with easy-to-read tables that are color-coded to show systems status. For example, when troubleshooting a data path problem; red indicates a failed path, black indicates an unknown status, and green indicates the data path is operating normally. These different colors appear on the status page troubleshooting tables displayed on the airplane's PFD, ND, or MFD, depending on the aircraft.

The more complex Garmin systems found on high performance aircraft typically provide

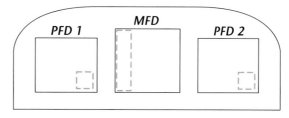

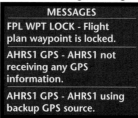

Figure 4-6-11. Typical fault and warning messages

greater information to the pilot or technician during a system failure. High performance aircraft will typically incorporate more processing power and/or additional LRUs in the integrated display system. Some aircraft may also incorporate a separate central maintenance computer that stores fault data for maintenance purposes. These systems will be discussed in Chapter 5.

The Garmin system, much like the newest avionics equipment produced by other manufacturers, employs microelectronic circuits requiring little power and is therefore very sensitive. The sensitivity comes with a cost in that magnetic and/or radio interference can easily create problems for the system. During engineering this interference is addressed for aircraft flight; however, interference should also be considered during troubleshooting. For example, the attitude and heading reference system can easily fail on the ground due to aircraft operations near metal structures like hangers. Cell phones or other devices using cell phone technologies in the aircraft cabin can create electromagnetic interference, even if the phone is in the monitor, or idle, mode. It is not uncommon for a technician to carry a cell phone. During maintenance, always suspect that this can cause a problem for sensitive avionics systems.

Other Modern Indication and Warning Systems

The last decade has seen numerous changes in flight deck display systems that now offer a variety of information to pilots, flight crew, and technicians. As systems became more inte-

1. *Primary aircraft cruising at 15,000 ft.*

2. *Target aircraft at slow closure rate. Short distance to CPA. Alarm triggered via tau criteria.*

3. *Target aircraft at high closure rate. Long distance to CPA. Alarm triggered via tau criteria.*

4. *Target aircraft at very slow closure rate. No CPA. No tau. Alarm triggered via DMOD criteria.*

NOTE: Examples for 15,000 ft (thresholds vary with altitude)

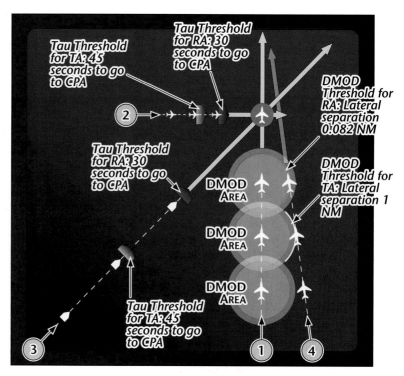

Figure 4-6-12. A typical TCAS tau and DMOD example: Primary Aircraft is heading north at 15,000 feet altitude. Aircraft No. 1: potential threat, slow closure rate, and short distance to CPA; Aircraft No. 2: potential threat, high speed aircraft, and long distance to CPA; Aircraft No. 3: potential threat, slow closure rate with no CPA, TA and RA triggered by DMOD.

grated, it was soon difficult to separate hardware and software according to a given function. This section of the text will examine three indicating and warning systems found on most modern aircraft. These systems, Enhanced Ground Proximity Warning, Traffic Alert and Collision Avoidance and SiriusXM satellite weather, are typically part of an integrated avionics package and will share data input/outputs, processing, and display functions with other equipment. Although the detailed examination of these systems is beyond the scope of this text, a brief discussion follows.

TCAS

The acronym TCAS has become the common name for a system designed to reduce the incidence of mid-air collisions between aircraft known as either the *traffic collision avoidance system* or the *traffic alert and collision avoidance system*. TCAS monitors the airspace around an aircraft for other traffic and warns pilots of a potential mid-air collision hazard. In order for TCAS to work independent of air traffic control, the system relies on the installation of an ATC transponder for all aircraft. Briefly described, a transponder is an electronic unit that continually broadcasts altitude, airspeed, and other critical information about that aircraft. For TCAS to work all aircraft in the general vicinity must be transponder equipped

and all transponders must communicate with each other. Transponder equipment used to provide advice to the pilot on potential conflicting aircraft is often referred to as secondary surveillance radar (SSR) systems. In the US and most other countries, TCAS is required for all civilian aircraft carrying 19 or more passengers.

In modern aircraft, the TCAS display may be integrated into the navigation display (ND), multifunction display (MFD), or electronic horizontal situation indicator (EHSI). In older aircraft and those with mechanical instrumentation, the TCAS display may be combined with the aircraft's vertical speed indicator (VSI). There are two basic levels of TCAS equipment, TCAS I and TCAS II. The system used mostly by general aviation is TCAS I. This system monitors approximately 40 miles of airspace and offers the pilot only limited information about a potential collision. If an aircraft is approaching a potential hazard, the synthesized audio voice will say, "TRAFFIC, TRAFFIC." It is the pilots job to determine what evasive maneuver, if any to take. TCAS II is a more elaborate system and provides more information in the event of a potential risk. This text will focus on TCAS II systems.

In 2006, the TACS version 7.0 was implemented and is now required on most commercial aircraft. TCAS II version 7.0 is the second and current generation of instrument warning TCAS. This version offers all the benefits of TCAS I, but also provides vocalized instructions to avoid danger, known as a resolution advisory (RA). Audio warnings include: "DESCEND, DESCEND," "CLIMB, CLIMB," or "ADJUST VERTICAL SPEED ADJUST" (meaning reduce vertical speed). Of course, a visual display will also be shown on the PFD showing the locations of any threat aircraft. In order for the system to issue a corrective action it is vital that both threat aircraft make corrections in the opposite direction. In other words, one aircraft must climb, while the other aircraft must descend. To ensure this is coordinated the TCAS on both aircraft must communicate their actions through the aircraft's transponders.

Theory of Operation

TCAS involves communication between all aircraft within a certain radius employing an operational transponder. The interrogation takes place via the 1,030 MHz radio frequency and the reply occurs on 1,090 MHz. These frequencies are digitally encoded with a variety of information allowing the TCAS processor to evaluate the potential threat of a mid-air collision. The interrogation/response occurs several times each second. The TCAS system builds a three dimensional map of airspace, incorporating the range, altitude, and bearing of all aircraft. Then, by extrapolating current range and altitude difference to anticipated future values, it determines if a potential collision threat exists.

TCAS and its variants are only able to interact with aircraft that have a correctly operating Mode C or Mode S transponder. A unique 24-bit identifier is assigned to each aircraft that has a Mode S transponder. This allows the system to keep track of each aircraft independent of ground operations. In effect the TACS equipment on each aircraft "talks" to each other and determines if a potential collision might occur. The next step beyond identifying potential collisions is automatically negotiating a mutual avoidance maneuver between the two, or more, conflicting aircraft. The TCAS software employs mathematical formulas to perform this task. The avoidance maneuvers are then communicated to the flight crew through one or more displays and by synthesized voice instructions.

In order to better understand the concepts of TCAS one must first understand some of the basic terminologies related to the system. During the following discussions please refer to Figure 4-6-12. *Tau* is a term used to define the protected area around the aircraft (called the primary aircraft, heading northbound, in this example). The limits of tau to the target aircraft are defined by the time to a potential impact or near miss. This point in space is calculated by TCAS and is known as the CPA (closest point of approach). If the tau minimum value is exceeded, TCAS will issue the appropriate warning. Of course tau changes depending on the speed of the two aircraft; faster aircraft require an advanced warning. In some cases tau may also change with altitude.

There are cases when a closure rate is so slow that minimums are never reached, yet the physical separation may be just a fraction of a mile. In such cases, the calculated closure rate is no longer useful for collision avoidance. Therefore, the TCAS design incorporates an additional warning parameter known as *distance modification* (DMOD). Time or closure rate are not a factor here. DMOD affects only the physical separation between the two aircraft. DMOD values, also, vary with altitude.

There are two alert levels associated with TCAS, the *traffic advisory* (TA) and the *resolution advisory* (RA). The TA is the first level of alert and typically given to the pilot in the form of the word TRAFFIC displayed in yellow on the ND or MFD, and the aural voice annunciation "TRAFFIC TRAFFIC." TCAS triggers a TA as soon as an intruder violates the specific DMOD or area and is used to call attention to a possible conflict. If no altitude data is available from the intruder aircraft, TCAS assumes the intruder's relative altitude is within 1,200 feet.

The RA is the highest TCAS alert level. Its purpose is to resolve a conflict by providing the pilot with aural and visual pitch commands. TCAS triggers an RA when a target enters the RA specific DMOD or area. A typical RA will include an audio alarm "CLIMB CLIMB" and TACS will show *TRAFFIC* in red on the display. If no altitude data is available from the target, an RA will not occur. Upon receiving an RA the pilot must disengage the autopilot immediately and execute the TCAS escape maneuver. Flight director commands as well as ATC advisories have to be ignored during this process; the TCAS pitch command has priority.

Figure 4-6-12 shows three different aircraft that pose a potential threat to Aircraft No. 1. Each aircraft has different criteria to define a TA or RA since they are all traveling at different speeds and headings. This Figure is drawn from the perspective TCAS of Aircraft No. 1; assuming the other aircraft have an active TCAS, they too would have DMOD and thresholds of their own.

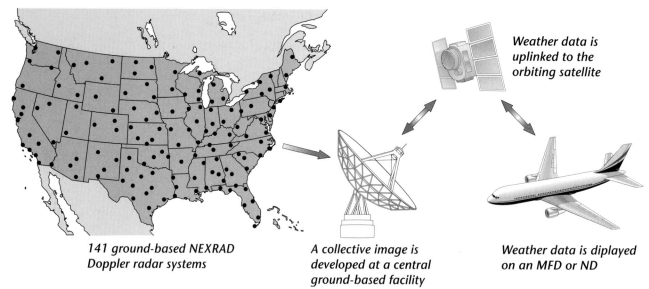

141 ground-based NEXRAD Doppler radar systems

A collective image is developed at a central ground-based facility

Weather data is uplinked to the orbiting satellite

Weather data is diplayed on an MFD or ND

Figure 4-6-13. Satellite weather system

SiriusXM Satellite Weather

XM Satellite Radio was a corporation started in the late 1980s and initially designed to provide pay-for-service radio, analogous to cable television, including a variety of music, news, sports, and other entertainment channels. Over time there have been corporate mergers and other changes to the service. In 2008, Sirius Satellite Radio bought XM forming a new company, SiriusXM Satellite Radio.

SiriusXM has the ability to broadcast weather data and automobile traffic information. The SiriusXM WX weather service, as it is known, has become extremely valuable to the aviation community. The service provides weather information sent through the SiriusXM satellite system to a receiver/processor onboard the aircraft. The versatility and relatively low cost of the SiriusXM satellite weather service, coupled with the multitude of hand-held and onboard display options, has made this weather system extremely popular to pilots, sailors, and the general public (Figure 4-6-14).

Unlike airborne weather radar, which relies on the aircraft's own equipment, the SiriusXM WX weather service can give the pilot information about weather anywhere in the country. The downside is that the various weather streams, radar, cloud coverage, lightning, etc., are transmitted about every five minutes, meaning the information is somewhat out-of-date by the time it is received. Airborne radar and lightning receivers return truly real-time information, but since these are stand-alone systems installed in each aircraft, they can cost thousands of dollars, add substantial weight, and are not easily suited to small aircraft. With the advent of SiriusXM WX service, weather data is now available to even light personal aircraft. Certain aircraft also integrate the SiriusXM radio service into the aircraft's audio system, thus, allowing passengers to listen to SiriusXM entertainment radio while flying. Today many high performance aircraft employ SiriusXM weather radio as well as airborne radar equipment, thus providing a very comprehensive weather display during flight.

The SiriusXM WX weather transmits information from the National Aviation Weather Radar (NEXRAD) through the satellite radio frequencies similar to the SiriusXM audio broadcasts. The NEXRAD system provides a snapshot of

Figure 4-6-14. A SiriusXM WX weather service display *Courtesy of Garmin International, Inc.*

the entire U.S. weather situation using a network of 141 ground-based Doppler radar stations (Figure 4-6-13). The NEXRAD image is updated every five minutes. This information is converted to satellite signals uploaded to the SiriusXM satellites in geosynchronous orbit above the earth and then transmitted to the aircraft. The airborne equipment processes the SiriusXM signal and converts the information into a weather display. The display is typically incorporated into other weather or navigational information and displayed to the pilot on the ND or MFD.

SiriusXM weather shows seven levels of precipitation across the country, revealing nested storms, and widespread, fast–changing convective activity. Unlike airborne radar, which often scans a limited range, NEXRAD shows a big picture providing a valuable tool for fast moving aircraft. Of course the airborne SiriusXM weather equipment allows the pilot to zoom in/out as desired. The SiriusXM system also provides the ability to "look through" large storm activity; this is a limitation of most airborne weather radar. The SiriusXM weather data also incorporates lightening reports. The system delivers lightening patterns depicting recent strikes, with the lightning strikes depicted over a five-minute time period so the pilot is presented a good view of storm intensity.

Enhanced Ground Proximity Warning Systems

In an effort to increase air safety, in the 1960s, the FAA began a program to reduce the number of accidents caused by controlled flight into terrain (CFIT). It was determined that CFIT could be minimized through use of an automated system to alert pilots of potential ground impact. The FAA defines the general concepts of this type of system as terrain awareness warning system (TAWS). Through the years a variety of different manufactures have developed different equipment to create a reliable TAWS. Today the most advanced systems are known as enhanced ground proximity warning systems (EGPWS). Although sometimes called TAWS, in actuality, EGPWS is a type of TAWS. Today the FAA and most foreign countries require some form of TAWS to be employed on all turbine-powered aircraft with a capacity of six or more passengers.

The EGPWS was introduced in the 1990s and has now become the standard for terrain warning systems. The system operates using mathematical formulas to determine the risk of ground impact. To do this the EGPWS processors must know the:

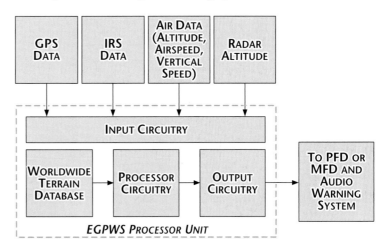

Figure 4-6-15. EGPWS processor unit

1. Aircraft altitude above the ground

2. Airspeed of the aircraft

3. Specific position of the aircraft on the earth's surface

4. Local terrain around the aircraft at all times

As seen in Figure 4-6-15 the EGPWS incorporates a large worldwide digital terrain database. This database is virtually a map of the earth's surface containing elevations of all terrain and human made obstacles. The database also knows the locations of airports; these locations are important so the EGPWS processor can adjust or eliminate any warning messages during aircraft landing and takeoffs. Aircraft altitude is provided by a number of sources including the radar altimeter that measures above the ground altitude. Other air data is sent to the processor as well. Position information is provided by the aircraft's global positioning system or inertial navigation system. The output data from the EGPWS is typically displayed on the instrument panel using the ND or MFD. EGPWS operates in conjunction with a variety of already available aircraft systems and therefore is a very cost effective means of improving air safety.

The EGPWS is typically connected to the aircraft's flight warning system. In the event that EPGWS determines a potential ground impact, the system creates both an audio and visual warning on the flight deck. In most cases EGPWS will provide an audio message related to the impending danger of the situation. For example, if the computer detects an impending ground strike, an automatic audio message "TERRAIN PULL UP" will be repeated until the condition is corrected. These warnings are of very high priority and in all most all cases the pilot must follow the recommendations of the EGPWS.

Chapter 5

INTEGRATED
test equipment

Section 1

Introduction

Aircraft designers have always been aware of the need for maintainability in the design of aircraft. In the past, maintainability was mainly a function of allowing technicians access to parts for replacement or inspection. Today's aircraft are so complex that design engineers must consider the ability to troubleshoot a system just as important as the ability to repair or inspect that system.

Troubleshooting a complex digital aircraft would be nearly impossible without self-diagnostic systems. Technicians have always been accustomed to using various tools and test equipment to troubleshoot aircraft. Simple items such as a voltmeter could be used to diagnose many of the electrical problems on early aircraft. As radio equipment and autopilot became popular, more sophisticated test equipment became necessary. Many of the tests were too complex and required shop, or at least, hanger maintenance. Other problems could be repaired on the flight line given the proper carry-on test equipment. To repair some systems, it required several pieces of test equipment and considerable time to troubleshoot the problem. Designers began to take advantage of computer technologies and added self-diagnostic circuits. Self-contained diagnostics used for electronic/avionics systems troubleshooting became known as built-in test equipment (BITE).

BITE systems were introduced on transport category and many corporate type aircraft. BITE systems came into practical use with the invention of digital electronic systems. The Boeing

Learning Objectives:

- Built-In Test Equipment (BITE)
- Central Maintenance Computer System (CMCS)
- Centralized Fault Display System

Left. Dispatch reliability is a critical part of good airline service. Digital avionics systems provide the technician with the tools to aid in identifying and quickly repairing component failures.

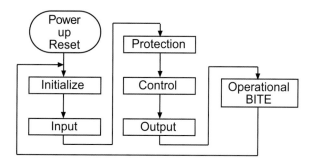

Figure 5-2-1. BITE flow diagram showing the sequence of circuit tests

Boeing 787 and Airbus A-380 can monitor systems, analyze faults, and automatically download maintenance information through the use of wireless systems such as cellular or satellite technologies. These automated troubleshooting systems improve aircraft maintainability and enhance safety. The design and operation of individual and centralized diagnostic systems found on modern aircraft are presented in this chapter. After studying this chapter the reader will be familiar with the majority of built-in troubleshooting systems found on today's aircraft.

757 and 767 were the first civilian aircraft to make extensive use of BITE. As with most aircraft systems, an evolution process took place during the development of BITE. Early systems were limited to troubleshooting of a particular component or LRU. More modern BITE could monitor input or output signals for a complete system. In general, early BITE systems were accessed through the specific LRU that housed the BITE circuitry. This meant early BITE systems were accessed in the electronics equipment bay or similar area.

The second generation of self-diagnostics equipment incorporates the use of a centralized monitoring system. The faults detected through several BITE systems could be monitored in one location. The built-in troubleshooting system found on Boeing aircraft is known as CMCS (central maintenance computer system). The diagnostic system used by Airbus S.A.S. is called CFDS (centralized fault display system). In general, the advanced systems are more easily accessible and understood than older systems.

Third generation systems have became more integrated and allow for wireless communications of troubleshooting data. Aircraft like the

Section 2

Built-In Test Equipment (BITE)

Built-in test equipment (BITE) became possible with the increased use of electronics and digital systems found on modern aircraft. In general, BITE consists of the circuitry needed to test various LRUs, power supplies, wiring and related switches, circuit breakers, and relays. The majority of BITE circuits are only a small portion of the circuitry contained within an LRU; however, on early systems there are a few LRUs dedicated to system tests. These LRUs are often referred to as BITE Boxes. On later generations of built-in test systems, the equipment became more integrated and performed additional functions. These integrated test circuits are typically known as central maintenance systems.

Built-in test equipment systems are used in conjunction with many digital circuits to aid in system troubleshooting. BITE systems are designed to provide fault detection, fault isolation, and operational verification after defect repair. Fault detection is performed continuously during system operation. If a defect is sensed, the BITE initiates an appropriate control signal to isolate any defective component(s). In order to repair the defective system, the technician can utilize the BITE to identify faulty components or wiring. The majority of the aircraft digital systems contain several LRUs. Defective LRUs may be quickly identified by the BITE system and exchanged during ground maintenance. Use of the LRU and BITE concepts greatly reduce aircraft maintenance down time. After the appropriate repairs have been made, the system should be run through a complete operational check. The BITE will once again monitor the system and verify correct operation if the system has been properly repaired.

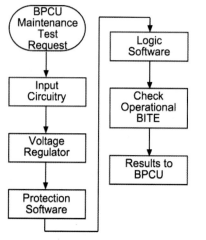

Figure 5-2-2. Bus power control unit BITE flow diagram

A typical commercial airliner may contain several BITE circuits used to monitor a variety of systems. A Boeing 757 or 767 aircraft, for example, utilizes built-in test equipment systems on approximately 50 LRUs located throughout the aircraft. Seven separate BITE units located in the aircraft's electrical equipment bay or aft equipment center are used to monitor electrical power, environmental control, auxiliary power, and flight control systems. BITE receives inputs from several individual components of the system being tested. Other individual systems also contain their own dedicated built-in test circuitry. These BITE systems are relatively simple and are contained within the LRU being monitored. Systems that employ dedicated BITE on the B-757 include the following:

- Engine indicating and crew alerting

- VHF communication radios

- HF communication radios

- ARINC communication addressing and reporting

- Selective calling

- Passenger address

- Weather radar

- ATC transponder

- Radio altimeter

- Automatic direction finder

- Inertial reference

- Air data computer

- Electronic flight instruments

- Flight management computer

- Radio distance magnetic indicator

- Lighting

- Fuel quantity

- Fire and overheat

- Antiskid autobrake

- Instrument landing

- VHF omnirange receiver

- Distance measuring equipment

- Window heat

- Proximity switch electronic unit

- Hydraulic management

- BITE test programs

A complex BITE system is capable of testing thousands of input parameters from several different LRUs. Typically a system performs two types of test programs: an operational test

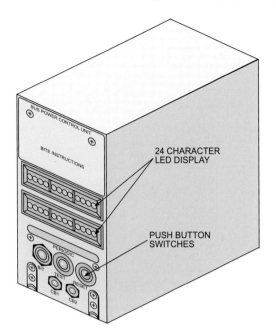

Figure 5-2-3. Typical LRU with access to BITE controls on the face of the unit

and a maintenance test. Normal operational checks start with initialization upon acquisition of system power (Figure 5-2-1). The operational BITE program is designed to check input signals, protection circuitry, control circuitry, output signals, and the operational BITE circuitry. During normal system operation, the built-in test equipment monitors an operational test routine 1 initiated by the BITE program. This test routine detects any hardware failure or excessive signal distortion that may create an operational fault. If the BITE program detects either of these conditions, it automatically provides isolation of the necessary components, initiates warning, caution, or advisory data, and records the fault in a nonvolatile memory.

The maintenance program of the built-in test equipment begins operation when the aircraft is on the ground and the maintenance test routine is requested. When initiated, the maintenance BITE will exercise all input circuitry and software routines of the system being checked. The corresponding output data are then monitored and faults are recorded and displayed by the BITE system. An illustration of the Bus Power Control Unit (BPCU) maintenance BITE routine is shown in Figure 5-2-2. This routine checks the input circuitry, voltage regulator circuitry, protection software, logic software, and operational BITE circuitry. The test results are returned to the BPCU for storage and display. The software, or operating programs, of the system are tested through utilization. That is, input data is initiated by the BITE and manipulated by the software program. The BITE program evaluates the corresponding output data in order to determine the system's

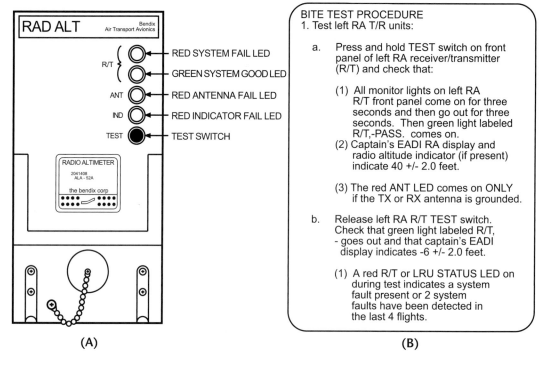

Figure 5-2-4. (A) Radio altimeter BITE - LRU front panel, (B) Radio altimeter BITE - Procedures for testing the left transmitter/receiver
Courtesy of the Boeing Commercial Airplane Company

performance. If a discrepancy in the output data is detected, the BITE system considers the system, or portion of the system, faulty and provides the appropriate indication.

Discrete digital signals are used as the code language for BITE systems. Built-in test equipment interprets the various combinations of digital signals to determine a system's status. If an incorrect input value is detected, the BITE system records the fault and displays the information upon request. Figure 5-2-3 shows an LRU containing a 24-character Light Emitting Diode (LED) display that can show a variety of fault information. The LED display is located on the face of the unit along with the appropriate test switches. The LRU is typically mounted in an equipment rack in the aircraft's electrical equipment bay. This type of BITE is used on aircraft built in the 1980s and early 1990s, and are typically considered the first generation of built-in test systems.

Troubleshooting with Built-In Test Equipment (BITE)

There are several versions of built-in test equipment in use today. Simple BITE systems may incorporate a go/no-go red or green LED on the LRU. More complex systems use a multi-character display and monitor more than one LRU. Some BITE can also test the associated wiring. Even more advanced BITE systems incorporate displays activated from the flight deck,

have paper printouts, and may have a means to transmit data from the aircraft to the maintenance facility during flight. These systems will be discussed later in this chapter.

A Bus Power Control Unit BITE

The BITE system incorporated into the early model B-757 aircraft Bus Power Control Unit (BPCU) can be seen in Figure 5-2-3. This system monitors the entire electrical power generation system, including the left, right, and APU generators, constant speed drives, and their related control units. The BIT button is depressed on this system to activate the manual BITE test function. Typically, this type of BITE system will display fault information in a coded message. The technician then decodes this message through the use of the aircraft's maintenance manual. The appropriate manual will inform the technician of any LRU to be replaced or circuit to be repaired and their respective locations within the aircraft. The fault information on this system is displayed for two seconds and then automatically advances to the next fault, if any. This type of BITE system will make an appropriate indication when all fault data has been displayed.

After the system has been repaired, the BITE should be reset, and an operational check performed. The repaired system should be run through a complete cycle of operation. In the case of the electric power generating system, the appropriate engine and AC generator should be subjected to a variety of operating parameters. The flight deck instruments and failure indicators are monitored during the test to detect any further problems.

After repair and operation of the system, the BITE fault display should be reactivated. This will initiate the readout of the nonvolatile memories and the BITE will display any remaining faults. If the system is found to be without fault, the BITE display will respond accordingly. Always be sure to reset the BITE prior to the verification test. The BITE fault messages are stored in memory and will remain unless erased by the reset function.

A Simple BITE Circuit

Most aircraft systems built in the 1980's or later incorporated some type of BITE circuitry. Independent systems, such as the radio altimeter shown in Figure 5-2-4A, often contained BITE circuitry in the main LRU of the system. This radio altimeter is one of three installed on an early model Boeing 757. Each Radio Altimeter Receiver/Transmitter (R/T) contains its own BITE. To run the radio altimeter BITE, press the TEST button and monitor the LEDs on the face of the R/T. The procedures for the BITE test are shown in Figure 5-2-4B. It should be noted that this BITE circuitry performs simple tests on the R/T, the antenna, and the radio altimeter display. Similar tests are extremely common on many LRUs found on transport category and corporate type aircraft.

Maintenance Control Display Units (MCDU)

The Boeing 757 and 767 incorporate a Maintenance Control Display Unit (MCDU) used to monitor and test the flight control computers, flight management computers, and the thrust management computers. The MCDU was the beginning step in the development of advanced integrated BITE systems found on today's state-of-the-art aircraft. MCDUs similar to the one presented here are very popular and found on hundreds of transport category aircraft. Through the MCDU, over 68 individual components, input signals, and systems are monitored.

System data is accessed in the main electrical equipment bay via the control panel located on the front of the MCDU (Figure 5-2-5). On some aircraft a carry-on MCDU controller is connected to the system on the flight deck. If the MCDU controller is used on the flight deck, the fault information is displayed on the

lower EICAS CRT. During a complete MCDU ground test, certain controls must be operated and various indicators observed, therefore, if the MCDU test mode were accessed from the equipment bay, two technicians would be required. Only one person can access fault data stored in memory from either location.

MCDU Operation

The MCDU operation is similar to the previously described BITE system. The MCDU receives digital data in an ARINC 429 format transmitted from the thrust management, flight control, and flight management computers along with various other system inputs. The MCDU monitors in-flight faults and performs ground test functions. In-flight faults are directly correlated to the various flight deck effects associated with in-flight problems. Flight deck effects are considered to be any EICAS display or discrete annunciator used to inform the flight crew of an in-flight fault.

Upon landing, the MCDU automatically records any in-flight faults, from the last flight, in its nonvolatile memory. To access this memory, the technician must first cycle the MCDU off and on again. This action will perform an internal test of the MCDU. After the internal test is complete, the technician should select the in-flight mode of operation by pressing the YES/ADV button on the MCDU when the display reads LAST FLT FAULTS (Figure 5-2-5). The unit will respond accordingly with faults listed in order of occur-

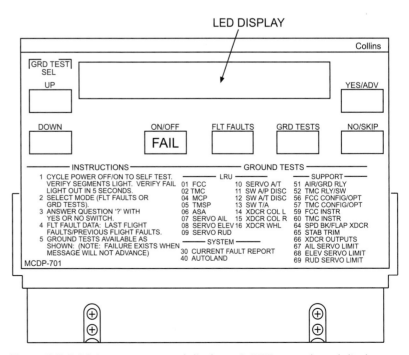

Figure 5-2-5. Maintenance control display unit BITE controls and display located on the face of the MCDU

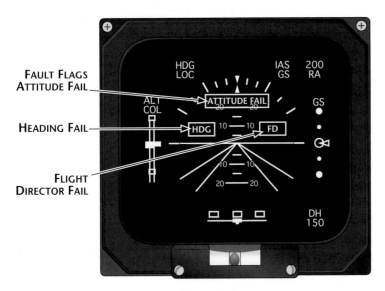

FAULT FLAGS
ATTITUDE FAIL

HEADING FAIL

FLIGHT
DIRECTOR FAIL

Figure 5-2-6. Typical EFIS display showing fault messages

rence. Press the YES/ADV switch to sequentially advance through all the faults. If no faults were found during the last flight, NO LAST FLT FAULTS will be displayed. Pressing the NO/SKIP button will skip to the next display.

At the end of the last flight fault data, the MCDU display will ask PREV FLT FAULTS? Answer yes to view faults which occurred during previous flights. To terminate the display of fault data press the ON/OFF or GRD TEST switch. The MCDU can store faults from a maximum of 10 previous flights.

MCDU Ground Tests

To run the ground tests function of the MCDU, the unit should be turned on and the self-test

DEC	HEX		MPR		
68	44	0100	0111	1001	
196	C4	0000	1001	1001	
69	45	0000	0000	0001	
197	C5	0000	0000	0001	
70	46	0000	0000	0100	
198	C6	0000	0000	0100	
71	47	0000	0000	0000	
199	C7	0000	0000	0000	
72	48	0010	0010	1001	
200	C8	0000	0010	1001	
73	49	0000	0000	0000	
201	C9	0000	0000	0000	
74	4A	0000	1110	0000	
202	CA	0000	1110	0000	
75	4B	0000	0000	0000	
203	CB	0000	0000	0000	

Figure 5-2-7. Typical BITS mode display

Courtesy of Rockwell International, Collins Avionics Divisions

must be valid. The ground test will begin when the GRD TEST button is pressed. At that time, the MCDU will run an initialization test and display any associated fault messages. If no initialization faults are found, the MCDU will begin testing the Flight Control Computer (FCC) when the YES/ADV switch is pressed. The remaining ground tests are activated in a similar manner. If you wish to skip to a given test press the NO/SKIP button until the appropriate test appears in the MCDU display. At that time, press YES/ADV. The UP and DOWN ground test select switches can also be used to move through the menu to a given ground test. To terminate the ground tests function, turn the MCDU off or press select for the flight faults option.

Built-In Test Equipment for Light Aircraft

Light aircraft electronic/avionic systems have also moved into an era of self-diagnostics. Modern avionics equipment found on light aircraft is often digital systems that incorporate BITE software programs. Many of the communication and navigation systems, as well as electronic flight instrument systems (EFIS), utilize BITE for internal and limited external testing. BITE systems are most often found on individual avionics equipment for corporate type aircraft. However, the trend of increased BITE systems is continuing to move into even the smallest aircraft. Today modern light aircraft incorporate integrated electronic systems which monitor digital data and troubleshoot faults down to the component level. These advanced systems will be discussed later in this chapter.

Single System BITE

The BITE systems found on single function avionics, like a VHF communications radio, typically employ only limited diagnostics. An LED indicator may be used to report the condition of the internal radio circuitry. If the technician encounters a radio problem, the test button on the transceiver would be pressed and the LED PASS/FAIL lights would be monitored.If the unit fails, the transceiver would be removed for bench repair. If the transceiver passed its test, the fault lies in other portions of the system, such as the antenna or system wiring.

BITE Systems for EFIS

Of all the BITE found on corporate type aircraft, the systems found on EFIS equipment are typically very advanced. Electronic flight instrument systems employ one or more CRTs or LCDs used to display flight and navigational data. Therefore, EFIS is interconnected to a multitude of other avionics equipment needed to determine the information to be displayed on the CRTs. The EFIS built-in test equipment often tests both the EFIS circuitry and the incoming signals from the monitored systems.

The Bendix/King EFIS 40/50 incorporates an extensive self-test during operation. The self-test begins with an analysis of internal circuitry and the data being sent to the EFIS processors. A more comprehensive test can also be initiated for maintenance purposes. If a failure is detected during operation, one or both of the EFIS CRTs will display a message related to the failure. Figure 5-2-6 shows an EFIS 40/50 with several fault messages shown on the display. Typically the most critical fault messages are displayed in red, less critical messages are shown in yellow.

Most electronic flight instrument systems for light aircraft allow the avionics technician to analyze input signals using the EFIS test functions. Input signal diagnostics help to troubleshoot the entire EFIS/autoflight system by providing a look at the data being analyzed by EFIS. In general, this type of data analysis requires reference to the system manuals. The Collins EFIS-85B utilizes a BITS Mode to provide troubleshooting data stored in a random accessory memory. The BITS mode presents a display similar to Figure 5-2-7. In this display the first two columns present the decimal (DEC) and hexadecimal (HEX) codes for the source of the data. The MPR (multiport RAM) data is shown in binary. The MPR can be decoded using the installation manual for the Collins EFIS. For a more detailed look at EFIS equipment and the related BITE systems, see Chapter 3.

Section 3

Central Maintenance Computer System (CMCS)

The Central Maintenance Computer System (CMCS) is found on a variety of modern Boeing aircraft. This system is designed to perform inflight and ground tests of virtually all aircraft systems, accessed from a central location. The CMCS is not a BITE system; the CMCS is a manager of the individual built-in test equipment, which are contained in the software of the various LRUs located throughout the aircraft. Although only one aircraft, the B-747-400 will be presented here, this material can be used as a good example of all CMCSs and will

Figure 5-3-1. Three control display units on a Boeing 747-400 aircraft
Courtesy of the Boeing Commercial Airplane Company.

apply to a variety of aircraft. The CMCS incorporates greater memory capability and better information access compared to previous BITE systems. This leads to more reliable troubleshooting using CMCS. The B-747-400 CMCS is typical of those found on many transport category aircraft. This type of built-in test equipment is often thought of as a second generation integrated test system.

System Description

The B-747-400 CMCS is accessed through one of four control display units (CDU). There are three CDUs located on the center console of the flight deck and one located in the main equipment bay. Figure 5-3-1 shows a photograph of the center console and three CDUs. The CDUs contain an alphanumeric keyboard and a CRT display used to access the CMCS data. This type display allows for a more descriptive message of faults that are directly correlated to flight deck effects created by the same fault. A CMCS printer is incorporated to provide a written report of the fault data, and a software data loader can be used to store faults on a computer disc. Aircraft equipped with ACARS are capable of transmitting fault data from the aircraft to a ground facility. ACARS will also answer all maintenance data requests from the ground facility.

The CDUs communicate directly with the central maintenance computers (CMC) of the B-747-400 computerized maintenance system. The CMCs monitor virtually every electronic system on the aircraft. If a fault occurs, the CMCs record that information that can be recovered

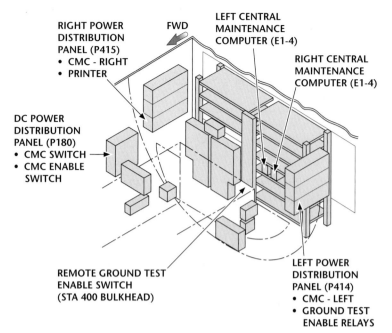

Figure 5-3-2. Locations of central maintenance computer equipment in the B-747-400 equipment bay

later through the CDUs. Figure 5-3-2 shows the location of the CMCs in the equipment bay. During flight, the CMCs receive fault data from the aircraft's EIUs (Electronic Interface Units) and other digital and discrete systems to record in-flight failures. The EIUs monitor system parameters and control the EICAS and EFIS displays. During flight, the CMCs also monitor the integrity of CMCS inputs and store any fault data.

Once on the ground, the CMC can be interrogated for fault history which may be stored on the computer's nonvolatile memory. Up to 500 faults can be stored in the CMC memory. From the failures detected by the CMC, one of 6,500 different fault messages can be displayed on the CDU. The CMC is also used to initiate various BITE tests of the aircraft's electronic LRUs. Each CMC weighs about 17 pounds and has a 300-pin connector on the rear of the unit. On the front of the CMC is an automatic test equipment (ATE) connector that is used to access CMC data during shop maintenance. The CMCs are electrostatic discharge sensitive so the technician must always take the necessary precautions during maintenance.

System Architecture

Figure 5-3-3 shows a block diagram of the CMC data inputs and outputs. There are two central maintenance computers (CMC) that receive up to 50 low-speed and six high-speed ARINC 429 data inputs. Most aircraft systems report to

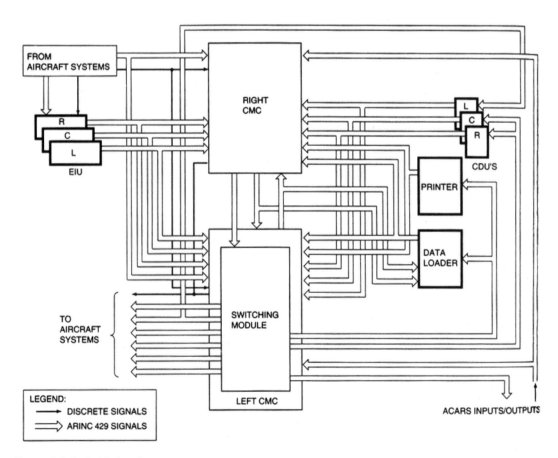

Figure 5-3-3. CMC data bus structure

Courtesy of Northwest Airlines, Inc.

DIGITAL INPUTS

These systems send ARINC 429 inputs to the CMCs

- Left control display
- Center CDU
- Right CDU
- Left EIU (high speed)
- Center EIU (high speed)
- Right EIU (high speed)
- Left flight control computer (FCC)
- Center FCC
- Right FCC
- Left VHF transceiver
- Right VHF transceiver
- Left HF transceiver
- Right HF transceiver
- Left air traffic control (ATC) transponder
- Right ATC transponder

- MAWEA (master monitor card A)
- MAWEA (master monitor card B)
- MAWEA (crew alerting card)
- MAWEA (left aural synthesizer card)
- MAWEA (right aural synthesizer card)
- Lower yaw damper module (high speed)
- Upper yaw damper module (high speed)
- Audio management unit
- Left weather RADAR transceiver
- Right weather RADAR transceiver
- Digital flight data acquisition card (DFDAC)
- Captain's clock
- Software data loader panel (high speed)
- Multiple input printer
- ACARS management unit

Table 5-3-1. ARINC 429 inputs to the CMCs

ARINC OUTPUT BUS 1 Connects to the following systems:	ARINC OUTPUT BUS 2 Connects to the following systems:	ARINC OUTPUT BUS 3 Connects to the following systems:
Pack temperature controller B	Electrical system card file	Pack temperature controller A
Cabin pressure controller A	Zone temperature controller	Cabin pressure controller B
Left flight control computer	Tire pressure monitor unit	Right flight control computer
Left stabilizer trim rudder ratio module	Center flight control computer	Right stabilizer trim rudder ratio module
Upper yaw damper	Center flap control unit	Lower yaw damper
Bus control unit 1	Ground flap control unit	Bus power control unit 2
Left flap control unit	Ground proximity warning computer	Right flap control unit
Left window heat controller	Center ILS receiver	Right window heat controller
Left VOR receiver	Center radio altimeter transceiver	Right VOR receiver
Left DME interrogator	Hydraulic quantity monitor unit	Right DME interrogator
Left ILS receiver	Center EFIS/EICAS interface unit	Right ILS receiver
Left ADF receiver	Proximity switch electronic unit	Right ADF receiver
Left range radio alitmeter transceiver	Brake system control unit	Right radio altimeter transceiver
Left ATC transponder	APU control unit	Right ATC transponder
Right weather radar transceiver	Left control and display unit	Left weather radar transceiver
Left EFIS/EICAS interface	Power supply – emergency lights	Right EFIS/EICAS interface unit
Left flight management computer		Right flight management computer
Left air supply control/test unit		Right air supply control test unit
Electrical system card file		Electrical system card file
ARINC OUTPUT BUS 4 Connects to the following systems:	**ARINC OUTPUT BUS 5** Connects to the following systems:	**ARINC OUTPUT BUS 6** Connects to the following systems:
Left VHF transceiver	Center VHF transceiver	Right VHF transceiver
Left HF transceiver	Fuel quantity processor	Right HF transceiver
	Fuel system card file	
	Digital flight data acquisition card	

Table 5-3-2. Connections for the CMC busses 1-6

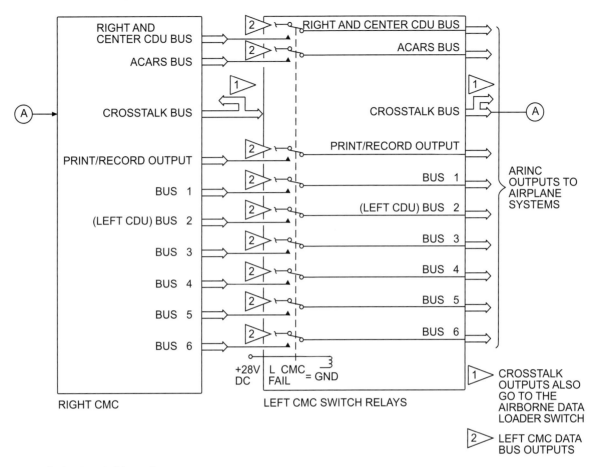

Figure 5-3-4. Left CMC switching relay

Courtesy of Northwest Airlines, Inc.

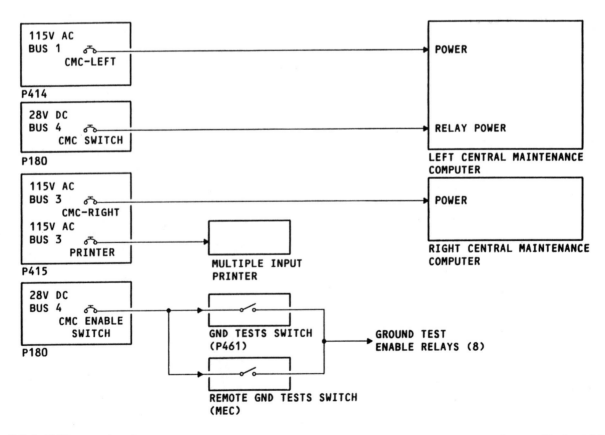

Figure 5-3-5. CMC power inputs

Courtesy of Northwest Airlines, Inc.

both the EIUs and the CMCs. The CMCs also receive information from the EIUs. The CMC compares fault information from the EIU and aircraft systems for fault verification. A list of ARINC inputs to the CMCs is shown in Table 5-3-1.

Although there are only 10 output busses from each CMC, the information is shared by many aircraft systems. Table 5-3-2 shows the systems connected to ARINC busses 1-6. The remaining four busses are connected to the CDUs; the ACARS management unit, the printer, the airborne data loader, and the opposite side CMC.

The right CMC has ten ARINC 429 outputs; one is a cross-talk bus to the opposite side CMC. The outputs are sent to the various aircraft systems through the left CMC switch relays (Figure 5-3-4). The switch relays are normally closed to the left CMC outputs. If the left CMC fails, a ground will be applied to activate the switch relay; hence, the right CMC outputs will be sent to the aircraft's systems. In other words, if the left CMC detects internal faults, it automatically passes output data from the right CMC directly through the switch relay. Remember, the CMCs are identical. The switch relay is simply inactive in the right CMC. In some cases the aircraft can be operated with only one CMC. In this situation, the CMC must be installed in the left slot.

Each CMC will receive up to 22 discrete inputs. Note that discrete inputs are typically analog signals. All input data is wired in parallel and therefore shared between the two CMCs. Up to 41 discrete parallel output signals are available from each CMC. Four discrete signals are also used for pin programming of each CMC. There are two spare program pins, one used for a parity bit, and one to determine the left/right installation of the CMC. If the CMC is installed in the right-hand rack, the pin should be grounded. If the CMC is installed in the left-hand rack, the pin is open. The parity bit employs odd parity to ensure the correct signals are connected to the program pins. A list of the discrete inputs and outputs can be found in Table 5-3-3.

The power to the CMCS is supplied through three different busses (Figure 5-3-5). The number 1 AC bus powers the left CMC. The number 3 AC bus powers the right CMC and the CMCS printer. The number 4 DC (28V) bus supplies the CMC switch relay and the ground test switches.

There are eight ground test enable relays that are used to ensure that certain LRU BITE tests are performed only on the ground and not during flight. The eight relays are activated by the ground test enable switch located on the flight

DISCRETE OUTPUTS
Equipment cooling test relay through CMC ground test enable relay 3
Left flight control computer (FCC)
Center FCC
Right FCC
Center air data computer (ADC) through CMC ground test enable relay 1
Right ADC through CMC ground test enable relay 1
Left ADC through ground test enable relay 2
R7421 left pitot probe heater test relay
R7422 right pitot probe heater test relay
Left inertial reference unit (IRU)
Right IRU
Center IRU
R7683 wing thermal anti-ice system relay through CMC ground test enable relay 8
MAWEA (left stall warning management card)
MAWEA (right stall warning management card)
MAWEA (configuration warning card)
Right ozone valve, left ozone valve, and ozone switch (1 output)
Fire test 1 through CMC ground test enable relay 3
Fire test 2 through CMC ground test enable relay 4
Fire test 3 through CMC ground test enable relay 5
Fire test 4 through CMC ground test enable relay 6
Fire test 5 through CMC ground test enable relay 7

DISCRETE INPUTS
Left radio communication panel
Center radio communication panel
Right radio communication panel
R7421 left pitot probe heater test relay
R7422 right pitot probe heater test relay
V474 valve ozone catalytic converter bypass right
V472 valve ozone catalytic converter left
V471 valve ozone catalytic converter left
Flight control electronics (FCE) power supply module 1 left (2 inputs)
FCE power supply module 1 right (2 inputs)
FCE power supply module 2 right (2 inputs)
FCE power supply module 2 left (2 inputs)
R7746 ozone converter command indication

HARDWARE PROGRAM PINS
L/R CMC
Spares (2)
Parity
The L/R CMC pin defines the position of the CMC. The left CMC's pin is grounded. The right CMC's pin is open.
The two spare pins are not connected
The parity pin shows odd parity for the program pins. The left CMC's pin in open. The right CMC's pin in grounded.

Table 5-3-3. CMCS discrete inputs and outputs.

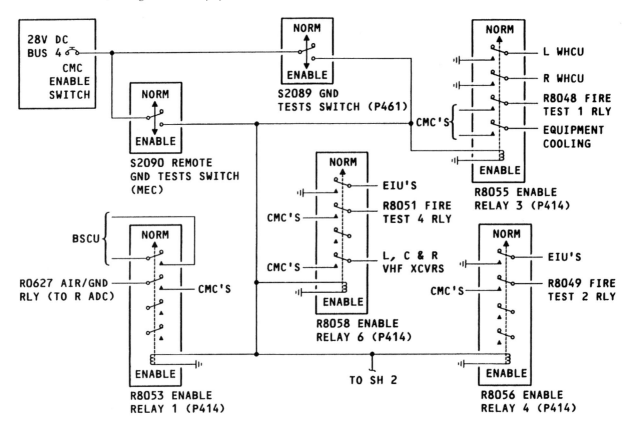

Figure 5-3-6. CMC enable switches and relays

Courtesy of Northwest Airlines, Inc.

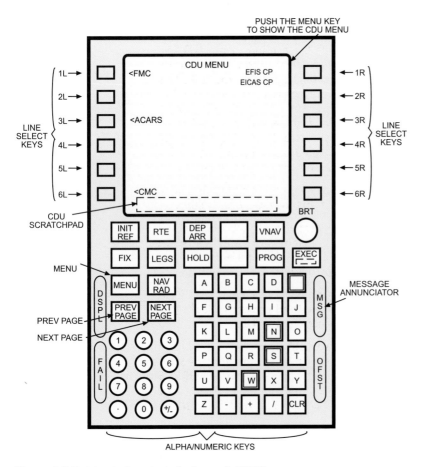

Figure 5-3-7. A typical control display unit (CDU)

Courtesy of Northwest Airlines, Inc.

deck or the remote ground test enable switch located in the main equipment bay. When energized, the relays send a discrete signal to the affected LRUs. Figure 5-3-6 shows the two enable switches and four of the relays. Relay position is monitored by the EIUs, and whenever energized the message GND TEST ENABLE is displayed on the auxiliary EICAS panel. The operation of the CMC ground test will be discussed later in this chapter.

CMCS Operations

Any of the four control display units can be used to access the data contained in the CMC. In most cases, the technician will use the flight deck CDU since the EICAS displays can also be viewed from that location. The CDU contains a CRT display located at the top of the unit between 12 line select keys (Figure 5-3-7). The line select keys (LSK) on the control display unit are used to select items for display or activate functions available on the CDU. Whenever a given line is active, a caret symbol (<) will be shown adjacent to the appropriate LSK. Other keys, which are used to access CMCS information, are the MENU, PREVIOUS PAGE, and NEXT PAGE keys.

To initiate CMC operations, the menu must be brought to the display screen by pressing the MENU button on the CDU (Figure 5-3-7). The line select key 6L, which is adjacent to <CMC on the display, should be pressed to select the CMC menu. The two page CMC menu will then be available (Figure 5-3-8). Page one of the CMC menu will be the initial display. Pressing the next page button on the CDU accesses the second page. As seen on the CMC menu, there are a total of seven options available from the CMC, displayed on the two pages:

1. Present faults

2. Confidence tests

3. EICAS maintenance pages

4. Ground tests

5. Existing faults

6. Fault history

7. Other functions

To make a selection, press the LSK adjacent to any of the seven functions on the display.

Present Faults

The present faults function of the CMC is used to display any fault and related flight deck effect that occurred during the present leg. The present leg is defined as the elapsed time between first engine start and last engine shut down. When the first engine is started for the next flight the present faults data is moved to the fault history memory. There are two basic formats presented under the present faults function: present leg faults and present leg messages. Present Leg Faults can be thought of as a general list of present leg faults. Present leg messages can be thought of as the details of each present leg fault. To view the present faults data, press the LSK 1L on the CMC menu page, the flight deck effect of the present leg faults will be displayed. For each flight deck effect, a related present leg message is available by pressing the appropriate LSK. A flight deck effect (FDE) is considered any EICAS message or parameter exceedance, any primary flight display (PFD) flag, or any navigational display (ND) flag. If no FDEs were reported during the present leg, the message NO FLIGHT DECK EFFECTS REPORTED DURING THIS FLIGHT will be displayed when the present leg faults page is accessed.

Present Leg Faults

Figure 5-3-9 shows the series of CDU displays for the present leg faults function of the CMCS. Refer to this figure during the discussion on present leg faults.

The following information is presented on the present leg faults data page(s):

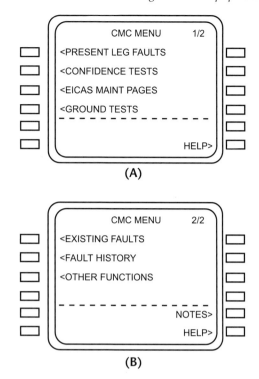

Figure 5-3-8. CMC menu display: (A) page 1 of 2 is mostly for line maintenance, (B) page 2 of 2 is for extended maintenance troubleshooting.

1. All present leg faults are displayed, listed by their FDE, in sequential order with the most recent fault at the top of the list.

2. The type of FDE, caution, memo, ND flag, etc., is given along with a flight reporting manual (FRM) code for the fault. This information is shown just above the FDE of each fault. The FRM code follows ATA specifications to establish a reference number for the fault. The FRM code can easily be referenced to the aircraft's maintenance and troubleshooting manuals.

3. The asterisk (*) shown next to the type of FDE means that the fault is still active.

4. The number of pages of fault data is shown in the upper right corner of the display. To access additional pages press the next page button on the CDU.

5. A category for NON-FDE FAULTS is listed to allow access to all present leg faults that did not create a given FDE.

6. If the word ERASE is shown next to a latched status message, pressing that LSK will remove that message from the EICAS status page. This should be done after repairing the fault that caused the status message.

7. The ERASE STATUS LSK is used to remove all latched messages from the EICAS status page.

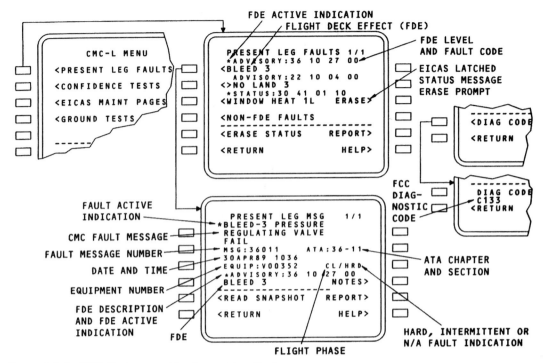

Figure 5-3-9. CDU displays used for Present Leg Faults

Courtesy of Northwest Airlines, Inc.

8. The RETURN LSK is used to change the display back to the previous menu.

9. The REPORT LSK is used to display the report menu.

10. The HELP LSK is available on most CMC page displays and is used to gain information concerning use of the CMC.

Present Leg Messages

After a technician has viewed the list of present leg faults, he/she may select the associated message for that fault by pressing the adjacent LSK. The CMCS present leg message page provides a detailed look at the fault that caused the FDE selected from the present leg faults page. In our example, Figure 5-3-9, the FDE BLEED

3 was selected. The present leg message page contains the following information:

1. The title of the FDE from the present leg faults page. The asterisk denotes the fault is still active.

2. The CMC fault message is listed beneath the FDE title.

3. The fault message number is used to find data about the fault in the aircraft's fault isolation manual (FIM).

4. The date and time at which the fault occurred is listed to help aid in troubleshooting. For example, if several messages occurred simultaneously they may be the result of the same defect. On the other hand, they may be nuisance messages

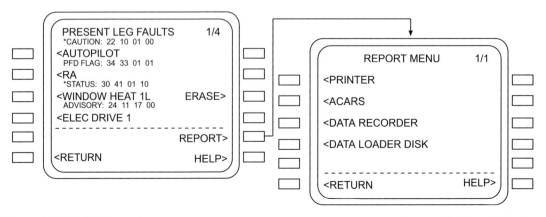

Figure 5-3-10. CMC report menu

Courtesy of Northwest Airlines, Inc.

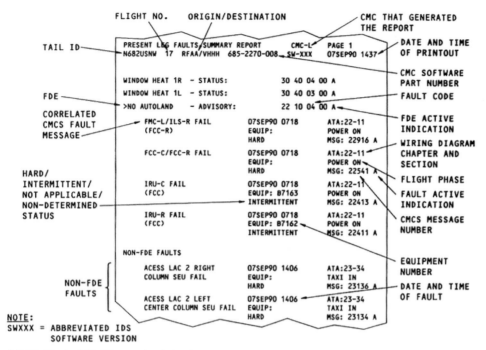

Figure 5-3-11. Typical CMC report printout

Courtesy of Northwest Airlines, Inc.

caused by a normal operating procedure, such as, an electrical power transfer.

5. The equipment number to which the fault occurred.

6. The FDE description.

7. The flight phase in which the fault occurred and the type of failure either hard or intermittent. This information is extremely useful during troubleshooting.

8. The ATA chapter and section related to the failed system.

9. In most cases, an auto snapshot of the system is taken at the time of failure. Pressing the LSK adjacent to READ SNAPSHOT on the display accesses this information. As stated during the discussions on EICAS, auto snapshots will record various system parameters, which can be recalled during troubleshooting.

Nuisance Messages

A nuisance message is considered any FDE or CMC message that is not caused by an actual fault. Nuisance messages may be caused by certain normal operating procedures, such as electrical power transfers or abnormal operation of a system during preflight. To help eliminate nuisance messages, the B-747-400 CMCS software incorporates flight phase screening. Flight phase screening is accomplished through correlation of any fault with the flight phase at the time the fault occurred. If the software determines that the failure occurred outside the normal operational flight region for that system the CMC will

ignore that failure. If the fault still exists when that system enters its normal operating region the CMC will then report the fault. Much of the flight phase screening is completed while the aircraft is still on the ground.

Flight Phases

Flight phases are recorded by the CMC to help the technician determine the aircraft's configuration at the time of a fault. The CMC uses data from the EIUs to determine the flight phase. There are 14 different flight phases, each with a two-letter designator as follows:

1. Power on (PO)

2. Preflight (PF)

3. Engine start (ES)

4. Taxi out (TA)

5. Takeoff (TO)

6. Initial climb (IC)

7. Climb (CL)

8. Enroute (ER)

9. Descent (DC)

10. Approach/Land (A/L)

11. Rollout (RO)

12. Taxi in (TI)

13. Go around (GA)

14. Engine shutdown (ES)

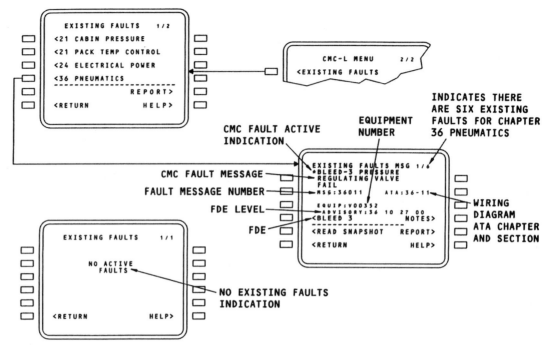

Figure 5-3-12. CMC existing faults display

Courtesy of Northwest Airlines, Inc.

Printing and ACARS

The CMCS report menu contains two options: printer or ACARS (Figure 5-3-10). The ACARS function will transmit the present leg fault data to the airline's ground facility. The print option will print a detailed present leg faults summary as shown in Figure 5-3-11. Printing the report is a handy way to document the fault information needed for troubleshooting or warranty verification.

Existing Faults

The existing faults consist of any real time faults monitored by the CMC that are present at the time of CMCS interrogation. Existing faults are often the only CMC function accessed during an aircraft turn around defined as the time between two flight legs, typically 30-60 minutes. Since existing faults show real-time failures, this information is very helpful in determining aircraft departure status. Faults

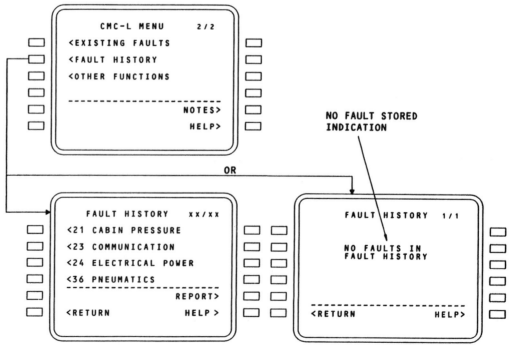

Figure 5-3-13. CMC fault history displays

Courtesy of Northwest Airlines, Inc.

that are currently active can be identified via the existing faults function and correlated with the aircraft's minimum equipment list (MEL). If approved by the MEL, the aircraft may be dispatched without repairing the fault. The existing faults memory may contain some data which is not stored in the present leg faults memory. This occurs since flight phase screening restricts certain faults from entering the present leg faults category and existing faults are recorded regardless of flight phase. These restricted faults often occur during ground activities.

To access the existing faults memory simply select the existing faults function from the CMC menu page on the CDU. As shown in Figure 5-3-12, the existing faults menu consists of one or more pages that list the title of the failed system according to ATA chapter. Lowest order chapters are listed first and only chapters that have a failed system will be listed. If no existing faults are present at the time of interrogation, the message NO ACTIVE FAULTS will be displayed.

To view the fault message(s) for a given system by ATA chapter, press the appropriate LSK. The first page of existing faults messages for that ATA chapter will be displayed on the CDU. Only one fault will be displayed per page. To access the next fault for that chapter press the NEXT PAGE key on the CDU keyboard. To access messages from other ATA chapters, press the LSK adjacent to RETURN. The information displayed on the existing faults message pages is nearly identical to the present leg faults messages. Compare Figure 5-3-9 and 5-3-12.

Fault History

To aid in the troubleshooting and repair of chronic problems, the CMCS fault history function should be accessed. There can be up to 500 faults from a maximum of 99 flight legs stored in the fault history nonvolatile memory. To access fault history, press the correct LSK on the CMC menu. All failure categories are listed according to ATA chapter as seen in Figure 5-3-13. If the fault history memory is empty, the terms NO FAULTS IN FAULT HISTORY will be displayed.

Pressing the appropriate ATA chapter LSK of the fault history menu page will display a fault history summary for each fault in that chapter. Each fault will be shown on a separate page. To access additional faults, press the next page key on the CDU. As seen in Figure 5-3-14, the fault history summary page contains the following information:

1. CMC fault message

2. Leg(s) in which the failures occurred

3. Type of failure (hard or intermittent)

At the top right corner of the fault history summary page is the page number and number of pages available in that summary.

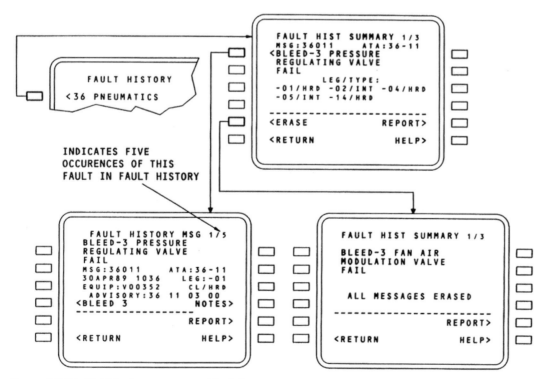

Figure 5-3-14. CMC fault summary and fault history message pages.   *Courtesy of Northwest Airlines, Inc.*

The CMCS fault history message page gives a detailed summary for each occurrence of the fault selected from the fault history summary page. To activate the fault history message page press the correct LSK on the fault history summary page (Figure 5-3-14). The data displayed on each page is listed in order of occurrence by flight leg. The first page displayed shows the most recent flight leg in which the selected fault occurred. Subsequent pages display additional legs in which the fault occurred. The fault history message page contains information that is almost identical to a present leg message page. The exception being that the fault history message page also contains the leg number for the fault occurrence. Present legs are considered leg

number 00, the fault previous to that is called flight leg -01, before that is -02, and so on.

EICAS Maintenance Pages

The EICAS maintenance page function of the CMC is used to access real-time data for 11 different aircraft systems. The EICAS maintenance pages are accessed through the CMC menu on the CDU; however, the data is displayed on the lower EICAS display. For more information on EICAS, see Chapter 3. The EICAS maintenance pages present three types of information: real-time data, auto event data, and manual event data. Real-time data presents information about the selected system as it currently exists. Real-time data is dynamic information and will change with changes in system parameters. Auto event data is information about the selected system, which was automatically recorded at the time of a system fault or exceedance. Auto event data is helpful in troubleshooting faults since it allows the technician to see the conditions of the system at the time of failure. Manual event data is a snapshot of the system parameters recorded by the flight crew or maintenance technician action. Table 5-3-4 shows a list of the 11 EICAS maintenance pages and the availability of manual and auto recorded events.

Accessing EICAS Maintenance Pages

Access to the EICAS maintenance pages is actually accomplished through four different types of LRUs: the CDU, CMCs, EIUs, and EICAS displays. Therefore, each of these systems must be operable to access the EICAS maintenance pages. Pressing the correct LSK from the CMC menu will activate the choice of EICAS maintenance pages. Figure 5-3-15 shows the maintenance page menu and the components involved in retrieving maintenance pages.

After a given system has been selected, the EICAS page control will be displayed (Figure 5-3-16). From this page the operator can select display, record, manual snapshot, erase system, auto snapshot, report, return, or help. The display function will present the real-time parameters for the previously selected system. The record function will take a manual snapshot of the parameters currently displayed on the EICAS CRT. Since the EICAS maintenance pages are shown on the lower EICAS CRT, the EICAS page control will be displayed simultaneously on the CDU.

The manual and auto snapshot functions will display a list of their respective snapshot data for the selected system (Figure 5-3-17). There

ATA	MAINTENANCE PAGE	MANUAL	AUTO
\multicolumn{4}{l}{**EICAS MAINTENANCE PAGES RECORDING SUMMARY**}
21	ECS	X	Ⓧ
24	Electrical	X	X
27	Flight Controls	X	None
28	Fuel	X	Ⓧ
29	Hydraulic	X	X
31	Configuration	N/A	N/A
32	Gear	X	None
49	APU	X	X
73	EPCS	X	▷1
73	Performance	X	X
73	Engine Exceedence	N/A	▷2

Fuel Pages

Main 1/Main 4
Main 2/Main 3   ECS Pages
Reserve 2/Reserve 3   Air Conditioning
Center Main/Stabilizer Air Supply

▷1 Store when a performance auto snapshot is taken

▷2 Stores automatically in cumulative manner – erased through CMC

Table 5-3-4. Manual and automatic event recording related to various EICAS maintenance pages

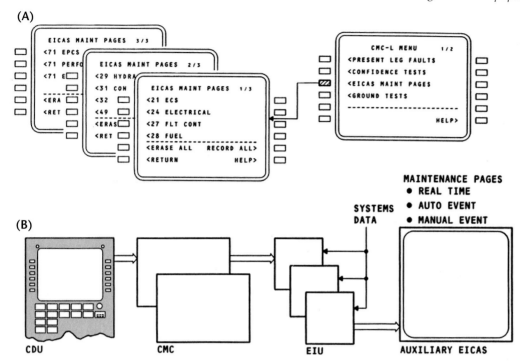

Figure 5-3-15. EICAS maintenance pages: (A) Typical EICAS maintenance menu pages, (B) components involved in the retrieval of EICAS maintenance pages
Courtesy of Northwest Airlines, Inc.

are a maximum of five auto and five manual snapshots available for each system. The list of snapshots is displayed showing the flight leg, date, and time of the recording. Pressing the LSK adjacent to a given snapshot will display the system's parameters on the lower EICAS display. Auto snapshots are also available from five other CMC functions: present leg faults, confidence tests, ground tests, existing faults, and fault history. Manual snapshots are only available through the EICAS maintenance page function.

Pressing the REPORT LSK on the EICAS page control menu will allow access to the print on ACARS function of the EICAS maintenance

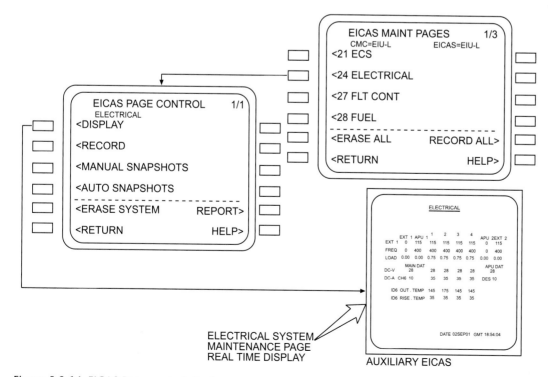

Figure 5-3-16. EICAS Page control display
Courtesy of Northwest Airlines, Inc.

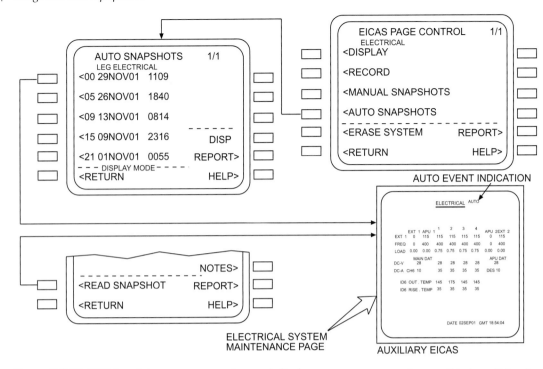

Figure 5-3-17. EICAS maintenance page auto event displays

Courtesy of Northwest Airlines, Inc.

pages. The print function will send the currently displayed EICAS maintenance page to the CMC. The CMC will relay that data to the flight deck printer. This feature is handy to allow for retrieval of the parameter data, which can be studied at a more convenient time or place. The ACARS function will send the currently displayed data to the ACARS management unit for transmission to the airline's ground facility. The ACARS function is only used during flight.

Of the 11 systems available for display by EICAS maintenance pages, seven have one page formats such as the display in Figure 5-3-18. Keep in mind that this one page can

display real-time, auto snapshot, or manual snapshot data. The four other systems available through EICAS maintenance pages are ECS, Fuel, Configuration, and Engine exceedance and have unusual page formats. The ECS (environmental control system) EICAS maintenance page format consists of two pages. One page displays conditioned air while the other shows supply air. The fuel system has four EICAS maintenance pages available for display. To allow access to each of these four pages an additional fuel menu page is displayed when requested using the CDU. The Configuration maintenance pages are used to access information on the integrated display system (IDS) software numbers and LRU pin programming. The configuration maintenance page does not have an auto or manual snapshot mode. The engine exceedance page consists of a single page format; however, not all engine parameters are always listed. Certain data is displayed as a function of an exceedance in that area. The engine exceedance page menu contains an erase function, which is used to clear the exceedance memory.

Confidence Tests

The confidence tests function of the CMCS allows for preflight testing of three aircraft systems: the stall warning/stick shaker, the take off configuration of the aircraft, and the ground proximity warning computer (GPWC). In general, confidence tests provide a quick GO/NO-GO status of these systems. Pressing the appropriate LSK activates the various tests.

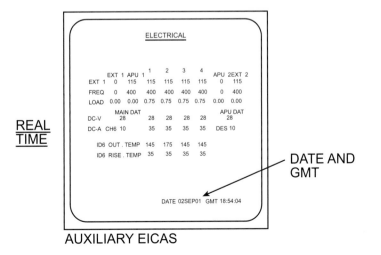

Figure 5-3-18. EICAS maintenance page, electrical, one-page format

Courtesy of Northwest Airlines, Inc.

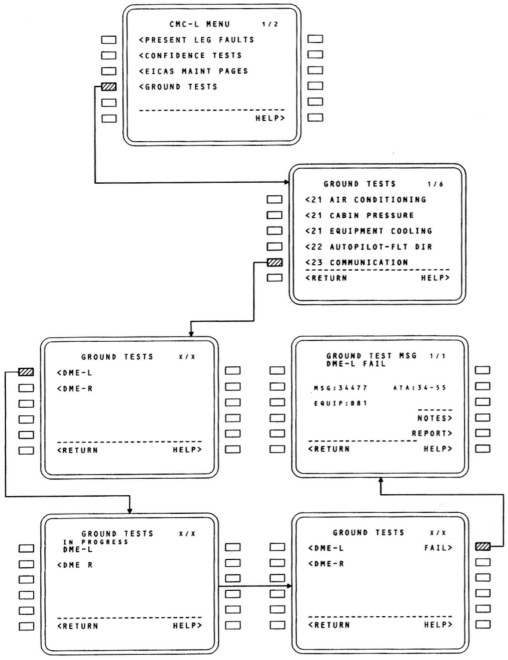

Figure 5-3-19. CMC Ground tests menu

Courtesy of Northwest Airlines, Inc.

ATA CHAPTERS AND MAJOR SUBSECTIONS

- **Chapter 21**
  Air Conditioning
  Cabin Pressure
  Equipment
  Cooling

- **Chapter 22**
  Auto Pilot
  FLT DIR
  Yaw Damper

- **Chapter 23**
  Communications

- **Chapter 24**
  Electrical Power

- **Chapter 26**
  Fire Protection

- **Chapter 27**
  Flaps Control
  Stall Warning

- **Chapter 28**
  Fuel

- **Chapter 29**
  Hydraulic Power

- **Chapter 30**
  Ice and Rain

- **Chapter 31**
  Indicating/
  Warning
  Recording

- **Chapter 32**
  Brake Control
  PSEU System
  Tire System
  Brake
  Temperature

- **Chapter 34**
  Air Data
  Inertial Reference
  Navigation
  Radios
  Flight
  Management

- **Chapter 36**
  Pneumatics

- **Chapter 45**
  Central
  Maintenance

- **Chapter 49**
  APU

- **Chapter 73**
  Engine Fuel and
  Control

Table 5-3-5. Ground test availability listed according to ATA chapter

Ground Tests

The ground tests function of the CMCS provides access to test functions of up to 80 different systems and LRUs. To initiate a ground test, the aircraft must be on the ground and the ground test enable switch must be placed in the enable position. When the ground test function is selected from the CMC menu, the list of ATA chapters is displayed on eight consecutive pages. The ATA chapters that have ground tests available are shown in Table 5-3-5. Selection of a given ATA chapter will display a list of LRUs or systems for testing. Up to five tests are available for each page of the menu. Figure 5-3-19 shows the series of CMC menu pages used to access the DME ground tests.

To start a given test, press the LSK adjacent to the LRU or system shown on the display. The CDU will initiate a digital signal to the CMC, which will in turn transmit a test signal to the appropriate LRU. The actual test will be performed within the LRU in question. During the test, the LRU will respond with a test initiation signal. Upon completion, the LRU will transmit the test results back to the CMC. The CMC will generate the appropriate message for display on the CDU.

At the completion of a ground test, the CDU will display one of the following messages:

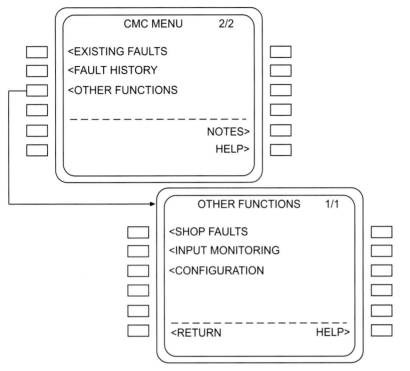

Figure 5-3-20. CMCS other functions menu · *Courtesy of Northwest Airlines, Inc.*

determine the test results. For example, a test of the flight deck annunciators would require the technician to visually identify if the lights illuminated at the appropriate time.

Shop Faults

Shop faults are typically used to further define faults found during other CMC tests. The other functions LSK on the CMC menu is used to access shop faults (Figure 5-3-20). The shop faults option will present a list of ATA chapters from which to make a selection. When a given ATA chapter is selected, the LRUs that can be tested are listed on the CDU. When a given LRU is selected, the BITE for that LRU is activated, and an internal test is performed. The results will be displayed as shown in Figure 5-3-21. During a ground test for example, a fault in the cabin pressurization may be detected. The use of shop faults may isolate the given defect to a specific LRU, or within a given LRU.

Input Monitoring

Input monitoring is accessible on the ground or in flight to display the ARINC 429 input data to various LRUs. Input monitoring data can be displayed in binary, hexadecimal, or engineering code formats. The CMC input monitoring function could be used to detect ARINC 429 bus problems similar to a carry-on data bus analyzer. As shown in Figure 5-3-22, input monitoring is found through the other func-

PASS, FAIL, or DONE. Pass simply states that the component or system test found no faults. A test fail will record a CMC message that is retrieved by pressing the LSK adjacent to the FAIL> prompt on the CDU (Figure 5-3-19). The message DONE indicates the end of the test for systems that require the technician to

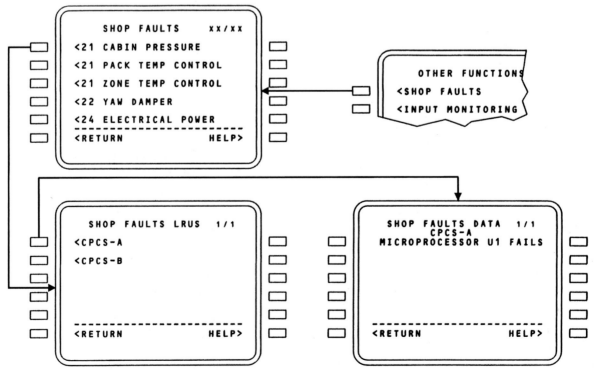

Figure 5-3-21. CMC typical shop fault

Courtesy of Northwest Airlines, Inc.

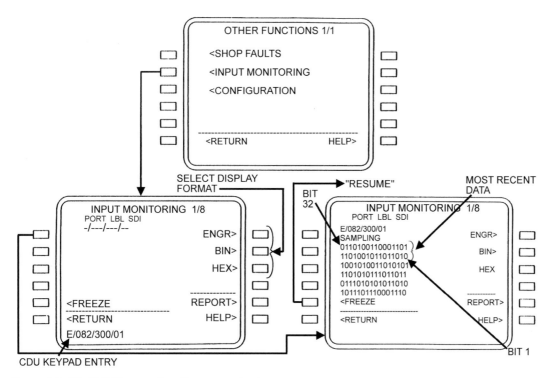

Figure 5-3-22. CMCS example of input monitoring

Courtesy of Northwest Airlines, Inc.

tions menu. Up to eight pages of input monitoring can be done at any given time. This is very handy when comparing the same data sent to several different LRUs.

To select a given system and bus to be monitored, enter the following four designators, using the CDU keypad:

1. The computer designator; enter either E for EIU, or C for CMC.

2. The specific port on the selected computer to be monitored; enter a three-digit identifier.

3. The label to be monitored; enter a three digit code. This label corresponds to the ARINC 429 specifications and identifies the information being transmitted to the selected CMC/EIU port.

4. The source destination indicator (SDI); enter a two digit code. The SDI is an ARINC specified number, which corresponds to the source transmitting the data.

The specific digits that should be entered are available from the aircraft's maintenance manuals, schematics manuals, or fault isolation manuals. When the code is initially entered, it appears on the scratch pad at the bottom left corner of the CDU (Figure 5-3-22). To select the number displayed in the scratch pad, press the top left LSK. To determine the format for the displayed data, press the LSK adjacent to the

engineering (ENGR), binary (BIN), or hexadecimal (HEX) label.

The data displayed on the CDU is real-time information for the system/bus selected. In Figure 5-3-22, the data is displayed in binary language and each of the 32 bit characters is a binary 1 or 0. Remember, ARINC 429 data consists of a 32-bit word. On the input monitoring display, each data word consists of two lines of 16 bits each. The bits read from left to right, the top left is bit number 32 and the bottom right is bit number 1. The top two rows of data are the most recent sampling of data. The next two lines are from the previous sampling, and so on. Up to six lines or three samples of data can be displayed at any given time. With each new sample of the data many of the binary digits may change from 0 to 1 and vice versa as the system being monitored changes parameters. The values displayed can be compared to the appropriate values given in the aircraft manuals to determine the validity of the data.

When the input monitoring function is updating data, the word SAMPLING is displayed on the CDU. During sampling the data will advance to the next sample approximately once a second. To study the data, press the LSK adjacent to the *FREEZE* prompt. When in the freeze mode, the data display remains stable and the message *FREEZE* replaces the word *SAMPLING* on the CDU display. The report function of the input sampling mode allows for

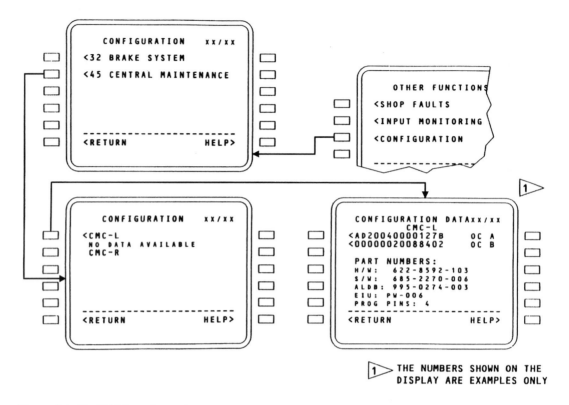

Figure 5-3-23. CMCS configuration page

Courtesy of Northwest Airlines, Inc.

printing of the data or transmission through ACARS to a ground facility.

Configuration

The configuration function of the CMC allows for verification of part numbers and programming options for the brake system control unit (BSCU) and the central maintenance computer (CMC). The configuration option is listed under other functions on page 2 of the CMC menu. As shown in Figure 5-3-23, selection of the *<45 CENTRAL MAINTENANCE* will activate the display of the left CMC data. As long as there is a left CMC installed, no data will be available from CMC-R. This is a normal condition because of the output switching relays discussed earlier. To access data from the right computer, the left CMC must be deactivated. To select the left CMC, press the LSK adjacent to the CMC-L prompt.

A typical data display for the CMC configuration is shown in Figure 5-3-23. This display shows the hardware (H/W) and software (S/W) part numbers, the airline database (ALDB) numbers, the EIU software configuration, and the CMC program pins. At the top of the display are the two airline option codes (OC A & OC B). To verify the CMC memory contains the correct option code, the hexadecimal form of the code is entered into the scratch pad. From the scratch pad, the code is moved to OC A or

OC B by pressing the corresponding LSK. The CMC will then compare codes and display the appropriate response.

Options Summary

The CMCS has a multitude of information available through the CDU. As a technician becomes familiar with the aircraft, it becomes second nature as to which function of the CMC will access the needed information. The CMC function accessed will be a direct result of the current needs of the technician and aircraft status. For example, if the aircraft is awaiting departure the existing faults function may be used; if the technician is troubleshooting a chronic fault, the fault history may be accessed. Table 5-3-6 shows the various functions of the CMC often used during maintenance and troubleshooting.

Troubleshooting Using the CMCS

Whenever troubleshooting any system, the ultimate goal is to repair the aircraft. This should be accomplished in the least time possible without sacrificing safety. The CMCS is designed to help isolate faults for troubleshooting and verify when the repair has been completed successfully using operational testing. For either of these operations, the CMCS cannot stand alone. The CMCS is always used in conjunction with the various aircraft maintenance manuals. For troubleshooting purposes the Fault Isolation Manual (FIM) portion of the maintenance manuals is extremely valuable. In many cases, the FIM is a completely separate manual or series of manuals. In some cases, the FIM data is included in the front of each chapter of the maintenance manuals.

The Fault Isolation Manual contains three sections, which are used extensively when isolating faults with the CMCS: the Fault Code Index, EICAS Messages, and CMCS Message Index. Each of these sections has a specific function, which may be used under different conditions to help isolate faults. These sections of the FIM will be presented in the upcoming examples of typical troubleshooting procedures. In all but a handful of situations, the fault isolation procedures begin with a discrepancy write-up in the logbook made by the flight crew. Between flights a technician will examine the log to decide if any problems have occurred which require immediate attention. In some cases, the fault repair can be deferred.

The log entry made by the flight crew typically consists of a description of the fault, the related EICAS message, if any, and the Fault Reporting Manual (FRM) fault code number. The FRM provides a fault code number for the various FDEs (Flight Deck Effects) caused by system faults. The FRM fault code number is identical to the Fault Code Number, which is used to identify faults in the FIM. On some aircraft, ACARS is used to transmit fault code data to the airline ground facilities. This allows the technician the opportunity to review the fault information prior to aircraft landing. This creates an obvious advantage when maintenance takes place during a short turnaround period.

It should be noted, that although the CMCS is very effective at fault isolation, these systems monitor and test only electrical circuitry, not mechanical devices. In some situations, the mechanical device can be monitored using electrical/electronic means; in other situations it cannot. For example, limit-switches can be used to determine the position of a pressurization air valve; however, there is no simple electrical means to measure the integrity of the cargo door seal. If the CMCS detects cabin pressurization is too low it will look for several conditions to determine the fault. If the cargo door seal has been damaged, the CMCS will not be able to determine a defect or the CMC may fault a different portion of the system. Although these situations are rare, one must always consider that the mechanical defects are often missed by the CMCS.

VARIOUS CMC FUNCTIONS	
CMC FUNCTION	PURPOSE AND TYPICAL USE
Present leg faults	Provides a list of faults and related fault data from the most recent flight leg.
	Typically used to confirm a fault reported by the flight crew during the last leg.
	Allows access of snapshots.
Existing faults	Provides data on all faults which currently exist (i.e. real time data).
	Used most often to examine aircraft status prior to dispatch.
Fault history	Provides a list of fault data for failures within the past 99 flight legs.
	Used for troubleshooting recurring problems.
Confidence tests	Tests certain critical systems and provides data on the aircraft's configuration.
	Typically used by the flight crew for preflight testing.
EICAS maintenance pages	Presents the real time parameters for 11 different aircraft systems.
	Used for dynamic testing during troubleshooting or repair verification.
Ground tests	Displays test results for various systems and LRUs.
	Often used to verify a fault or as a test after repair.
Shop faults	Displays data on specific LRUs or software failures.
	Typically accessed only for in-depth troubleshooting.
	Often used to pinpoint a defect after identifying a system fault through other tests.
Input monitoring	Functions as a data bus analyzer for ARINC 429 data transmitted to the CMCs or EIUs.
	Used during in-depth troubleshooting of a system.
Configuration	Presents specifics as to which hardware and software components are installed in the CMCs and and BSCU.
	Used to verify the installation of the correct components.

Table 5-3-6. Various CMC functions

System Troubleshooting

Example 1

This example will provide a general guideline for troubleshooting a system fault. A specific detailed sequence will be provided in the second system troubleshooting example. The following steps should be used to isolate a typical system fault:

1. Access the flight log to determine the problem. The log should give a description of

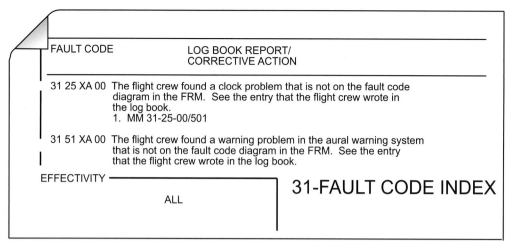

FAULT CODE | LOG BOOK REPORT/ CORRECTIVE ACTION

31 25 XA 00 The flight crew found a clock problem that is not on the fault code diagram in the FRM. See the entry that the flight crew wrote in the log book.
1. MM 31-25-00/501

31 51 XA 00 The flight crew found a warning problem in the aural warning system that is not on the fault code diagram in the FRM. See the entry that the flight crew wrote in the log book.

EFFECTIVITY

ALL

31-FAULT CODE INDEX

Figure 5-3-24. Fault code data from the Boeing 747-400 Fault Isolation Manual (FIM)

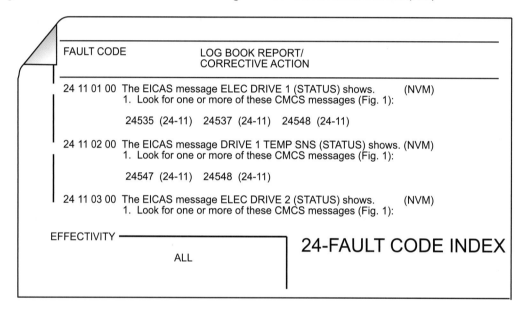

FAULT CODE | LOG BOOK REPORT/ CORRECTIVE ACTION

24 11 01 00 The EICAS message ELEC DRIVE 1 (STATUS) shows.   (NVM)
1. Look for one or more of these CMCS messages (Fig. 1):

24535 (24-11)   24537 (24-11)   24548 (24-11)

24 11 02 00 The EICAS message DRIVE 1 TEMP SNS (STATUS) shows. (NVM)
1. Look for one or more of these CMCS messages (Fig. 1):

24547 (24-11)   24548 (24-11)

24 11 03 00 The EICAS message ELEC DRIVE 2 (STATUS) shows.   (NVM)
1. Look for one or more of these CMCS messages (Fig. 1):

EFFECTIVITY

ALL

24-FAULT CODE INDEX

Figure 5-3-25. Fault code index message 24 11 01 00, 24 11 02 00, and 24 11 03 00

EICAS MESSAGE	LEVEL	DESCRIPTION	FAULT CODE
AURAL SYNTH CARD	(STATUS)	AURL SYNTHSIZER CARD LEFT AND/OR RIGHT (IN THE MAWEA) HAS FAILED	31 51 01 00
AURAL WARN SPKR	(STATUS)	FAILURE OF THE MAWEA AURAL WARNING LEFT AND/OR RIGHT SPEAKER	31 51 02 00
>CONFIG FLAPS	(WARNING)	LE AND/OR TE FLAPS NOT IN TAKEOFF POSITION FOR A TAKEOFF AND TAKEOFF THRUST SELECTED ON ENGINES	31 51 03 00
>CONFIG SPOILERS	(WARNING)	SPEED BRAKE HANDLE NOT IN DOWN DETENT FOR A TAKEOFF AND TAKEOFF THRUST SELECTED ON ENGINES 2 OR 3	31 51 07 00

INDICATING/RECORDING SYSTEMS – EICAS MESSAGES

EFFECTIVITY

31 EICAS MESSAGES

ALL

Figure 5-3-26. EICAS messages from the Fault Isolation Manual

the problem, the fault code from the FRM, and the FDE. In some cases, a fault code number is not available for a given defect. For this situation, begin troubleshooting using the fault description.

In some cases, only the EICAS message is used to initiate troubleshooting. In this situation, the EICAS message portion of the FIM should be used to determine the fault code number.

2. Access the Fault Code Index found in the FIM or the appropriate chapter of the maintenance manual.

If the flight log entry did not include a fault code, use the general list at the beginning of the index. Figure 5-3-24 shows a portion of the Fault Code Index for a clock problem. In this case, the isolation procedures would be found in the maintenance manual (MM) 31-25-00/501.

If the flight crew reported a specific fault code and FDE in the flight log, the fault code number should be found in the fault code index. As shown in Figure 5-3-25, the fault codes are listed in numerical order in the Fault Code Index. For the fault code number 24 11 01 00, the EICAS message ELEC DRIVE 1 (STATUS) is given along with three possible fault message numbers: 24535, 24537, and 24548. Adjacent to the fault message numbers, listed in parenthesis, are the corresponding ATA chapters and sections.

3. If an EICAS message is being used to initiate the fault isolation procedures, the EICAS message pages from the FIM or MM must be used to find the corresponding fault code number. Then the Fault Code Index is used to find the fault message number as in step two above. An example of an EICAS message page is shown in Figure 5-3-26. EICAS messages are arranged in alphabetical order and

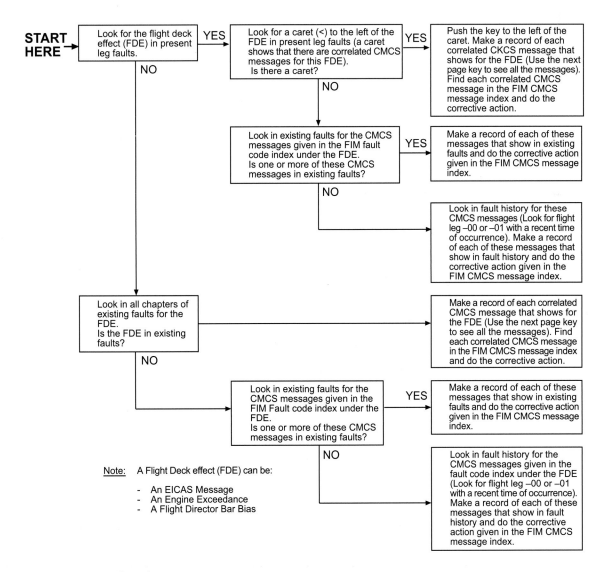

Figure 5-3-27. CMCS access flow chart

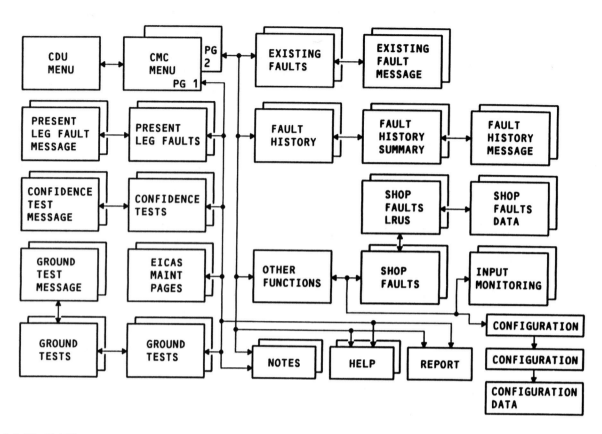

Figure 5-3-28. CMCS menu tree

Courtesy of Northwest Airlines, Inc.

provide the message level (status, advisory, caution, or warning), description, and the fault code number.

4. Once the fault message number has been found through the aircraft's manuals, it must be correlated to the actual CMCS message. To do this, the technician must log on to the CMC using the CDU. A CMC message that related to the fault in question should be located in the CMC nonvolatile memory. In most cases, the search will begin in the Present Leg Faults memory. If the correct CMC message is not found there, the Existing Faults and/or Fault History function of the CMC should be accessed. Figure 5-3-27 shows a flow chart diagram from the B-747-400 Fault Isolation Manual, which provides guidance to CMC access.

A large portion of working your way through CMC data is simply knowing how to access different functions of the system. The CMC menu tree shown in Figure 5-3-28, begins CMC access in the upper left corner of the diagram. The CDU menu is used to access the two-page CMC menu, from here the various fault nonvolatile memories and system tests are accessed. The bottom center portion of the menu tree shows that the help and report portion of the CMC program can be accessed from

almost any function. These options become very handy if used properly.

5. The proper correlation is made between the CMC message and the flight log or EICAS message in order to find the correct fault code needed for the repair. Correlation helps to ensure the work performed is for the fault in question. After correlating the CMC message and the recorded fault, use the CMC Message Index portion of the FIM or MM. The CMCS Message Index will provide the necessary suggestions for repair and/or further fault isolation. As seen in Figure 5-3-29 the fault codes are listed in numerical order in the CMCS Message Index. The index includes one or more CMCS messages for the related fault code, the EICAS message/type, and the corrective action. The corrective action portion of the CMCS Message Index identifies all necessary references, such as the maintenance manual (MM) or wiring diagram manual (WDM).

6. After finding the correct fault code, CMCS message, and related FDE, it would be wise to check the Boeing Service Tip information. The service tips provide helpful information to aid in the troubleshooting and repair of the aircraft. A typical service tip is shown in Figure 5-3-30.

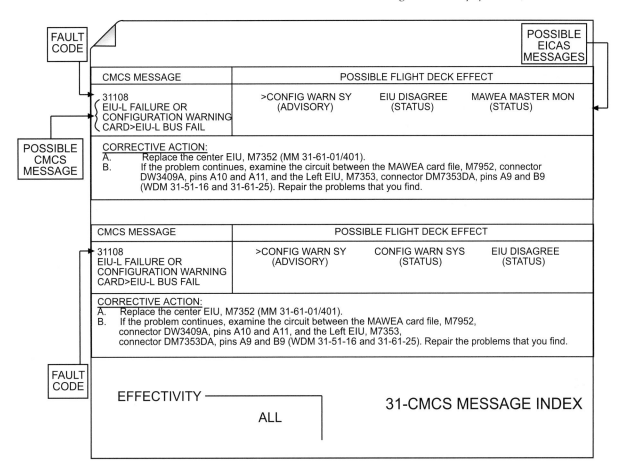

Figure 5-3-29. CMCS message index

Courtesy of the Boeing Commercial Airplane Company

7. Access all necessary manuals as suggested in the CMCS Message Index and perform the necessary corrective actions. Be sure to always verify the repair through the correct testing procedures. In many cases, the CMCS can be used for ground test or review of existing faults of the repaired system. The correct testing procedures will be listed in the manuals under the appropriate page block. As discussed in Chapter 1, the page blocks of the maintenance manuals define the materials covered in that section. Page blocks typically used for troubleshooting and repair of a system include: 001-099 Description and Operation, 101-199 Troubleshooting, including electrical schematics, 201-299 Maintenance Practices, 401-499 Removal and Installation, and 501-599 Adjustment and Test.

In most cases the troubleshooting process begins with a reported system failure. The CMCS is then used for fault isolation of the reported problem. During the initial troubleshooting, always correlate reported faults to CMCS and EICAS messages. Correlation may require a thorough search of the CMCS information. Figure 5-3-31 provides a flow chart that can be used during investigation of the CMCS.

If several improper FDEs are reported at once, suspect the same faulty condition may have caused several items to create an EICAS message. To determine if one failure is to blame for several FDEs, look for CMC messages that may have occurred at the same time. Also consider the time at which the faults occurred; did they correspond to a normal flight crew action? This

_____ SERVICE TIPS_____

1. General
   A. EICAS message ELEC TR UNIT (X) may be displayed if the Split System Breaker (SSB) is open and the right and left sync buses voltages differ by 2.5 volts AC.
   B. TRU currents may be at or near zero if the SSB is open.
   C. This may be a nuisance message.
   D. The 747-422 FIM 24 – FAULT CODE INDEX will be updated with this information.
   E. A future BCU modification will reduce the occurrence of this nuisance message.

2. Recommended Action
   A. Disable power source to one SYNC bus half:
      (1) Set one APU (If on APU power) or one EXT (If on external power) switch to AVAIL position (the split system will close).

      NOTE: This may cause automatic load shedding

   B. If the ELEC TR UNIT (x) messages disappear and/or the TRU output current return to normal, the message is a nuisance.
   C. If the message still exists replace the TRU.

Figure 5-3-30. Typical service tips

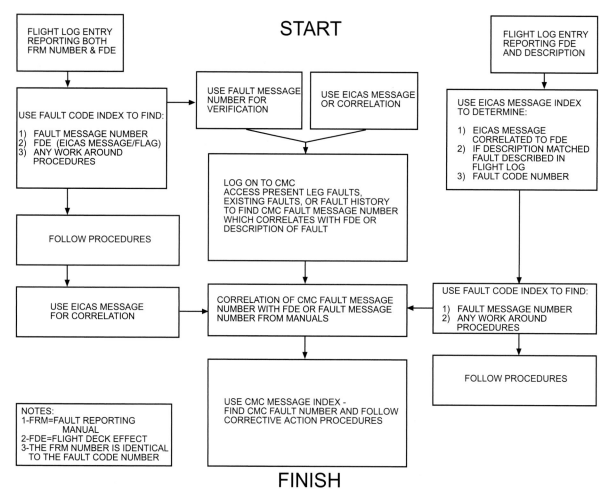

Figure 5-3-31. CMCS flow chart

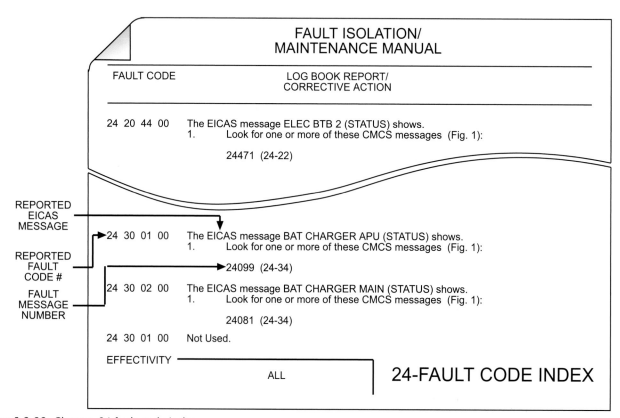

Figure 5-3-32. Chapter 24 fault code index

FAULT ISOLATION/
MAINTENANCE MANUAL

EICAS MESSAGE	LEVEL	DESCRIPTION	FAULT CODE
BAT CHARGER APU	(STATUS)	ONE OR MORE OF THESE PROBLEMS OCCURRED: (1) THE APU BATTERY CHARGER HAS A FAILURE (2) THE INPUT POWER TO THE APU BATTERY CHARGER IS OFF OR HAS A FAILURE (3) THE APU BATTERY CHARGER INTERLOCK IS OPEN (4) THE APU BATTERY IS IN AN OVERHEATED CONDITION	24 30 01 00

Figure 5-3-33. EICAS message for fault code 24 30 01 00

normal action may have caused the FDEs. If so, the reported FDEs and related CMCS messages are nuisance messages and most likely require no further action.

Example 2

In this example we will look at an actual fault and follow it through from logbook entry to repair and testing. The following steps should be taken to repair a problem with the APU battery system.

1. The technician at the gate received a flight logbook entry written as follows: Fault number 24 30 01 00, EICAS status message BAT CHARGER APU appeared during cruise. This message occurred during the last flight of the day; therefore, time was available for the repair. If the message appeared earlier in the day between flights, the repair would most likely have been deferred. To determine if the repair could be deferred the technician would have to consult the aircraft's minimum equipment list (MEL).

2. The troubleshooting process begins with the Fault Code Index. As seen in Figure 5-3-32, the fault code number 24 30 01 00 (as recorded in the flight log) is used to locate the CMCS message number (fault message number). In this case, the fault message number is 24099.

If the FRM code number was not reported in the flight log, the EICAS message pages of the maintenance manual could have been used to find the fault code number. The EICAS message page for this fault is shown in Figure 5-3-33.

3. At this time, the technician should log onto the CMC to find the CMC message number for the fault in question. The first place to look would be in Present Leg Faults. If there were no faults that corre-

lated to the one in question, move to the Existing Faults function. If that specific fault message number cannot be found in Existing Faults, access the Fault History function. When in Fault History, look at leg 00 or leg -01 for faults with a recent time of occurrence. When the correlation of the CMC message number and the fault message number (found in the Fault Code Index) is complete, move to the next step.

4. Use the CMCS Message Index to determine the corrective action (Figure 5-3-34). Since the message number 24099, EICAS message BAT CHARGER APU, the CMCS message, and the flight log entry all correlate; this is the correct location in the CMCS Message Index. In this example, the corrective action is a procedure that requires several tests to further identify the problem. Assume the following:

   a. The resistance of the APU battery charger connector was measured. The reference for this test is given as WDM 24-31-21.

   b. The measured value was 424 ohms; therefore, the instructions in part A apply. If after a one hour cooling period the second measurement was 424 ohms. The battery should be replaced.

5. Replacement of the battery would begin with reference to the maintenance manual (MM) 24-31-06, p. 401. This section of the MM includes general information, additional reference materials, locations, and procedures for complete removal and installation of the battery.

6. After completing the installation of the new battery, the system should be tested. Section 24-31-06 p. 501 of the MM should be referenced during testing (Figure 5-3-35).

7. If the APU operates correctly, any latched EICAS messages related to the fault should be erased. The proper mainte-

CMCS MESSAGE	POSSIBLE FLIGHT DECK EFECT
24099 APU BATTERY CHARGER FAIL (BCU-2)	BAT CHARGER APU (STATUS)

CORRECTIVE ACTION:

NOTE: This CMCS message will also show if the APU battery becomes too hot. If you have used the APU battery heavily, let at least 2 hours go by before you do more trouble-shooting. Alternatively, you can replace the APU battery, M7432, (AMM 24-31-06/401) and continue trouble shooting.

NOTE: The APU battery charger is disabled when the battery voltage drops below 4 volts. This is a protective function to prevent charger operation if a battery is not connected to the system. If the battery charger is faulty, and battery voltage has dropped below 4 volts, a new battery should be installed in addition to the battery charger.

 (1) Measure the resistance between pins 11 and 12 of the APU battery charger, M7431, connector DM7431 (WDM 24-31-21).

A. If the resistance is less than 575 ohms, do the steps that follow:
 (1) Let the battery cool for at least 1 more hour or replace the battery (AMM 24-31-06/401).
 (2) If you have let the battery cool and the resistance is still less than 575 ohms, replace the APU battery, M7432 (MM 24-31-06/401).

B. If the resistance is more than 574 ohms, do the steps that follow:
 (1) Measure the voltage at pins 4,7, 10 of the APU battery charger, M7431, connector DM7431 (WDM 24-31-21).
 (2) If the voltage on the pins is 115 volts ac, do the steps that follow:
 (3) Replace the APU battery charger, M7431 (MM 24-31-07/401) and APU battery, M7432 (AMM 24-31-06/401).
 (4) If the problem continues, examine the circuit between the APU battery charger, M7431, connector DM7431, pins 1, 3, 11, 12, and the APU battery, M7432, connector DM7432B, pins 1,3, 11, 12 (WDM 24-31-21). Repair the problems that you find.
 (5) If the problem continues, examine the circuit between the APU battery charger, M7431, connector DM7431, pin 9, and the No. 2 BCU, G11, connector DG11CA, pin B-H6 (WDN 24-34-22). Repair the problems that you find.

C. If the voltage on the pins is not 115 volts ac, do the steps that follow:
 (1) Make sure that the pins that follow are not shorted to electrical ground.
 (2) The APU battery charger disable relay, R7218, connector DR7218, pin X2 (WDM 24-31-21)
 (3) Electrical system control module, M7307, connector DM7307F, pin 35 (WDM 24-31-21).
 (4) APU battery, M7432, connector DM7432B, pin 8 (WDM 24-31-21).
  (a) If the APU battery, M7432, connector DM7432B, pin 8 is shorted to electrical ground, replace the APU battery, M7432 (MM 24-31-06/401 and WDM 24-31-21).
  (b) If the problem continues, examine the circuit between the APU battery charger, M7431, connector DM7431, pins 4, 7, 10 and the APU BATTERY CHGR circuit breaker, C805 (WDM 24-31-21). Repair the problems that you find.

EFFECTIVITY ———————

ALL

24-CMCS MESSAGE INDEX
01F.1   PAGE 33

Figure 5-3-34. CMCS message number 24099

Courtesy of the Boeing Commercial Airplane Company

MAINTENANCE MANUAL

APU BATTERY – ADJUSTMENT/TEST

1.   General
     A.   This procedure contains a task to do the operation test of the APU battery (M7432).
     B.   The APU battery is installed on the E-33 equipment rack with the APU battery charger. The E-33 equipment rack is on the left side of the aft passenger cabin. Access to the E-33 equipment is through the E-33 access door.

     TASK 24-31-06-705-001

2.   Operational Test of the APU Battery
     A.   References
          (1)   24-22-00/201, Manual Control
          (2)   45-24-00/201, MCS – Electrical Power
          (3)   IPC 24-31-06 Fig. 1
          (4)   WDM 24-31-21
          (5)   SSM 24-31-02
     B.   Access
          (1)   Location Zone
                271 Passenger Cabin, LH
     C.   Do a Test of the APU Battery

     S 865-002
     (1)   Supply electrical power (AMM 24-22-00/201)

     S 715-003
     (2)   Do a test of the APU battery:
          (a)   Get access to the maintenance page for the electrical system
                (AMM 45-24-00/201).
          (b)   Make sure the DC-Volts for the APU battery is more than 24 volts.
          (c)   Make sure the DC-Amps is in the positive CHG mode.
          (d)   Set the STANDBY POWER switch, on the P5 panel, to BAT.
          (e)   Make sure the main EICAS display operates correctly.
          (f)   Make sure the advisory EICAS message BAT DISCH APU shows on the main EICAS screen.
          (g)   Set the STANDBY POWER switch, on the P5 panel, to AUTO.

     S 865-004
     (3)   Remove electrical power, if it is not necessary
           (AMM 24-22-00/201.)

EFFECTIVITY
          ALL

24-31-06
PAGE 501

Figure 5-3-35. APU battery adjustment and test procedures   *Courtesy of the Boeing Commercial Airplane Company*

nance log entry should be made and the aircraft returned to service.

Troubleshooting the CMCS

The CMCS is an extremely reliable system seldom requiring maintenance. If the CMCS should fail, in most cases, the CMC can pinpoint the defect and the repair can be made swiftly and accurately. The CMCS runs a self-test program whenever the system begins operation. If a defect is found, the appropriate message will be displayed on the CDU scratch pad. If the CMC is operable, the fault will be recorded in the CMCS nonvolatile memory.

If the left CMC fails, the right CMC will take control of the system. In this case, when you select the CMC from the CDU menu, the message CMC-L FAIL will appear on the CDU scratch pad. Also, the amber fail light (MSG) on the CDU keyboard will illuminate to indicate there is a message in the scratch pad. In most cases, the aircraft can be dispatched with one inoperative CMC. The repair can then be made at the next maintenance opportunity.

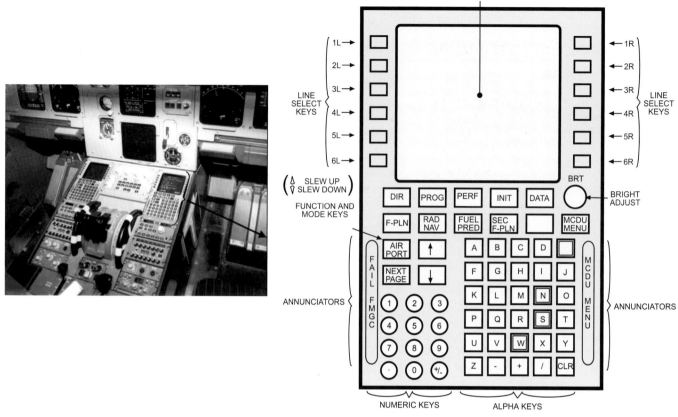

Figure 5-4-1. An A-320 multipurpose control and display unit (MCDU)

There are two types of failures that can be reported during the CMCS self-test: priority one messages and priority two messages. Priority one messages are the most important CMCS fault messages and have priority over priority two messages. Only one message can be displayed at a time in the CDU scratch pad. To cycle through all messages, press the clear (CLR) button on the CDU keyboard. There are a total of 11 priority one messages. Some of the priority one messages are: *CMC-L/R fail*, indicating a CMC internal failure; *CMC-L/R program pin fail*, indicating a program pin

error detected by the parity program pin; *SW part number disagree*, meaning there is different software installed in two CMCs. A part number error would only occur if one of the CMCs was recently changed or the software recently upgraded.

There are six priority two messages. For example, fault history disagree indicates that the fault history stored in the CMC memory differs between the left and right CMC. Both priority one and two fault messages are real-time and are displayed unless canceled. If the fault clears itself and then fails again, the message will reappear. If either a priority one or two is displayed on the CDU, the aircraft's maintenance manual can be used to determine the correct actions to repair the fault.

There are several EICAS messages used to show different types of CMC failures. If EICAS is active, the EIUs monitor the health of the CMCs. If a fault occurs, the message will be displayed using a standard fault message. A typical fault message might be *EIU-L-CMC-L BUS FAIL*, meaning there is a failure of the data bus between the left EIU and left CMC. The repair procedures would be outlined in the CMC message index. To isolate the fault, the technician should use standard troubleshooting procedures as outlined in this chapter.

FLIGHT DECK

2 MULTIPURPOSE MCDUs   **MULTIPURPOSE PRINTER**

Figure 5-4-2. Location of MCDU and multipurpose printer on an A-320

Section 4

Centralized Fault Display System

Airbus S.A.S. aircraft use the centralized fault display system (CFDS) for fault isolation and system analysis. The Airbus A-320 utilizes an advanced CFDS, which incorporates an integrated system used to access almost every computer on the aircraft. The A-320 CFDS will be presented in this portion of the chapter to describe a typical centralized fault display system employed by Airbus. Since this system is typical of a second generation CFDS the concepts discussed here can be applied to a variety of Airbus aircraft.

System Description

As the name implies, the centralized fault display system can access various BITE systems throughout the aircraft and display the information in one central location. Previous generation aircraft required that each individual system be tested independently. The CFDS utilizes a multipurpose control and display unit (MCDU) to input commands and monitor replies of the CFDS. The MCDU used on the A-320 is nearly identical to the control display unit (CDU) used on the B-747-400. Figure 5-4-1 shows the alphanumeric control panel and CRT of a typical MCDU.

The CFDS offers greater standardization, better data interpretation and simplification of technical documentation over previous individual BITE systems. The CFDS also provides a more in-depth analysis of a system failure. This helps to reduce maintenance time and improve aircraft reliability. Documentation of data is simplified through the use of an onboard printer, which can be used to print most of the data that is displayed on the MCDU. The MCDU and printer are located on the flight deck in the center pedestal area (Figure 5-4-2). The MCDU gets the term multifunction since it is used by maintenance technicians for accessing CFDS data as well as the flight crew to access the flight management system.

Centralized Fault Display Interface Unit

The Centralized Fault Display Interface Unit (CFDIU) is the main computer that manages the CFDS information. There is only one CFDIU on the A-320; however, a CFDIU backup channel is used to provide redundancy. As seen in Figure 5-4-3, the CFDIU is in direct communications with the aircraft systems, flight warning computers (FWC), MCDUs, and the airborne communication addressing and reporting system (ACARS) management unit. The ACARS system can be used to transmit information directly from the CFDIU to the airline ground facility. This capability allows the aircraft technician to begin troubleshooting the system even before the aircraft has landed.

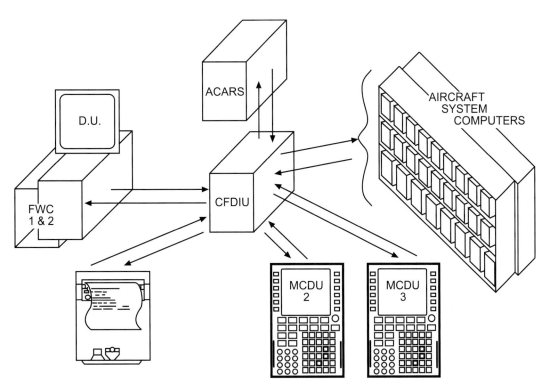

Figure 5-4-3. Pictorial diagram of CFDIU communications interface *Courtesy of Airbus S.A.S.*

TYPE OF DATA TRANSMITTED/ RECEIVED TO/FROM THE CFDIU	SYSTEMS TRANSMITTING SPECIFIED DATA TO THE CFDU	SYSTEMS RECEIVING SPECIFIED DATA FROM THE CFDIU
Flight number and city pair	Flight augmentation computer (FAC)	
Aircraft identification information and class 2 failures	Flight data interface units (FDIU)	
Flight phase and ECAM warning information	Flight warning computer (FWC)	
DMU class 2 failure information	Data management unit (DMU)	Flight warning compuer (FWC)
Engine serial number	Display management computer (DMC)	Engine vibration monitoring unit (EVMU)
Time and date information	Aircraft clock	
Flight phase and time/date information		All type 1 systems
Aircraft identification and city pair information		ACARS management unit (MU)
City pair		Data management unit (DMU)

Table 5-4-1. Various systems monitored by ECAM during normal operations

Virtually every system which creates an ECAM message is also fed to the CFDIU. The failure or exceedance that caused the ECAM message is stored in memory and can be accessed on the ground through the MCDU. The various systems monitored by the CFDS are in Table 5-4-1. It should be noted that not all LRUs within a system are directly connected to the CFDIU. In most cases, only one LRU reports BITE information for any given system.

System Architecture

The CFDS employs both discrete and digital data signals to communicate with the various aircraft LRUs. Figure 5-4-4 shows a generalized CFDIU interface diagram. On the A-320, 49 different systems report to the CFDIU with BITE information. It is important to remember that the CFDS is not a BITE system; the CFDS

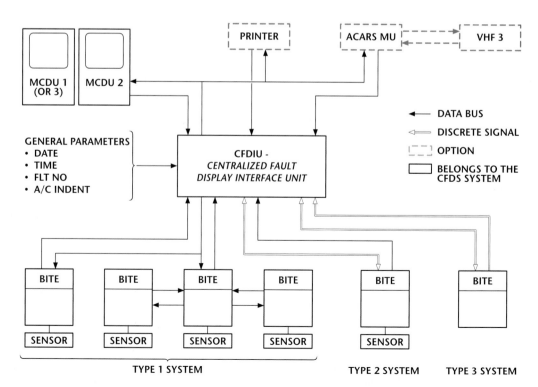

Figure 5-4-4. CFDIU interface diagram

is a manager of the BITE system related to the specific LRUs. The actual built-in test circuitry is contained in the computer software for the individual systems. Most LRU BITE software can store failure data for up to 64 previous flight legs. This capacity for storage can be helpful when troubleshooting reoccurring problems.

System Types

The CFDS architecture can be further divided according to the type of systems that report to the CFDIU. There are three system types, which encompass all LRUs monitored by CFDS: type 1, type 2, and type 3. The system type dictates the BITE memory and the interconnections between the CFDIU and the respective system LRU. Most of the LRUs monitored by the CFDS belong to type 1 systems (Figure 5-4-5). The type 1 systems transmit BITE data to the CFDIU through an ARINC 429 bus. The CFDIU also communicates to the type 1 LRUs using an ARINC 429 data bus. The data transmitted from the CFDIU is needed for BITE interrogation and to provide flight/ground information to the LRUs. Type 1 systems contain the most memory for storage of fault data, typically up to 10 faults per flight for a maximum of 64 flight legs.

Type 1 systems. In the centralized fault display system there are three categories of type 1 systems: single computer, multi-computer, and duplicated systems (Figure 5-4-5). A type 1, single computer system will always communicate directly to/from the CFDIU. Single computer systems are relatively simple LRUs, which perform a specific function. Examples of a type 1, single computer system are the VHF 1, 2, or 3 transceivers. Each of these units contains their own BITE that communicates directly to the CFDIU using a 429 data bus.

The multi-computer systems require more than one computer to perform their operations. These systems typically employ one designated computer to coordinate the BITE data and report to the CFDIU. More than one computer in the multi-computer system will most likely have BITE capabilities; however, only one computer will communicate directly to the CFDIU (Figure 5-4-5). Duplicated systems are LRUs that contain two different subsystems within one computer. For duplicate systems, each subsystem will communicate on a dedicated bus to/from the CFDIU.

Type 2 systems. Type 2 systems can only store fault data for the last leg; previous leg fault data is automatically erased. The BITE data for the type 2 LRUs is transmitted to the CFDIU through an ARINC 429 data bus. To initiate the BITE test, or retrieve the last failure, the CFDIU

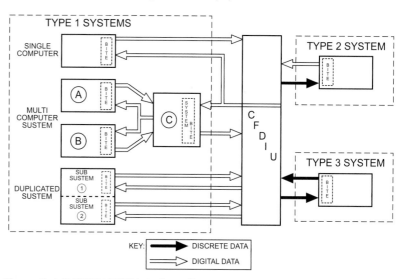

Figure 5-4-5 CFDIU/BITE interface diagram

Courtesy of Airbus S.A.S.

sends a discrete signal to the LRU. All memorized type 2 systems fault data is lost at the next engine start.

Type 3 systems. Type 3 systems have no capability of memorizing fault data. Therefore, type three data is considered real-time information. Type 3 systems are typically LRUs, which perform switching or simple monitoring tasks. For example, transformer rectifier (TR) units and ice detection sensors are type 3 devices. Type 3 LRUs receive a discrete signal from the CFDIU to initiate a test. Type 3 LRUs transmit information to the CFDIU using a discrete pass/fail signal.

Internal and External Failures

The CFDS has the capability to distinguish the difference between internal and external failures. An internal failure is defined as a failure that occurs to a component (LRU) that causes the system (or part of the system) to fail. For example, if the angle of attack sensor fails in the air data system, the air data system will report an internal failure (Figure 5-4-6). Other systems that use air data information will report an external failure. An external failure can be defined as a failure of a component outside of the failed system. In Figure 5-4-6, system A, B, and C, will detect a component failure outside of their system and report that the air data computer created an external failure in their system. In this example, the CFDIU would receive four fault messages; one internal and three external. All four of these faults were caused by the failed angle of attack sensor. A faulty component can cause a failure of its own systems (internal failure) or failure of a related system (external failure). Any system reporting an external failure will also report what system caused the problem. The system that caused an

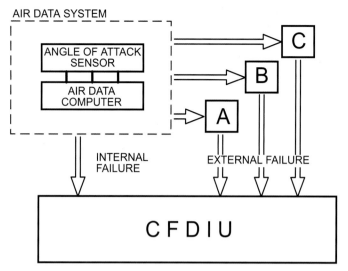

Figure 5-4-6. Example of internal and external failures reported to the CFDIU
Courtesy of Airbus S.A.S.

external fault would be listed as an identifier of the fault.

CFDIU Architecture

There are five main functions of the CFDIU: memory, management, correlation, monitoring, and detection. The memory section of the CFDIU is used to store ECAM warning and system failure information for later display by the CFDS. The management section of the computer is used to add information, such as time and date, flight phases, and flight legs to any failures or ECAM reports, which are stored in the CFDIU memory. The correlation function of the CFDIU is used to relate one system failure to the failures of other systems. For example, if the number 1 flight augmentation computer (FAC 1) failed, that system would not be able to send data to several other systems. The correlation software of the CFDIU will record the FAC 1 failure; but, it will ignore messages that state no data from FAC 1. The CFDIU will present only the initial failure for the last leg report and the IDENT function will present the systems affected by the failure.

The monitoring function of the CFDIU is used to detect a total or intermittent input bus failure. As the name implies, the monitoring software continuously monitors each input bus. The detection circuitry of the CFDIU is used to determine what type of failure occurred in the system. Different failure types include internal, external, intermittent, and class III.

CFDIU Interfaces

The CFDIU interfaces with several different aircraft systems (Figure 5-4-7). The FWC (flight

warning computer) is used to transmit flight phases and ECAM warnings to the CFDIU. The FAC transmits flight number and route to/from city pair to the CFDIU. This information is then sent to the management unit (MU) of the ACARS and the data management unit (DMU). The clock is used as the central time/date base for the CFDIU management activities. The flight data recorder interface unit (FDIU) transmits aircraft identification to the CFDIU, which relays the data to the FWC. The CFDIU receives the engine serial numbers from the display management computer (DMC) and transmits that information to the engine vibration monitoring unit (EVMU).

CFDIU Backup Functions

The backup section of the CFDIU provides redundancy for two basic functions: backup of the system CFDS report/test function and backup of the aircraft's clock. If the aircraft's clock fails while the CFDIU is powered, the CFDIU's backup clock software will provide time and date information to the CFDS and related systems. The backup channel of the CFDIU contains a completely independent power supply and software for backup in order to provide the report/test function in the event the CFDIU main channel fails.

CFDS Operations

To understand the CFDS operations, the system failure classification should be discussed. There are three failure classifications for the A-320 systems: Class 1, Class 2, and Class 3. In general, systems having the highest priority to flight safety produce a Class 1 fault when they fail; lowest priority failures produce Class 3 faults.

Class 1 failures will have the following effects:

1. Display a discrete caution or warning annunciator and a message on the ECAM display

2. Will have a direct consequence on the operation of the current flight

3. May have dispatch consequences for the next flight

4. Must be reported by the pilots

5. Will provide maintenance information available through the CFDS at the end of each flight leg

Class 2 failures have the following effects:

1. Display a message on the ECAM status page only

2. Will have no direct consequence on the operation of the current flight

3. Have no dispatch consequences

4. Must be reported by the pilots

5. Maintenance information will be available through the CFDS at the end of each flight leg

Class 3 failures will have the following effects:

1. Have no message or status page information displayed by ECAM

2. No operational consequence on the current flight

3. No dispatch consequence

4. Maintenance information is available through the CFDS

5. Correction of these faults can be left until the next regularly scheduled maintenance opportunity

CFDS Operating Modes

There are two modes in which the CFDS will display fault reports: in flight and on ground. The in flight mode of the CFDS allows the flight crew to access current leg reports and current leg ECAM reports.

Using the on ground mode of the CFDS, the following six reports can be assessed:

1. Last leg

2. Last leg ECAM

3. Previous legs

4. Avionics status

5. System report/tests

6. Post flight

For both *in-flight* and *on-ground* modes the reports are activated using the MCDU (Figure 5-4-8).

MCDU Operation

The MCDU operation is relatively simple and typically requires operation of only a few select keys for CFDS access. Figure 5-4-9 shows the MCDU and highlights some of the CFDS controls. There are two MCDUs installed on the A-320; however, only one MCDU can be used to access CFDS information at any given time. The opposite side MCDU can be used at that time for functions other than accessing the CFDS. The MCDUs turns on automatically when aircraft electrical power is available. The first step to select any report on the MCDU is to press the MCDU MENU button.

The current leg and last leg reports are virtually identical. The difference is a function of the time at which the report is taken. The current leg report is available in the flight mode only, the current leg report data is moved to the last leg memory after landing. The same scenario just described is also true for the current leg ECAM and last leg ECAM reports.

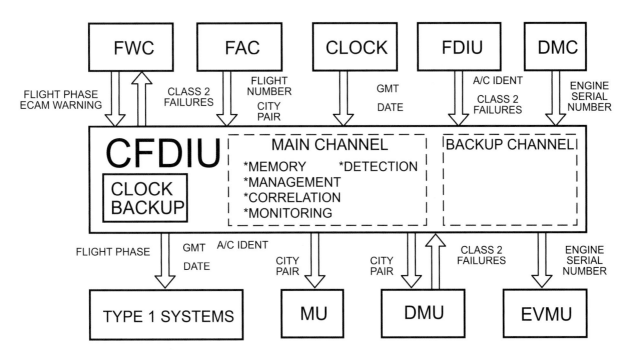

Figure 5-4-7. Block diagram of CFDIU interface with other aircraft systems

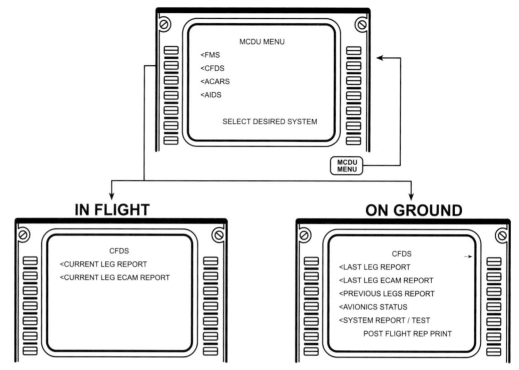

Figure 5-4-8. Example of access to in-flight and on-ground reports using the MCDU

Courtesy of Airbus S.A.S.

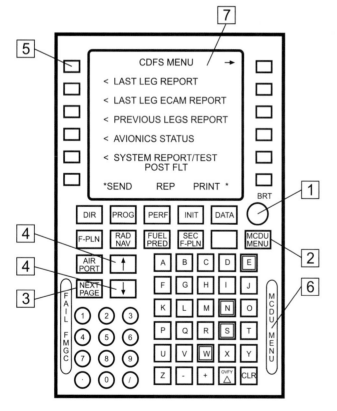

Figure 5-4-9. MCDU control panel: (1) Brightness knob, (2) MCDU key, used to begin access of the MCDU functions, (3) NEXT PAGE key gives access to various pages of the display sequencing, (4) Scroll up/down keys,used to access data, requires more than one page for the entire display, (5) Twelve LINE SELECT keys provide access to the MCDU function adjacent to the selected key, (6) MCDU MENU indicator illuminates when a system connected to the MCDU requests information to be displayed, (7) CRT display contains a maximum of 14 lines each having 24 characters

Courtesy of Airbus S.A.S.

Last (and Current) Leg Report

The last (and current) leg reports can store up to 40 lines of fault information sent to the CFDIU from the individual BITE systems. These reports list only Class I and II failures. For each fault, the time at which the fault occurred is listed along with the associated ATA chapter for quick maintenance manual reference (Figure 5-4-10). It should be noted that the time could be displayed under UTC (Universal Time Constant) or GMT (Greenwich Mean Time). The last leg reports can also display the fault identifiers that incurred a related external failure (Figure 5-4-10). If the report consists of more than one page an arrow is displayed in the upper right corner of the first page. Pressing the MCDU next page key will bring consecutive report pages to the display.

To access the current/last leg reports, press the appropriate line select key on the CFDS menu page (Figure 5-4-9). The title of the report (LAST LEG REPORT) and the date will appear at the top of the display. Figure 5-4-11 shows the sequence of displays accessible through the last (and current) leg reports. The report shows the LRU failure and its associated functional identification number (FIN) for each fault. The report can be printed if the print option is displayed in the lower left corner of the display. The < symbol displayed next to a fault indicates that source identifier information is available for that fault.

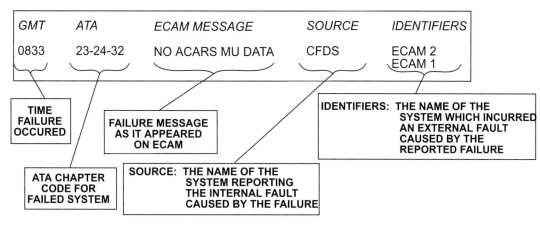

Figure 5-4-10. CFDS last leg report format

Last (and Current) Leg ECAM Report

The last (and current) leg ECAM report is used to access the actual ECAM messages that have been displayed to the flight crew. This data is sent to the CFDIU from the flight warning computers for storage of the data. Up to 40 lines can be stored for each report in one or more pages. Figure 5-4-12 shows the typical page displays for the last (and current) leg ECAM report. The last (and current) leg ECAM report also displays the time that the ECAM message was initiated, the flight phase number (PN) at the time of the ECAM message, and the ATA chapter and section. Pressing the line select key labeled print * will print all pages of the ECAM report.

Previous Legs Report

At the beginning of each new flight (start of the first engine) the last leg report currently in memory is transferred to the previous legs report memory. The previous legs report is therefore comprised of a series of past last leg reports. A maximum of 63 flight legs and a total of 200 failures can be accessed through the previous legs report. The report contains the same information as the last leg report along with a two-digit number used to show the flight leg when the fault was recorded (Figure 5-4-13). The designation (INTM) means the fault has occurred intermittently during that flight leg. The previous flight report is only available on the ground and the print feature will only

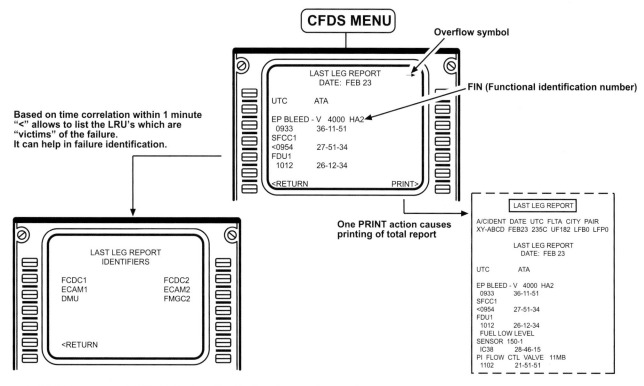

Figure 5-4-11. Sequence of MCDU displays for the Last Leg or Current Leg reports

Courtesy of Airbus S.A.S.

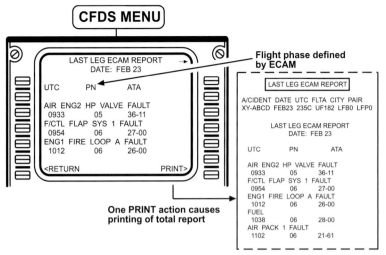

Figure 5-4-12. Sequence of MCDU display for the last leg or current leg ECAM reports *Courtesy of Airbus S.A.S.*

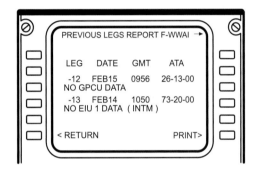

Figure 5-4-13. Example of a Previous Legs Report *Courtesy of Airbus S.A.S.*

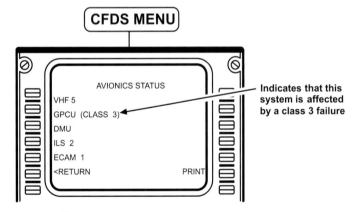

Figure 5-4-14. Typical avionics status page *Courtesy of Airbus S.A.S.*

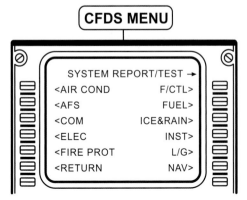

Figure 5-4-15. Example of system report/test menu *Courtesy of Airbus S.A.S.*

print the page currently on the MCDU display. This printing arrangement is necessary since the previous legs report could be several pages long.

When analyzing the previous legs report, leg 01 is actually two flight legs old. After engine shut down, the most recent flight (current leg) becomes the last leg, the flight before that is the previous leg 01. Each leg moves back in the memory sequence at the start of the first engine prior to a flight. If you are required to start an engine for troubleshooting purposes be sure to keep track of where the flight leg data is currently being stored.

Avionics Status

The CFDS avionics status data presents a list of systems that are affected by a current failure. This list can be used during troubleshooting to help determine the departure status of the aircraft or to determine the aircraft's current auto flight capabilities. The avionics status information is real time and is therefore continuously updated. During an avionics status inquiry many systems must be powered, and in some cases set to a normal operating condition, or they will be displayed as failed. As seen in Figure 5-4-14, if the system is affected by a Class 3 failure it will be identified in the avionics status data. If the indication Class 3 appears that system may be affected by a Class 2 or Class 1 failure also.

System Report/Test

The CFDS system report/test function is used for an in-depth look at the faults that occurred in a particular system. System report tests are only available during ground operations. The system report/test will allow for access of fault and troubleshooting data for any system monitored by the CFDIU. As seen in Figure 5-4-15, the main menu for the system report/test identifies the monitored systems listed in order of ATA chapter. The actual information available in the system report/test mode is a function of the type of system being monitored. Type 1 systems will provide the most in-depth analysis; type 3 systems offer the least information. The data presented during system report/tests is retrieved directly from the individual system's BITE, not from the CFDIU memory. These reports/tests are therefore interactive between the CFDIU and the respective LRUs.

Type 1 system reports. Type 1 systems are connected to the CFDIU via an ARINC 429 data bus, and provide a menu specific to the selected system. Figure 5-4-16 shows the sequence of

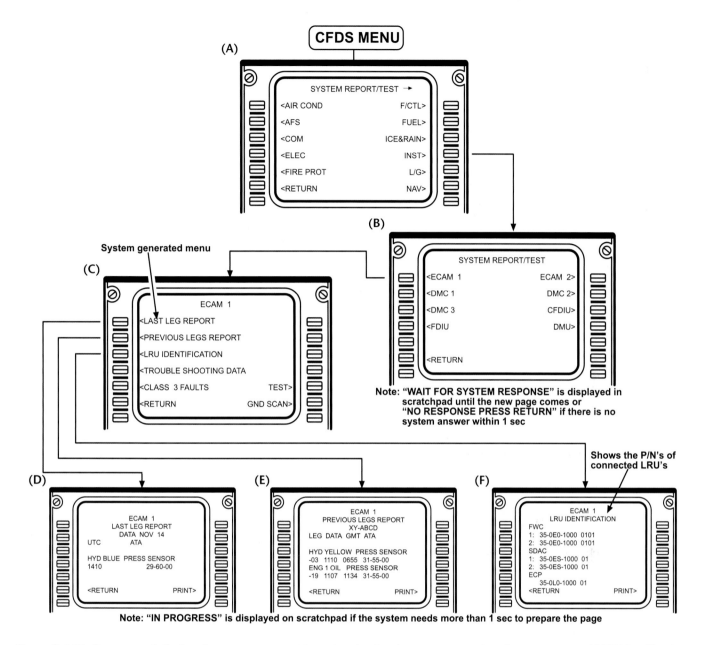

Figure 5-4-16. Sequence of displays for a system report/test showing last leg reports, previous legs reports, and LRU identification functions

Courtesy of Airbus S.A.S.

displays for a typical type 1 system. The displays are described as follows:

1. The system report/test menu is retrieved from the main CFDS menu. From this display or following pages, a given ATA chapter is selected. In this case, the technician presses the instrument (INST) key.

2. The CFDIU displays a list of the systems contained under this selection. The technician selects ECAM 1 for interrogation.

3. The CFDIU interrogates the ECAM 1 BITE, which sends a list of available reports back to the CFDIU. This display

shows the type of reports and tests available for ECAM 1.

4. The last leg report shown here is similar to the last leg report previously discussed; however, in this case, the report only contains data on the ECAM 1 system.

5. This previous legs report contains only the information that deals with the ECAM type 1 system.

6. The LRU identification page is used to display the specific part numbers of all electronic LRUs connected to the ECAM 1 system.

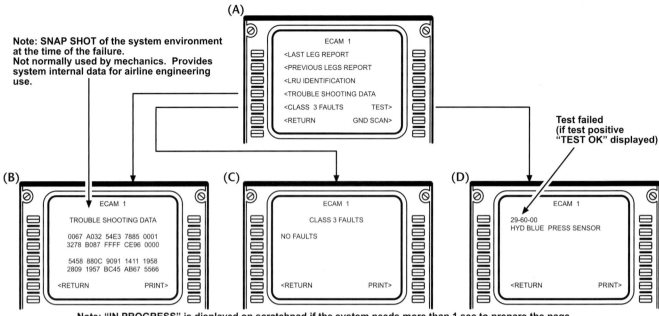

Figure 5-4-17. Sequence of displays for a system report/test (type 1 systems) showing troubleshooting data, class 3 faults, and test functions

Courtesy of Airbus S.A.S.

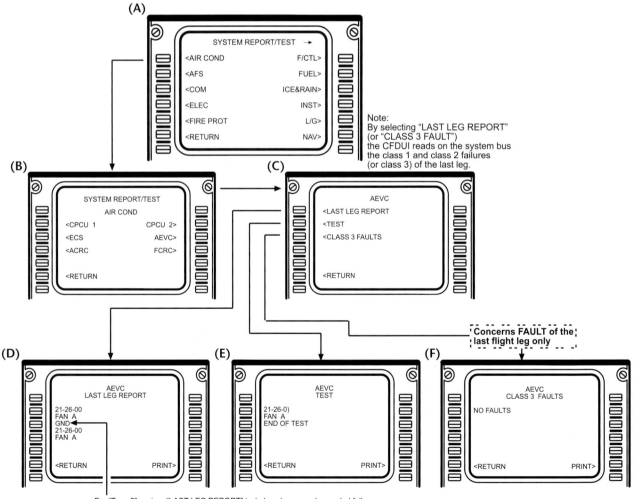

Figure 5-4-18. Sequence of displays for a system report/test (type 2 systems)

Courtesy of Airbus S.A.S.

The remainder of the potential ECAM 1 report/tests is shown in Figure 5-4-17. The displays are described as follows:

1. This is the ECAM 1 system report/test menu. The technician will make a selection from this display in accordance with his/her current requirements.

2. This page, troubleshooting data, provides access to the digital data transmitted between the various LRUs in the system. The data is displayed in a hexadecimal format and can provide an in-depth look at the system operations. Typically this data is only accessed if previous troubleshooting methods fail. The hexadecimal codes are typically sent to the airline engineering department to be analyzed. Engineering then relays information on a potential repair back to the technician. Troubleshooting data will be discussed in more detail later in this chapter.

3. Any Class 3 faults are listed in this display. Remember, Class 3 faults are not critical repair items.

4. Pressing the TEST line select key will cause the ECAM 1 system BITE to perform a real-time system test. The results of that test will be presented on this display. The associated ATA chapter will be displayed adjacent to the failed component.

Type 2 system reports. Type 2 systems are monitored by the CFDIU and allow a maximum of three options in the system report/test mode; last leg report, test, and Class 3 faults. The sequence of displays for a type 2 system is shown in Figure 5-4-18. Using the air conditioning system as an example, the displays are described as follows:

1. From the air conditioning menu, a second subsystem selection is made. In this case, the avionics equipment ventilation computer (AEVC) was selected.

2. The options for the AEVC are displayed here. The last leg report is displayed as an option for all type 2 systems. A list of Class 3 faults and a real-time test are also possible options on some type 2 systems.

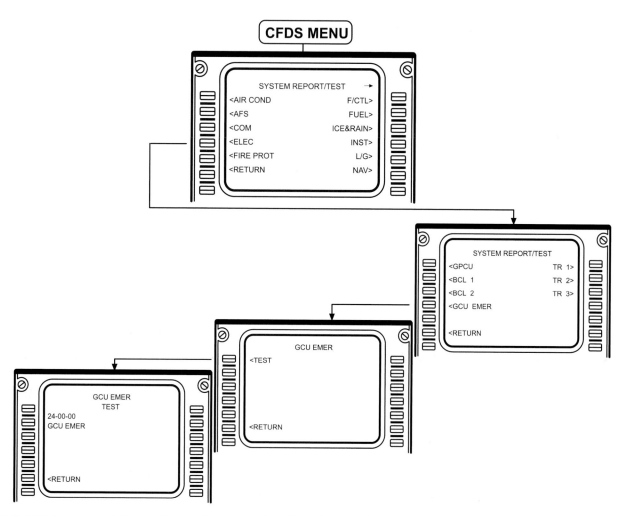

Figure 5-4-19. Sequence of displays for a system report/test (type 3 systems)

Courtesy of Airbus S.A.S.

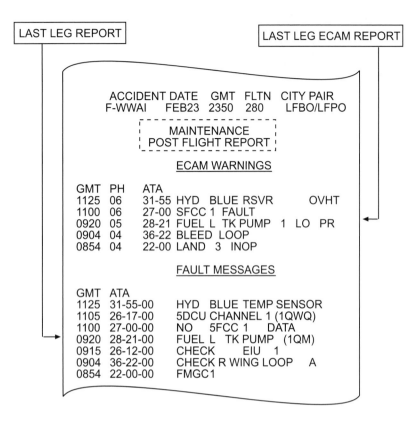

LAST LEG REPORT

LAST LEG ECAM REPORT

```
ACCIDENT DATE   GMT   FLTN   CITY PAIR
F-WWAI    FEB23   2350   280    LFBO/LFPO

          MAINTENANCE
        POST FLIGHT REPORT

          ECAM WARNINGS

GMT   PH   ATA
1125  06   31-55  HYD  BLUE RSVR        OVHT
1100  06   27-00  SFCC 1 FAULT
0920  05   28-21  FUEL L TK PUMP  1  LO PR
0904  04   36-22  BLEED  LOOP
0854  04   22-00  LAND  3  INOP

          FAULT MESSAGES

GMT   ATA
1125  31-55-00   HYD  BLUE TEMP SENSOR
1105  26-17-00   5DCU CHANNEL 1 (1QWQ)
1100  27-00-00   NO  5FCC 1   DATA
0920  28-21-00   FUEL L TK PUMP  (1QM)
0915  26-12-00   CHECK      EIU  1
0904  36-22-00   CHECK R WING LOOP   A
0854  22-00-00   FMGC1
```

Figure 5-4-20. Typical post flight report   *Courtesy of Airbus S.A.S.*

3. The last leg report is similar to the last leg reports previously described; however, ground faults are also included in this display. If a ground fault is recorded, the term *GND* will be displayed adjacent to the fault.

4. The test feature will run a real-time test for the displayed system. The test results will be shown at the end of the test.

5. Any Class 3 faults that occurred during the last flight leg will be displayed.

Type 3 system reports. Type 3 systems have only one function or test and therefore have no associated menu. An example of a type 3 system display sequence is shown in Figure 5-4-19.

Post flight reports. The CFDS post flight report is a combination of the last leg and last leg ECAM reports available from the CMCS printer. The post flight report is a handy means of accessing fault data from the last flight leg. This data must be accessed through the CFDS menu page and is available in print form only. An example of a typical post flight report is shown in Figure 5-4-20. Be sure to print any desired post flight reports prior to the first engine start for the next flight. At that time the last leg and last leg ECAM reports move to the previous flights memory and are not available in the post flight report.

Clock Initialization

In the event the aircraft's central clock fails, appropriate repairs must be made and the clock time/data must be reset. The second page of the CFDS menu allows access to the clock software. The clock option of the menu will be displayed only in the event of a clock failure and a CFDIU power interruption. In that case, the MCDU alphanumerical keyboard can be used to enter in the new time and date.

CFDIU Backup Mode

As discussed earlier, the CFDIU has a backup channel in the event of a primary system failure. The backup mode can be accessed during a partial CFDIU failure using the second page of the CFDS menu. The backup mode will be the only system operational in the event of a serious primary system failure. As seen in Figure 5-4-21, the backup mode allows for access of the system report/test function only. The backup mode is operational on the ground, not in flight.

ACARS/Print Program

Most modern Airbus aircraft are equipped with ACARS and transmit CFDS information directly from the CFDIU to the ground. The ACARS/print program is used to determine what CFDS data will be sent via ACARS and what data will be printed onboard the aircraft. Dependent on the specific airline configuration, only certain data can be selected for ACARS transmission or onboard printing. ACARS can be an extremely valuable troubleshooting tool when linked to the aircraft's centralized fault display system. This combination can provide automatic transmission of system fault information to the airline ground facility. This allows the ground technician to prepare for aircraft repair even prior to the aircraft landing. This can greatly reduce aircraft delays due to maintenance and enhance airline productivity.Many airlines limit the use of ACARS for transmission of aircraft systems fault data. Limits are imposed for maintenance information since ACARS frequencies are already overcrowded. At the time the A-320 was introduced, it was anticipated that transmission of fault data via ACARS will become commonplace. With third generation fault detection systems, ACARS as well as satellite and wireless communications can be used for the transmission of maintenance data. These systems will be discussed later.

Figure 5-4-22 shows a typical display for the ACARS/print program. In the SEND column on the left side of the display, the selection of

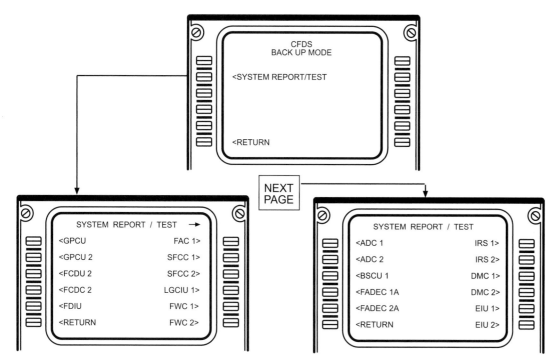

Figure 5-4-21. CFDS backup mode allows access only to system report/test function

Courtesy of Airbus S.A.S.

yes or no can be made to determine what information will be sent by ACARS. In the PRINT column (right side) the selection is made for the information to be printed. Pressing the corresponding line select key will change the label from yes to no and vice versa. If the label (yes or no) is displayed in blue, it could be changed manually. Labels displayed in green cannot be changed.

Level 1 Troubleshooting

There are three levels of troubleshooting that can take place using the central fault display system. On the A-320, level 1 troubleshooting is performed at the gate during a quick turn around and is relatively limited in scope. Typically the technician will be under very tight time constraints to finish any necessary repairs and therefore will try to defer most maintenance. Level 1 troubleshooting will always begin with a flight crew log entry concerning a failed component or system. In many cases, the post flight report will allow the technician to access the correct repair using the aircraft maintenance manuals. An example of a CFDS message which would result in level 1 troubleshooting is *FAC 1 FAIL*. In this case the technician would quickly determine that the FAC (flight augmentation computer) number 1 should be replaced. If the repair cannot be made quickly and the problem cannot be deferred, the technician will study the problem further and make the repair prior to dispatching the

aircraft. This will require the use of level 2 or 3 troubleshooting techniques.

Level 2 Troubleshooting

The A-320 level 2 troubleshooting employs techniques of moderate complexity and depth that often require several hours or longer to correct the fault. An example of a CFDS message requiring a level 2 troubleshooting procedure is check pins FAC 1. In this case, the technician would be required to check the pin programming of the number 1 FAC (flight augmentation computer). This type of troubleshooting is always deferred if possible. If deferment is not possible because of other system failures, weather, or the system is simply too critical to fly in an inoperative condition, the aircraft must be repaired before takeoff.

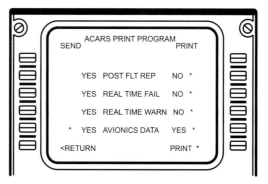

Figure 5-4-22. Typical ACRS print program display

Courtesy of Airbus S.A.S.

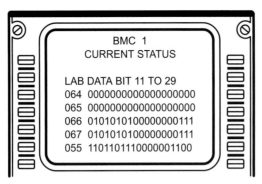

```
            BMC  1
        CURRENT STATUS

    LAB DATA BIT 11 TO 29
    064  0000000000000000000
    065  0000000000000000000
    066  0101010100000000111
    067  0101010100000000111
    055  1101101110000001100
```

Figure 5-4-23. Typical CFDS operational test display, Data bit at the far right is bit #11 and data bits at the far left is bit #29

Level 3 Troubleshooting

The A-320 level 3 troubleshooting is an in-depth study of the failed system, typically involving access of the binary or hexadecimal fault code data. As previously discussed, fault code data is found under troubleshooting data accessed through the system report/test of the CFDS. Fault code data is also available through some operational tests and, in some cases, displayed in last leg reports. Level 3 troubleshooting is typically performed only when standard level 1 and 2 troubleshooting fails to find the defect. In most cases, the information needed

Status Type	0	1		
P+	Loss of voltage	Applied voltage		
P-	Applied ground	Loss of ground		
LABEL = 066				
BIT N°	PARAMETER DEFINITION	STATUS 0	STATUS 1	COMMENTS
11	Eng2 OPV position	Fully open	Not fully open	
12	Eng2 FAV position	Fully open	Not fully open	
13	Eng2 FAV position	Fully closed	Not fully closed	
14	Eng2 HPV position	Fully open	Not fully open	
15	Eng2 HPV position	Fully closed	Not fully closed	
16	Eng2 PRV position	Fully open	Not fully open	
17	Eng2 PRV position	Fully closed	Not fully closed	
18	Eng2 Bleed Push Button position	Normal	Pushed	
19	Eng 2 Fire Push Button postion	Normal	Pushed	
20	PRV2 Low regulation	No	Yes	
21	Starter Valve Eng2	Closed	Not closed	
22	Valve closure (for engine start)	No	Yes	
23	CFM validity	No	Yes	
24	IAE Validity	No	Yes	
25	Eng2 HPV Solenoid	Not energized	Energized	Used by IAE only
26	Eng2 PRV Closure Control	No control	Closure control	Output state
27	Eng 2 Precooler Inlet Pressure status	12PSIG<P<COPSIG	COPSIG<P<12PSIG	
28	ENG2 Bleed Fault	Fault	No fault	Self maintained signal except switch upon off
29	Eng2 PRV PWR Supply	No power	Power	

Table 5-4-2. Decoding information for label 066

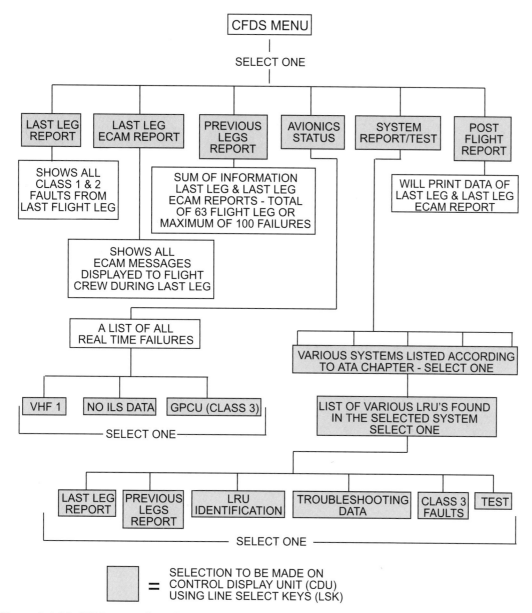

Figure 5-4-24. CFDS access flow chart

to analyze the fault code data is only available through the airline's engineering department. The specific fault codes can be printed using the CFDS printer, hand written, or downloaded to a floppy disc using the aircraft's data loader.

Figure 5-4-23 shows a CFDS page as displayed during an operational test of the engine bleed air system. In this example, the bleed-air monitor computer (BMC) current status fault code data is displayed. This information is displayed in binary language. Some codes are presented in hexadecimal numbers. The label (LAB) number 064, 065, 066, 067, or 055 bits and data bits (11-29) can be analyzed using the appropriate reference. Table 5-4-2 shows the decoding reference information for label 066. Comparing the binary bits to the appropriate columns of the reference information will provide valuable troubleshooting information.

In most cases, all three levels of troubleshooting begin with a logbook report from the flight crew concerning a given fault. That information may lead to the correct troubleshooting manual or may require further interrogation of the CFDS. The flow chart in Figure 5-4-24 shows the various reports and system tests available through the CFDS. In this chart it is easy to see the depth of information available through the CFDS for troubleshooting purposes.

A Typical Troubleshooting Sequence

The troubleshooting sequence can involve many different steps that all lead to the common goal of repairing the aircraft. If the fault occurs during normal operations of the aircraft, the flight crew will initiate a logbook

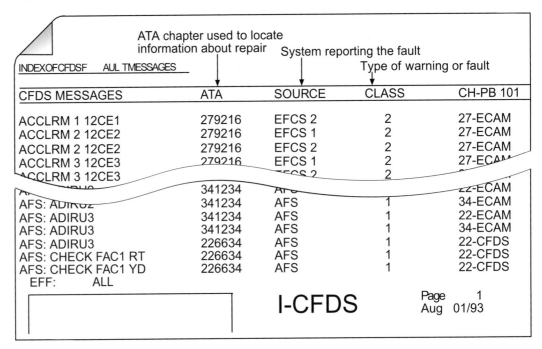

CFDS MESSAGES	ATA	SOURCE	CLASS	CH-PB 101
ACCLRM 1 12CE1	279216	EFCS 2	2	27-ECAM
ACCLRM 2 12CE2	279216	EFCS 1	2	27-ECAM
ACCLRM 2 12CE2	279216	EFCS 2	2	27-ECAM
ACCLRM 3 12CE3	279216	EFCS 1	2	27-ECAM
ACCLRM 3 12CE3		EFCS 2	2	
AFS: ADIRU2	341234	AFS	1	22-ECAM
AFS: ADIRU2	341234	AFS	1	34-ECAM
AFS: ADIRU3	341234	AFS	1	22-ECAM
AFS: ADIRU3	341234	AFS	1	34-ECAM
AFS: ADIRU3	226634	AFS	1	22-CFDS
AFS: CHECK FAC1 RT	226634	AFS	1	22-CFDS
AFS: CHECK FAC1 YD	226634	AFS	1	22-CFDS

Figure 5-4-25. An example of the troubleshooting manual index of CFDS fault messages

entry to inform ground crews of the need for maintenance. The fault information may have been transmitted, either verbally or using ACARS, during flight to the airline ground facility or by a post-flight report printed by the flight crew and handed to maintenance personnel. In either case, troubleshooting begins with referencing the aircraft's troubleshooting manuals. For the A-320, the troubleshooting manual has an index of warnings and malfunctions which can be directly referenced to the pilot's malfunction logbook

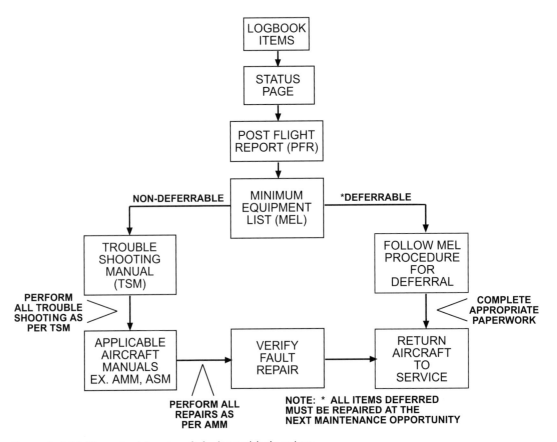

Figure 5-4-26. Flow chart for aircraft fault troubleshooting

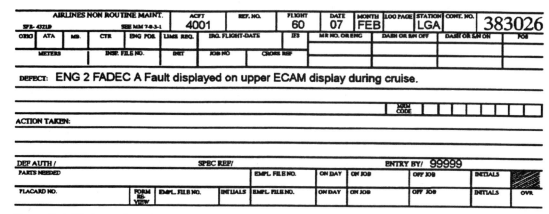

AIRLINES NON ROUTINE MAINT. SFR- 4321P SEE MM 7-8-3-1	ACFT 4001	REF. NO.	FLIGHT 60	DATE 07	MONTH FEB	LOG PAGE	STATION LGA	CONT. NO. 383026

OBO	ATA	MB.	CTR	ENG POS.	LIMS REQ.	ENG. FLIGHT-DATE	IFB	MR NO. OR ENG	DASH OR SN OFF	DASH OR SN ON	POS

METERS	INSP. FILE NO.	INIT	JOB NO	CROSS REF			

DEFECT: **ENG 2 FADEC A Fault displayed on upper ECAM display during cruise.**

MM CODE | | | | | | |

ACTION TAKEN:

DEF AUTH /	SPEC REF/			ENTRY BY/ 99999			
PARTS NEEDED		EMPL. FILE NO.	ON DAY	ON JOB	OFF JOB	INITIALS	
PLACARD NO.	FORM RE-VIEW	EMPL. FILE NO.	INITIALS	EMPL. FILE NO.	ON DAY	ON JOB	OFF JOB	INITIALS	OVR.

Figure 5-4-27. Examples of a typical logbook entry

entry. The index of warnings and malfunctions is divided into four categories: ECAM (ECAM warnings), EFIS (EFIS fault flags), LOCAL (local warnings), and OBSV (crew observations). The left hand column of each index is a list of the message, flight deck effect, or observation related to the fault. The right hand column shows ATA chapter and section of the troubleshooting manual (TSM) to be referenced next. All TSM references are to page block 101 (PB101).

The TSM also contains an index of CFDS fault messages. If the exact CFDS message were known this index would be referenced (Figure 5-4-25). The CFDS message index provides the most information to aid the repair. This index should be accessed if the CFDS message is available. Any TSM index will lead the technician to the proper procedures to effectively troubleshoot and repair the system.

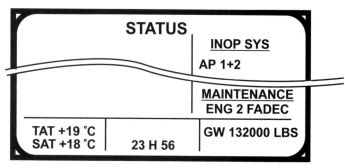

Figure 5-4-28 Confirmation of a logbook entry using the CFDS status page

In many cases, troubleshooting/repair of the system is deferred until the next maintenance opportunity. Deferring fault repair can be done if approved by the aircraft's Minimum Equipment List (MEL). The following paragraphs provide an example of the typical trou-

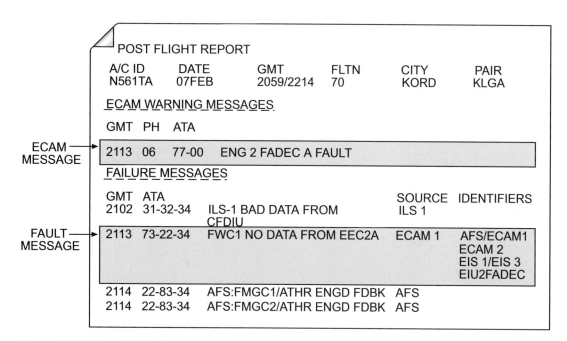

Figure 5-4-29. Identify CFDS message using Post Flight Report.

MINIMUM EQUIPMENT LIST

Fault(s) indicated by FADEC Maintenance Status

ECAM message
  ENG 2 FADEC FAULT
Fault message
  FWC1 : NO DATA FROM EEC2 A

No.	DEFERRED POS.	PLCD#
7171B	ENG NO. 1.....................	1485
7171A	ENG NO. 1 AND NO. 2...	1484
7171F	ENG NO. 2.....................	1557

SPEC NOTES:
A. These maintenance status messages are displayed on lower ECAM.
B. CFDS interrogation not required.

MAINT:
A. Install DEFERRED. placard adjacent to lower ECAM.

OPS PLACARD: "___ FADEC MAINT STS MESSAGE DISPLAYED ON ECAM DOES NOT AFFECT DISPATCH."

Figure 5-4-30. Example of a minimum equipment list entry

bleshooting steps that follow the flow chart of Figure 5-4-26.

1. Review the logbook entry received from the flight crew. (Figure 5-4-27). In this example, the entry states that the message ENG 2 FADEC was displayed by ECAM during flight.

2. Verify that the fault is still active by accessing the status page on the MCDU. If the fault no longer exists, it might be inter-

mittent or related to a given flight phase. In this example, we will assume the fault is confirmed (Figure 5-4-28).

3. Identify any CFDS message related to the fault by reviewing the post flight report (Figure 5-4-29). The appropriate fault message is found and a related ECAM message is generated.

NOTE: *Repairing the FADEC may take several hours. In most cases, this would be too long for a gate turn around. Therefore,*

Fault Related ECAM Message	ATA Chapter & Section	Troubleshooting Manual Reference
ECAM	PFR ATA CH/SE	TSM ATA CH-PB 101
EDW WARNING(S)		
AUTO FLT A/THR OFF	22-00	22-ECAM
AUTO FLT A/THR OFF		77-ECAM
...AB OFF	22-00	
AUTO...	22-00	22-ECAM
ENG 1 FADEC FAULT	77-00	77-ECAM
ENG 2 BEARING 4 OIL SYS – HI PRESS	77-00	77-ECAM
ENG 2 BEARING 4 OIL SYS – SCAVENGE		
VALVE FAULT	77-00	77-ECAM
ENG 2 EIU FAULT	77-00	77-ECAM
ENG 2 FADEC A FAULT	77-00	22-ECAM
ENG 2 FADEC A FAULT	77-00	77-ECAM
ENG 2 FADEC B FAULT	77-00	22-ECAM
ENG 2 FADEC B FAULT	77-00	77-ECAM
ENG 2 FADEC FAULT7	77-00	77-ECAM

EFF :   ALL

ECAM MESSAGE FROM POST FLIGHT REPORT

I-ECAM

Figure 5-4-31. Example of a troubleshooting manual index (I-ECAM)

this type of repair would typically take place during an overnight stay or during the next scheduled maintenance check if authorized by the MEL.

4. Compare fault message against MEL (Figure 5-4-30). In this example, the MEL states that the repair can be deferred if a placard is installed adjacent to the lower ECAM display. The required placard states FADEC maintenance status message displayed on ECAM does not affect dispatch.

5. Defer maintenance and return aircraft to service after the appropriate paperwork had been completed per MEL.

When time permits, the FADEC repair would be accomplished in the following sequence:

1. Reference the aircraft's TSM index as in Figure 5-4-31. In this example, the ECAM message from the post flight report was used along with the ECAM index of the TSM. The index states that Chapter 77-ECAM of the TSM should be used to find the potential fault. The warnings/malfunctions, CFDS fault messages, and fault isolation procedures references are listed in the TSM (Figure 5-4-32).

2. The TSM task number 73-20-00-810-811, page 211 is now referenced (Figure 5-4-33). This portion of the TSM offers specific details on fault confirmation (paragraph 3) and fault isolation (paragraph 4). Once the fault has been repaired according to the aircraft maintenance manual (AMM) or aircraft schematics manual (ASM), the fault confirmation procedures should be repeated to verify that the repair was successful.

False and Related Failure Messages

ECAM and the CFDS report all system malfunctions regardless of the cause. For example, during power transfer from ground power to engine-driven generator power the electrical system may go out of limits for an instant, which could trigger several CFDS messages. Many of these false fault messages may be totally unrelated to the electrical system; however, the power transfer caused a momentary power glitch and many systems incurred an instantaneous fault. That fault was then recorded by the CFDS. In this case, the problem is not aircraft related, but the CFDIU identified the fault and memorized the data anyway. The last leg, last leg ECAM, and post flight reports would all show these false fault messages.

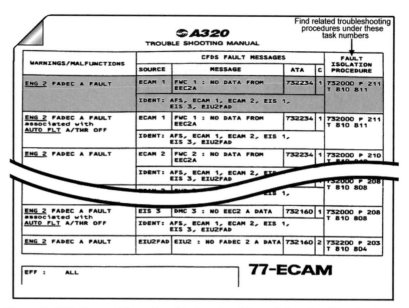

Figure 5-4-32. Excerpt from Troubleshooting Manual showing Warning/Malfunction CFDS Fault Messages, and Fault Isolation Procedures

Courtesy of Airbus S.A.S.

To verify these faults as nuisance messages, check the time at which the fault occurred on the CFDS report. If the faults all happened at the same time, and if that time coincides with an engine start, the problem is most likely related to a power transfer. It can also be verified to see if the faults still exist. To do this, look at any real-time display, such as ECAM.

To help eliminate the problem with false fault messages, troubleshooting procedures should be instigated by a pilot report of the problem. The pilots will not report nuisance messages that occur during power transfer or similar situations. A similar fault message can also be misleading if the time of occurrence is not considered. For example, if two CFDS messages occur at the same time it is likely they are caused by the same LRU failure.

Section 5

Third Generation Integrated Test Equipment Systems

During the past decade, advancements to aircraft electronic systems commonly known as Integrated Modularized Avionics (IMA) have created significant improvements in fault monitoring and recording systems on modern aircraft. As stated in previous chapters, IMA is a design concept which allows for advanced software technologies and improved data bus communications between aircraft systems.

A320
TROUBLE SHOOTING MANUAL

TASK 73-20-00-810-811

Loss of the Output 1 Bus on SEC 2 Channel A

1. **Possible Causes**

- EEC (4000KS)
- wiring of EEC A OUTPUT 1 signal from the EEC 2 (4000KS) to the first terminal block

2. **Job Set-up Information**

A. **Referenced Information**

REFERENCE	DESIGNATION

AMM 31-50-00-710-001   **Operational Test of the Central Warning Systems**
AMM 73-22-34-000-010   **Removal of the Electronic Engine Control (EEC) (4000KS)**
AMM 73-22-34-400-010   **Installation of the Electronic Engine Control (4000KS)**
ASM 73-25/10

3. **Fault Confirmation**

A. **Test**

1) Do the operational test of the central warning systems (FWC) (Ref. AMM TASK 31-50-00-710-001).

4. **Fault Isolation**

A. If the test gives the maintenance message FWC1 : NO DATA FROM WEC 2A:
- replace the EEC (4000KS), (Ref. AMM TASK 73-22-34-000-010) and (Ref. AMM TASK 73-22-34-400-010).

   (1) If the fault continues:
     - do a check and repair the wiring of EEC A OUTPUT 1 signal from the EEC 2 (4000KS) to the first terminal block, (Ref. ASM 73-25/10).

B. Do the test given in Para. 3.

73-20-00

EFF : ALL

Figure 5-4-33. Excerpt from troubleshooting manual showing: (1) possible causes, (2) job setup procedures, (3) fault confirmation, (4) fault isolation procedures

Courtesy of Airbus S.A.S.

Many new aircraft also utilize wireless communication systems such as local area network, satellite, or airborne cell, for the downloading of system performance data. Many of these advanced concepts were introduced with the B-777 aircraft and enhanced with the newer Airbus A-380 and the Boeing 787. Many corporate aircraft also employ these concepts of improved aircraft monitoring. Today, third generation aircraft systems employ built-in test equipment to check system status; a maintenance data storage area onboard the aircraft, and a communications link to the airline's ground facility for real-time downloads of any problems detected during flight. Wireless communications between the aircraft and ground facilities are also used to upload pertinent data related to flight or software configuration changes to the aircraft.

The real-time downloads of system status also provides a means of condition monitoring. The ability to continually monitor the status of various systems onboard the aircraft is one of the

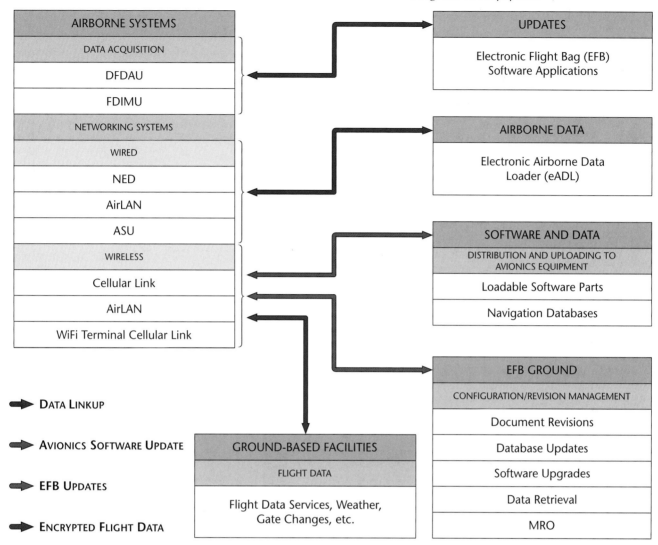

Figure 5-5-1. Example of a typical data monitoring system using wireless connectivity

major advancements found on third generation central maintenance systems. The capability to record and analyze trends in a real-time format allows airlines to "see problems" before they happen. Trend monitoring of critical flight systems should considerably reduce maintenance costs and may turn out to be one of the best safety improvements to reach the aviation industry in recent history.

Many of the latest aircraft, such as the B-787 and A-380, also employ Electronic Flight Bags (EFB) which play an integral part of the aircraft's central maintenance system. As discussed in Chapter 3, EFBs found on transport category aircraft employ integrated flat panel display units that can access and process a variety of aircraft information. The EFB offers pilots a variety of data related to flight and navigation functions, as well as maintenance information such as the Minimum Equipment List (MEL). Prior to any flight the pilot can easily check the current status of aircraft systems and determine dispatch capability using the EFB. The MEL is used to determine the MIMIMUM number of operational systems or maximum failures allowable to dispatch the aircraft according to FAA regulations. On these modern aircraft, pilots have instant access to all maintenance and dispatch information using the EFB. Since all EFB information can be updated instantly using wireless technologies the pilots will always be using the latest information when making critical decisions.

In recent years, many manufactures have began to offer a data monitoring service for airframe, engines, and/or various airborne systems. The services are offered by a variety of international corporations such as Teledyne Technologies, Honeywell, Rolls Royce, and Rockwell International. A variety of programs are offered which provide remote connectivity to airborne systems and integration to web-based maintenance, analysis, and trouble-shooting tools. The functional components of a typical data monitoring system are shown in Figure 5-5-1. Whenever a system exceeds a set

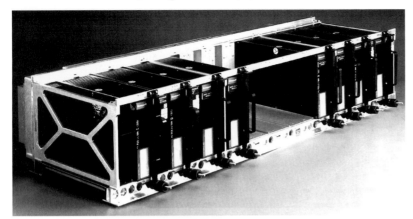

Figure 5-5-2. AIMS interface cabinet   *Courtesy of Honeywell, Inc.*

limit, or a failure occurs, the central maintenance system records the data and also sends critical information to a ground facility using a wireless local area network, cellular, and/or satellite technologies. The service provider monitors incoming data and uses the web to communicate with a variety of maintenance authorities that may include the airline, the system, and/or aircraft manufacturers, and other technical providers. The maintenance monitoring service can automatically make recommendations to the airline for any needed repairs and coordinate maintenance at the next airport facility for that aircraft.

In general, the concepts of a data monitoring service is to improve aircraft efficiencies by reducing the downtime required for typical maintenance activities. Reduced aircraft downtime is mainly achieved through trend monitoring and the use of wireless connectivity. Trend monitoring is the process where small changes in a system are detected, recorded, and analysed in order to find problems and repair systems before they fail. For example, if a turbine engine records a slight increase in exhaust gas temperature over the last 100 flight hours, trend monitoring will detect this change, and maintenance can be scheduled before the problem creates a major engine failure. The use of wireless connectivity has helped to make this type of trend monitoring practical. Since the aircraft systems can now export vast quantities of diagnostics data automatically the system analysis and troubleshooting time can be reduced substantially and critical systems can be examined more closely.

With the miniaturization of electronic systems and the improvements to wireless technologies, data monitoring can be found on virtual all aircraft systems. Data monitoring and trend analysis is now used on light business jets and even small personal aircraft. Today operators of small rental fleets equipped with many of the latest Garmin avionic systems have the

capability to monitor system trends. The collection of data needed for this type of analysis is often called Data Mining. The data can be transferred from the aircraft using wireless technologies or using solid-state memory cards and flash drives. In some cases, the data can be analyzed by the owner/operator of the aircraft or through the use of a data monitoring service.

Many of the new data mining techniques are also used for a variety of flight applications. The use of electronic flight bags linked to wireless communications has made it possible for independent service providers to offer flight deck management applications to airlines and corporate operators. In this case, an independent company would take responsibility for routine updates of charts and flight manuals, as well as offering services related to flight planning and aircraft performance calculations. Since these systems are completely interactive with web-based weather services, real-time airport weather can be used to determine takeoff performance and help minimize delays and improve safety.

Boeing 777 Central Maintenance System

The Boeing 777 aircraft could be considered the first aircraft to employ a third generation central maintenance system. As discussed in previous chapters, the B-777 employs an integrated architecture designed by Honeywell called the Airplane Information Management System (AIMS). The concept behind AIMS is simple; integrate various systems by sharing common functions/components and replace larger LRUs with line replaceable modules (LRM). AIMS actually integrates data collection, computing functions, power supplies, and output functions for several subsystems. By sharing both hardware and software components AIMS enhances reliability, improves redundancy, and creates a substantial weight savings as compared to older systems.

Boeing and Honeywell designed AIMS to incorporate fault tolerant software. Fault tolerance along with proper redundancy of critical systems hardware allows AIMS software to detect a fault and reconfigure the system for uninterrupted operations. In many cases, the flight crew would not even be aware of a system failure. In theory, this type of design will allow the B-777 to continue flying with the failed system until the next convenient maintenance opportunity. Typically, the repair would be performed during night maintenance not during the short time period available between flights, or in some cases, delayed until the next phase check.

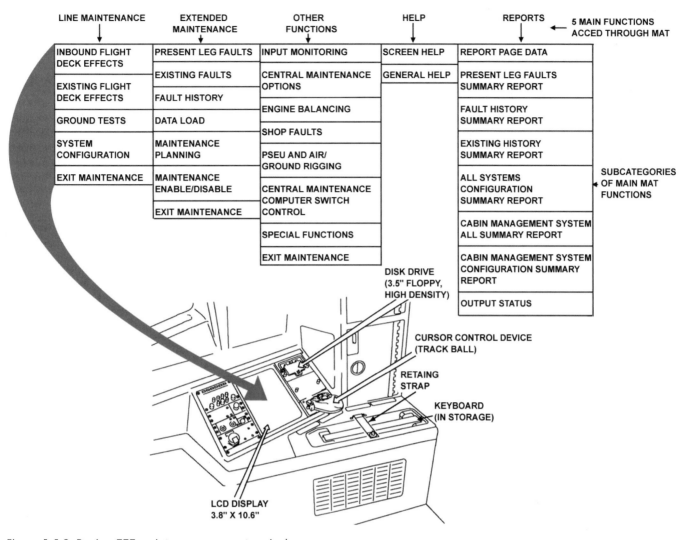

Figure 5-5-3. Boeing 777 maintenance access terminal

On the B-777, there are two AIMS cabinets, each with eight LRMs. Each cabinet contains four Input Output Modules (IOM) and four Core Processing Modules (CPM). The IOMs transfer data to/from the AIMS cabinet and within the four CPMs. The four CPMs perform the calculations for the various avionics systems serviced by AIMS. Each LRM is designed for quick removal and installation. Without the AIMS concept, several LRUs would be required to perform the same operations. The LRMs are smaller than the conventional LRUs since they share several functions with other modules within AIMS. Figure 5-5-2 shows a typical AIMS cabinet and LRMs.

Troubleshooting the B-777 avionics is achieved through the central maintenance computing system (CMCS). The major computing functions for this system will be contained in the AIMS processors. On the B-777 the technicians will access the central maintenance system data using the Maintenance Access Terminal (MAT) located on the flight deck just behind the first officer's station. The technicians can also access maintenance data supplied by AIMS through the local area network (LAN) and a portable access terminal. The LAN is connected to AIMS using a fiber optic data transfer connection following the ARINC 629 data bus standard.

As seen in Figure 5-5-3, five different functions will be accessed through MAT: Line Maintenance, Extended Maintenance, Other Functions, Help, and Reports. Each of these functions is used to access a specific operation of the CMCS. The terminal consists of an 8 x 10 inch LCD display, a keyboard that is stowed during flight, a disk drive, and a cursor control device. The system is menu driven and designed to operate similar to a personal computer.

The A-380 Integrated Test Equipment

The Airbus A-380 is one of the newest commercial passenger jets flying today. This aircraft employs all of the advanced concepts of IMA

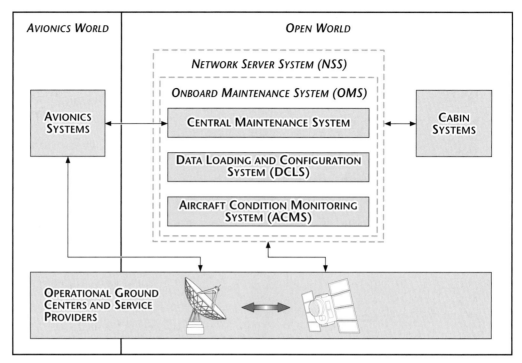

Figure 5-5-4. The A-380 Onboard Maintenance System (OMS) interface diagram

(Integrated Modular Avionics), which provides improved communications between all critical systems on the aircraft. The modern concepts of wireless connectivity are also employed on the A-380 allowing for easy communications between the airplane and airline facilities even during flight. Once connected to ground facilities, virtually all manufacturers, suppliers, vendors, regulatory agencies and, of course, the airline facilities can communicate through the World Wide Web. So in effect, the central maintenance system of the A-380 is connected to every worldwide agency that might help to analyze and troubleshoot maintenance problems.

The integrated central maintenance system found on the A-380 is called the Onboard Maintenance System (OMS). The OMS is an integrated system used to provide support for scheduled and unscheduled maintenance, aircraft servicing, aircraft condition monitoring, and configuration and reconfiguration of sys-

tem software. As seen in Figure 5-5-4, the OMS communicates to other systems through the Network Server System (NSS) allowing for in-flight transmission of data to ground facilities. The OMS is comprised of three major subsystems: the Central Maintenance Systems (CMS), the Data Loading and Configuration System (DLCS), and the Aircraft Condition Monitoring Systems (ACMS). To ensure the integrity of critical systems, the CMS receives data from all avionics directly through a secure data bus. The CMS is similar to second generation central maintenance systems like the A-320 and B 474-400. The A-380 CMS identifies, centralizes, and stores fault data. The CMS receives fault information from individual components and systems. The components and/or systems are tested by BITE circuitry within specific LRUs located throughout the aircraft.

The ACMS (Aircraft Condition Monitoring System) is used to provide support for preventative maintenance and in-depth analysis of maintenance data. The ACMS can transmit trend data to ground facilities for real-time analysis during flight, or the data can be downloaded after landing. The DLCS (Data Loading and Configuration System) is designed to manage all data related to system configuration including all data uploads/downloads. The download of all CMS fault reports is therefore controlled through the DLCS.

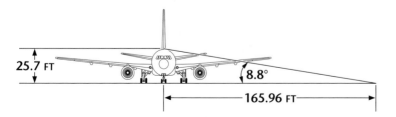

Figure 5-5-5. The TMLU antenna is located on the top of the fuselage 25.7 feet above the ground and operates via line of site. The minimum distance of 165.96 feet occurs with an angle of 8.8 degrees to meet this line of sight requirement to the ground.

The OMS (Onboard Maintenance System) is hosted on the main aircraft Network Server System (NSS), which allows virtually all systems to communicate to the OMS. As can be seen in Figure 5-5-4 the cabin systems, including passenger entertainment, connects to the OMS through the NSS. The more crucial avionic systems communicate directly to the CMS through a secure communications channel. The secure and direct communications route to the CMS provides an extremely reliable connection for all flight critical systems. During flight the NSS uses satellite technologies to communicate to ground-based operational centers and service providers. These facilities monitor aircraft data for trend analysis and any faults which may occur.

Whenever a system fault is detected, the CMS classifies the problem and creates a standard or customized report. The reports are sent to the flight deck where they can be consulted by the flight crew and transmitted to the operational ground centers via satellite. On the ground the maintenance personnel can access the CMS data Servicing Reports. The servicing reports contain a list of items that require service according to current aircraft status and aircraft specific guidelines. The CMS also provides electronic links to maintenance documents as needed. Using the CMS the technician can also access the specific BITE for any given system and create standard or customized reports for that system. Of course the CMS can also be used for system verification after a repair has been completed.

The technician can access the CMS through either the Onboard Maintenance Terminal (OMT) or a Portable Maintenance Access Terminal (PMAT). The OMT is located on the flight deck behind the co-pilot's seat and, of course being a wireless device, the PMAT can be used anywhere in or near the aircraft. Maintenance personnel can also access OMS functions from the flight deck using the Onboard Information Terminals (OIT). Flight personnel primarily use the OIT as discussed in Chapter 3.

The Boeing-787 Central Maintenance Computing System

The B-787 was still under development at the time this text was written, yet much is known about the aircraft and the functionality of the central maintenance system. As discussed in previous chapters the aircraft will employ an integrated architecture using an ARINC 664 data bus structure. On the B-787 many of the crew information and maintenance functions are combined since they both provide similar functions (communications of system operations and performance). As with the A-380, the B-787 will perform traditional central maintenance functions as well as condition monitoring for trend analysis in order to prevent any

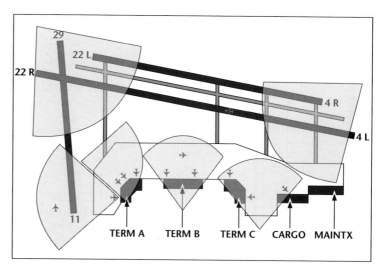

Figure 5-5-6. Coverage of multiple wireless antennas

potential future faults. The majority of trend monitoring is performed through screening routines for engines and data bus systems. The software for the trend monitoring and central maintenance computing (CMC) functions will be hosted in the common core network. The B-787 also employs wireless communications to download trend and maintenance data before the aircraft ever reaches the gate.

Data loading and configuration management are separate functions also provided by the aircraft's central maintenance system. The data loader function controls the insertion of operational software into the appropriate avionics LRUs. The configuration management function of the CMC tracks the current version of software and hardware installed in the B-787 avionics. Configuration management on a complex aircraft such as the B-787 becomes an extremely difficult task due to the variety of systems; and therefore, configuration management software is vital to flight safety.

In preparation for the B-787's commercial operations Boeing has worked with airlines and major airports in an effort to establish an operational ground wireless system to support the airplanes advanced communication systems. As seen in Figure 5-5-5 a wireless local area network antenna will be mounted on the top of the aircraft with line-of-sight transmission/reception. The system will begin to authenticate ground access points with weight-on-wheels at touchdown. Since the radio signal requires line of sight with the ground antenna, there will typically be several ground antennas located at different locations on the airport. With the ground antennas properly located the aircraft can download maintenance data anytime after touchdown. The eventual goal for this system is to allow all authorized maintenance personnel access

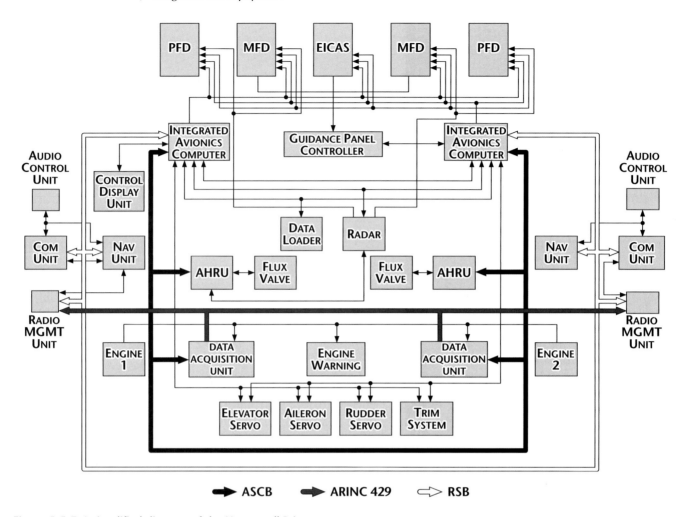

Figure 5-5-7. A simplified diagram of the Honeywell Primus system

to the aircraft's crew information and maintenance system through a laptop computer equipped with a WiFi card or to allow the appropriate ground facility access to CMC data before the aircraft is parked at the gate. In most cases the airport facility will have several wireless hot spots as seen in Figure 5-5-6. Each of these ground stations employs a directional antenna covering a certain section of the airport.

Light Aircraft Built-In test Equipment

The miniaturization of electronic systems and advanced processing power found in today's avionics equipment has made it possible for smaller, less complex aircraft to employ centralized maintenance systems. Aircraft such as the Cessna Mustang or Citation Ultra, very light jets such as the Embraer Phenom, and even small piston engine aircraft such as the Cirrus RS-20 contain some type of built-in test equipment. The systems found on these aircraft are typically not as elaborate as those found on transport category aircraft, how-

ever with modern technologies virtually any aircraft with integrated avionic systems can incorporate centralized maintenance functions.

As discussed in previous chapters, the Honeywell Primus 1000 or 2000 model systems are integrated avionics suites installed on many modern corporate and commuter type aircraft. As seen Figure 5-5-7, Primus is a completely integrated avionics system that includes everything from instrument display panels to control surface servos. Each LRU, such as the data acquisition unit (lower right portion of Figure 5-5-7) contains its own BITE circuitry. The LRUs can therefore perform system monitoring and fault detection. As with larger aircraft, the individual BITE circuits report to a centralized location where all fault data can be accessed. Of course, any fault information critical to flight safety is displayed to the pilots on one or more flat panel display units.

In general, the Primus BITE will monitor functional performance of both hardware and software during four distinct phases of operation:

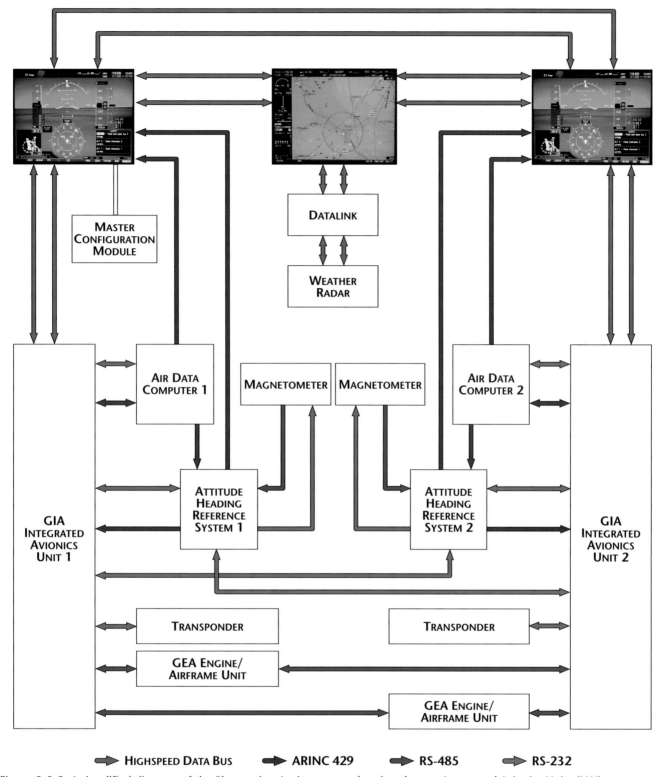

Figure 5-5-8. A simplified diagram of the Phenom's avionics system showing the two Integrated Avionics Units (IAU)

power-up, continuous monitoring, pilot activated, and ground maintenance tests. When power is first applied to the system or LRU, the BITE software initiates a series of tests that would interfere with normal operation and therefore can only be performed at startup or during ground maintenance. Power-up BITE will test all critical systems, such as, alternate power switchover, volatile and non-volatile memory, and the input/outputs of the data busses. Even signals to/from autoflight control surface servos are tested. Continuous monitoring is performed any time the system is operational and does not interfere with normal operation. Pilot activated and ground maintenance tests are performed on request; typically to verify system operation or check the status of a system or LRU.

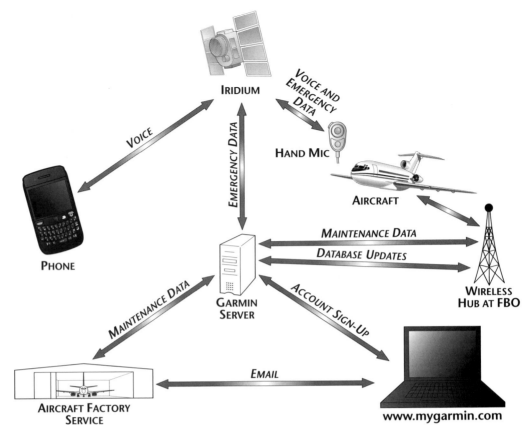

Figure 5-5-9. The Phenom Datalink Management System

The Primus BITE has three classifications: critical, noncritical, and maintenance. Critical failures are those that would affect primary functions, causing erroneous outputs, such as an incorrect pitch commands to the primary flight display or autopilot servo. Critical failures will have a distinct annunciation on the flight deck and typically limit the capabilities of the aircraft. A noncritical fault is one which will not affect a primary flight function; such as heading angle. Noncritical faults will typically have a flight deck annunciation, but will have little or no effect on aircraft capabilities. Maintenance faults are those which are stored in the BITE memory for maintenance use only. These faults are accessed during ground maintenance tests.

A typical light business jet which employs an advanced integrated maintenance system is the Phenom made by Embraer. This is a new style of aircraft typically constructed mostly of composite materials and containing small yet powerful turbine engines and advanced integrated avionics. This type of aircraft is commonly known as a Very Light Jet (VLJ) since it is light weight and fuel efficient. The Phenom's advanced avionics equipment is produced by Garmin International and is similar to the Garmin systems discussed previously.

As shown in Figure 5-5-8 there are two main processing units for the Garmin system called the Integrated Avionics Units (IAU). As with other aircraft, the IAUs contain software for the central maintenance functions; however, the actual BITE software is contained within the system's individual LRUs. For example, on this aircraft, system tests for the AHARS (attitude heading reference system) are performed by the AHARS BITE; the IAU central maintenance computer function is designed to access, store, and manage all AHARS fault data. The central maintenance system can identify faults down to the LRU part number and stores this data for download to technical service personnel.

The Phenom aircraft also has the ability to perform trend monitoring of critical systems. Trend monitoring is an extremely useful tool for predicting potential system failures, especially on turbine engines. The system can monitor fuel flow, temperature, and vibrations related to engine components virtually every second. Analyzing these trends has proven to be an excellent predictor of a turbine engine's health and ability for continued flight safety. Collection of this data is useless until it is analyzed; it is therefore critical that IAU data be downloaded at regular intervals.

As seen in Figure 5-5-9 this aircraft has the capability to employ wireless connectivity for the download/upload of data. During flight the data link system employs the use of satellite communications used for voice or emergency fault data. There is also a wireless, WiFi type, connection which is automatically enabled once the aircraft is on the ground and close enough to the appropriate wireless hub. When a wireless data connection is established a status page showing all connection settings will be displayed on the MFD. Through the satellite or the WiFi connection, data is shared with the aircraft's FBO (fixed base operator), Embraer, and Garmin's maintenance data center. If the aircraft experiences a critical fault the data will be analyzed and directed to the maintenance FBO as soon as possible. If the data is to be used for trend monitoring, the Embraer and Garmin facilities store the data, utilize computer analysis, and inform the aircraft owner/operator as needed. Trend analysis reports are also generated periodically or upon request. At the end of each flight a status report can be transmitted automatically or through a pilot's command. The report will contain all crew alert system (CAS) message reports, engine trend and exceedance reports, and CMC message reports.

Trend monitoring analysis and wireless connectivity are all costly to implement. For example, Embraer has a for-hire service known as AHEAD (Aircraft Health Analysis and Diagnostics) that can be purchased by owners of the Phenom jet. The service is designed to monitor the aircraft fault and trend data and provide feedback even enroute as the aircraft travels from airport to airport. This service can be extremely valuable since it helps to eliminate aircraft downtime and flight delays. If the aircraft owner decides not to participate in the AHEAD program or use wireless connectivity, the maintenance data can also be offloaded from the aircraft using an SD card installed into the MFD bezel. Use of the SD card was discussed in Chapter 4, and of course, system status reports containing all fault data are easily viewed on the MFD after landing.

Chapter 6

AUTOPILOT
and autoflight systems

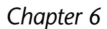

Section 1

Introduction

Autopilots have been installed on aircraft for several decades. The systems have been proven reliable and, for the most part, more accurate than human pilots. The variety of autopilot systems is almost as vast as the variety of airplanes. Light aircraft may have simple autopilots installed; while transport category aircraft often incorporate complex systems with full autoland capabilities. Most twin-engine aircraft incorporate some type of autopilot system and many corporate aircraft employ complex systems similar to large passenger jets. Autopilot technologies will also play a large role in the upcoming decades as airspace becomes more crowded due to the increased number of flights globally. The initiative known as NEXTGEN, or Next Generation Air Transportation system, relies on modern technologies to enhance safety and improve capacity.

Autopilots were first developed to relieve the pilot/co-pilot from constantly having to handle the aircraft controls. On long flights this became especially important on older transport category aircraft like the Boeing 707. These aircraft were difficult to control and were physically exhausting, especially in bad weather. As technology improved, and systems became lighter and smaller, autopilots began to filter into the light aircraft market. In the early 1980s, autopilot technology advanced to the point where the machine became more efficient than the human; flying with advanced autopilot systems saved both time and fuel. Automated flight systems developed in the 1990s and early 21st century, improved safety while increasing the number of aircraft in the authorized airspace. Today a

Left. Modern autoflight systems are a far cry from the original autopilots of the 1930s. Today's systems, with features like autoland, can almost operate the aircraft from the departure gate to the arrival gate.

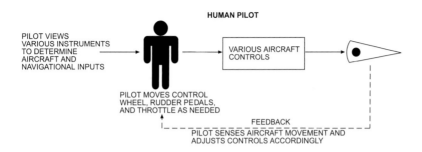

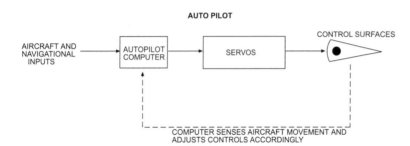

Figure 6-2-1. Comparison of human pilot and autopilot

lot, and presents several modern systems in depth. Many modern autopilots are known as autoflight systems, both terms can be considered synonymous; however, it is generally accepted that autopilots are relatively simple and autoflight systems are more complex and capable of more functions.

Section 2

Basic Autopilot Theory

By definition, the autopilot is designed to perform pilot duties automatically. The autopilot must first interpret the aircraft's current attitude, speed, and location. Second, if adjustments are necessary, the autopilot must move the appropriate control surface, and throttles on advanced systems. Third, the autopilot must anticipate the aircraft's movement and reposition the control surfaces, and/or throttles, to prevent the aircraft from overshooting the desired course and attitude.

wide range of autopilots is available for almost any type of aircraft.

Simple autopilots provide guidance along only the longitudinal axis of the aircraft. These systems, often found on light single-engine aircraft, were called *wing levelers* because they were used to keep the wings level. More complex systems provide total control for attitude and navigation. Many modern transport category aircraft incorporate systems that provide aircraft control, yaw damping, navigational guidance, thrust control, and Category III landing capability. This chapter examines the operational theory of autopilot systems, explains individual subsystems of the autopi-

To perform the functions just described, the autopilot must monitor various aircraft parameters including airspeed, altitude, pitch, roll, and yaw (Figure 6-2-1). Navigational aides are also monitored to provide course data. Next, the autopilot will analyze the data to decide if adjustments are needed. An autopilot computer is used to analyze the data and output the necessary control information. If adjustment is required, servos are used to move control surfaces and reposition the aircraft. *Servos* are devices used to move the flight controls, or throttles, in accordance with autopilot com-

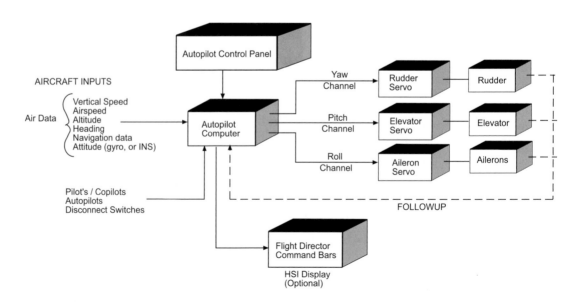

Figure 6-2-2. Basic autopilot block diagram

mand signals. Hence, older autopilot computers were sometimes referred to as servo amplifiers. A follow-up system is used to inform the autopilot computer that the control surface has changed position. The follow-up system allows the computer to anticipate when the control surfaces should be returned to the neutral position. Since aircraft move in three axes (pitch, roll, and yaw), many autopilot systems typically contain three channels or subsystems.

The autopilot must be capable of performing all of the necessary functions in a safe and reliable way. Two major safety considerations that every autopilot system must have, is the ability to be quickly and positively disengaged electronically, and the ability to be overridden manually by the pilot, if necessary, to regain control of the aircraft.

The autopilot block diagram (Figure 6-2-2) shows the inputs and outputs for a typical three-axis autopilot. The autopilot computer receives aircraft inputs from:

1. Air data sources supplied by pitot static pressures, or electronic signals from an air data computer

2. Heading sources provided by the aircraft's compass system

3. Navigational inputs, such as ILS, DME, or VOR

4. Attitude information from an inertial reference system (IRS) or attitude sensors

A quick disconnect push-button is on the control yoke to force a quick disconnect of the autopilot.

Each autopilot system will incorporate some means by which the flight crew can input commands. The two panels shown in Figure 6-2-3A are used to control an autopilot for a typical corporate jet aircraft. The top panel is used to engage the autopilot and to control the manual pitch and roll functions. This panel also contains the *AP XFR*, for transfer of a dual autopilot system, and turbulence mode (**TURB**) push-button. The panel shown in Figure 6-2-3B is used to select the different modes of the autopilot system. From this panel the pilot can select several different modes of operation including navigation (*NAV*) or vertical navigation (*VNAV*). The *navigation mode* allows the autopilot to fly the selected lateral navigational course (i.e., control of north, south, east, and west directions). The *vertical navigation mode* allows the autopilot to fly the selected altitude or glide path.

The main autopilot outputs include the three signals used to control the pitch, roll, and yaw servos. As the respective control surfaces

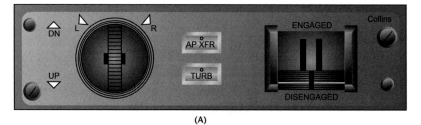

(A)

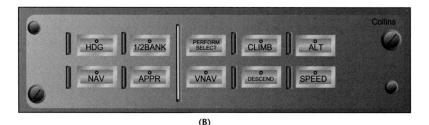

(B)

Figure 6-2-3. Typical autopilot control panels from a corporate-type aircraft: (A) Used to engage the autopilot and for manual pitch and roll functions, (B) Used to select the different modes of the autopilot system

Courtesy of Rockwell International, Collins Divisions

move according to autopilot signals, a follow-up signal is transmitted back to the autopilot computer. The autopilot computer may also have an output dedicated to the flight director control. The *flight director* presents a visual aid to the pilot that is used for manual control of the aircraft. As seen in Figure 6-2-4, the flight director display is typically incorporated in the attitude director indicator (ADI) or integrated into the primary flight display (PFD). Since the flight director utilizes many of the same inputs as the autopilot, the two systems often share components. In the example of Figure 6-2-2, the autopilot computer drives the command bars for the flight director display.

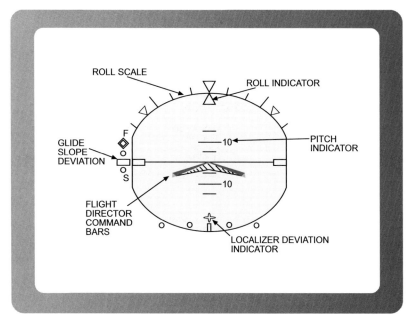

Figure 6-2-4. A typical flight director display incorporated on an attitude director indicator (ADI)

Yaw Damping

Virtually all high-speed aircraft are designed with swept back wings. The aerodynamics of a swept back wing causes a stability problem known as *Dutch roll*. Dutch roll is basically a slow oscillation of the aircraft about its vertical axis. Correct application of the rudder can prevent dutch roll; however, it requires constant repositioning of the rudder. This process becomes almost impossible for the pilot.

The yaw damper system is designed to control rudder and eliminate dutch roll. The *yaw damper* is basically an autopilot component dedicated to rudder control. If the system detects a slip or skid of the aircraft, the rudder is activated to correct this condition; hence, eliminating dutch roll. On most aircraft, the yaw damper is considered independent of the autopilot system, although they may share the same control panels, sensors, and computers. During flight, the autopilot and yaw damper may both be used to position the rudder. The autopilot positions the rudder to coordinate turn activity; the yaw damper positions the rudder to eliminate dutch roll. On most aircraft, the yaw damper can be engaged independent of the autopilot.

Section 3

Autopilot Components

A variety of components are incorporated into every autopilot system. Many of these components or subsystems are not actually part of the autopilot, but are essential to autopilot operation. Older autopilots employ analog systems and mechanical sensors. Newer autopilot components are typically digital systems and communicate through data bus cables. The following presents many of the basic elements of an autopilot system.

Gyroscopic Sensors

There are two common types of gyroscopic sensors used in modern autopilots: rotating mass gyros and ring laser gyros. Most autopilots use gyro sensors to detect movement of the aircraft. Gyro outputs are also used for reference on certain navigation systems. Gyro systems are both fragile and expensive; it is very important the technician becomes familiar with the system before performing maintenance on gyros.

Rotating Mass Gyros

Rotating mass gyros have a tendency to stay stable in space. This effect allows a rotating mass gyro equipped with a rate sensor to detect aircraft motion (Figure 6-3-1). The output signal from the rate sensor can be sent to an autopilot computer and/or used to stabilize a gimbal platform.

Gimbal platforms. Rotating mass gyros are often used to stabilize acceleration sensors mounted on gimbal platforms. A gimbal platform is made rigid in space, parallel to the earth's surface regardless of the aircraft's attitude. This is accomplished by mounting three rotating gyros to the platform as shown in Figure 6-3-2. One gyro is needed for each axis of the gimbal. If the aircraft's attitude changes, the gyro rate sensor produces an output signal that is proportional to the amount of attitude change. This signal is monitored and ampli-

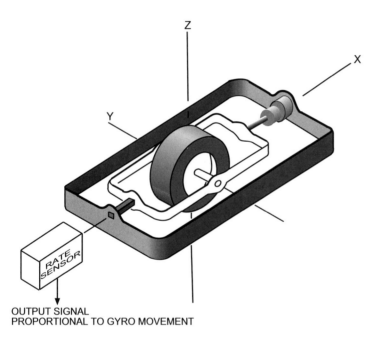

Figure 6-3-1. Diagram of a basic rotating mass gyro

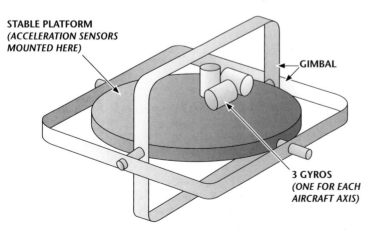

Figure 6-3-2. Diagram of a typical gimbal platform

fied by the autopilot computer, then sent to the torquer units on the stable platform. The torquer produces the counter-force, typically a magnetic field, needed to stabilize the platform. With the acceleration sensor mounted on a stable platform, an accurate acceleration can be monitored by the system. If the acceleration sensors were not mounted to a gimbal platform, the sensors would measure attitude changes as well as aircraft accelerations. Due to advancements in modern technologies, rotating mass gyros and gimbal platforms are quickly being replaced by more reliable and accurate systems.

Laser Gyros

The ring laser gyro (RLG) is actually an angular rate sensor and not a gyro in the true sense of the word. Conventional gyros generate a gyroscopic stability through the use of a spinning mass. The gyro's stability is then used to detect aircraft motion. The RLG uses changes in light frequency to measure angular displacement.

The term *laser* stands for "light amplification by stimulated emission of radiation." The RLG system utilizes a helium-neon laser; that is, the laser's light beam is produced through the ionization of a helium-neon gas combination. A typical RLG is shown in Figure 6-3-3. This system produces two laser beams from the same source and circulates them in a contra-rotating triangular path. As shown in Figure 6-3-4, the high voltage potential, approximately 3,500 V, between the anodes and cathodes produce two light beams traveling in opposite directions. The laser is housed in a glass case, which is drilled with precise holes to allow travel of the light beam. Mirrors are used to reflect each beam around an enclosed triangular area. A prism and detector are installed in one corner of the triangle. The prism reflects the light to allow both laser beams to be measured by the detector.

The resonant frequency of a contained laser is a function of its optical path length. When the RLG is at rest, the two beams have equal travel distances and identical frequencies. When the RLG is subjected to an angular displacement around an axis and perpendicular to the plane of the two beams, one beam has a greater optical path and the other has a shorter optical path. Therefore, the two resonant frequencies of the individual laser beam change. This change in frequency is measured by photosensors and converted to a digital signal. Since the frequency change is proportional to the angular displacement of the unit, the system's digital output signal is a direct function of the angular rate of rotation of the RLG.

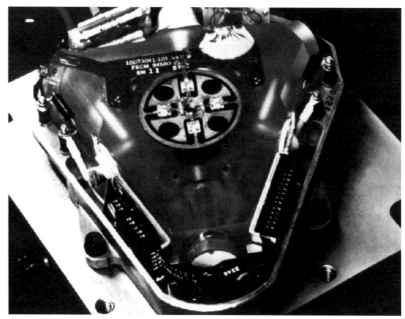

Figure 6-3-3. A ring laser gyro

Courtesy of Honeywell, Inc.

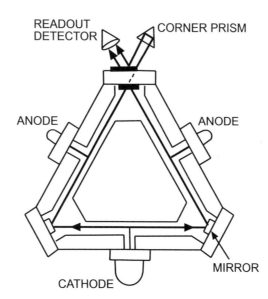

Figure 6-3-4. Pictorial diagram of a ring laser gyro

The RLG system is typically coupled to a complete navigation system. The digital signals from the RLG can be used to control inertial reference and navigation systems and/or autopilot functions. Each inertial sensor assembly contains three triangular lasers. In Figure 6-3-5, two of the lasers can be seen mounted within the sensor assembly. RGL technology provides a new era of aircraft safety. Since the RLG has no moving parts, it has a much greater reliability than the conventional rotating mass gyro system.

No matter which type of gyro is employed, all three axes of the aircraft must be monitored for a fully functional autopilot. Pitch, roll, and yaw can only be monitored by sensors aligned in the correct position. On most systems, alignment

Figure 6-3-5. An inertial sensor assembly containing three ring laser gyros

Courtesy of Honeywell, Inc.

units are interchangeable between locations on the aircraft. Pin programming is used to identify the specific installation location.

The most modern ring laser gyros operate on the same principles as those found in 10-year -old aircraft. The main improvements in modern RLGs is the miniaturization of components allowing for the construction of lighter and smaller gyros, which in turn has allowed manufacturers to integrate gyro components in a single housing. For example, air data and inertial reference units are now commonly combined into an air data initial reference unit (ADIRU). Another advancement found in modern laser gyro systems is the ability to electronically align the unit. In these systems, when installing the gyro, all final alignment is performed through electronic circuitry during initial set-up operations. The technician must simply follow the required steps and the RLG software will ensure that the system is aligned with the aircraft and the vertical, longitudinal, and horizontal axes.

Maintenance and Troubleshooting

In general, laser gyro systems are relatively maintenance free and rotating mass gyros seem to be subject to frequent failures. Every gyro provides some type of electrical output signal. The simplest way to detect the operation of a gyro is to check the output signal. On the most modern systems, the output is probably a digital signal. On many rotating mass gyros, the output is a three-phase AC signal. Many rotating mass gyros are combined with rate sensors. The output signal from the rate sensor is typically a single AC voltage.

On any gyro system, the output signals are relatively low current; therefore, a poor connection can easily create a loss of output. If a gyro's output is inaccurate or missing, check the electrical connections to the unit. Rotating mass gyros all produce a humming noise as the gyro rotates. This noise should be present any time the gyro is active. If the gyro is completely quiet, the unit is defective or there is no power to the gyro assembly. If the gyro assembly makes an unusually loud rumble noise, the gyro bearings are probably worn beyond limits and the unit should be replaced.

All gyros require some warm-up period. Laser gyros must reach a given temperature to stabilize; rotating mass gyros must reach a certain RPM, and digital control circuits often perform test functions prior to operation. In many gyro systems, it is important that the aircraft remain stationary during this warm-up or initialization period. Be sure to provide sufficient warm-up time to all gyro systems whenever performing maintenance.

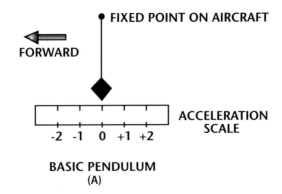

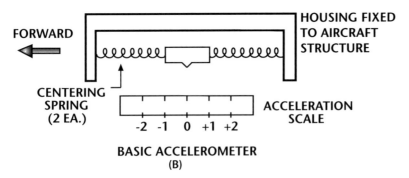

Figure 6-3-6. Simple pendulum used to measure acceleration: (A) A basic pendulum, (B) A basic aircraft accelerometer

becomes very important during the installation, removal, and replacement of rate sensing LRUs. Each unit must be oriented in the correct position with respect to the rest of the aircraft. For example, if a unit is installed 180° out of alignment, the system could react exactly opposite of what is necessary. Be sure the unit is installed according to the manufacturer's instructions. Many gyro

Accelerometers

An accelerometer is a device that senses aircraft acceleration. Acceleration is a vector force and therefore is measured in both magnitude and direction. Since an aircraft can accelerate in three directions, a minimum of three accelerometers are used on most installations.

Figure 6-3-6A depicts how a basic accelerometer resembles a simple pendulum. The pendulum will swing to the right as the aircraft moves forward and to the left as the aircraft decelerates. Figure 6-3-6B shows the arrangement of a simple aircraft accelerometer. This design incorporates two springs to center the pendulum. As the aircraft moves, the indicator moves in the opposite direction, relative to the aircraft.

An actual accelerometer incorporates a pick-off device to convert pendulum movement into an electric signal (Figure 6-3-7). The electric signal is amplified and, in some cases, converted to a digital format. Some accelerometers use a secondary current flow to keep the armature centered over the pick-off coils. A torquer current is used to produce a magnetic field that centers the armature in the null position. This produces an armature that is very stable and increases accelerometer accuracy.

Modern accelerometers are often small *microelectromechanical systems* (MEMS) consisting of little more than a cantilever beam with a small mass at the end. This mass is often called the *proof mass*. Under the influence of external accelerations the proof mass deflects from its neutral position. An electronic element is used to sense the deflection and produce an analog or digital signal. This method is simple, reliable, and inexpensive. Micromechanical accelerometers are available in a wide variety of measuring ranges. The designer must make a compromise between sensitivity and the maximum acceleration force that can be measured.

Most micromechanical accelerometers operate in-plane, that is, they are designed to be sensitive to accelerations in only one plane. By integrating two devices perpendicularly on a single die, a two-axis accelerometer can be made. By adding an additional out-of-plane device, three axes can be measured. Such a combination always has a much lower misalignment error than three discrete models constructed independently and mounted into an assembly.

MEMS accelerometers are designed in a manner similar to microelectronic circuits, such as, microprocessors. MEMS devices are constructed of silicon using a photo-etching process, which has proven to produce high quality components with great reliability at low cost. A complete MEMS accelerometer measures typically in a range of 20

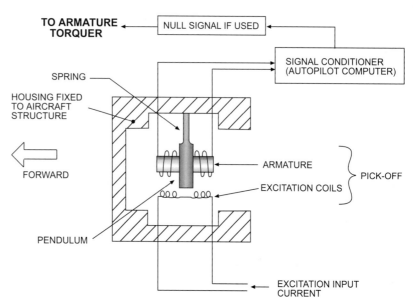

Figure 6-3-7. An aircraft accelerometer showing the pick-off unit

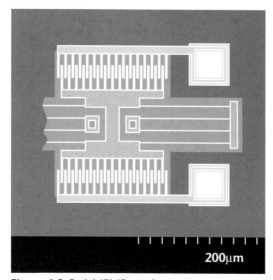

Figure 6-3-8. A MEMS accelerometer

microns (20 millionths of a meter) to as large as a millimeter. Figure 6-3-8 shows components of a typical MEMS accelerometer. MEMS accelerometers are found in a variety of applications on aircraft and other consumer products. For example, automobile airbag controls use a MEMS device to measure sudden decelerations, smart phones use MEMS to determine if the display is horizontal or vertical and laptop computers use MEMS accelerometers to instantly park the hard drive if a fall is detected.

Modern accelerometers typically contain one of two electronic elements: piezoelectric or capacitive components, which convert mechanical motion into an electrical signal. Piezoelectric accelerometers rely on crystals to produce a voltage/current when pressure is applied to the crystal. The voltage/current

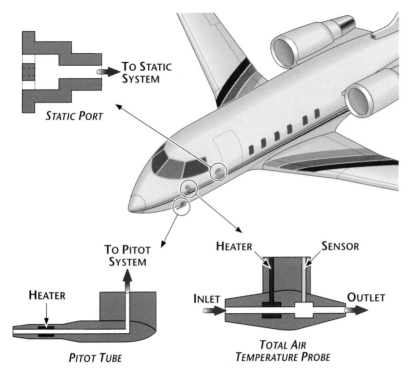

Figure 6-4-1. Typical locations for installation of pitot tubes, static ports, and temperature probes

converted to a digital signal, which represents the acceleration force.

Capacitive accelerometers typically use a silicon micromachined sensing element. Like all capacitors, the capacitance value will change according to the distance between the plates. The sensor is constructed of an extremely small capacitor and, like the crystal accelerometer; an acceleration force easily affects it. As the aircraft accelerates the microcapacitor plates bend changing the distance between the plates. The plates on this capacitor would be measured in microns and thus invisible to the naked eye. The miniature size makes the capacitor vulnerable to even the slightest acceleration force. The accelerometer circuitry measures this change in capacitance, which is directly proportional to the acceleration and creates the appropriate output signal.

On many transport category aircraft, there are at least two accelerometers to monitor the acceleration of each aircraft axis. The outputs of both accelerometers are combined to provide an extremely accurate measurement. The accelerometers for each axis may also be mounted at opposite ends of the aircraft (i.e., the tail and the nose). This also improves accuracy.

Accelerometers are used in a variety of applications on modern aircraft. Originally accelerometers were found in autopilot circuits used for navigation and to determine aircraft position. Eventually as aircraft systems became more integrated and accelerometers became smaller, lighter, and increased sensitivity, engineers found additional uses for these sensors. For example, today modern aircraft employ accelerometers to measure sudden position changes caused by turbulence. An accelerometer can detect a sudden lift allowing a computer to activate flight controls, such as spoilers, and change wing efficiency lift. This creates a smoother ride for passengers and decreases stress on the aircraft structure.

produced is directly proportional to the force applied to the crystal, within the limits. If the crystal is made small enough, an acceleration force "pressing" on the crystal will create an electrical output. The electrical output is directly proportional to the pressure created by the acceleration. This is an extremely weak analog signal that is typically amplified and

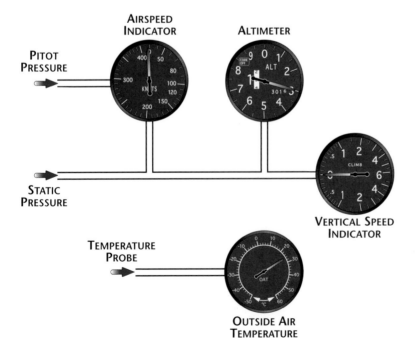

Figure 6-4-2. A pneumatic pitot-static air data system

Section 4

Air Data

To fly safely, it is essential to know the aircraft's airspeed and the current altitude. These items are difficult for the pilot to sense and are best-determined using flight instruments. To measure these basic flight parameters, the air mass surrounding the aircraft must be monitored. The measurement of this air mass is known as air data. The three air data elements typically measured are temperature, static pressure, and pitot pressure.

There are two different temperatures typically measured by an air data system: static air temperature (SAT) and true air temperature (TAT). *Static air temperature* is the temperature of the undisturbed air surrounding the aircraft. *True air temperature* is a measure of the air as it is compressed by the moving aircraft. Temperatures are an important air data reference used to improve the accuracy of other parameters and enhance the efficiency of modern autoflight systems.

Static pressure. Static pressure is the absolute pressure of the air that surrounds the aircraft. Static pressure varies inversely with the altitude of the aircraft and also changes with the general atmospheric conditions of the area. On a standard day, 59°F at sea level, the static pressure is 29.92 Hg (1013 mb). Static air pressure should be measured in undisturbed air, which is difficult to find near a moving aircraft; therefore, correction factors that are calculated by the air data computer are often employed when determining static pressure. Static pressure is used to determine the aircraft's altitude and vertical speed.

Pitot pressure. Pitot pressure is an absolute pressure of the air that enters the pitot tube. With the aircraft at rest the pitot pressure is equal to static pressure. Since the opening of the pitot tube faces the direction of aircraft travel, as the aircraft increases speed pitot pressure will increase. The difference between pitot pressure and static pressure is often referred to as *dynamic pressure*. Dynamic pressure is used to determine the aircraft's airspeed. Figure 6-4-1 shows the installation locations of typical static ports and pitot probes. Both pitot and temperature probes are typically heated to prevent ice formation. Redundant probes may be installed on the opposite side of the aircraft to avoid errors caused by aircraft yaw.

Types of Air Data Systems

There are basically three types of air data systems currently in use: pneumatic, electropneumatic, and electronic. Each of these systems can be connected to an autopilot. In general, it might be said that newer aircraft designs incorporate more electronic air data systems while older and less complex aircraft employ pneumatic systems. A pure pneumatic system relies solely on the static and pitot pressures to drive the altimeter and vertical speed indicators (Figure 6-4-2). Pneumatic air data systems can be used only on simple autopilots.

The electropneumatic air data system employs both electronic circuitry and pneumatic functions (Figure 6-4-3). The pitot and static air pressures are sent directly to an altimeter and vertical speed indicators that are often used as backup instruments. Both pitot and static pressures are also sent to the electronic air data unit along with signals from the air temperature sensors. The electronic air data unit converts the pressure data into an electrical signal and then sends that information to one or more displays and the auto pilot computer. On large

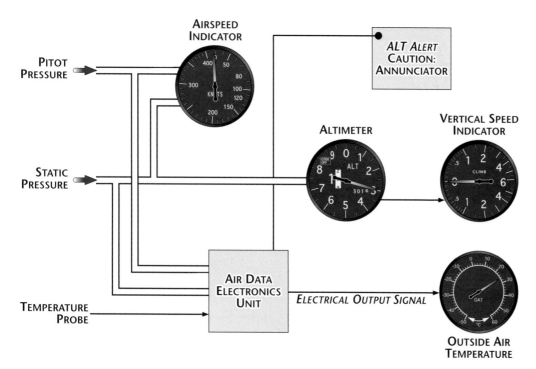

Figure 6-4-3. Block diagram of a typical electropneumatic air data system

PRIMARY FLIGHT DISPLAY (PFD)
-*SHOWING AIRSPEED, ALTITUDE, VERTICAL SPEED, AND OTHER AIR DATA*

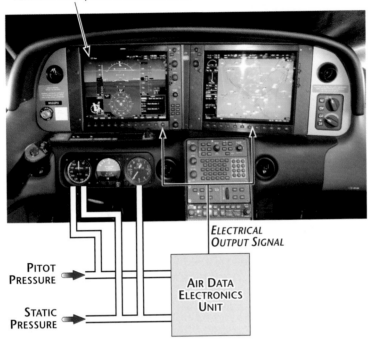

Figure 6-4-4. A modern air data system interfaced with an electronic flat-panel display

corporate or transport aircraft, electronic signals may also be transmitted to other aircraft systems, such as the flight data recorder or central maintenance computer. Figure 6-4-4 shows a simplified example of how a modern light aircraft might employ an electropneumatic air data system.

Electronic air data systems are employed on the newest aircraft and typically convert all air pressure values to an electrical signal for distribution. There are no pneumatically operated instruments on this type of system; even the backup altimeter, airspeed, and vertical speed indicators are electronic instruments (Figure 6-4-5). This design concept greatly reduces weight, especially on large aircraft, since all plumbing lines used to distribute pitot and static pressures are kept to a minimum. Electrical wiring or data bus cable is used to distribute the air data information to all end users. The electronic unit employed in this type of system is typically called an air data computer (ADC).

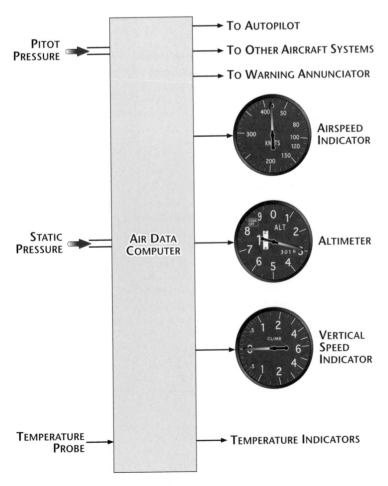

Figure 6-4-5. Distribution of air data signals to the various aircraft systems

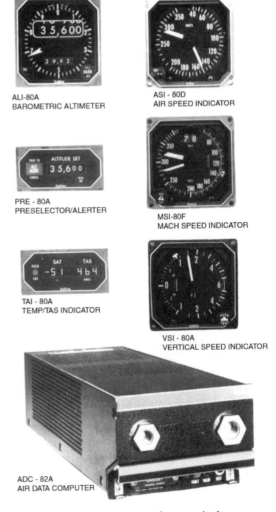

Figure 6-4-6. Components for a typical corporate aircraft air data system

Courtesy of Rockwell International, Collins Divisions

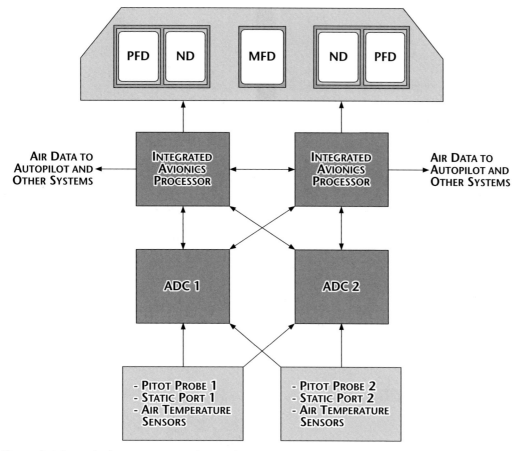

Figure 6-4-7. An air data system interface with a typical integrated display system

Air Data Computer Systems

An ADC system monitors pitot pressure, static pressure, and air temperature to determine various parameters. The air temperature input is most likely an analog electric signal produced by a temperature transducer. For most systems, the ADC would receive pitot, static, and air temperature inputs from redundant sources. The ADC will transform the pressure inputs to an electrical signal using pressure transducers. The electrical signals are then sent to the processing circuitry where the data is manipulated into useful information, such as indicated airspeed, vertical speed, altitude, etc. The ADC outputs whether either analog or digital depending on the system are sent to the autopilot, flight instruments, flight warning systems, central maintenance computer, and other systems that require air data information. The components of a typical air data system employed in corporate-type aircraft are shown in Figure 6-4-6. This system would be found in an aircraft employing individual electromechanical instruments. Aircraft with integrated display systems would send digital air data signals to some form of display management computer where they are processed and sent to the pilot's and co-pilot's PFD (Figure 6-4-7). As discussed in Chapter 3, these types of systems are often highly integrated,

employ digital data transfer systems, and many of the components share software and/or operational functions.

Maintenance and Troubleshooting

Modern air data systems incorporate built-in diagnostics that can be accessed through the aircraft's central maintenance computer system or through a specific LRU. If electromechanical instruments are used to display air data, they often employ a self-test button. Pressing the self-test button will cause a specific indication on the instrument. If this indication is not displayed, the instrument should be replaced. If the indicator passes the self-test, then there is most likely something wrong with the electronic air data unit, related wiring, or the pitot/static plumbing. On aircraft equipped with central maintenance systems, air data faults would be stored and accessed through the central maintenance computer.

In many cases, built-in diagnostics can detect the fault on the first attempt. However, even on modern systems there are many mechanical faults to the system plumbing that cannot be detected by the diagnostics. Faults with partially

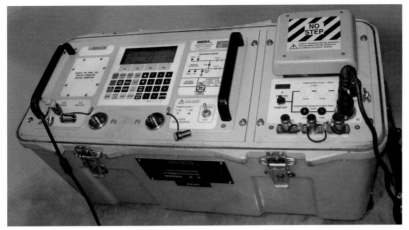

Figure 6-4-8. An air data test unit

Figure 6-5-1. A flux detector typical of those found on corporate aircraft

clogged pitot tubes or static ports or leaking plumbing lines can give inaccurate readings on air data instruments. Static ports are especially prone to clogs since their openings are relatively small. Bent or misaligned pitot tubes will also create accuracy problems. Water that may have entered the system can also cause problems. On dual systems, if only one probe is damaged the ADC may report a disagree message on the fault diagnostics. Any time an air data system malfunctions, be sure to inspect the static and pitot probes for damage or clogs.

Modern airspace around the world has become more crowded and regulatory agencies have reduced separation minimums between aircraft to help improve traffic control. In the United States, the FAA has a standard known as Reduced Vertical Separation Minimums (RVSM) that are required for most high altitude flights. The accuracy of air data instruments is critical in order to safely reduce vertical clearance. RSVM aircraft are required to perform air data accuracy checks according to the FARs every 12 calendar months. There are several commercial air data test units available from various manufacturers. A typical air data test unit is shown in Figure 6-4-8. In general, each tester will ensure accurate airspeed, vertical speed, altitude, and air temperature data. They will also test for pitot/static leaks and integrity. Many test systems will ensure the interface reliability between air data equipment and various end users.

Section 5

Compass Systems

Any autopilot that performs basic navigation functions must receive data from the aircraft's compass system. The magnetic compass is too inaccurate; therefore, the autopilot system must rely on an electrical/electronic compass system. The fluxgate Compass system employs one or more remote sensors to produce an electric signal that can be used to determine the aircraft's position relative to magnetic north. The newest aircraft integrated units, such as Honeywell Aerospace's Attitude Heading Reference System (AHRS), supply inertial reference data, magnetic heading information, and air data combined into one unit. Honeywell's AHRS employs MEMS technology accelerators, and integrates attitude sensors and digital laser gyros; eliminating the need for a fluxgate compass and performing in-flight alignment verification using a GPS cross reference. An AHRS unit of this type weighs about 10 pounds and uses approximately 20 watts of electrical power. Another common integrated system is known as the ADIRU (air data inertial reference unit). This system combines air data sensors and the components of a typical IRU to provide a variety of critical flight data including airspeed and heading information. In most cases each of these integrated systems work in conjunction with a complete flight management and/or autoflight system.

The remote sensor used in a fluxgate compass system is often called a flux detector or flux valve (Figure 6-5-1). The flux detector receives a constant input of 115 VAC, or 28 VAC, at 400 Hz. The output voltage is a function of the alignment of the detector with the earth's magnetic field. The sensing unit in the flux detector is the flux valve. The flux valve is comprised of a three-spoked frame with an output winding on each spoke (Figure 6-5-2). An excitation winding is located in the center of the flux frame. The frame is typically suspended in a sealed case on a universal joint, which allows it to pivot and remain relatively stable at different aircraft attitudes. The unit

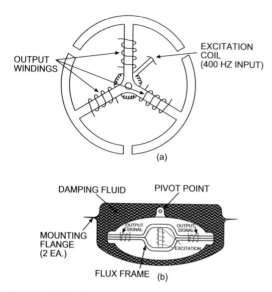

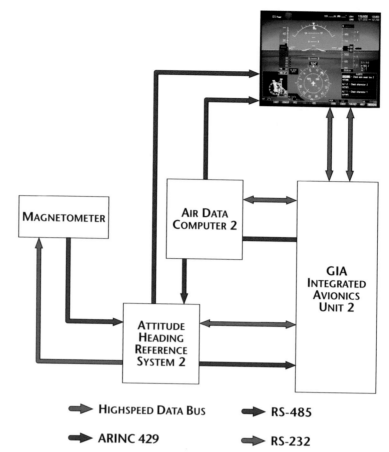

Figure 6-5-2. Flux detector internal components: (A) The flux frame, (B) The flux frame and housing assembly

is surrounded by oil to dampen the flux frame movement.

The operation of the flux detector relies on the interaction of the earth's magnetic field and the magnetic field induced in the flux frame by the 400 Hz excitation coil. Without the earth's magnetic field, each output coil of the flux frame would produce an equal voltage. As the aircraft moves with respect to magnetic north, different legs of the flux frame become saturated with magnetism. As the saturation of the frame changes, different voltages will be induced in the three output coils. Therefore, the output coils produce a three-phase AC voltage that changes characteristics relative to the aircraft's heading.

In older systems the output signals from the flux detector can be sent directly to a remote slaved compass such as those found in horizontal situation indicators (HSI). This requires a relatively large current flow in order to move the HSI; therefore the fluxgate itself must also be large. On more modern systems, the flux detector output signals may be sent to an electronic circuit where they are amplified and distributed to various systems that will use the data. This allows for the design of a smaller and more sensitive fluxgate. Systems that monitor flux detector signals include the autopilot system, the flight data recorder, the radio magnetic indicator (RMI), and/or the electronic flight instrument systems (EFIS). On some aircraft, the output of the flux detector is sent to a compass coupler. The compass coupler contains a mechanical servo/synchro combination. The output from the compass coupler is directed through relays to the RMI or HSI. The relays are used for reversionary switching to select

Figure 6-5-3. The magnetometer signal is converted into digital information on most modern systems.

which source would be used to drive which RMI or HSI.

As with many modern aircraft systems the remote compass or magnetometer, is often combined with other functions. As seen in Figure 6-5-3, the magnetometer in the Garmin G-1000 system converts all information to a digital data format. The magnetic information is sent to the AHRS unit in an RS-485 format; AHRS converts the signal to ARINC 429 data and sends compass signals to the PFD and the integrated avionics unit (IAU). The IAU uses compass data for a variety of navigation and autopilot functions.

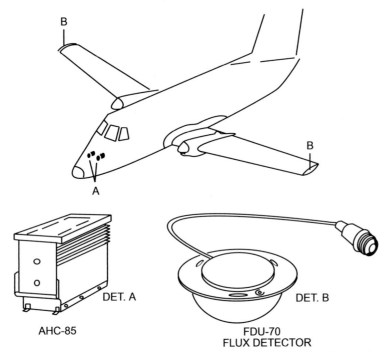

Figure 6-5-4. Typical installation locations for a flux detector and attitude heading computer
Courtesy of Rockwell International, Collins Divisions

Maintenance and Troubleshooting

In most installations, the flux detector is placed in the aircraft's wing to isolate it from magnetic interference caused by the other electrical systems (Figure 6-5-4). The mounting structure for the flux detector contains an adjustment that is used to ensure the unit is correctly aligned. This alignment becomes an important maintenance item when replacing the unit.

In general, alignment of the flux detector is done by placing the aircraft facing a known compass point and moving the flux detector until the remote compass on the HSI or PFD reads the correct compass heading. This is a simplified explanation of the procedures since most systems contain two flux detectors. Always follow the manufacturer's recommended procedures for flux detector alignment.

To isolate a defective component in the compass system, test all LRUs containing built-in test equipment. Troubleshooting the RMI or HSI typically becomes a remove and replace procedure. That is, the suspect component is swapped with a known operable unit and the system is tested. The flux valve unit is a nonrepairable item and must be replaced if found to be defective. Keep in mind the flux valve (magnetometer) creates a relatively weak electrical signal. Dirty, worn, or loose electrical connector pins can easily affect weak signals. Be sure all connectors are in good condition whenever troubleshooting compass systems.

CAUTION: *When testing any fluxgate compass system, be sure the aircraft is away from items that may interfere with the earth's magnetic field. The test must be done outside the hangar and away from other aircraft, cars, railroad tracks, and power cables. Also be aware that many metal items are not visible, such as buried power lines, fuel tanks, or concrete reinforcing rods. Placing the aircraft on a compass rose is the best way to test the system. These general practices must also be observed by the pilot during a preflight test. If a problem occurs with the fluxgate system on the ground, be sure to retest the compass system in a known environment.*

Compass system tests should be performed with various electrical equipment both on and off. If the slaved compass is affected by the operation of certain electrical equipment the problem must be fixed. First check to see if all wires near the flux detector are properly shielded. If proper shielding fails to produce the desired results, reroute electrical wires or equipment away from the flux detector to illuminate the error.

Section 6
Inertial Reference Systems

An inertial reference system (IRS) is a combination of laser gyros and accelerometers used to sense the aircraft's angular rates and accelerations. IRSs are relatively expensive and typically found only on corporate, transport, or military-type aircraft. The LRUs of a typical inertial reference system are shown in Figure 6-6-1. The laser gyros and accelerometers are installed in the inertial reference unit (IRU), which is typically installed in the aircraft's equipment bay. The IRU also contains the computer circuitry for signal processing and system interfacing.

The data produced by an IRS is used in conjunction with a total autoflight system. The IRS data is typically combined with air data outputs to compute:

1. Attitude (pitch, roll, yaw)

2. Angular rate changes (pitch, roll, yaw)

3. Aircraft velocity

4. Course track angle

5. Inertial altitude

6. Linear accelerations

7. Magnetic heading

8. True heading

9. Position (latitude and longitude)

10. Vertical speed

11. Wind speed

12. Wind direction

The output data from the IRS is a primary input for a modern autoflight system. IRS outputs are also sent to electronic flight instrument systems for display of attitude and navigational data. IRS data is sent to the flight data recorder along with other aircraft systems.

Many of the latest IRSs are so accurate, the need for a fluxgate compass is eliminated. For example, aircraft such as the B-757/767, the B-747-400, and the A-320 use the IRS for magnetic heading data. The IRS sends magnetic compass data to the RMI and/or EFIS for display to the flight crew.

Some state-of-the-art inertial reference systems integrate IRS with magnetic heading and air data functions. The advanced Attitude Heading Reference System (AHRS), by Honeywell Aerospace, integrates attitude sensors and laser gyros. This system uses MEMS technology accelerators and advanced digital circuitry to miniaturize component size and save weight. The AHRS unit eliminates the need for a fluxgate compass and performs in-flight alignment verification using a GPS cross reference. The Honeywell system weighs approximately 10 pounds and uses approximately 20 watts of electrical power. Another common integrated system is known as an air data inertial reference unit (ADIRU). This system combines air data sensors and the components of a typical IRU to provide a variety of critical flight data including air speed and heading information. In most cases, integrated systems work in conjunction with a complete flight management and/or auto flight system.

Initialization

Since an IRS can only measure changes in position, the unit must be given a starting reference point. The procedure used to provide the IRS with initial latitude and longitude is called *initialization*. Initialization typically occurs at the aircraft gate before the first flight of the day. If the aircraft has not been moved overnight, the position in memory can be used. If the aircraft has been towed to a new location, the crew must enter the correct latitude and longitude into the IRS, typically using a multifunction alphanumeric keyboard.

During initialization, the IRS accelerometers measure the direction of the earth's gravity

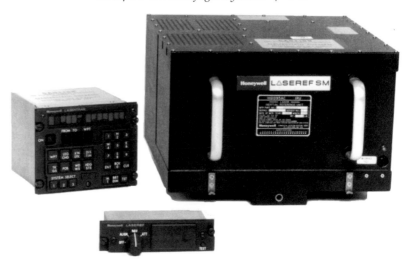

Figure 6-6-1. Components of a typical laser inertial reference system

Courtesy of Honeywell, Inc.

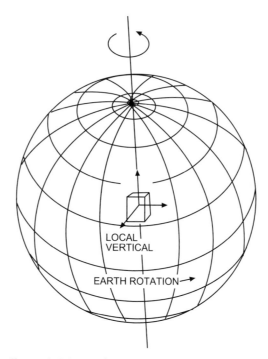

Figure 6-6-2. Local vertical is measured between the aircraft's location and the rotational axis of the earth.

force to determine the aircraft's local vertical. *Local vertical* is a direction perpendicular to the rotational axis of the earth that intersects the aircraft's position (Figure 6-6-2). During initialization, the IRS rate sensors measure the speed and direction of the earth's rotation relative to the aircraft. This, along with the latitude, longitude, and local vertical allows the system to determine true north. At the completion of the initialization process, the IRS computer contains the necessary data to compute the aircraft's current position and heading. Initialization takes approximately five to ten minutes and the aircraft cannot be moved during this time.

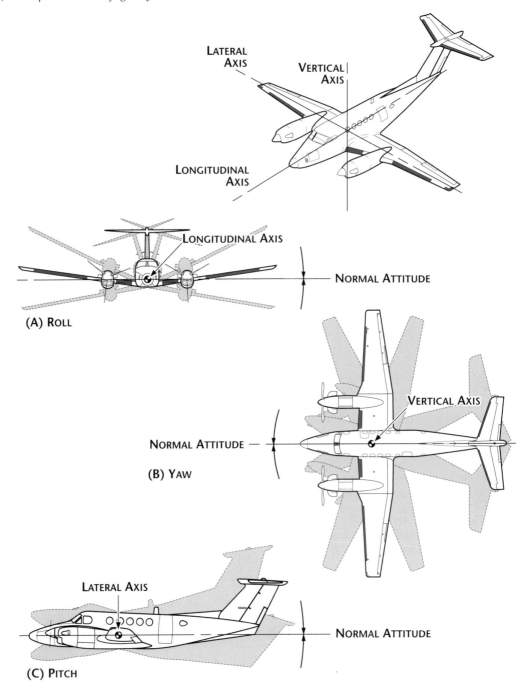

Figure 6-6-3. The three axis of the aircraft; one IRS unit must be aligned with each axis

Theory of Operation

Each IRS unit is made up of three laser gyros and three accelerometers. One each of these units is aligned with the pitch, roll, and yaw axis of the aircraft (Figure 6-6-3). Figure 6-6-4 shows the three laser gyro assemblies and accelerometers within a typical IRU. The three gyros measure angular displacement about their respective axis (pitch, roll, and yaw). The accelerometers are used to measure the rate of acceleration about each axis. Each of the three axes must be monitored since the aircraft travels in three-dimensional space. Also, most aircraft will contain two or three

IRUs, each with the capability to monitor all three axes of the aircraft. Multiple IRUs provide the redundancy needed for safety and reliability.

Once the IRU has been initialized, the system knows where it is located in all three dimensions and current heading. As the aircraft moves in any direction from its initial position, the IRS will sense the movement and compute the new location and heading using a high speed processor. Using Figure 6-6-5 as an example, assume the aircraft is stationary at a given point in space (Point A). If the longitudinal accelerometer measures

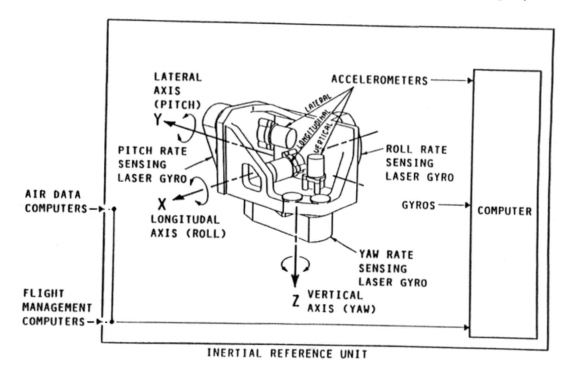

Figure 6-6-4. The typical configuration of an inertial reference unit.

Courtesy of Northwest Airlines, Inc.

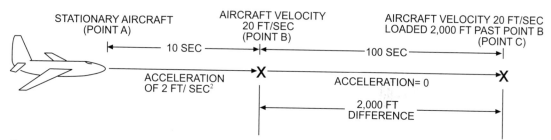

Figure 6-6-5. The IRS measures acceleration and time to calculate the aircraft's change of position.

an acceleration of 2 ft/sec²; this mean the aircraft is accelerating forward. After ten seconds the aircraft would be flying at a velocity of 20 ft/sec (10 x 2 ft/sec² = 20 ft/sec). Assume the aircraft stops accelerating and the velocity remains constant at Point B. If the aircraft continues to fly with a velocity of 20 ft/sec for 100 seconds the aircraft's new location (Point C) is 2,000 feet from Point B. Distance equals velocity multiplied by time (20 ft/s x 100 = 2,000 ft).

The IRS computer performs similar calculations for the angular rate changes measured by the laser gyros. Assuming the IRS detects a yaw rate of 5° per second for 15 seconds, the computer would determine that the heading has changed 75° from the aircraft's original heading. The IRS computer continuously performs acceleration and angular rate calculations for all three axes. By measuring both accelerations and angular rates, the IRS can provide a constant update on the aircraft's location and heading. Heading and location information are constantly being compared on multiple IRU systems to ensure accuracy.

Several other factors that can affect the accuracy of the IRS are the:

1. Earth's rotation at approximately 15.04° per hour

2. Spherical shape of the earth meaning aircraft do not travel in a straight line over the surface

3. Laser gyro is subject to drift over time. This drift is much less than rotating mass gyros; however, it is still important to consider

To compensate for these inherent errors, the IRS software is programmed to make the necessary corrections while processing the data. On many systems, another means of ensuring accuracy is to periodically cross-reference location data with other navigational aids, such as GPS.

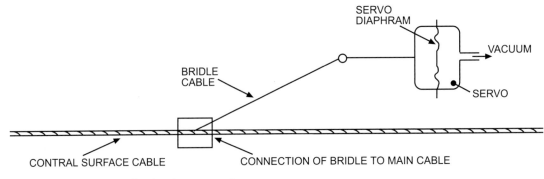

Figure 6-7-1. Diagram of a simple pneumatic servo

Maintenance and Troubleshooting

Most aircraft that employ an IRS also contain some type of centralized maintenance computer system (CMCS). IRS troubleshooting is typically accomplished using this system. Faults are stored in a nonvolatile memory and displayed when requested by the technician. In most cases, the aircraft will contain two or three IRUs. Each of these units is interchangeable and can be swapped to help identify a defective unit, or reversionary switching can be used to swap units from the flight deck. Whenever removing any IRU, make sure to handle the unit with care; the IRUs are fragile and can be severely damaged if dropped. Any unit that has been dropped is not airworthy. Whenever shipping an IRU, be sure to use the appropriate shipping container to help protect the unit.

Whenever testing an IRS, be sure to allow proper time for the aircraft to initialize. The aircraft must remain stable during the initialization process. Large wind gusts or maintenance being performed on the aircraft may upset the initialization procedure. In this case, the procedure should be repeated on a "quiet" aircraft. Never condemn an IRS that will not align on the first attempt, it may be caused by a moving aircraft.

Installation of the IRS in the correct position with respect to aircraft axes is also an important consideration. For example, if the IRS unit was installed with the gyros and accelerometers out of alignment, the system could not produce accurate data. This is typically not a problem for removal and replacement since the LRU is installed in a rack that is permanently mounted to the aircraft. However, if the rack should become bent, cracked, or somehow misaligned, the IRS will not work properly. Some modern IRUs can be electronically aligned. In this case, the physical alignment is less critical and the fine adjustment is made using software corrections.

Section 7

Inertial Navigation Systems

A modern inertial navigation system (INS) uses airborne equipment for aircraft navigation without relying on external radio signals. Laser gyros and accelerometers provide three dimensional navigation capabilities. This system is employed mainly by the military. The major advantage of the system is that it requires no external navigation aids. All the equipment for worldwide navigation is contained in the INS.

Many older transport category aircraft, such as the DC-10, early B-747s, and the L-1011 employ an INS which utilizes rotating gyros, gimbal platforms, and accelerometers to sense aircraft position. This system is called INS since it is responsible for coordinating the aircraft's navigational parameters, including flight plan and waypoint selection. The INS can be interfaced with the autopilot or flight director system to steer the aircraft. The difference between and INS and IRS is that the INS provides waypoint and flight plan capabilities; a modern day IRS must work in conjunction with a flight management system (FMS) to provide these functions.

Servos

A servo is a device used to apply a force to the aircraft's control surface in response to an autopilot command. There are basically three types of servos: pneumatic, electrical, and hydraulic. There are also hybrid servos; these are typically combinations using hydraulic activators combined with an electric motor. Each servo must incorporate some type of mechanism so the pilot can override the autopilot command. They also typically contain a feedback system that provides a return signal to the autopilot computer.

Pneumatic Servos

Pneumatic servos are vacuum actuated units used on simple autopilots for light aircraft. As seen in Figure 6-7-1, the pneumatic servo operates using a vacuum applied to the servo diaphragm. The autopilot computer controls the vacuum. Two servos are required for each control surface. A bridle cable is used to connect the servo to the control surface cable. Pneumatic servos offer limited range of travel and provide a relatively weak actuating force; therefore, pneumatic servos have limited use. Today pneumatic servos are only found on older aircraft using simple autopilot systems.

Electric Servos

Electric servos utilize an electric motor and clutch assembly to move the aircraft's control surface according to autopilot commands (Figure 6-7-2). Due to their reliability and excellent torque production, electric servos are commonly found on all types of aircraft, including trainers, personnel, corporate-type turbine and turboprop aircraft. A bridle cable is installed between the servo's capstan assembly and the control surface cable. The capstan is used to wind/unwind the bridle cable; hence, moving the control surface. Figure 6-7-3 shows a capstan with the test fixtures installed in preparation for adjustment of the slip clutch assembly. This figure also shows the torque adjustment nut for the slip clutch. The slip clutch gives the pilot the ability to overpower the unit in the event of a servo malfunction.

Most electric servos use some type of clutch assembly to connect the servo motor to the capstan. During manual operation of the controls, the clutch is disengaged and the capstan moves freely. During autopilot operation the clutch is engaged and the capstan is connected to the servo motor. The clutch engage/disengage is accomplished using an electromagnet simulator to a solenoid. The clutch is engaged when the electromagnet is energized and the servo is active. The autopilot computers and pilot activated controls are used to energize the servo clutch. Virtually all autopilot systems employ an autopilot disconnect switch that is typically a push-button located on the control wheel. This gives the pilot the ability to instantly disconnect the autopilot function and manually fly the aircraft. The electric motor and clutch assembly usually operate on direct current; however, some transport category aircraft may employ AC motors. In order to keep capstan r.p.m. relatively low while the motor operates at high speed, the unit contains a gear reduction assembly. The gear assembly is normally self-lubricating and requires no regular maintenance.

Hydraulic Servos

Hydraulic servos are the most powerful type of servo actuator; hence, these units are typically used on transport category aircraft. Ever since the B-727 and DC-9 took to the air, transport category aircraft have employed hydraulically operated control surfaces. In brief, the systems operate using an engine-driven hydraulic pump and employ a control valve to route hydraulic fluid to a control surface actuator. The control surface actuator is mechanically linked to the control surface. On most transport category aircraft, the control surface actuator is linked to the control wheel and rudder pedals through control cables. On the newest aircraft, like the A-380 and B-787, the control wheel or side stick controller and rudder pedals are connected to the control surfaces via electrical wiring and computer circuits.

The Airbus A-320 hydraulic servos operate in two different modes: active and damping. As seen in Figure 6-7-4, the active mode is employed when the servo valve is pressurized and the Elevator Aileron Computer (ELAC) energizes the solenoid valve. In the active mode, hydraulic fluid is controlled by the servo valve and directed through the mode selector valve to the aileron actuator.

In the damping mode, the actuator follows control surface movement as hydraulic fluid is allowed to flow through the restricting orifice. The servo is in damping mode whenever the solenoid is de-energized or hydraulic pressure is not supplied to the servo valve. In this

Figure 6-7-2. A typical electric servo

Courtesy of Rockwell International, Collins Divisions

HOLD DOWN SCREW

CAPSTAN HOLDING FIXTURE

CAPSTAN LOCKING PEN

CAPSTAN

CAPSTAN TEST FIXTURE

TORQUE WRENCH NUT

TORQUE ADJUSTMENT NUT

Figure 6-7-3. An electric servo capstan mounted to the capstan test fixture

Courtesy of Rockwell International, Collins Divisions

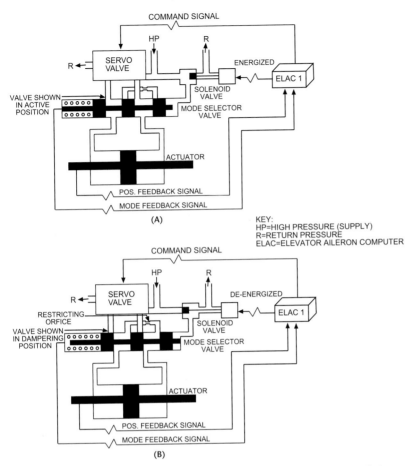

Figure 6-7-4. Diagram of a typical hydraulic servo and related control circuitry: (A) Active mode, (B) Damping mode

situation, the mode selector valve moves to the right via a spring force and connects the actuator hydraulic fluid to the restricting orifice. The mode selector valve and the main aileron actuator both produce a feedback signal to the computer (ELAC 1).

Although there are a large variety of hydraulic servos for transport category aircraft, they all operate in a similar fashion. Each hydraulic servo will contain some type of servo or control valve, typically actuated by an electric solenoid. The control valve will move hydraulic fluid to the main actuator in order to move the control surface. Every unit will also contain some type of bypass or damping mode in the event of system failure. On fly-by-wire aircraft, the flight deck controls are moved via an artificial force produced by the autoflight computer(s). On more traditional aircraft, the control wheel and rudder pedals move via a cable linked to the actuators.

Hybrid Servos

Some modern aircraft, such as the B-787 or A-380, employ servos that combine hydraulic actuators and electric drive motors. These electro-hydraulic servos are self contained units which do not require a connection to the aircraft's main hydraulic system. An independent servo actuator is more reliable and allows the aircraft manufacturer to reduce or illuminate the plumbing needed for a central hydraulic system. This reduces overall aircraft weight. This type of servo receives an electronic signal which controls an electric motor contained within the actuator assembly. The electric motor is directly coupled to a hydraulic pump which supplies the pressurized fluid; the actuator moves accordingly. This type of actuator is independent of the central hydraulic system, but requires a large supply of electrical current to drive the pump motor.

Other types of hybrid servos employ combinations of electric drive motors and actuators connected to a central hydraulic system. The main hydraulic system powers the actuator under normal flight conditions, however; if a failure occurs the electric motor is available for back up. All hybrid servo actuators are electronically controlled; often with a digital signal sent trough a high speed bus. Therefore, modern servos often contain the circuitry necessary to decode digital data in order to operate the actuator and produce the correct control surface movement.

Servo Feedback Systems

As mentioned earlier, all servo systems must contain some type of feedback circuit to inform the autopilot computer that the control surface has moved. The feedback system produces an electrical signal that is directly proportional to the movement of the servo actuator. There are two common devices used to generate the feedback signal: an AC synchro and a differential transducer. Synchros are typically employed on electric and hydraulic servos used in conjunction with analog autopilot systems. State-of-the-art digital autopilot systems often employ differential transducer feedback systems.

Synchro Systems

The most common autopilot feedback synchro is a transformer-like device that monitors angular displacement using a stationary primary winding and pivoting secondary winding. As shown in Figure 6-7-5, the primary winding receives an input voltage of 26 VAC 400 Hz. The output voltage of the secondary is a function of the angular position of the secondary winding. Figure 6-7-6; position number one, shows the secondary winding in the null position or perpendicular to the primary winding. In this position, no voltage is induced in the secondary. As the secondary rotates clockwise, the voltage induced in the secondary increases until the secondary is parallel to the primary (position number four). The voltage then decreases as the rotor continues to turn clockwise. A second null is reached when the rotor becomes horizontal to the primary once again (position number seven). The secondary voltage is in phase with the primary voltage for rotor positions two through six.

As shown in Figure 6-7-6, when the secondary winding rotates past the second null (position number 7), the output voltage is 180° out of phase with the primary voltage. The voltage value continues to change as the secondary continues to rotate clockwise. The out-of-phase condition exists until the rotor reaches the first null position once again. The output voltage is 180° out of phase with respect to the input voltage in positions eight through twelve.

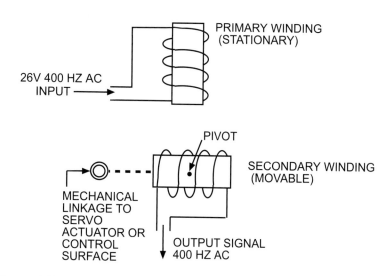

Figure 6-7-5. Components of a typical autopilot feedback synchro

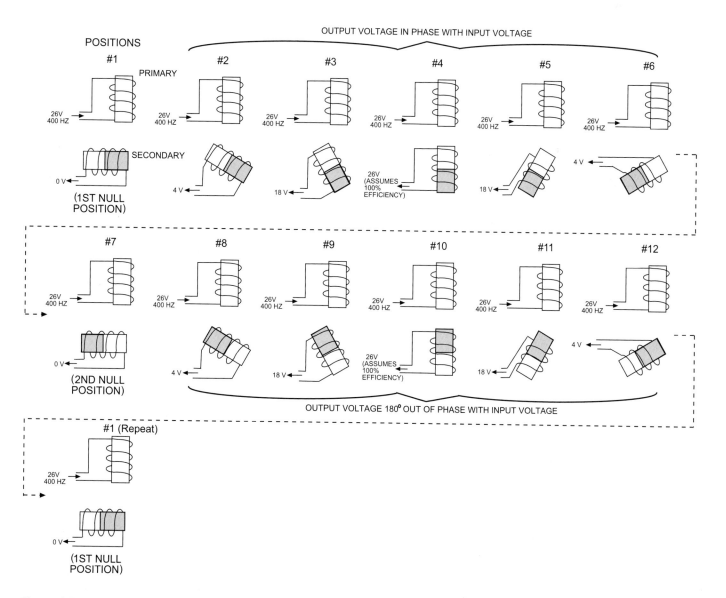

Figure 6-7-6. Voltage and phase relationship of an autopilot feedback synchro as it rotates 360°

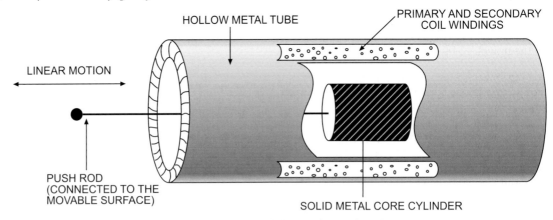

Figure 6-7-7. Cut-a-way of a LVDT (linear voltage differential transducer)

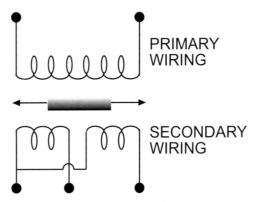

Figure 6-7-8. Wiring diagram of a LVDT

AC synchros provide excellent feedback signals for many autopilot systems. The phase shift principle, discussed above, allows for accurate measurements of even small control surface movement. When placed in the null position, any movement clockwise or counter clockwise is easy to measure due to the phase shift and voltage change. The most accurate measurements using a synchro are therefore obtained near the null positions. On most systems, the synchro rotor is connected to the servo output or control surface through a mechanical linkage, hence the synchro rotor moves in unison with the control surface.

Troubleshooting Synchro Systems

Most synchro systems are fairly reliable. The electrical components are simply wire coils and therefore seldom fail. The secondary pivot bearing can fail or become worn which causes inaccurate feedback signals. Likewise, if the mechanical linkage connecting the synchro becomes worn or binds during movement, inaccurate signals will result. The mechanical linkage and pivoting secondary coil are critical and must have free movement. On many systems, adjustment of the synchro to the null position is critical for proper operation. Many electric servos drive the synchro to the null

position prior to engaging the servo clutch. This ensures the synchro starts in the null each time the autopilot is engaged.

The electrical system of a synchro can be tested for the proper input voltage to the primary. In most cases, the input is 26 VAC 400 Hz. The synchro primary and secondary coils can be tested for continuity and shorts to ground. When measuring continuity, it is critical that the coil resistance be within specifications. A change in resistance of just a few ohms can create inaccurate readings most likely resulting from a breakdown of the coil's insulation. If an insulation breakdown is suspected, be sure to monitor the system closely during the next several hours of operation.

Differential Transducers

The differential transducer is typically used to provide a feedback signal from hydraulic servos. There are two common types of transducers used for autopilot feedback systems: the linear voltage differential transducer (LVDT) and the rotary voltage differential transducer (RVDT). LVDTs and RVDTs produce a relatively weak electrical signal and are found on modern digital autoflight systems. Since they are used in conjunction with hydraulic servo systems, both LVDTs and RVDTs are typically found on transport category and high performance corporate-type aircraft. LVDTs and RVDTs are also found in other non-autopilot systems to measure position or rate of motion. For example, the Airbus A-320 employs an LVDT on the turbine engine to measure stator vane position. The nose wheel steering system of the A-320 employs RVDTs to measure nose wheel position.

The increasing popularity of LVDTs and RVDTs stems from the simplicity of their design. As seen in Figure 6-7-7, the LVDT consists of a hollow metallic tube and a solid metal cylinder that is allowed to slide inside the tube. Around the tube are two electrical windings, a primary and secondary, similar to a transformer. A push rod is used to connect the solid metal cylinder to the movable object that is being monitored.

An LVDT or RVDT is a mutual inductive device. The primary winding is flanked by two secondary windings as shown in Figure 6-7-8. The secondary windings are wired to form a series opposing circuit. The primary receives an alternating current. This AC will induce a voltage into both secondary windings. If the core material is exactly centered, the output signals from the secondary windings will cancel. As the core is displaced from center, the output signal increases in amplitude. The phase of the AC output signal, with respect to the input signal, is determined by the direction of core displacement. Hence, the transducer can measure both direction and magnitude of any movement. Figure 6-7-9 shows the relationship between core position and the secondary output signal.

An RVDT operates similarly to a LVDT except it is designed to detect rotational movement. The RVDT contains a heart-shaped core material that rotates within a hollow tube (Figure 6-7-10). As the core is rotated, it changes the output voltage and phase of the secondary. The rotational movement measured by an RVDT is typically 120° or less, and the highest resolution is obtained in the first 40° of rotation. An RVDT contains two bearing assemblies, required to support the input shaft.

Both LVDTs and RVDTs are always used in conjunction with some type of electronics circuitry. The circuit is used to interpret the output signals of the LVDT or RVDT. On many aircraft, the secondary's output signal is sent to an LRU, which converts the AC voltage and phase relationships into usable data. Figure 6-7-11 shows a hydraulically operated yaw damper servo containing an LVDT. The output signal

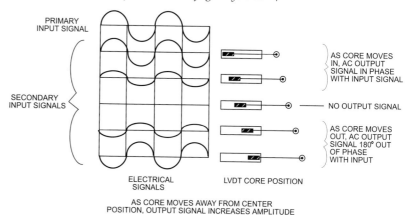

Figure 6-7-9. The relationship between core position and LVDT output

from the LVDT is sent directly to the Flight Augmentation Computer (FAC).

Troubleshooting LVDTs and RVDTs

Both LVDTs and RVDTs are relatively maintenance free. The LVDT contains only one moving part, the core, which is typically supported so there is no contact between the core and the coil housing. The RVDT core is supported by two bearings that have extremely long life due to the light loads on the input shaft. Both LVDTs and RVDTs therefore have virtually infinite mechanical life unless damaged by some external force.

Electrically, the primary or secondary winding of the transducer may fail due to an open or shorted circuit. In many cases, an ohmmeter can be used to detect these failures. Simply disconnect the transducer from the aircraft wiring and perform a continuity test of the coils. Opens most often occur due to induced stress on the windings caused by vibration, failed solder, or crimped connection. However, even

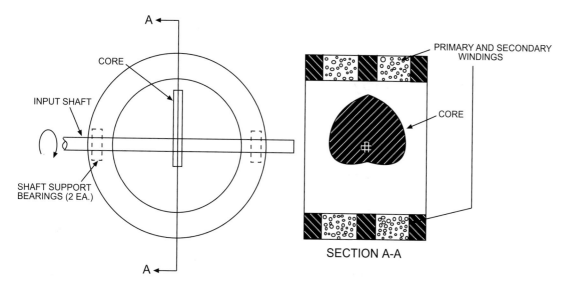

Figure 6-7-10. Internal components of a typical RVDT

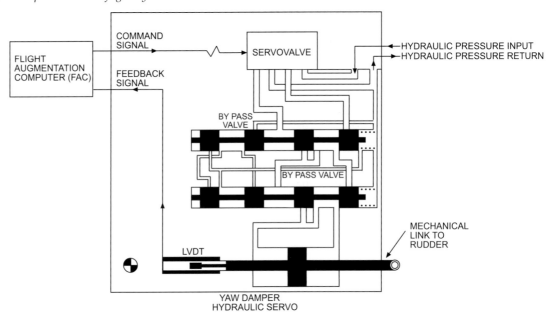

Figure 6-7-11. An LVDT used in a yaw damper servo

these failures are rare and the MTBF (mean time before failure) of a typical aircraft quality transducer is over one million hours.

Another means of troubleshooting an LVDT or RVDT is to measure the input and/or output voltage of the transducer. The input signal must be an AC voltage within the limits established by the manufacturer. A high impedance voltmeter must be used for this test to ensure the meter does not distort the signal being measured. The output signal from the secondary winding is determined by the position of the core. A dual channel oscilloscope can be used to show the voltage and phase relationship of the input and output signals. A voltmeter can be used to measure the transducer output voltage as the core changes position. This test is typically sufficient since it is virtually impossible to change the

phase relationship of the output signal once the transducer is installed correctly.

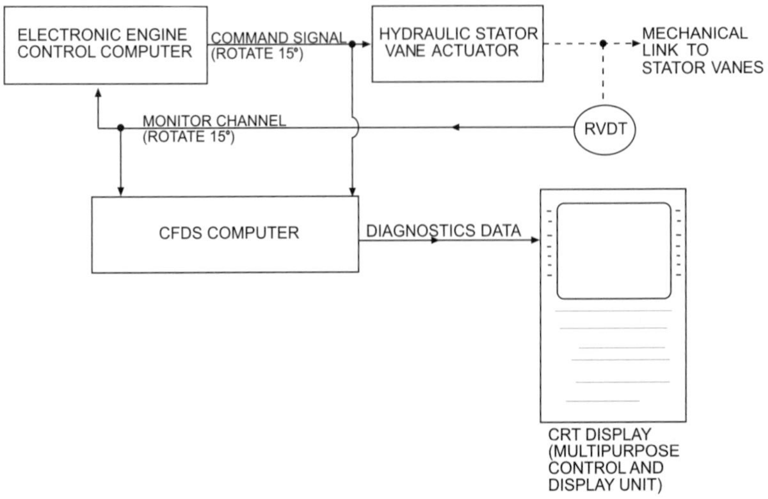

As noted earlier, the output voltage of the secondary should be zero when the core is exactly centered (i.e., located at the null position). This condition rarely exists in the real world since the excitation voltage often contains high-level harmonics, which induce stray voltages into the secondary. Whenever measuring the output voltage at the null position, remember that a small AC signal, approximately 0.25 percent of maximum, may be acceptable. If the null position output voltage exceeds this amount, be sure to check the purity of the input voltage to the primary.

On transport category aircraft, the troubleshooting process for LVDTs and RVDTs often becomes simplified through the use of built-in test equipment (BITE) or central maintenance systems. The integrated test equipment can be used to monitor the output of the transducer as the unit is moved through its operating range. On many aircraft, a comparison is made of input and output signals to verify correct operation. For example, on the A-320, the Centralized Fault Display System (CFDS) monitors engine stator vane position (Figure 6-7-12). The CFDS checks the input signal to the stator vane actuator and compares that to the output signal from the RVDT, which monitors the stator vane position. Using the CDFS, the technician can read command channel and monitor channel signals on a CRT display. The stator vane's position is measured in angular degrees. On this system, a 1° tolerance is allowable. The CFDS also performs continuous fault monitoring of the stator vane positioning system. If a fault were detected during flight,

Figure 6-7-12. RVDT output signals monitored by the centralized fault display system

the CFDS would take the necessary corrective actions and record the failure in memory for later recall. A transducer system found on the Boeing 767 is used to monitor aileron travel as specified by the autoflight computer. On this aircraft, the CMC (Central Maintenance Computer) monitors signals from three LVDTs. The technician can display transducer output on a CRT for comparison purposes.

Whenever troubleshooting an LVDT or RVDT, always remember that these units are extremely reliable. In most cases, the associated wiring or electrical connectors are more likely to fail than the transducer itself. However, if a transducer is to be replaced in the field, take caution to ensure the proper installation of the new unit. Note any markings on the core and/or housing assembly. Be sure the core is installed in the correct configuration. Some LVDTs and RVDTs can only be replaced at an overhaul facility. In this case, if the transducer has failed, the entire autopilot servo assembly must be replaced in the field. Whenever changing any transducer or servo assembly, always follow the manufacturer's instructions on installation and rigging very carefully. Some installations may also require a flight test of the autopilot system.

Tachometer Generators

Tachometer generators, or tach generators, are often used in electric servo systems as rate sensors. The tach generator measures the rotational speed of the electric motor and provides feedback to the servo amplifier or autopilot computer. This feedback signal is typically used to regulate and limit motor speed. The tach generator consists of a permanent magnet and armature assembly. The generator spins in direct relationship to the servo motor. Hence, the electrical signal produced by the generator is directly proportional to the motor's movement. A typical tach generator installation is shown in Figure 6-7-13. Tach generators can easily be tested for correct operation by measuring the output voltage as the generator spins. If the generator produces inadequate voltage it must be replaced.

Section 8

Collins APS-85 Autopilot System

The APS-85 is a typical digital autopilot system found on high performance corporate-type aircraft. The system includes both autopilot and yaw dampening capabilities. The components

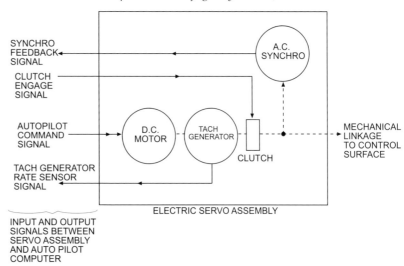

Figure 6-7-13. An electric servo assembly containing a tach generator

include a mode select panel, (two panels are used for a dual system), a flight control computer, an autopilot panel, three primary servos, and three servo mounts (Figure 6-8-1). The APS-85 Flight Control Computer (FCC) is a dual channel system that provides redundancy for the autopilot. The FCC, located in the equipment rack, is cooled with forced-air. The three servos are mounted throughout the aircraft in appropriate locations for their respective control surfaces. The control panels are located on the flight deck and are accessible to both pilots.

Autopilot panel. The autopilot panel (APP) shown in Figure 6-8-2 contains the main controls for the system. The autopilot and yaw damper switches are guarded levers and must be raised to engage the respective systems. On some aircraft, the autopilot may be engaged independent of the yaw damper; on other aircraft, both systems must be engaged simultaneously. Prior to activating the autopilot/yaw damper, the FCC monitors the system for faults. If a fault is detected, the FCC will not activate the autopilot/yaw damper. Whenever the autopilot/yaw damper is engaged, the appropriate message is displayed by EFIS.

The APP *pitch wheel* is a spring-loaded rotary switch (Figure 6-8-2). Moving the pitch wheel up or down modifies the vertical reference being flown by the autopilot. As a new vertical reference is entered, the value is displayed by EFIS. The *turn knob* is a bidirectional switch used to initiate a roll mode and define a given roll rate. The turn knob is inoperative while the autopilot is in the approach mode. The autopilot transfer (*AP XFR*) switch is used on systems that employ a dual flight guidance system. The transfer switch is used to shift from the left to the right FCC. The turbulence (**TURB**) push-button switch is used to "soften" the ride when fly-

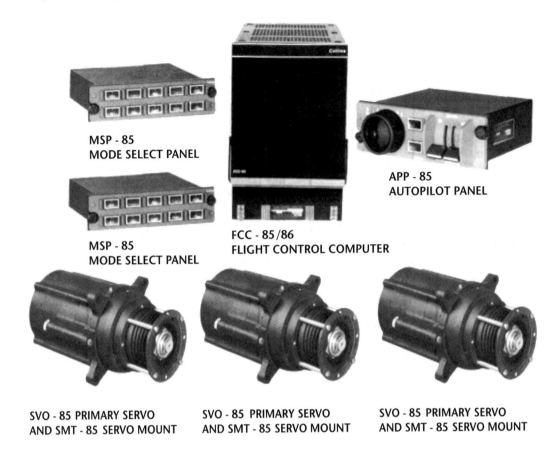

Figure 6-8-1. System components of the APS-85 autopilot   *Courtesy of Rockwell International, Collins Divisions*

ing through rough air. When in the turbulence mode, the FCC lowers autopilot gain signals. This degrades the intensity of control surface movement.

Mode Select Panel

The mode select panel (MSP) consists of ten push-button switches used to control the various autopilot modes (Figure 6-8-3). The various mode select push-buttons are somewhat self descriptive. The **HDG** (heading) switch commands the autopilot to steer a given heading. The **1/2 BANK** mode reduces all bank angles to approximately 13.5 degrees. The 1/2 bank mode is inoperative during approach to land. The **NAV** (navigation) mode causes the autopilot to follow the navigation source currently displayed on EFIS. The **APPR** (approach) mode is used when the pilot wishes to navigate via a localizer and glide slope signal during an approach to land. The NAV and APPR modes can be armed without actually "capturing" the mode. Capture of the mode occurs only when valid navigation signals are available. If no valid navigation signal (such as the localizer) is received the approach mode will be armed (not captured). The autopilot will fly the aircraft's last heading until capture occurs.

Vertical mode switches include the:

- **CLIMB** button is used to activate a given climb rate (IAS or MACH) according to FCC software. Three different rates can be selected (low, medium, or high speed) using the **PERFORMANCE SELECT** button.

- **ALT** (altitude) button is used to maintain the current barometric altitude of the aircraft.

- **VNAV** (vertical navigation) mode is used to fly a vertical profile established by the flight management system.

- **DESCEND** mode commands the autopilot to fly a preprogrammed descent rate. The APP pitch wheel can be used to increase or decrease that rate.

- **SPEED** mode will cause the autopilot to fly a given speed by adjusting aircraft pitch accordingly.

Flight control computer. The flight control computer (FCC) is a dual channel unit designed to receive input data, process the information, and send the appropriate outputs to the autopilot servos and the electronic flight instrument system. While the autopilot is engaged, the FCC controls aircraft attitude through the control surface servos. With the autopilot engaged

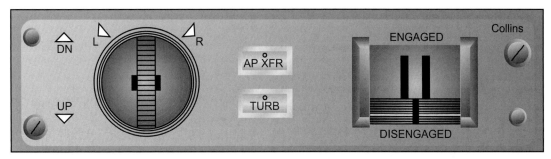

Figure 6-8-2. The APS-85 autopilot panel

Courtesy of Rockwell International, Collins Divisions

Figure 6-8-3. The APS-85 mode select panel

Courtesy of Rockwell International, Collins Divisions

or disengaged, the FCC controls the V-bar position on the EADI. As mentioned earlier, the V-bars are part of the flight director system that provides visual reference to the pilot.

Servos. The APS-85 uses three electrically actuated servos. Each servo is equipped with an engage/disengage clutch that allows for quick response time of the servo mechanism. The servos also employ a slip clutch that is used as a backup for manual override. The APS-85 is designed to interface with the aircraft's trim motor assembly and therefore the aircraft controls do not require additional trim servos.

Theory of Operation

Refer to the block diagram in Figure 6-8-4 of the APS-85 system during the following discussion on theory of operation. The FCC contains two channels (A and B), which receive identical inputs for data processing. The dual channels share the same FCC housing, yet operate completely independent. The FCCs perform system monitoring to ensure autopilot reliability and present all diagnostic data through the aircraft's EFIS displays. The FCC outputs control servo operation and flight director displays on the EADI.

A voter circuit is contained within each channel of the FCC (Figure 6-8-5). The voter circuit determines which channel calls for the least servo movement and sends that signal to the motor. A torque limiter is used as a current limiting device, which allows the pilot to manually

overpower the motor in the event of a servo runaway.

Flight Control Computer Interface

Inputs to the FCC include:

1. Digital data in CSDB format from the two MSPs, the air data system, the Attitude Heading System (AHS), and EFIS. EFIS supplies all navigation inputs to the FCC. The FCC will also accept AHS data in ARINC 429 format.

2. Discrete inputs are received from the annunciator test switch, the autopilot disconnect switch, the go-around switch, the flaps switch, configuration strapping, options strapping (Lower center of Figure 6-8-4) as well as the pilot's and co-pilot's sync switches.

3. The APP sends analog data to the FCC for pitch and roll commands and autopilot/yaw damper engage commands.

4. The three servo units send analog servo rate data to both channels of the FCC (Figure 6-8-5).

The FCC outputs include:

1. A two wire 28 VDC analog signal to each servo motor.

2. A CSDB output is sent to EFIS with flight director information. The same data is sent to the air data system.

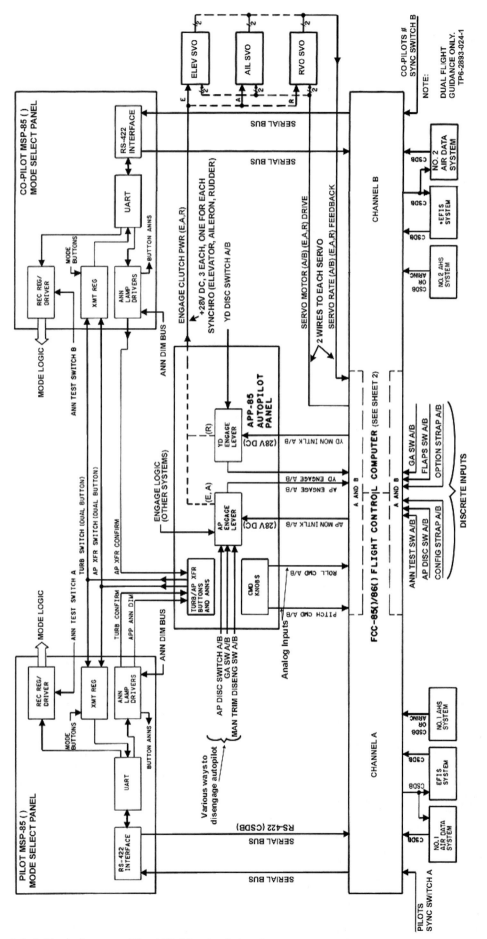

Figure 6-8-4. Block diagram of the APS-85 autopilot system   *Courtesy of Rockwell International, Collins Divisions*

3. A CSDB output is sent to both MSPs. This signal informs the MSP of the current FCC operating mode(s).

Attitude Heading System

The APS-85 is designed to interface with a unique attitude heading reference system, the AHS-85. The AHS-85 measures angular rates and accelerations along all three aircraft axes using piezoelectric sensors (Figure 6-8-6). The AHS-85 system employs an Attitude Heading Computer (AHC) containing the piezoelectric sensors. The piezoelectric sensors replace the accelerometers, gyros, and rate sensors found on conventional attitude heading systems (Figure 6-8-7).

The AHS-85 AHC uses a conventional flux detector. The flux detector is needed to provide a magnetic heading reference to the system. Since each aircraft has slightly different magnetic characteristics, a compensator unit must be used to correct for magnetic errors and flux detector misalignment. Remember, the compensator unit corrects for a specific aircrafts magnetic error, and therefore must stay with the aircraft when changing other attitude heading system components.

The AHC contains a dual sensor assembly that houses two rotating wheels mounted at 90° angles from each other (Figure 6-8-8). The spinning wheels, which rotate at a constant 2,500 r.p.m., contain four piezoelectric crystals called benders. One pair of benders measures acceleration, the other pair of benders measures rate changes (Figure 6-8-9). A conventional rotating mass gyro spins at 20,000 r.p.m. The slower rotational speed of 2,500 r.p.m. on the AHC

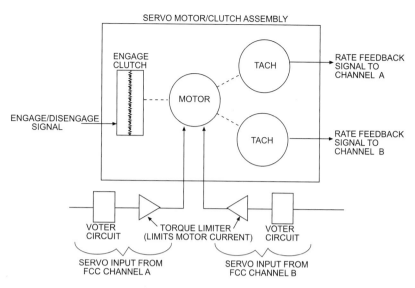

Figure 6-8-5. Diagram showing the connections be a two-channel FCC and a control surface servo

Figure 6-8-6. A piezoelectric sensor from the AHS-85 attitude heading system
Courtesy of Rockwell International, Collins Divisions

Figure 6-8-7. Attitude/heading system components: (A) Early version systems with several independent components—gyros, flux detector rate sensors, and accelerometers, (B) Newer version system containing all the necessary components in two units, the attitude/heading computer and the flux detector
Courtesy of Rockwell International, Collins Divisions

Figure 6-8-8. A dual sensor assembly

Courtesy of Rockwell International, Collins Divisions

sensor assembly makes this unit much more reliable than a conventional gyro.

When pressure is applied to a piezoelectric material, a voltage is produced. As the rotating piezoelectric crystals are subject to an acceleration or rate change, the materials bend. Bending the material applies a pressure to the crystals; therefore, the benders produce a voltage. The direction in which the crystal bends, determines the polarity of the output voltage (Figure 6-8-10). The rotating wheel contains a timing mark. This mark provides a reference point for AHC. The rotating wheels also contain two transformer primary coils. These transformers are used to induce the signal from the benders to the stationary portion of the sensor assembly.

It is important that the Collins AHC-85 be oriented in the correct position for the system to operate properly. AHC configuration strapping is used to tell the computer which direction within the aircraft the AHC is fac-

ing. Once the computer knows the magnetic heading, determined by the flux detector, and the gravitational reference, determined by the rotating sensors, any changes in angular rate or acceleration can be easily converted into changes in aircraft position. These changes in position are transmitted to the autopilot FCC.

Like the inertial reference system discussed earlier, the AHS must be initialized prior to use. This process requires that the aircraft be parked in an area free of magnetic interference. This allows for proper flux detector operation. The aircraft must also remain still during the initialization process. This system also has the capability to initialize during smooth, straight and level flight.

Inspection and Maintenance

The APS-85 requires very little routine maintenance. Operational tests should be performed in accordance with the aircraft's approved maintenance schedule. The servos are time-limited components and require regular maintenance. Every major aircraft overhaul or every 10,000 flight hours, the main control surface servos and servo mounts should be inspected by an authorized repair facility. It is recommended that any trim servos be inspected in as little as 1,000 flight hours.

On the aircraft, the servo capstan should be inspected for wear, security, and proper cable alignment. The servo unit should be operated through its entire range and observed. If a cable binding or fatigue is present, the problem must be corrected. If the servo unit makes a grinding or rubbing sound, the servo and servo mount should be removed, inspected, and repaired. At regular intervals, the servo slip clutch must be tested. The test procedures are outlined in the service manual.

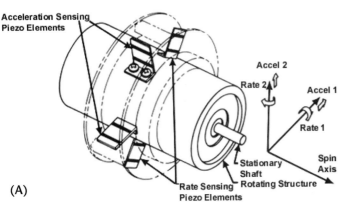

Figure 6-8-9. Sensor wheel assembly: (A) Diagram showing piezoelectric crystals (benders), (B) An expanded view of the sensor wheel

Courtesy of Rockwell International, Collins Divisions

Troubleshooting Procedures

The APS-85 is designed to operate in conjunction with the Collins electronic flight instrument system (EFIS). This allows the autopilot diagnostics to be displayed on the EFIS primary flight display or multifunction display. The APS-85 diagnostics operate in three modes: input, report, and output mode. The input mode displays various input parameters that report to the FCC. The input mode can therefore be used to determine the operational outputs of systems, such as AHS, air data, and navigational aids. The report mode presents data on various systems that are monitored by the FCC. If a fault flag is displayed during flight, the report mode should be accessed prior to turning off electrical power. The output mode is used to display FCC software outputs. In general, the report mode is used for primary troubleshooting, and the input/output modes are used for detailed fault isolation.

The pilot(s) of any aircraft using an APS-85 should be made aware of the autopilot diagnostics. In the event of an autopilot problem, the pilot should enter the diagnostics mode on EFIS and record all fault codes prior to power shut down. This will help the troubleshooting process since the report codes give technicians important diagnostics data. The report codes are decoded using the autopilot maintenance manual. Report mode data includes categories such as *REPAIR CODE* (general fault data), *AP DIS CODE* (faults causing an autopilot disengage), *STEER CODE* (flight director steering faults), and *RAM ERRORS* (FCC memory faults). Diagnostic codes of 000000 indicate a system with no faults detected. Other fault codes must be decoded using the autopilot maintenance manual. The diagnostic procedures for using the Collins EFIS-85 were explained in Chapter 3 of this text.

Autopilot and Flight Director Problems

Whenever troubleshooting any autopilot system, it is sometimes difficult to determine if the fault lies in the autopilot or the flight director. Keep in mind that the:

1. Autopilot computer typically controls the flight director indications, and

2. Autopilot and the flight director typically receive different inputs from the autoflight computer.

If the flight director indicator is an electromechanical type, test the indicator if possible,

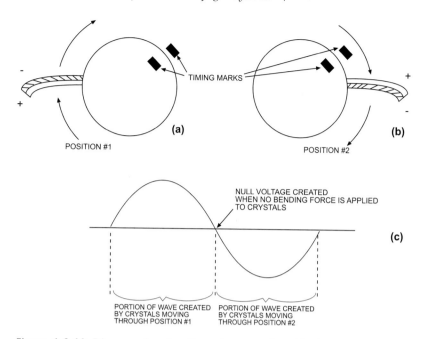

Figure 6-8-10. Diagram showing the voltage produced by bending crystal sensors: (A) As the crystal sensors rotate from straight down through position #1 to straight up the positive portion of the sine wave is produced, (B) As the crystal sensors rotate from full up through position #2 to straight down the negative portion of the sine wave is produced, (C) The sine wave is produced as the crystal rotates and the aircraft is under an acceleration force. The amplitude of the sine wave changes proportionally to the acceleration force of the aircraft.

using the appropriate test switch. If an EFIS is used, check for an EFIS fault using the built-in diagnostics. If either the electromechanical instrument or EFIS show faults, the defect is in the indicator and the autopilot is most likely OK. If the indicator tests OK, suspect an autopilot problem.

In general, when the autopilot is engaged, the flight director and autopilot functions are isolated. This is the best time to troubleshoot the two subsystems. If the pilot commands a left turn and the flight director responds, but the autopilot does not, the fault is most likely in an autopilot component, perhaps a defective aileron servo. If the pilot commands the same left turn and the autopilot "flies" the aircraft into the turn, but the flight director does not respond, the fault is in the flight director, perhaps a fault in the ADI or EFIS interface. If both the flight director and autopilot fail to respond, the problem is in a component common to both subsystems, perhaps a defective APP or FCC.

Troubleshooting: Helpful Hints

The following are several troubleshooting techniques that may help isolate faults on the APS-85, as well as other autopilot systems.

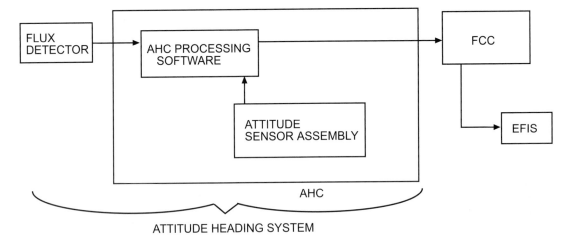

Figure 6-8-11. Block diagram of an attitude heading system

1. Many autopilot problems come from the subsystems that feed the autopilot computer. One of the most complex and frequently failed subsystems is the attitude heading system. This is especially true for attitude heading systems that employ rotating mass gyros. The AHS-85 contains both the flux detector for heading information, and the attitude sensors for attitude data. As shown in Figure 6-8-11, the attitude sensors are part of the Attitude Heading Computer (AHC). Attitude and heading data are each processed and distributed through the AHC, which means:

   - If the EFIS displays a heading flag (*HDG*), the flux detector is most likely at fault

   - If the EFIS displays an attitude flag (*ATT*), the attitude sensors are at fault and the AHC must be replaced

   - And if both flags (*ATT* and *HDG*) are displayed, the AHC processor software is faulty and the AHC should be replaced. Always become familiar with any subsystems that feed the autopilot, as this will aid you in troubleshooting.

2. If a YELLOW message such as *HDG* (heading) appears on EFIS, this normally means a dual system disagreement. That is, two redundant subsystems that feed the autopilot are not transmitting the same data. In this case, determine which unit is faulty by operating the subsystems independently. Independent operation can be done through reversionary switching or by opening circuit breaker(s) to one of the subsystems.

3. Any time a heading inaccuracy problem occurs, consider that one of the fluxgates may be too close to a metal object. Move the aircraft and see if the problem corrects itself. If a heading disagree fault is stored in the diagnostics memory, check the time of occurrence. If the fault occurred shortly after starting the engines, the fault is most likely caused by the aircraft taxiing too close to a metal structure.

4. While operating in autopilot mode, if the aircraft consistently changes altitude while in a banked turn, the fault is most likely a misalignment of the attitude heading system. It is very important that the AHS-85 mounting tray be aligned correctly. An alignment fixture can be used to verify alignment of the AHS mounting tray. The tray can be shimmed to adjust the alignment if necessary. On any attitude heading system, if a misalignment occurs, inspect the mounting structure for cracks, bends, or looseness.

5. Any autopilot system is only as good as the related control surface elements. If the control surfaces are improperly installed, loose, or poorly balanced, the autopilot will most likely be unable to hold a steady attitude. If the control surface cables are too loose, the aircraft will oscillate or porpoise while in the autopilot mode. This will be especially evident at capture of a given attitude or heading.

The Phenom VLJ Autoflight System. One of the most modern aircraft designs at the time this text was written is the category of aircraft known as very light jets (VLJ). These aircraft are typically constructed using composite materials in order to save weight and employ advanced integrated avionics which would typically include automatic flight control systems. The Embraer Phenom VLJ, introduced in 2008, has a capacity of four to six passengers, can be flown with one pilot, and has been well accepted by the industry. Since the Phenom employs an avionics package constructed by

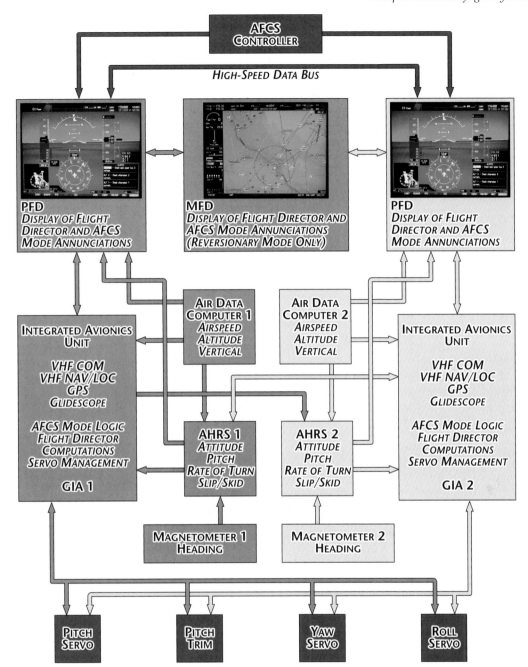

Figure 6-8-12. IAUs process data from aircraft systems and pilot commands for control of the pitch, roll, and yaw servos

Garmin International, the autoflight functions are similar to many small high performance aircraft also using Garmin avionics. The Phenom employs a Garmin Prodigy system similar to the Garmin G-1000 found in the Cessna Mustang and other aircraft.

As discussed earlier in this text, the Garmin integrated avionics system incorporates two PFDs and one MFD. These display units contain the circuitry for various software functions that deliver information to/from the Phenom autoflight system. The two main processors in the aircraft are called the Integrated Avionics Units (also known as the GIA, Garmin Integrated Avionics). These computers receive and process a variety of the data from aircraft systems as well as pilot commands from the flight deck. (Figure 6-8-12) The IAUs send information to the control surface servos for control of pitch, roll, and yaw. The servo actuators in the Phenom employ DC electric motors connected through various cables and mechanical systems to move the control surface. There is also a dedicated electrically operated pitch trip servo that moves a trim tab not a main control surface.

The autoflight system found on the Phenom is called the Flight Guidance and Control System

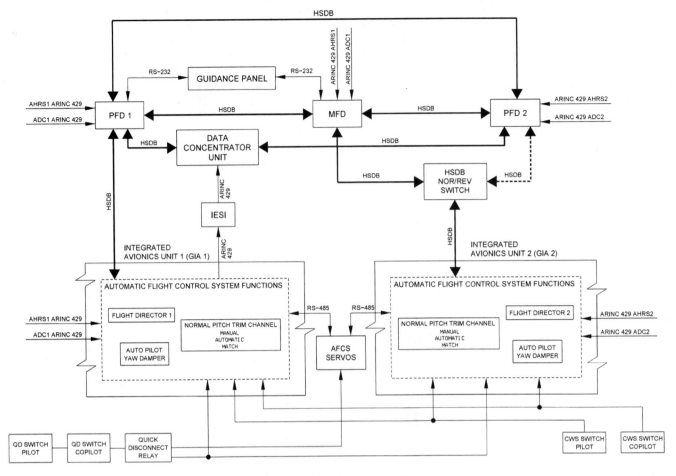

Figure 6-8-13. The Phenom Flight Guidance Control System interface connections

Courtesy of Embraer

(FGCS) and is divided into four major functions:

1. Flight director (FD)
2. Automatic pilot (AP)
3. Yaw damper/turn coordinator (YD)
4. Automatic pitch control.

The two GIAs compute the flight director and automatic pitch command functions. The three servo units each contain processing software responsible for automatic pilot and yaw dampening.

Figure 6-8-13 shows the various interface connections of the FGCS on the Phenom. The number 1 and 2 GIAs each contain identical software for flight director, yaw damper, and automatic pitch trim functions. Only one GIA performs FGCS calculations depending on the pilot's selection; the other GIA is ready in standby mode. The guidance panel, located top left of the diagram contains most of the flight deck controls needed for the FGCS. According to pilot commands, the guidance panel (GP) sends RS-232 data to the PFD and MFD; these units process, convert, and send the GP data to

the GIAs via a high-speed digital bus (HSDB). The HSDB is a Garmin proprietary bus used on the G-1000 and similar Garmin systems. The GIA software performs the automatic flight control functions and outputs data to the appropriate servo actuators via a RS-485 bus.

The servo units on this aircraft receive inputs from the GIAs and the pilot and co-pilot's control wheel autopilot and quick disconnect switches. These are considered "smart servos" since they contain software circuitry and process the incoming information prior to taking any servo action. Each servo contains two RS-485 transceivers and two processor circuits, providing redundancy. The servo processors exchange data, perform validity checks, and then control the servo motors as needed. The processors also transmit motor speed, torque, current, and voltage values to the GIAs as a feedback signal.

Air data, attitude, and heading information is created by the ADC (air data computers) and the AHRS (attitude heading reference system). This information is sent directly to both PFDs and to GIAs via an ARINC 429 data bus. Using independent busses provides redundancy and allows for validity checks. The ADC and AHRS

FMC for internal FMC functions. The first officer's 115 VAC transfer bus supplies autothrottle servo excitation power along with tachometer generator excitation. The 28 VDC battery bus powers the FMS warning circuits. The first officer's (F/O) 115 VAC transfer bus powers the internal functions of the right FMC.

Maintenance and Troubleshooting

The FMS continually monitors itself using BITE systems programmed into the FMC software. The BITE is initiated at every power-up of the FMC. The BITE can also be initiated through the central maintenance computer system or using the *INITIATE TEST/LAMP TEST* switch on the front of the FMC (Figure 6-9-14). During this 15 second test, the main and auxiliary EICAS, the PFD, and ND each present specific test messages. During the test, the master caution and warning lights and aural tones sound for a short period. On the FMC, the red FAIL lamp illuminates while the FMC test switch is held in, or at the end of the test if the FMC BITE fails. The TEST IN PROCESS light illuminates any time the test is in progress.

Two major subsystems of the FMS can be accessed through the central maintenance computer system (CMCS): the FMC and the FMC servo loop. Both of these systems can be accessed from either the right or left FMS. The CMCS tests for the FMS can only be performed on the ground since the FMS is inoperative during CMC interrogation. FMS fault data stored in the CMCS memory can be accessed through the CMC existing faults or present leg faults page.

Autopilot Flight Director System

The B-747-400 autopilot flight director system (AFDS) receives inputs from various systems and sensors throughout the aircraft, and provides steering commands for automatic and/or manual control. For manual steering, the flight director provides the interface between the AFCS (automatic flight control system) and the pilots. During automatic steering, the aileron, elevator, and rudder servos provide an interface between the AFDS and the control surfaces. The autopilot is capable of pitch control to maintain a given airspeed, altitude, vertical speed, or vertical navigation including glide slope. Roll commands can maintain a given heading, track, lateral navigation, or attitude including localizer. The autopilot yaw function provides control for adverse yaw, and crab angle.

Controls

The mode control panel (MCP) is the main interface between the flight crew and the AFDS. The mode control panel is located on the glare shield, cooled by forced air, and connected to the system through three connec-

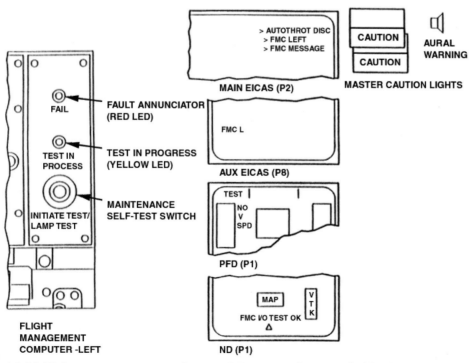

Figure 6-9-14. Flight management control system BITE test can be controlled by the maintenance self-test switch found on the FMC.

Courtesy of Northwest Airlines, Inc.

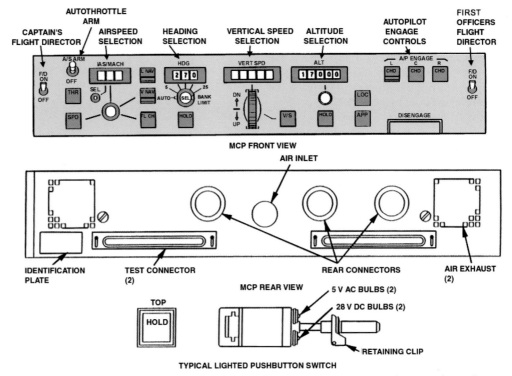

Figure 6-9-15. Boeing 747-400 Autothrottle Flight Director System (AFDS) mode control panel

Courtesy of Northwest Airlines, Inc.

tor plugs located on the rear of the unit. In Figure 6-9-15, a lighted push button assembly is removed from the face of the unit for lamp replacement. Each lamp assembly contains four bulbs, two powered by 5 VAC, and two powered by 28 VDC.

Refer to Figure 6-9-15 during the MCP control explanation in this paragraph. The captain's flight director is activated by the toggle switch on the far left of the MCP; the first officer's flight director toggle switch is located on the right of the panel. The autothrottle engage switch is located just right of the captain's flight director switch. Indicated airspeed **(IAS)**, or mach speed, can be selected from the speed mode of the autothrottle function. Lateral navigation **(L NAV)** or vertical navigation **(V NAV)** can be selected using the appropriate lighted push button switch. Pressing the flight level change **(FL CH)** switch will engage both vertical and lateral navigation. The HDG control can be used to select a given heading for the autopilot or flight director. Vertical speed is entered into the MCP using the vertical speed thumb wheel. A given altitude can be selected and displayed in the ALT window. The autopilot engage push buttons allow the pilot to select the left, center, or right FCC for command of autopilot/flight director functions.

An autopilot disengage switch is located on both the captain's and first officer's control wheel. These switches are removed by a screw located on the front of the switch plate (Figure 6-9-16). The switch wiring is fed through the control wheel to a terminal block. The autopilot go-around switches are located on the number 2 and 3 thrust levers (Figure 6-9-17).

Architecture

The three FCCs interpret data and provide the necessary calculations for the autopilot and flight director functions. The pilot

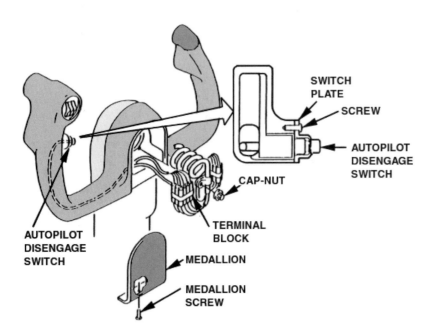

Figure 6-9-16. Autopilot disengage switch   *Courtesy of Northwest Airlines, Inc.*

selects inputs to the FCC through the MCP, the heading reference switches, the disengage switches, and go-around switches (Figure 6-9-18). The FCC receives three types of system inputs: navigational, airplane configuration, and triple redundant sensors. Navigational inputs are provided by the FMC and ADC. Airplane configuration sensors monitor items necessary for autoflight, such as hydraulic status and flap position. The triple redundant inputs are those needed for autoland functions. Triple redundant sensors include: ILS, IRU, and radio altimeter data.

The three FCCs each control a separate servo, one each for the ailerons, elevator, and rudder. The servos use electrical signals from the FCCs to control the flow of hydraulic fluid, which in turn controls the position of the related control surfaces. The FCC outputs display data to the EFIS/EICAS interface units (EIU). As seen in Figure 6-9-19, all three FCCs send a parallel data signal to each of the EIUs. The FCCs send a discrete warning signal to the modularized avionics and warning electronic assembly (MAWEA) for annunciation of warning data. Caution information is sent from the FCCs to the three EIUs.

The FCCs communicate to each other via a cross-channel data bus for exchange of health monitoring, and to provide redundancy for servo engage data. The ability to cross talk between

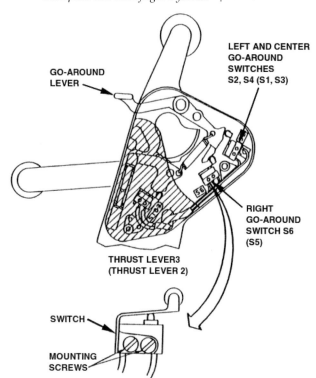

Figure 6-9-17. Autopilot go-around switches   *Courtesy of Northwest Airlines, Inc.*

FCCs improves system safety by allowing the comparison of information between computers. If any FCC detects a failed FCC or critical system out of tolerance, the autoland capability will not be available.

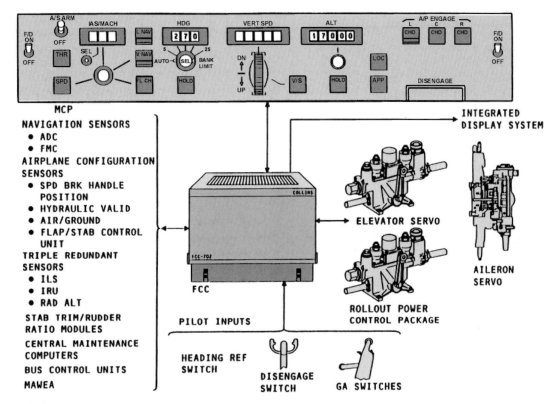

Figure 6-9-18. Interface diagram of the flight control computer (FCC) and various aircraft systems

Courtesy of Northwest Airlines, Inc.

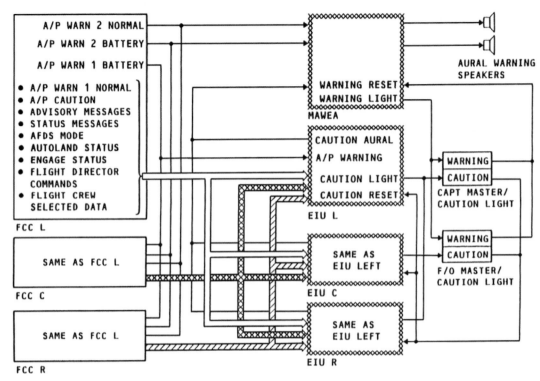

Figure 6-9-19. FCC/EIU/MAWEA interface diagram   *Courtesy of Northwest Airlines, Inc.*

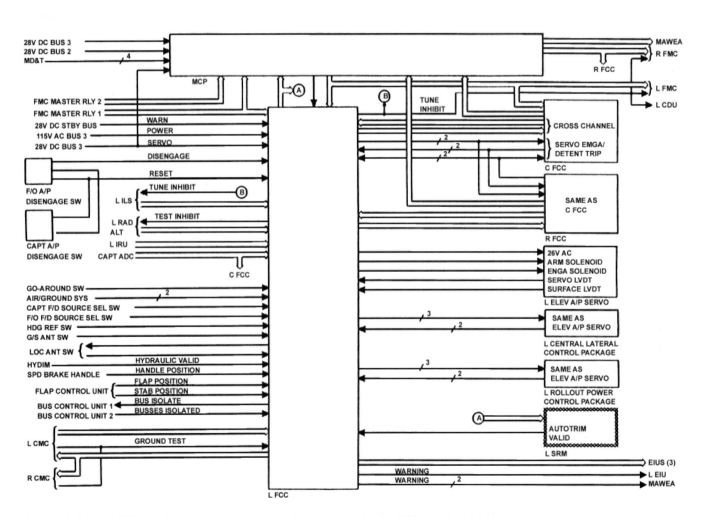

Figure 6-9-20. Left FCC interface diagram, note that the center and right FCCs receive similar inputs   *Courtesy of Northwest Airlines, Inc.*

Figure 6-9-20 shows the interface of the FCC and various aircraft systems. In the upper-right portion of the diagram are the cross channel busses for communication between the left/right/center FCCs. The data busses to the central maintenance computers (CMC) are shown in the lower left portion of the diagram. Discrete data, represented by a single line on the interface diagram, comes from a variety of other aircraft systems to the FCCs. In the top left portion of the interface diagram are the power inputs to the FCC and MCP. To operate this autoflight system, there are a total of nine different circuit breakers fed from seven different power distribution busses. Whenever troubleshooting the system, make sure power is available to all necessary circuits.

Maintenance and Troubleshooting

The B-747-400 Autopilot Flight Director System (AFDS) contains BITE circuits that continuously monitor the health of the FCCs and related systems. The BITE circuits are located within each FCC and report all autoflight failures to the central maintenance computer. The flight crew is made aware of failures by a flag on the PFD or ND, an EICAS message, and/or a discrete annunciator and audio tone. EICAS will always display a warning, caution, advisory, or status message for the various AFDS faults. The *A/P DISCONNECT* message is the only EICAS warning applicable to AFDS. Remember, warnings are the most serious EICAS message and require immediate crew action. This message will display on EICAS for either a manual or automatic disconnect. In the event of a manual (pilot activated) disconnect, the CMC will not store the message as a fault.

Any fault that is sensed by the BITE circuitry is automatically recorded in the CMC nonvolatile memory. The technician can access current or previous failures through the Existing Faults or Fault History pages of the CMCS. Ground tests can also be performed using the CMCS. To access AFDS test functions, go to the Ground Tests page of the CMC menu (Figure 6-9-21), select chapter 22 AUTOPILOT FLT DIR and choose the appropriate test from the menu. Table 6-9-1 is a list of the tests available through the AFDS ground tests menu.

Several of the tests have preconditions that must be met before the tests can take place. Preconditions are listed on the control display unit after the test selection has been made (Figure 6-9-21).

CAUTION: *Whenever performing operational tests on any autopilot be sure the aircraft is clear of personnel and machinery.*

TESTS AVAILABLE THROUGH THE AFDS GROUND TEST MENU	
1. L/R/C FCC	Tests the FCCs and the systems/sensors that interface the the FCCs
2. MCP test	Tests the displays, switches, and control of the MCP
3. Aileron servo	Command and engage signal are sent to the aileron servo, the FCCs monitor the response
4. Elevator servo	The elevator servo is tested same as aileron servo test
5. Rudder servo	The rudder servo is tested same as the aileron servo test
6. Autopilot disconnect switches	Tests function of A/P disconnect switch
7. Go-around switches	Tests function of G/A switches
8. Autoland unique test	Tests the operation of several functions critical to the autoland function
9. Air ground relay	Tests that all three FCC receive air/ground data
10. FCC configuration	Shows pin configuration of FCCs
11. FCC instrument	Monitors the interface between the FCCs and the integrated display system
12. Speed brake transducer	Monitors function and interface of S/B transducers
13. Flap transducer	Monitors function and interface of flap transducers
14. Stabilizer trim	The autopilot sends a given signal to the trim system and the FCCs monitor the response
15. Surface limit	Test to ensure each FCC has the same control surface travel limits (separate test conducted for aileron, rudder, elevator)
16. Tranducer output	Tests the stabilizer, aileron, rudder, elevator, speed brake, and flap transducer outputs

Table 6-9-1. Autopilot flight director system (AFDS) ground tests

Many of the autopilot tests will operate various control surfaces and/or thrust reversers. These control surfaces could cause damage to the aircraft or bodily injury to unsuspecting individuals. Also, be sure that other maintenance being performed on the aircraft will not adversely affect the autopilot tests and create a potential hazard. For example, if another technician is servicing the hydraulic system, the autopilot functional test should not be performed.

Yaw Damper

The B-747-400 yaw damper system provides dampening for Dutch roll prevention, turn coordination, and suppression of structural

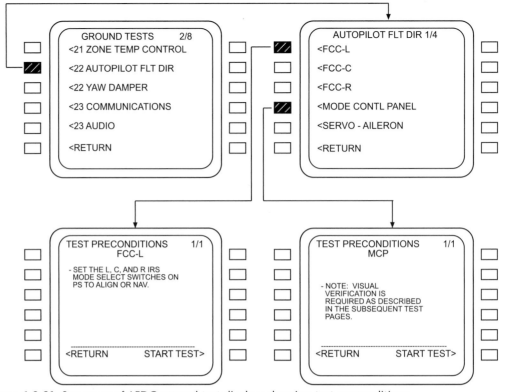

Figure 6-9-21. Sequence of AFDC ground test displays showing test preconditions

Courtesy of Northwest Airlines, Inc.

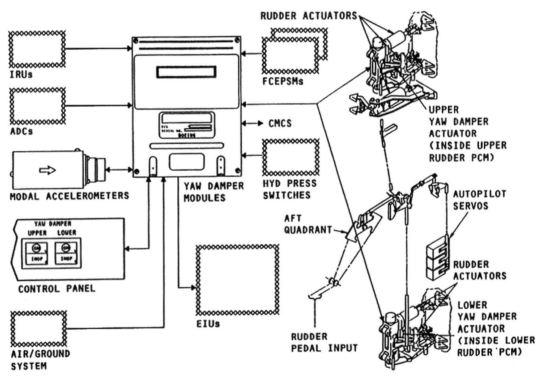

Figure 6-9-22. Block diagram of the yaw damper system

Courtesy of Northwest Airlines, Inc.

modal oscillations. *Structural modal oscillations* are an undesired effect created by turbulence, which causes bending of the fuselage around the wing area. There are two redundant yaw damper systems, each containing a yaw damper module powered by the FCEPSM (flight control electronics power supply modules). As seen in Figure 6-9-22, each yaw damper module receives inputs from the IRUs, ADC, dedicated modal accelerometers, the yaw damper control panel, air/ground systems, the CMCS, hydraulic pressure switches, and a feedback signal from the yaw damper actuators.

The main yaw damper module outputs go to the yaw damper actuators. Output signals also go to the EFIS/EICAS interface units (EIU) and the central maintenance computer system. The yaw damper control panel also receives an output from the yaw damper modules to verify the current operation of the yaw damper system.

Section 10

Fly-by-Wire

The basic concepts of Fly-by-Wire (FBW) are simple; replace cables, pulleys, and pushrods with electrical wiring as a means to connect pilot inputs to aircraft control surfaces. In a traditional system, the pilot moves a control wheel, or yoke, and a stainless steel cable is used to transfer this motion into control surface movement. On most large aircraft, the pilot would move the control wheel, the cable would move a hydraulic actuator, and the hydraulic actuator would move the control surface. As seen in Figure 6-10-1 in a FBW system the pilot would move the control wheel, an electrical signal would be sent to an electronics control unit computer, and the control unit would send an electrical signal to the hydraulic actuator that moves the control surface. A FBW system must also employ a feedback system to provide a "realistic feel" back to the pilot through the flight deck controls (wheel, yoke, or rudder pedals).

FBW is not actually a new concept. For years many aircraft have employed electrical circuits to operate certain control surfaces. For example, Cessna light aircraft, like the 172, have employed electric flap actuators for several decades. The pilot would simply select a flap position using a switch on the instrument panel, the signal would be sent to the flap motor and the flaps would move to the desired position. On some aircraft there was even a rudimentary feedback system, which would move an indicator to inform the pilot of flap position. The difference between this simple electric actuator and modern FBW systems is that today's aircraft use electrical signals to move primary flight controls, such as elevators and ailerons. Primary flight controls require constant repositioning by the pilot and therefore require a much more complex system. A modern FBW design permits a more efficient aircraft structure through the use of computer-aided controls. This technology allows the airplane to meet strict safety requirements while decreasing weight and

increasing fuel efficiency. Modern FBW aircraft require a complex flight control system employing several computers, digital data transfer, and multiple actuators for dozens of flight controls.

B-777 Automatic Flight Control System

The Boeing B-777, placed in service in the mid 1990s, was the first transport category aircraft designed to incorporate a fly-by-wire primary flight control system. The FBW design had been employed on some military aircraft prior to the B-777 release; and is also used on the newer B-787 and A-380 aircraft for its reliability, efficiency, and weight savings.

There are three distinct segments of the B-777 automatic flight control system:

1. Flight management computing system (FMCS)

2. Autopilot flight director system (AFDS)

3. Flight controls and related mechanisms

The B-777 flight controls are actually divided into two separate systems: the primary flight control system (PFCS) and the high lift control system (HLCS). Like most large aircraft, the B-777 employs dozens of independent flight controls (Figure 6-10-2). As the name implies the PFCS is used as the primary control system providing both automatic and manual operations.

The PFCS monitors a variety of inputs, employs various computers, and determines how and when to move control surfaces. The PFCS calculates commands to control surfaces using sensor inputs from control wheel, control column, rudder pedals, speed brake lever, and the pitch trim wheel. All three axes (pitch, roll, and yaw) are provided stability augmentation and envelope protection by the PFCS. Envelope protection is used to ensure the aircraft never exceeds the operational limits and enters into an unsafe configuration, such as a stall condition. The PFCS controls two ailerons, two flaperons, and fourteen spoilers for roll control; two elevators and a movable horizontal stabilizer for pitch control; and a segmented (tabbed) rudder for yaw control.

The HLCS is used to increase aircraft lift during takeoff and landing (low speed flight). The high lift control surfaces include one inboard and one outboard trailing edge flap on each wing. The B-777 also employs seven leading edge slats and one Krueger flap on each wing. A Krueger flap differs from an ordinary flap or

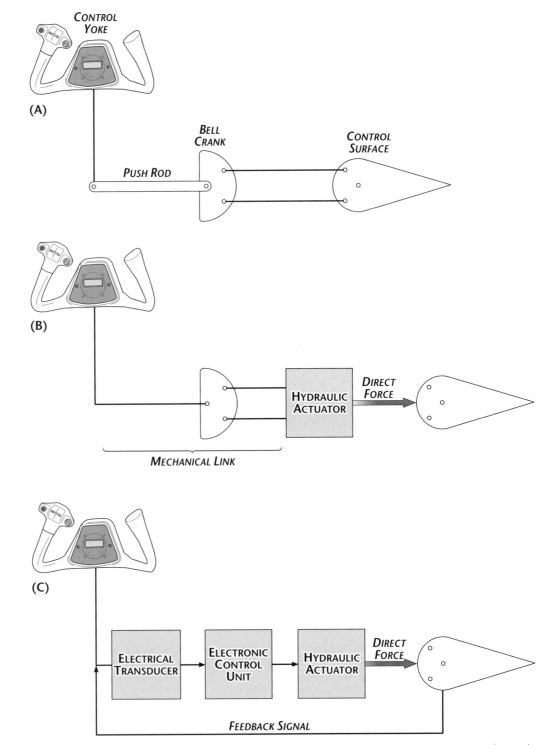

Figure 6-10-1. Operation of flight controls: (A) A mechanical flight control system, (B) A mechanical-hydraulic flight control system, (C) A fly-by-wire electronic-hydraulic flight control system

a slat in that the Krueger flap deploys from the leading edge of the wing and hinges from the front edge downward to increase lift.

As seen in Figure 6-10-3, as the flight crew moves the wheel/yoke assembly, rudder pedals or other flight deck controls the movement is converted into an electrical signal by position transducers. A transducer changes mechanical motion into an electrical voltage. The electrical signal is sent to the Actuator Control Electronic units (ACE). The ACEs convert the analog signal from the transducers into a digital format and send that data to the primary flight computers (PFC) through an ARINC 629 data bus. There are three 629 buses dedicated to flight control data.

The PFCs receive data from other systems, including the airplane information manage-

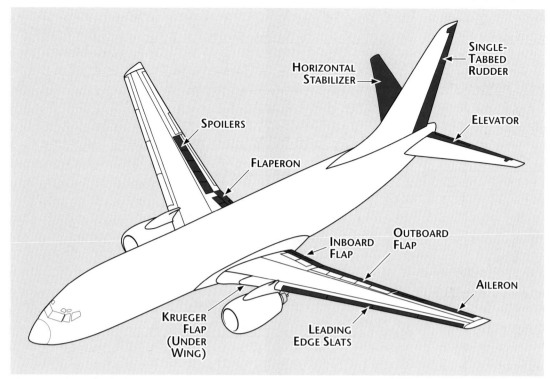

Figure 6-10-2. The flight controls of a B-777 aircraft

ment system, the air data and inertial reference units, and secondary air data sources. The PFCs consider the input data and employs control-law software to calculate augmentation and envelope protections. The PFCs then send digital command signals back to the ACEs. The ACEs convert the digital signals into analog signals for command of the power control units (PCU). Each PCU contains an electrically operated servo-valve that controls hydraulic actuators to move the control surface. Each control surface will be connected to one, two, or three PCUs depending on load demands.

The PCUs also contain a position transducer, which sends a feedback signal to the ACEs. The feedback signal is used to determine when the PCU should stop control surface movement. When in autopilot operations the AFDCs will initiate all signals for control surface movement. The primary flight control system responds in the same manner as if the pilot manually activated the flight deck controls. When in autopilot mode a signal is sent to a backdrive actuator in order to move the necessary flight deck controls into the appropriate position as commanded by the autoflight computer.

Since this aircraft is highly reliant on electrical power for flight control operations, interruption of electrical power could be catastrophic. As discussed in Chapter 2 of this text, there are multiple engine-driven permanent magnet generators, a ram air backup generator, and other power sources including batteries available for flight control operations. Figure 6-10-4

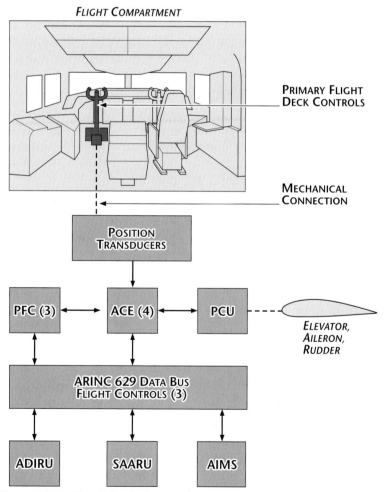

Figure 6-10-3. The flight deck control movement is converted into an electrical signal by position transducers and converted into a digital signal for use by the autoflight system.

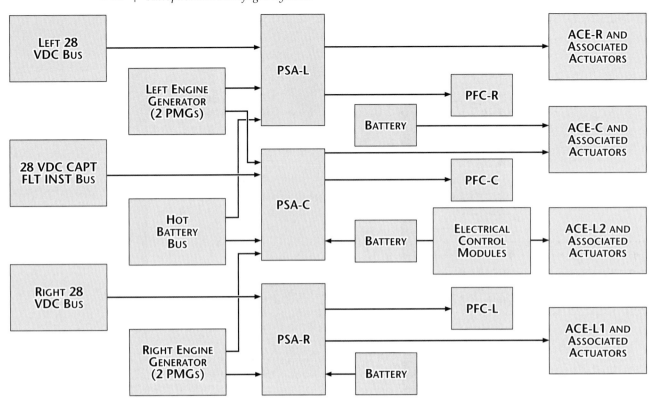

6-10-4. Redundant power sources provide safety for the B-777 flight control system.

shows the various sources available to power the ACEs and PFCs in the system. It is critical that these LRUs continue to receive power in order to maintain manual flight control.

The primary flight control system can operate in three distinct modes: normal, secondary, and direct; all of which are dependent on the health of system sensors, computers, and control devices. The system will automatically switch to a less automated mode if certain components fail and flight safety cannot be ensured. The *normal mode* of operation provides all envelope protections including stall warning, over-speed, over-yaw, and bank angle. The autopilot is also fully functional in the normal mode. If there are one or more critical failures the system software will switch to *secondary mode* that limits some of the automated protections. If additional failures occur, the systems must operate in *direct mode* and only manual pilot commands are accepted. The PFCs are non-operational in the direct mode.

A big consideration for any fly-by-wire system is redundancy. For the most part, each flight deck control contains up to three pressure transducers to ensure pilot commands create the correct electrical signals. Of course, there are also multiple computers and redundant software functions allowing the PFCS to fail fully active under most conditions. Of course if a failure is too extensive, the system will switch

operational modes from normal to secondary or direct.

Fly-by-wire flight control systems also require a relatively complex feedback system to provide each pilot with the correct feel on the flight deck controls. This is important when pilots fly the aircraft manually. The B-777 incorporates centering mechanisms, which returns the control (wheel, yoke, and/or rudder pedals) to the neutral position when appropriate. As seen in Figure 6-10-5 the system also incorporates electrical actuators, which return pressure as the pilot pulls/pushes or rotates a flight deck control. This return pressure will change with aircraft speed and as the aircraft reaches flight envelope limits. Each of these feedback systems rely on a variety of inputs and several computer functions to provide pilots a "natural" feel when controlling the aircraft.

B-777 Autopilot Flight Director System

The autopilot flight director system (AFDS) is an integral part of the B-777 flight controls. The AFDS has three channels that can each operate independently to provide redundancy. When activated, the autopilot function of the AFDS will control the aircraft on its selected vertical and horizontal flight path and selected airspeed. The flight director portion of the AFDS

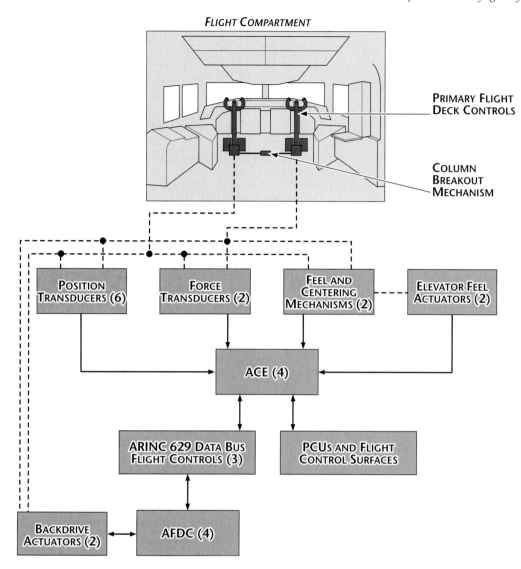

6-10-5. Two backdrive actuators position the flight deck controls during autoflight operations.

is operational whenever aircraft systems are powered that provides visual guidance commands on the aircraft flat panel displays giving the pilots all the data needed to manually fly the aircraft.

The AFDS has three major components: the mode control panel (MCP), three autopilot flight director computers (AFDC), and the various sensors, switches and transducers, which provide input signals to the system. Figure 6-10-6 shows the relationship of the AFDS components. Here it can be seen that AFDS receives pilot commands through the MCP and the miscellaneous control switches. The MCP is used to select the operational mode of the AFDS. The pilot may select various operations, such as, lateral navigation (LNAV), vertical navigation (VNAV) and others. The miscellaneous control switches include: the takeoff, go-around (TOGA) switches located on the throttle levers and the disconnect switches on the pilot's and co-pilot's control wheel.

The AFDS monitors the various pilot activated inputs using three AFDCs. Each computer calculates the necessary response and sends output signals through three ARINC 629 data bus cables to the ACEs and PFCs. Similar to the manual flight operations the ACEs and PFCs activate the appropriate control surface. As a control surface is moved the PFC software calculates the backdrive commands, which are sent to the AFDCs. The AFDCs then send the backdrive signals to the appropriate backdrive actuators that reposition the rudder pedals and/or control wheel/ yoke as needed. Most autopilot commands are redundant and the computers analyze multiple inputs. Software functions known as *mid-value selection* and *voting* are used to determine the validity of all data prior to moving any control surface.

To inform pilots of the current operating status, the autopilot flight director system will send data to the aircraft instrument display

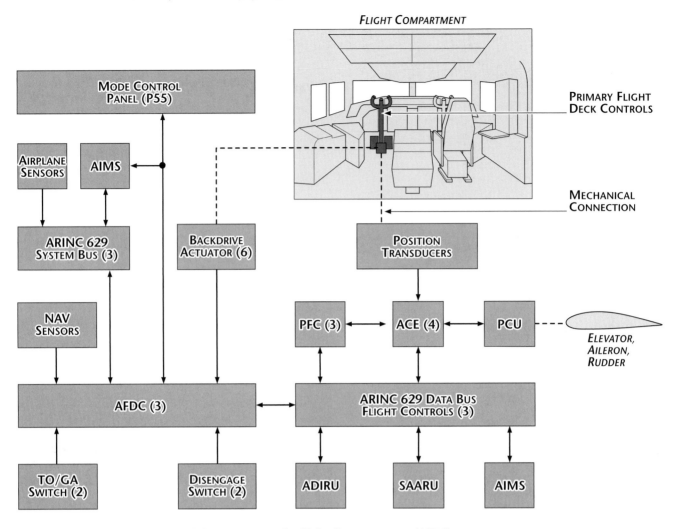

Figure 6-10-6. Simplified diagram of the B-777 autopilot flight director system (AFDS)

system. The aircraft's PFDs, EICAS, and MFD will each show AFDS displays and annunciations as needed. The PFDs show flight modes as well as autoland and autopilot indications. All warning and caution data is sent to EICAS and the MFD shows AFDS status.

B-777 Flight Management Computing System

To help reduce pilot workload, a flight management computing system (FMCS) is used to provide vertical and lateral guidance for all phases of flight excluding takeoff and landing. The FMCS will also automatically tune all radios and provides navigational data on the flight deck displays. The FMCS software, known as the flight management computing function (FMCF) is located in the two AIMS cabinets. The B-777 airplane information management system was discussed in Chapter 3. One of the flight management computing functions operates in active mode while the other is ready in standby in the event of a failure. The flight management system operates

in conjunction with the autopilot flight director system to provide complete navigation and autoflight functions. A simplified diagram of the FMCS is shown in Figure 6-10-7; please reference this diagram during the following discussions.

The flight crew interface for the FMCS is through the three control display units (CDU) located on the flight deck. The CDUs are mounted on the center pedestal between the two pilots and contain the traditional alphanumeric keys and liquid crystal display. The pilots enter all flight planning data on the CDUs and this information is sent to both AIMS.

The FMCF has four basic elements: navigation, flight planning, performance management, navigation radio tuning. The FMCF contains a large navigational database with all necessary navigational aids, waypoints, flight plans, and other necessary information. Flight planning functions use flight crew inputs to create the desired flight plan. The performance management function employs

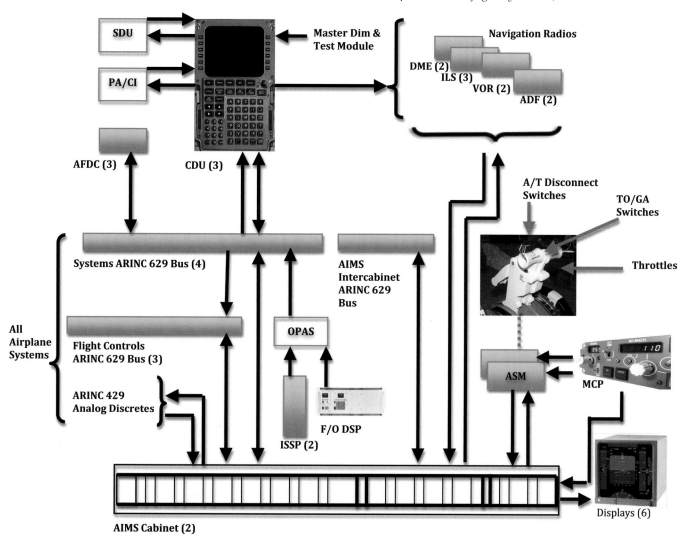

Figure 6-10-7. A simplified diagram of the B-777 Flight Management Control System (FMCS)

aerodynamic models and flight crew selections to calculate the most economical flight path and engine power settings. The navigation radio tune function sets all radio frequencies and settings necessary for complete navigation for the entire flight. The FMCF software is updated at regular intervals in order to ensure currency.

B-777 Thrust Management Computing System

Thrust management is an independent function also contained in the two AIMS cabinets. Both AIMS contain redundant systems to ensure fail active operations. The thrust management computing function (TMCF) is basically software used to send auto throttle commands to the throttle servo motors and engine trim commands to the engine electronic controllers (EEC). The servo motors are used for large throttle adjustments and the EECs make fine adjustments to keep both engines at peak efficiency for various flight conditions.

A-380 Flight Control System

The Airbus A-380 is a large four-engine transport category aircraft employing state of the art integrated electronics and the AFDX data transfer system. AFDX was discussed in Chapter 2. The aircraft uses an advanced flight control system called the *Auto Flight System* (AFS). The A-380 AFS can be divided into three distinct elements: flight guidance (FG), flight management system (FMS), and the flight controls. Due to the size and complexity of the A-380, the flight control system contains nearly 50 separate control surfaces activated automatically or manually by pilot commands (Figure 6-10-8).

The A-380 employs a fly-by-wire-type system with all flight deck inputs converted to electrical signals, routed through one or more processor circuits and eventually sent to an electrically controlled hydraulic actuator assembly (Figure 6-10-9). The flight controls are divided into two distinct categories: pri-

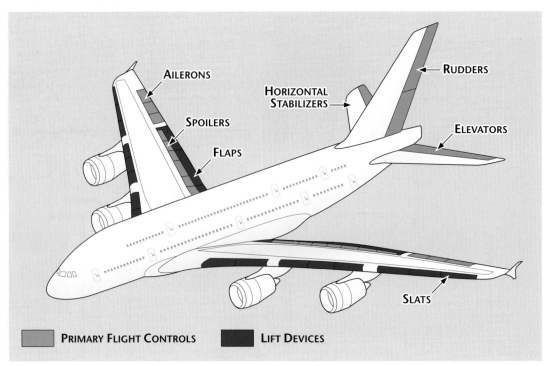

Figure 6-10-8. An A-380 flight control system contains nearly 50 separate control surfaces.

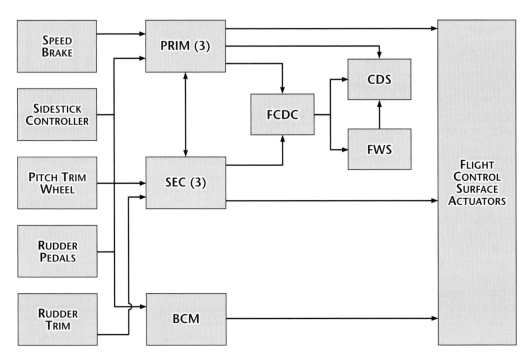

Figure 6-10-9. A simplified diagram of the A380 fly-by-wire system

mary flight controls and the slats and flaps. The primary flight controls are used for control of pitch, roll, and yaw during normal, direct, or alternate flight configurations. The slats and flaps are each considered a high-lift device used for low speed flight during take-off and landing.

The primary flight control system employs three primary computers (PRIM), which provide flight control, flight guidance, and envelop pro-

tection functions. Envelope protection is provided by system software to prevent exceedance of certain flight parameters, such as, excessive bank angle. The system also employs three secondary computers (SEC). Each computer, PRIM and SEC, can perform two functions: command computations and command executions.

Command computations convert pilot or autopilot commands into control surface deflection signals according to flight parameters and

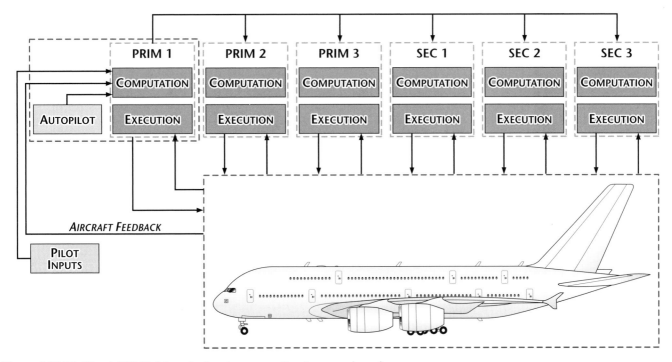

Figure 6-10-10. The A-380 flight control system operating in normal mode

envelop protection limitations. In addition, command computations analyze and compare servo actuators feedback signals in order to ensure proper control surface movement has been achieved. The command execution function of PRIM and SEC computers send the necessary electrical control signals to the servo actuators in order to create control surface movement.

The flight control system operates in three distinct modes: normal, direct and alternate. In normal mode one primary computer performs all computation functions and sends command signals to the other computers (Figure 6-10-10). All three PRIM and SEC computers perform the execution functions for their assigned control surfaces. As the diagram shows, the PRIM master computer also performs self-monitoring of the control surface feedback signal to ensure the systems respond accurately.

If a malfunction is detected in the master PRIM all computation functions will be passed to another PRIM. If all PRIMs are lost due to failures, each SEC will perform computation and execution functions as needed. At this point the flight control system will automatically downgrade to direct mode. Direct mode occurs whenever the system has degraded dramatically due to several failures and the normal mode of operation is not available. When operating in direct mode the auto trim function is no longer available and all envelope protections are lost. Warning information displayed on the flight deck such as, over speed or stall warn-

ings, inform the pilot of any potential envelope exceedance.

If all PRIM and SEC computers are lost due to system failures, the aircraft flight controls are operated in alternate or backup mode. The backup system is totally segregated from the normal system with dedicated sensors and transducers in the flight deck controls. At least one hydraulic system and backup electrical power source must be available for alternate mode operations. If all engines fail power is provided by the ram air turbine. In alternate mode only flight controls used for basic maneuvers and safe landing are available. Figure 6-10-11 shows the control surfaces that are operable in the direct mode of operation.

The A-380 flight control system employs three types of servo actuators to move the flight control surfaces and high lift devices. The actuators are combinations of electronic controllers, electric motors, and hydraulic actuators. Since this is a fly-by-wire aircraft, each servo is electrically controlled from one or more computers. There are three types of servo actuators: conventional actuators, electro-hydrostatic actuators (EHA), and electrical backup hydraulic actuators (EBHA)..

As seen in Figure 6-10-12 conventional servos employ an electrically controlled servo valve that regulates the flow of hydraulic fluid into the actuator and determines control surface movement. The servo valve can also be used to reverse the flow of hydraulic fluid, which changes the actuators direction of travel. In

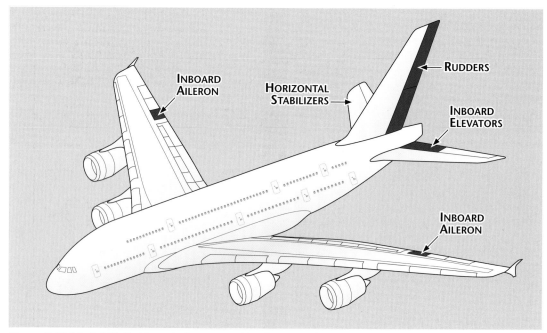

Figure 6-10-11. A-380 flight controls operated in direct mode

order to operate, conventional servos must have a supply of pressurized hydraulic fluid. The A-380 employs multiple centralized hydraulic systems to provide redundancy for the actuators. Each system is named by color, green or yellow.

The electro-hydrostatic actuators are hydraulic units that have their own electric motor and self-contained hydraulic system. The servo receives an electronic signal to its electric motor located within the actuator assembly. The electric motor is directly coupled to a hydraulic pump that supplies the pressurized fluid to move the actuator. This type of actuator is independent of the central hydraulic systems, but requires a supply of electrical current to drive the pump motor.

The electrical backup hydraulic actuator is a combination of the conventional and electro-hydrostatic actuators. This unit is connected to the central hydraulic system using a servo valve for electronic control and employs a self-contained electric motor/pump assembly to produce an independent supply of hydraulic pressure. The backup actuators can therefore operate using electric or hydraulic power.

A-380 Auto Flight System

The Airbus A-380 Auto Flight System (AFS) is comprised of two distinct elements; the flight guidance (FG) and the flight management system (FMS). The FG system provides short-term lateral and vertical guidance based on the flight parameters selected by the flight crew or the FMS. The FMS provides long-term guid-

ance by sending targets, such as waypoints, airports, and navigation aids, to the FG system. The AFS works in conjunction with the three PRIM computers for autopilot, flight director, and auto throttle functions. There are two complete FMS functions, which operate using one of three FMCs, providing redundancy to ensure that systems can fail and the FMS still remains operational. A basic auto flight system interface diagram is shown in Figure 6-10-13.

The flight crew can interface with the AFS using the AFS control panel or the three MFDs can provide back up for the control panel. The MFDs operate in conjunction with the KCCU (keyboard and cursor control unit) as discussed in Chapter 3. The PFDs provide visual feedback to the pilots regarding the AFS operations. The NDs show all navigation data related to AFS. Of course there are discrete controls such as autopilot disconnect switches on the side stick control and auto throttle disconnect switches on the throttle quadrant.

The FG function of the AFS is designed to follow short-term instructions and provide guidance and speed controls. The FG functions include:

1. Autopilots one and two (AP1 and AP2) provide calculations for pitch, roll, and yaw

2. Flight directors one and two (FD1 and FD2) provide guidance commands on the PFDs. This enables the flight crew to manually fly the aircraft or to monitor guidance orders during autopilot controls

3. Auto thrust (A/THR) controls engine thrust through the FADEC (full authority

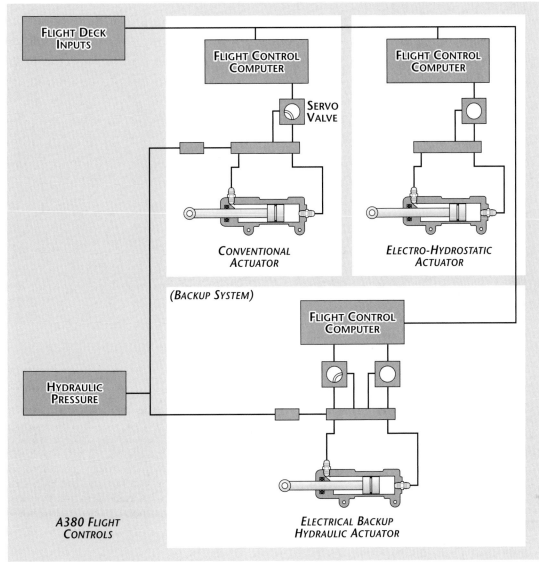

Figure 6-10-12. The A-380 employs three types of servo actuators, conventional, electro-hydrostatic and electrical backup hydraulic

digital engine control). The FADEC system analyzes the thrust command and operates electrical servo actuators that change engine power settings and position the engine throttles accordingly

The pilot interface to the AFS is through the traditional flight deck controls and instrument display system found on the A-380. It is important that fly-by-wire aircraft retain a "traditional feel" so a pilot can easily transition from one aircraft to another. The three main inputs to the AFS are the autoflight system control panel (AFS CP), the MFD and KCCU, the thrust levers, and the side stick control. Each of these input devices creates an electric signal, which connects to the PRIM, FMC, or FCU back up computers (Figure 6-10-13). The computers also send feedback signals to the flight deck controls which are employed to provide feedback to the pilot. The PFD, ND, and MFD provide indications as to the AFS system status.

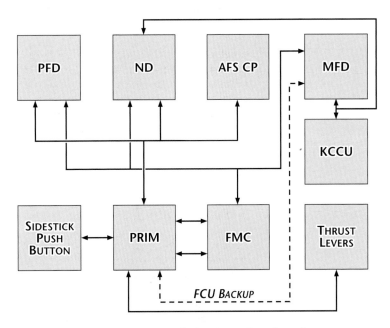

Figure 6-10-13. A basic A-380 autoflight system interface diagram

Chapter 7

COMMUNICATION
and entertainment systems

Section 1

Introduction

Even during the early days of aviation, communication between the flight and ground crew personnel was extremely important. The first aircraft were started by hand as the pilot simply yelled to his ground crew. Hand signals were often used for communication during times of adverse noise conditions. As aircraft began to carry passengers, they too needed to be informed of certain flight details. Early crew to passenger communications, were simply a matter of loud conversations. As aircraft grew in size and complexity, it became evident that better communications between flight and ground crews, as well as between crew members and passengers were necessary.

Today's transport category aircraft contain a variety of systems, all dedicated to communications. The flight crew can communicate with passengers, ground crew, air traffic control, and cabin crew. Some systems allow communication between airline operations, maintenance facilities, and the aircraft central maintenance system. Gate changes, passenger lists, or other pertinent data can be transmitted and printed using an onboard printer.

Many modern transport category aircraft also include an extensive passenger entertainment system. Entertainment systems include multichannel audio and video programs. All of these passenger systems must be linked to the flight crew in the event the pilot or flight attendant needs to make an announcement.

Transport category aircraft employ a sophisticated audio system to coordinate all this com-

Learning Objectives:

- *Airborne Communication and Address Reporting System*
- *Interphone System*
- *Passenger Entertainment and Communication System*
- *Advanced Cabin Entertainment Service System*

Left. Modern transport aircraft, both commercial and business, have added extensive in-flight electronic entertainment options for their passengers.

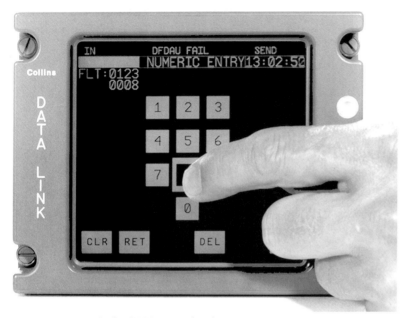

Figure 7-2-1. A typical ACARS control unit
Courtesy of Rockwell International, Collins Avionics Divisions

munication. This chapter presents an overview of communication and entertainment systems by examining the Boeing B-747-400 and Airbus A-380 aircraft. Most of the systems found on these aircraft are similar to those of other large aircraft. Remember, this chapter will present general concepts; specific information for each aircraft must be accessed prior to performing any maintenance activities.

Section 2

Airborne Communications Addressing and Reporting System

Airborne Communications Addressing and Reporting System (ACARS) is a digital air/ground communications service designed to reduce the amount of voice communications on the increasingly crowded VHF frequencies. ACARS allows ground to aircraft communications, in a digital format, for transmitting operational flight information, such as fuel status, flight delays, gate changes, and departure and arrival times. ACARS can also be used to monitor certain engine and system parameters and downlink relevant maintenance data to the aircraft operator. Prior to the use of ACARS, this information was transmitted using voice communications. ACARS can be thought of as e-mail for the aircraft. Since the message is transmitted in digital format via

ACARS, it occupies much less time on a given frequency than conventional voice communications. ACARS is an automatic system; transfer of information requires little to no flight crew involvement.

There are two major corporations that provide ACARS services worldwide: ARINC Incorporated and a French organization known as SITA. As stated earlier in this text, ARINC Incorporated is a global corporation based in the United States with primary stockholders consisting of various U.S. and international airlines and aircraft operators. ARINC provides services related to a variety of aviation communication and navigation systems, one of these services is called GLOBALink. In general, airborne equipment designed to operate using GLOBALink will operate with other ACARS services throughout the world.

ACARS Theory of Operation

The airborne components of ACARS connect to sensors throughout the aircraft. These sensors are used to detect various parameters to be transmitted by ACARS. For example, transport category aircraft transmit OOOI (Out, Off, On, and In) data. *Out* stands for out-of-the-gate. ACARS uses a parking brake or similar sensor to determine out-of-the-gate time. *Off* means aircraft off-the-ground. *On* refers to the aircraft on-the-ground; a landing gear sensor can detect these conditions and sends the information to ACARS. *In* is determined when the aircraft is in-the-gate. ACARS could transmit *In* data when the parking brake is set. ACARS operates by transmitting digital data. A code of 1s and 0s is transmitted to deliver all information. If one were to listen to ACARS, it might sound similar to a modem connecting to the Internet.

In the United States, ACARS transmits all information using VHF frequencies. The airborne equipment transmits through a VHF transceiver (typically located in the aircraft's equipment bay). HF frequencies are also available for ACARS transmissions during intercontinental flights. However HF transmissions are quickly being replaced with satellite equipment. At the time this text was written, some aircraft employ satellite transmissions of ACARS. Satellite usage will improve worldwide coverage and enhance ACARS performance. In future, more data will be transmitted through some form of digital data and it will most likely be broadcast through satellites.

ACARS is a digital system that initiates all U.S. transmissions on the VHF frequency 131.55 MHz. The ACARS airborne equipment contains a control unit, typically located on the flight deck, a Management Unit (MU), and the necessary VHF transceiver located in an electronics or avionics equipment bay. A typical ACARS control unit is shown in Figure 7-2-1. The ground-based equipment contains antennas and VHS transceivers located at various sites, a data link via telephone lines to one or more ACARS control facilities, and a data link to the various airlines (Figure 7-2-2). In some cases, the communications between ACARS ground facilities and the airlines are accomplished through microwave transmitters or satellite links. Of course, the airline element of ACARS is an elaborate system of command and control subsystems used for maintenance, crew scheduling, gate assignments, and other day-to-day operations.

The ACARS system is designed to use a variety of VHF frequencies from 129.00 MHz to 137.00 MHz. In North America, all ACARS transmissions begin on 131.55 MHz. The ACARS ground facilities may then assign a different VHF frequency (between 129.00 and 137.00 MHz) to the aircraft. The assigned frequency will be a function of the aircraft location and the particular VHF frequencies used in that area. The ACARS ground facilities will then reassign new VHF frequencies to the aircraft ACARS as needed. Any frequency changes are totally automatic and therefore unknown to the flight crew. It is very likely that ACARS will change frequencies several times during a given flight.

Each aircraft using the ACARS system is given a specific address code. This code is used by the ground-based facility whenever calling the aircraft. The airborne equipment will monitor all ACARS data transmissions on their assigned frequency and accept only those with the correct address code.

Since the VHF frequencies assigned to aircraft are relatively crowded, all ACARS messages must be as short as possible. To achieve a short message, special digitized code blocks using a

maximum of 220 characters are transmitted in a digital format. If a longer message is needed, more than one block will be transmitted.

In general, the ACARS system operates in two modes for data communications: the demand mode and the polled mode. The demand mode allows the flight crew or airborne equipment to initiate communications. To transmit, the airborne Management Unit (MU) determines if the ACARS channel is free from other communications. If the frequency is clear, the message is transmitted; if the frequency is busy, the MU will wait until the frequency is available. The ground station will send a reply to the message transmitted from the aircraft. If an error message or no reply is received, the MU will continue to transmit the message at the next opportunity. After six unsuccessful attempts, the airborne equipment will notify the flight crew.

In the polled mode, ACARS operates only when interrogated by the ground facility. The ground facility will routinely uplink "questions" to the aircraft equipment and when a channel is clear, the MU will respond with a transmitted message. The MU organizes and formats all data prior to transmission. Upon request, the flight information is transmitted to the ground facility. Information for ACARS is collected from several aircraft systems, including the flight management system (FMS), and the central maintenance computer system (CMCS).

Architecture

The various systems that interface with the MU communicate via an ARINC 429 data bus. As shown in Figure 7-2-3, the Boeing 747-400 ACARS management unit interfaces with the:

1. Modularized Avionics and Warning Electronics Assembly (MAWEA), for chime tones to announce calls,

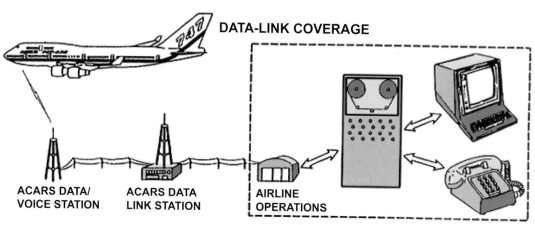

DATA-LINK COVERAGE

ACARS DATA/ VOICE STATION   **ACARS DATA LINK STATION**   **AIRLINE OPERATIONS**

Figure 7-2-2. Boeing 747-400 ACARS interface diagram

Courtesy of Northwest Airlines, Inc.

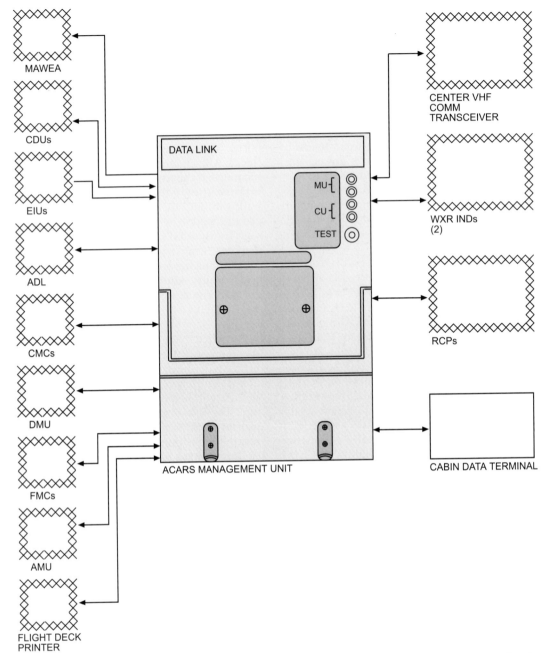

Figure 7-2-3. Boeing 747-400 ACARS interface diagram

Courtesy of Northwest Airlines, Inc.

2. Control Display Units (CDU), for basic operational controls

3. EFIS/EICAS Interface Units (EIU), for aircraft identification and current flight status

4. Airborne Data Loader (ADL), for the loading of ACARS MU software

5. Central Maintenance Computer (CMC), for transmission of systems fault data

6. Data Management Unit (DMU), for transmission of aircraft condition monitoring parameters

7. Flight Management Computer (FMC), for transmission of various flight plan parameters

8. Audio Management Unit (AMU), coordinating voice call lights

9. Flight deck printer, for receiving printed data

10. Center VHF Communications Transceiver, for transmission and reception of ACARS radio signals

11. Weather Radar Indicators (WXR IND), for display of received messages

12. Radio Control Panels (RPC), used for voice or data mode selection

13. Cabin Data Terminal, for downlink or uplink of nonverbal data

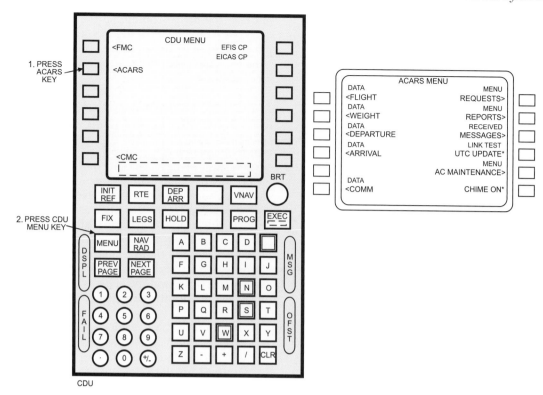

Figure 7-2-4. A Control Display Unit (CDU) showing access to the ACARS menu

Courtesy of Northwest Airlines, Inc.

On the B-747-400, there is typically only one ACARS MU, which is located in the main equipment center. There are two power sources required by the ACARS MU: 115 VAC from AC bus 3 and 28 VDC from the hot battery bus. The power from the hot battery bus is used by the MU's internal clock. The MU performs all of the following coordination functions of the airborne ACARS:

1. Controls operational modes

2. Stores and decodes messages

3. Encodes messages and coordinates data transmissions

4. Provides timing using an internal clock

5. Provides all interface activities between airborne ACARS subsystems

6. Monitors ACARS signals and accepts data transmitted to the specific aircraft

Controls and Displays

The ACARS control is managed through any one of the three Control Display Units (CDU) located on the flight deck. ACARS is accessed through the CDU menu where a selection is made from the ACARS menu (Figure 7-2-4). Pressing the appropriate **LINE SELECT KEY (LSK)** will access the different functions for ACARS. The flight crew can then input data as needed.

All data entered will be displayed on the CDU display for visual reference. The message can then be sent automatically or manually using the ACARS menu. For example, pressing the LSK adjacent to RECEIVE MESSAGES> will allow ACARS to display any received message(s).

To activate the ACARS voice mode using the CDU, press the **LSK** adjacent to <COMM, then follow the prompts on the display. A request for voice communications can also be made from the ACARS ground network. The request

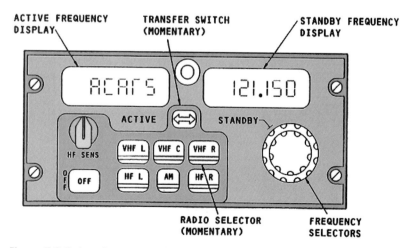

Figure 7-2-5. A radio communication panel showing the display of ACARS in the active frequency selection and 121.150 MHz in the standby frequency selection

Courtesy of Northwest Airlines, Inc.

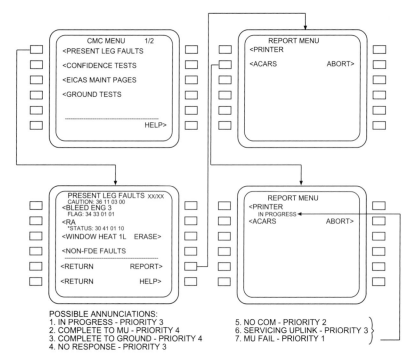

POSSIBLE ANNUNCIATIONS:
1. IN PROGRESS - PRIORITY 3
2. COMPLETE TO MU - PRIORITY 4
3. COMPLETE TO GROUND - PRIORITY 4
4. NO RESPONSE - PRIORITY 3
5. NO COM - PRIORITY 2
6. SERVICING UPLINK - PRIORITY 3
7. MU FAIL - PRIORITY 1

Figure 7-2-6. Sequence of CMC displays used to send the Present Leg Faults through the ACARS transmitter   *Courtesy of Northwest Airlines, Inc.*

will contain an assigned frequency. When the network is ready for voice communications, a call light on the audio control panel will illuminate and a soft chime will sound.

The B-747-400 Radio Communications Panel (RCP), located on the flight deck, is used for radio selection and tuning of the HF and VHF radios. As seen in Figure 7-2-5, the RCP displays the active and standby VHF/HF frequencies. If the center VHF radio is selected and ACARS is

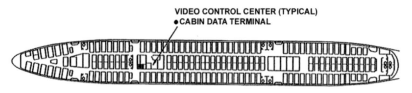

Figure 7-2-7. Location of the cabin ACARS access unit on the B747-400
Courtesy of Northwest Airlines, Inc.

Figure 7-2-8. ACARS cabin access terminal   *Courtesy of Northwest Airlines, Inc.*

in use, the active communications frequency is replaced by the term ACARS. This indicates that ACARS is transmitting. The active frequency automatically returns when ACARS transmission is complete.

Various EICAS maintenance pages and CMC fault reports can be sent to ground facilities using ACARS. This process becomes very helpful to maintenance personnel at line facilities. When aircraft have short turnaround times, it is essential that maintenance is performed as quickly as possible. If the technician has received a copy of the CMC fault data, the troubleshooting process can begin while the aircraft is still in flight. In many cases, the CMC fault data can provide the technician with enough information so replacement LRUs, the necessary paperwork, and the appropriate tools can all be waiting for the aircraft's arrival.

To downlink a CMC fault report or EICAS maintenance page, the crew would access the desired fault data on the CDU. In the example of Figure 7-2-6 the pilot wishes to send the present leg faults page. To send the report, press the **REPORTS** > line select key, then select <*ACARS*. If all systems are operational, the message *IN PROGRESS* will be displayed. The message should change to *COMPLETE TO MU* and then, *COMPLETE TO GROUND*. If a failure has occurred or ACARS is processing an uplink message, one of the following messages will be displayed: *NO RESPONSE, NO COM, MU FAIL,* or *SERVICING UPLINK*.

On many of the newest aircraft, such as the A-380, the central maintenance system can automatically downlink maintenance data. These aircraft employ automated systems which require no crew actions for maintenance downlinks. This data can be sent using an ACARS system or a specific system dedicated to a given airline or aircraft manufacturer. These systems were discussed in Chapter 5.

On the B-747-400, the ACARS data terminal is typically located at the cabin video control center in the main cabin of the aircraft (Figure 7-7). The unit is installed in the wall and folds down for use. The terminal contains a standard alphanumeric keyboard and liquid crystal display, resembling a typical laptop computer (Figure 7-2-8). This unit is typically used by the flight attendants to access ACARS data transmission/receive modes.

ACARS data can also be displayed on either of the two weather radar indicators located on the lower right and left sides of the instrument panel. Keep in mind weather radar data is typically displayed on the navigational display; therefore, accessing ACARS data does not inhibit the viewing of weather information. To display ACARS data, the WXR indicator control switch must be in the auxiliary (AUX) position. ACARS data is displayed in up to 12 lines of 32 characters each. Data can also be printed using an airborne printer.

Maintenance and Troubleshooting

For ACARS operations on the B-747-400, the aircraft battery power must be available to the hot battery bus. If the battery is removed from the aircraft during maintenance, the ACARS unit cannot be tested and the ACARS MU internal clock must be reset. The clock reset is accessed through the ACARS menu on the CDU.

A BITE circuit within the MU monitors the ACARS MU and related airborne subsystems. The BITE circuit continuously monitors the health of ACARS components and reports all failures to the central maintenance computer system. The MU test switch, located on the face of the MU, can also be used to initiate an ACARS test (Figure 7-2-9). During the test, all four lights should illuminate for three seconds, all lamps should extinguish for the next three seconds. After six seconds, the appropriate lamp should illuminate; green means 'OK', red means failed system. Red and green indicators are available for both the MU and CU.

To test the ACARS link between airborne and ground based equipment, the system must be configured for a link test. The link test is performed by accessing the ACARS menu on the CDU and pressing the **LSK** adjacent to **UTC UPDATE** (Figure 7-2-4). The link test will activate the MU transmitting, receiving, and system interfacing capabilities. For this test to take place, the aircraft must be within range of an ACARS ground station. Keep in mind; VHF transmissions with the aircraft on the ground are very limited.

Another system to consider when troubleshooting ACARS is the center VHF transceiver. Since ACARS will only transmit using the center VHF radio, that system must be working for ACARS operations. Of course, different aircraft may use a different VHF transceiver for ACARS.

Satellite Communications

Some commercial and most business aircraft are now equipped with some type of satellite communications system. Satellite communications operate using high frequency radio signals, which transmit from the aircraft to an orbiting satellite and then to a ground facility (Figure 7-2-10). The process reverses when com-

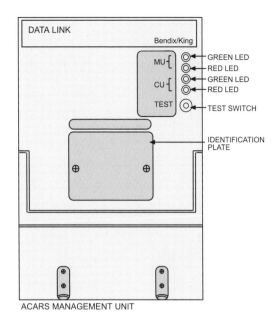

Figure 7-2-9. ACARS management unit showing the MU test switch   *Courtesy of Northwest Airlines, Inc.*

municating from the ground to the aircraft. Most systems can handle both voice and data transmissions following one or more ARINC specification standards.

Inmarasat is an international corporation providing two-way communication services to a range of governments, aid agencies, media

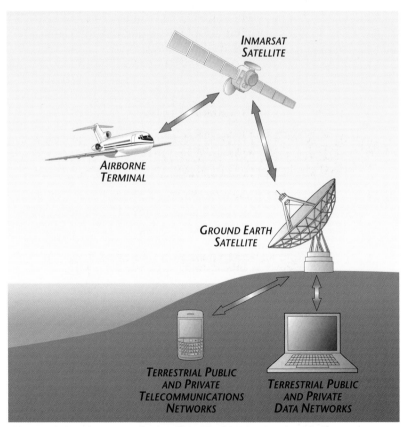

Figure 7-2-10. Elements of a Satellite Communication System

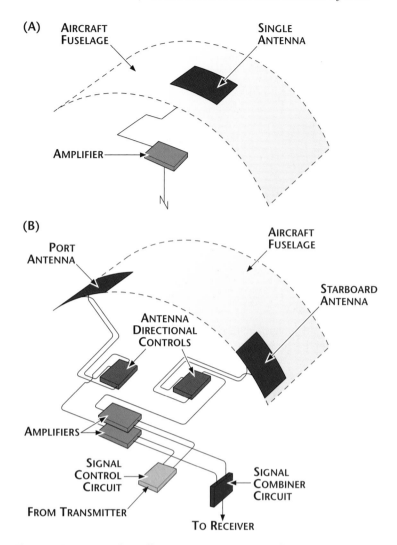

Figure 7-2-11. Typical Satellite System Antenna Installation: (A) Low-gain antenna, (B) High-gain antenna

2,400 bit/s. Different services require airborne equipment designed for that specific purpose. Inmarsat fees also change according to the level of service.

The system consists of three distinct system elements; the airborne equipment, the celestial element, and the terrestrial components (Figure 7-2-10). The airborne equipment consist of the following:

1. A satellite communications transmitter/receiver unit, typically installed in the electronics equipment bay,

2. An interface device, which may be a hand-held phone-type unit, a facsimile machine, laptop, or ACARS equipment, and

3. High- or low-gain antenna mounted on top of the aircraft (Figure 7-2-11). The low gain antenna is used for low speed transmissions and is a fixed unit. The high gain antenna is electronically steerable and employs more than one antenna element. The high gain system is used for exchange of high-speed data (typically about 10 KBps).

Section 3

Interphone Systems

The interphone system is found on transport category aircraft to provide communications between flight crew, ground crew, flight attendants, and maintenance personnel. Due to the size of transport category aircraft, communications between different areas of the aircraft, or inside and outside of the plane are nearly impossible without the aid of the interphone system. Considering that most airports are noisy places to begin with, the interphone system becomes nearly essential to perform any maintenance or ground service activities.

The B-747-400 interphone system is divided into four basic subsystems:

1. Flight interphone
2. Service interphone
3. Cabin interphone
4. Crew call system

The flight and service interphones, as well as the crew call system, all operate in conjunction with the audio management unit. The cabin interphone system is not controlled directly by the audio management unit and will be discussed in Section 4 of this chapter.

outlets, and businesses with a need to communicate in remote regions or where there is no reliable ground based network. Since aircraft often fly in remote locations, especially when crossing oceans, Inmarasat offers services to both airlines and corporate operators.

Inmarasat employs a series of satellites orbiting near the equator and provide coverage to a specific portion of the earth's surface. Collectively the Inmarasat satellites cover nearly the entire earth excluding the poles. The service provided to aircraft allows for the transmission/reception of voice, fax, and data signals.

There are three levels of Inmarasat service currently available for airborne use, Aero-L, Aero-H, and Aero-I. The Aero-L (Low-gain antenna) is used primarily for packet data that includes ACARS services. The Aero-H (High-gain antenna) is a service providing medium quality voice and fax/data at up to 9,600 bit/s. Aero-I (Intermediate-gain antenna) is used for low quality voice and fax/data at up to

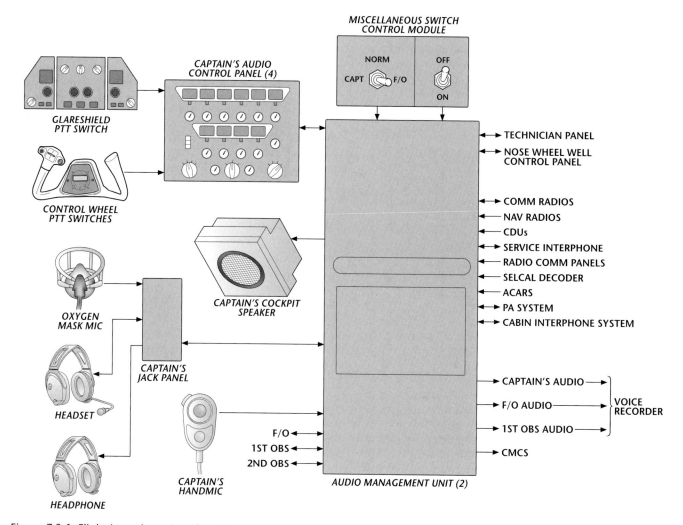

Figure 7-3-1. Flight interphone interface diagram

Flight Interphone

The B-747-400 flight interphone system is used to provide communications between flight crew members and/or other aircraft operations personnel. The flight interphone system is typically used for communications between flight crew members; however, the system may be interconnected to the service and cabin interphones when needed.

Flight Interphone Architecture

Refer to the flight interphone system interface diagram in Figure 7-3-1 for the following discussion. The Audio Management Unit (AMU), located in the main equipment bay, is used to process and control all audio signals required by the flight crew. The AMU receives input from the captain's and/or first officer's Audio Control Panel (ACP), which receives commands from the control wheel and glare shield Press-To-Talk (PTT) switches. The audio signals to the AMU can be sent from the captain's or

first officer's hand-held microphone (MIC), the oxygen mask MIC, or the headset boom MIC. The audio outputs can be sent to the headset, headphones, or flight deck speakers. The miscellaneous switch control module contains the switch to connect the service interphone with the flight interphone system. If the captain's or first officer's audio system fails, the observer audio system switch allows for transfer of the failed system to the observer's station. Locations of the flight interphone system components are shown in Figure 7-3-2.

Since the AMU controls all voice transmissions to/from the flight crew, the AMU must interface with the various radio systems on the aircraft, the passenger address system, and the cabin interphone system. The AMU also sends all audio signals to the aircraft's voice recorder. The AMU is comprised of four separate sections to provide redundancy: the captain's, first officer's, first observer's, and second observer's cards (Figure 7-3-3). There are three separate circuit breakers that feed the AMU and each card is internally fused. Diodes inside the AMU ensure that a shorted card will not

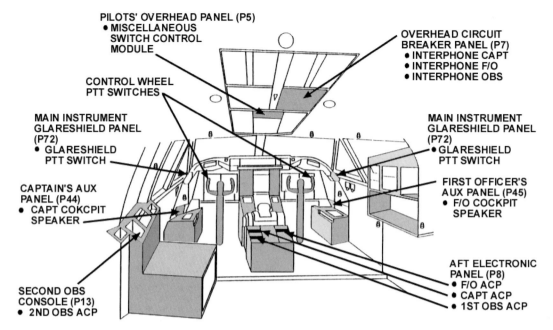

Figure 7-3-2. Location of flight interphone system components

Courtesy of Northwest Airlines, Inc.

draw power from an operational portion of the AMU. As seen in Figure 7-3-4, each card of the AMU is dedicated to a specific group of inputs and outputs.

Microphones and headsets. There are several options for each flight crew member regarding communication microphones (MIC) and speakers/headsets. The captain and first officer can choose from the following (Figure 7-3-5):

1. Headset (MIC/earphone combination)

2. Hand-held MIC

3. Oxygen mask MIC

4. Headphones

5. Cabin speaker

The headset contains an acoustic tube MIC that carries sound waves to the transducer. The

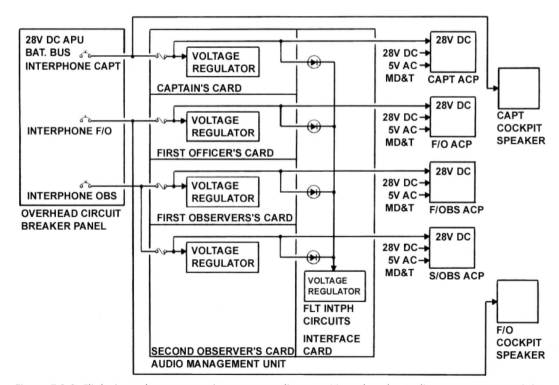

Figure 7-3-3. Flight interphone system input power diagram. Note that the audio management unit is subdivided into different cards to provide redundancy.

Courtesy of Northwest Airlines, Inc.

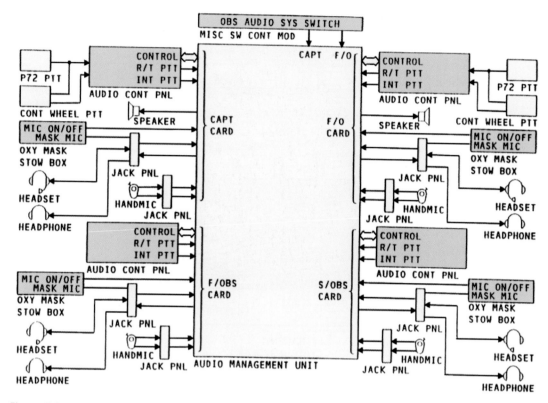

Figure 7-3-4. Audio management unit interface diagram

Courtesy of Northwest Airlines, Inc.

transducer converts the sound waves into an electrical signal, which is sent to the preamplifier located in the headset cord. The headset MIC is keyed by one of three Press-To-Talk (PTT) switches. The headset earphone is a standard earphone assembly that allows the crew member to monitor communications.

The oxygen mask contains a carbon-type MIC used for communications when the flight crew requires oxygen. The oxygen mask MIC is keyed by any of the PTTs. The hand-held MIC assembly contains a preamplifier and microphone assembly and keyed by a switch on the side of the unit. The crew member can select the headphones or flight deck speaker to monitor communications when using the oxygen mask or hand-held MIC.

Controls. The Audio Control Panels (ACP) provide command signals to the AMU for control of the communication and navigation audio signals. Each of the four ACPs (captain, first officer, first observer, and second observer) is operated independently and connects to the AMU via an ARINC 429 data bus. The Receive/Transmit (R/T) and Interphone (INT) PTT switches send a discrete signal to the AMU.

There are 13 audio receiver controls located on the ACP. Each receiver control is dedicated to a specific communication system. The receiver controls are a push on/off switch-volume con-

trol combination. When a receiver control is in the on position, the associated green indicator illuminates.

As seen in Figure 7-3-6, there are ten transmitting switches located on each of the four identical ACPs. Each of the transmit switches are dedicated to one of the following systems:

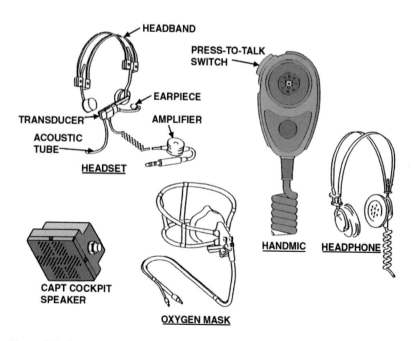

Figure 7-3-5. Various microphones/earphone combinations available to the flight crew

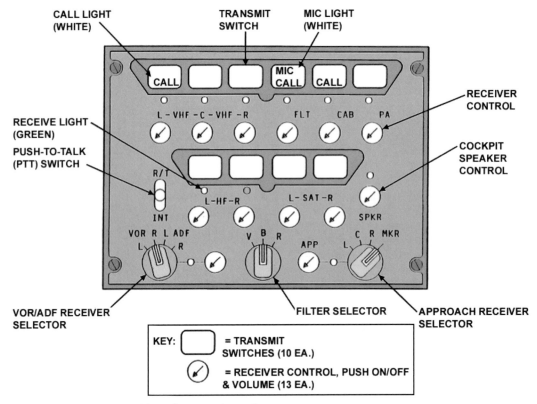

Figure 7-3-6. Boeing 747-400 audio control panel   *Courtesy of Northwest Airlines, Inc.*

left/center/right VHF, flight interphone (FLT), cabin interphone (CAB), passenger address (PA), and the left/right HF, left/right satellite communications (SAT). The transmit switches are lighted push-button switches divided into two sections: top and bottom. The bottom section displays a CALL light when the related system has an incoming message for the flight crew. When the **TRANSMIT** switch is pressed, the CALL light extinguishes and the MIC light illuminates on the top half of the switch. If a given system is already selected for transmission, the CALL light extinguishes when a **PTT** switch is pressed.

The ACP controls related to the interphone system are labeled FLT, CAB, and INT. The ACP flight interphone (FLT) call light illuminates whenever a request for communications is made by the ground crew via the flight interphone connection in the nose wheel well. The ground crew interphone connection in the nose wheel well is shown in Figure 7-3-7. Pressing the transmit switch allows the flight crew member to talk on the service interphone system via the hand-held MIC. Pressing the **INT PTT** switch will key the headset or oxygen mask MIC to the interphone system. The FLT volume control adjusts the volume for all flight interphone messages received at that station. The cabin interphone (CAB) call light illuminates when a communication request is made by the flight attendants via the cabin interphone system.

The lower controls on the ACP are used for navigation radio audio reception (Figure 7-3-6). The flight crew can choose to listen to the L/R VOR, L/R ADF, or the L/C/R marker beacon (MKR). The filter selector is used to choose what type of VOR audio will be transmitted to the crew member. In the *V* position, only voice frequencies will be passed. In the *R* position only the station range code identification will be available. In the *B* position, both the VOR range and voice frequencies will be sent to the flight crew.

Both the captain's and first officer's control wheel contain a two-position press-to-talk switch. As seen in Figure 7-3-8, the switch control is a rocker trigger-type assembly. Pressing the upper portion of the trigger keys activates the oxygen or headset MIC to the selected communication system. Pressing the lower portion of the trigger will activate the interphone system regardless of the communication system selected on the ACP.

Service Interphone

The service interphone system permits ground crew communications through various access points located throughout the aircraft. Technicians often use this portion of the interphone system during various maintenance operations. The service interphone may also be used during ground service operations.

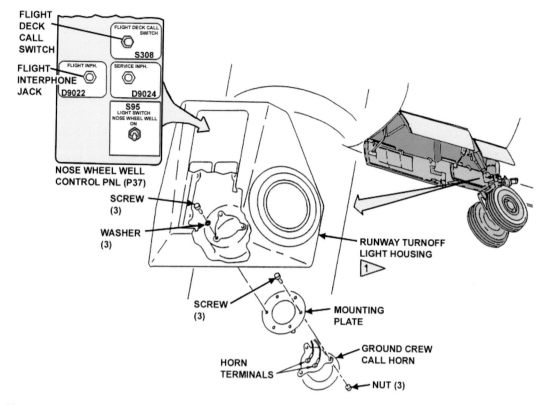

Figure 7-3-7. Wheel well flight interphone connection

As seen in Figure 7-3-9, on the B-747-400 there are 19 different service interphone connection points. The interphone connection points are combined into three groups: forward, mid, and aft. Each input connection of the group is paralleled and sent to the audio management unit.

The Audio Management Unit (AMU) receives input signals from the service interphone jacks. As discussed earlier, the AMU coordinates interphone system communications (Figure 7-3-10). The service interphone inputs are amplified by the AMU and sent back to the service interphone jacks or flight interphone system, if the service interphone switch is on.

Crew Call System

The crew call system is used to alert ground crew personnel of an interphone message from the flight crew, and to alert the flight crew of a message from the ground crew. A horn, along with the interphone control panel is located in the nose wheel well (Figure 7-3-7). Pressing the **FLIGHT DECK CALL** switch sends a signal to the Modularized Avionics and Warning Electronics Unit (MAWEA). MAWEA activates a chime through the flight deck speakers and illuminates a light on the audio control panel. The Cabin Interphone Controller (CIC) is used to interface the various subsystems of the crew

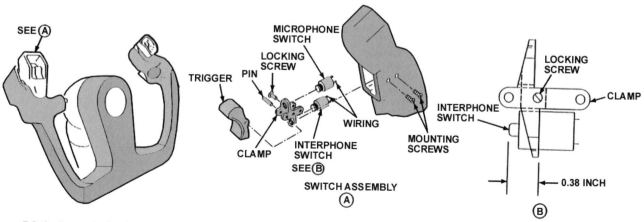

Figure 7-3-8. Control wheel press-to-talk switches

Courtesy of Northwest Airlines, Inc.

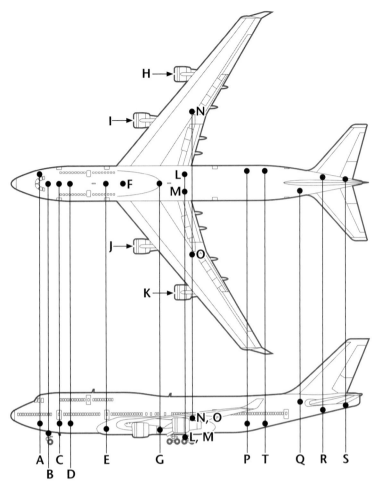

A - FORWARD EQUIPMENT CENTER (INTERNAL)
B - NOSE WHEEL WELL PANEL (EXTERNAL)
C - MAIN EQUIPMENT CENTER (INTERNAL)
D - FORWARD CARGO COMPARTMENT (INTERNAL)
E - CENTER EQUIPMENT CENTER (INTERNAL)
F - AIR CONDITIONER EQUIPMENT BAY (EXTERNAL)
G - GROUND AIR INLET (EXTERNAL)
H - ENGINE 4 NACELLE (EXTERNAL)
I - ENGINE 3 NACELLE (EXTERNAL)
J - ENGINE 2 NACELLE (EXTERNAL)
K - ENGINE 1 NACELLE (EXTERNAL)
L - RIGHT WHEEL WELL (EXTERNAL)
M - LEFT WHEEL WELL (EXTERNAL)
N - RIGHT REFUELING STATION (EXTERNAL)
O - LEFT REFUELING STATION (EXTERNAL)
P - AFT CARGO EQUIPMENT PANEL (INTERNAL)
Q - AFT EQUIPMENT CENTER (INTERNAL)
R - TAIL CONE (INTERNAL)
S - APU
T - BULK CARGO DOOR (INTERNAL)

Figure 7-3-9. Boeing 747-400 service interphone locations

call system. The CIC will be discussed later in this chapter.

The ground crew call horn sounds any time a flight deck interphone call is made to the

ground crew. The same horn is also used to alert the ground crew of an equipment cooling failure, or if the Inertial Reference (IR) unit is on and AC power is not supplied to the aircraft. If the IR is allowed to operate on battery power only, the battery will quickly discharge.

Interphone Maintenance and Troubleshooting

Maintenance and troubleshooting of the interphone system typically consists of wiring connector problems related to the various system's components, or removal and replacement of various LRUs. The AMU contains BITE circuitry that communicates with both Central Maintenance Computers (CMC). The CMCs record messages related to an AMU failure, ACP failure, or ARINC 429 interface bus failure.

MICs and headsets are items that require regular maintenance. These items are constantly being moved, bounced, and dropped about the cabin. This results in wires shaking loose causing intermittent audio transmissions or to break completely. The easiest way to troubleshoot MICs and headsets is to operate the system; shake the suspect unit and/or related wiring while listening for problems. Then, swap the suspect unit and try the system a second time. Service interphone connection jacks also seem to be problem components. Many of these jacks are exposed to the outside environment and are often treated roughly. In most cases, the jack becomes loose causing an intermittent connection. Be sure to inspect these components carefully when dealing with interphone faults.

Section 4

Passenger Entertainment and Communication Systems

In-Flight Entertainment Systems

Today the cabin area of many modern aircraft has evolved into a hi-tech audiovisual playground for passengers. There are a variety of options from which passenger can choose to work, play, or communicate during a flight. In general, these various systems are known as IFE (In-Flight Entertainment) systems.

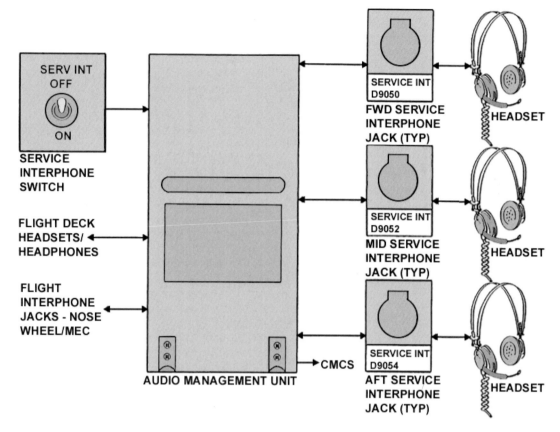

Figure 7-3-10. Block diagram of the service interphone system

Courtesy of Northwest Airlines, Inc.

Manufacturers such as Panasonic Avionics Corporation and Thales Group offer systems for commercial airliners and small business jets. In many cases, the airlines have contracted with a management company as a provider that can maintain the system and manage the passenger options.

Passenger entertainment first became popular in the 1980s as wide-body passenger jets were equipped with audio and video options for the passengers. These aircraft typically employed CRT displays mounted in various locations in the cabin (commonly in the ceiling). The video signal came from a videotape unit similar to a home VCR. Audio is transmitted to the passenger seat and through individual headphones. Today almost all commercial passenger jets are equipped with audiovisual systems, which employ LCD displays and multiplexed digital audio. Some systems incorporate individual video displays, most commonly mounted in the seatback directly in front of each passenger (Figure 7-4-1). This type of system allows for individual choices for each passenger, but requires extremely large quantities of data being transmitted throughout the aircraft.

Safety is a major concern with aircraft systems, and passenger entertainment is no exception. In order to ensure that the IFE system will not create safety hazards, there are two basic design characteristics, which must be followed. First, the IFE must be completely isolated from any systems that are flight critical. The IFE system will typically receive all digital data signals as well as electrical power through a dedicated IFE data bus or power distribution bus. Second, all IFE equipment, wiring, and instal-

Figure 7-4-1. Seatback video display for individual passengers

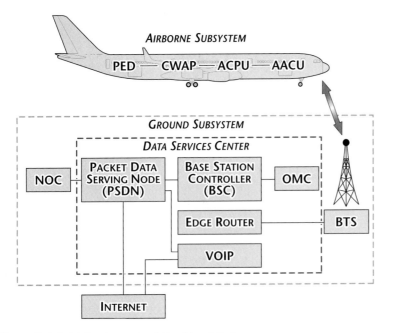

Figure 7-4-2. Main elements of the Aircell System

texting, Internet browsing, emailing, and even cell phone use, in some areas, can be provided. Since these systems often work in conjunction with passenger entertainment, they are often referred to as IFEC (In-Flight Entertainment and Connectivity or In-Flight Connectivity and Communications). The IFC system offered by Aircell will be discussed later in this chapter.

Airborne Broadband Internet

Today's wireless technologies have become almost a necessity for business operations as well as personal activities. Modern public demand for connectivity to the Internet has driven many airlines and most corporate aircraft owners to install an Airborne Broadband System (ABS) for passenger convenience. The ABS technologies allow laptops, WiFi enabled PDAs (personal digital assistants), and other wireless devices a means of connection to the Internet during flight. Of course the ABS requires an elaborate system independent of the aircraft using radio signals, satellites, and ground-based units to connect the aircraft to the Internet.

lations must be easily identified as well as FAA certified.

Video on demand, SiriusXM radio, and satellite TV services are also being offered on some aircraft. However, for the most part these features are limited and the providers are currently taking a close look at the cost effectiveness of this type of IFE. It is important to remember that as more systems are added to the aircraft, the airplane becomes heavier and less efficient; this eventually adds cost to each flight.

Another feature gaining popularity is the availability of wireless Internet (WiFi) service during flight. This is often referred to as In-Flight Connectivity (IFC). Services such as

Aircell broadband service is a corporation, which provides much of the wireless internet currently available to aircraft. Aircell is a private company started in the early 1990s and the only company in the United States authorized by the FCC and FAA to use frequencies in the 800 MHz band for in-flight communications. The broadband spectrum licensed to Aircell is 894 Megahertz (MHz) to 895.5 MHz for the air to ground frequency and 849 MHz to 850.5 MHz for the ground to air frequency. Aircell currently supplies broadband service to over one thousand commercial aircraft operated by various airlines in the United States.

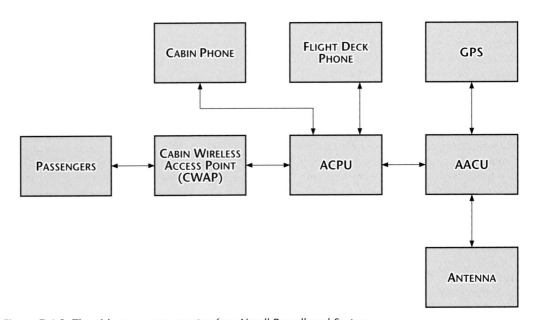

Figure 7-4-3. The airborne components of an Aircell Broadband System

At the time this text was written, the Aircell network is based on Third Generation (3G) Evolution-Data Optimized (EV-DO) Code Division Multiple Access (CDMA) cellular technology, which allows all passengers to be on a single broadband frequency. The speed of data transmission is dependent on the number of passengers using the system within the aircraft, as well as the number of planes sharing a cell site. As technology changes, the specifics of the Aircell network will no doubt improve. The ABS system consists of an airborne segment and a ground-based segment. As seen in Figure 7-4-2 the airborne segment consists of various line replaceable units (LRU) installed on the aircraft and the ground-based segment consists of a network of cell sites and the related circuitry for terrestrial-based connections to the Internet.

Figure 7-4-3 shows the components of the typical Aircell system installed on a commercial airliner. The airborne system is composed of four primary in-aircraft components:

1. Cabin Wireless Access Point (CWAP)

2. Controller Processor Unit (ACPU)

3. Air-to-Ground Communications Unit (AACU)

4. Cabin phone and flight deck handset

The CWAP provides the interface between passenger/crew wireless devices on the aircraft. This is often called the "hot spot" and employs two small CWAP antennas to capture wireless data from the passenger. Depending on the size of aircraft, more than one CWPA and their associated antennas may be located throughout the aircraft in order to provide adequate connectivity.

Once a wireless signal is received, the CWAP sends digital data to the ACPU. The ACPU is the central processor and control unit for the airborne system. In the ACPU, passenger data is authenticated, stored in buffers as needed, and prepared for transmission to the ground facility. To provide the air-to-ground link, the ACPU sends passenger data to the AACU (Air-to-Ground Communications Unit). The AACU employs the circuitry and two external antennas, mounted on the bottom of the aircraft, needed for connection to the terrestrial subsystem. The AACU also employs an external GPS antenna that is used to determine aircraft position. This position data allows the Aircell system to determine which ground-based antenna will be utilized as the aircraft flies across country. Of course, the GPS antenna is located on the top of the aircraft. Figure 7-4-4 shows a diagram of the antennas needed for a fully functional airborne broadband system.

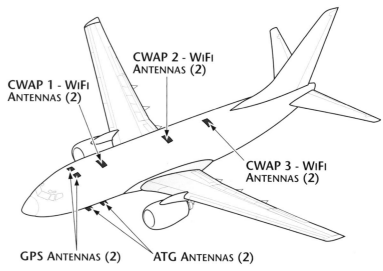

Figure 7-4-4. Broadband antenna locations for a typical commercial airliner

One disadvantage of the Aircell system is that the aircraft must be flying above 10,000 feet altitude and within a few hundred miles of a land-based antenna for proper operation. This means the system requires hundreds of ground-based cell antennas located throughout the US. Other systems have been developed for aircraft, which utilize satellite internet connections. These systems, although less popular and more expensive, provide connectivity in remote areas of the world.

Section 5

Advanced Cabin Entertainment Service System

The Advanced Cabin Entertainment Service System (ACESS) is used on the B-747-400 to control five major subsystems: the passenger entertainment audio system, the cabin interphone system, cabin lighting, the passenger service system, and the passenger address system. In short, ACESS controls and distributes passenger audio and lighting. ACESS is software driven and allows for improved cabin flexibility. That is, ACESS can easily reconfigure cabin audio and video signals to compensate for changes in cabin configuration. Through ACESS software, audio and video signals can be changed for virtually any seating arrangements, while older aircraft would have required changes in system wiring and components.

ACESS is a multiplexed digital system with the majority of ACCESS information transmit-

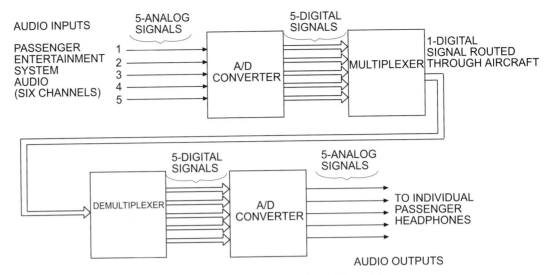

Figure 7-5-1. Typical configuration for the transmission of multiplexed data

ted on a serial data bus. Transmission of serial digital data provides a great reduction in wiring; however, multiplexer and demultiplexer circuits are required. As shown in Figure 7-5-1, multiplexers/demultiplexers are required at the beginning and end of each audio transmission route. Figure 7-5-1 also shows that the audio input signals go through an A/D converter circuit and the audio output goes through a D/A converter circuit. Since the initial audio signals are analog, they must be converted to digital signals prior to multiplexing. Multiplexing is the process of converting parallel digital data into serial digital data. After multiplexing, the data is then transmitted. After transmission, the data is demultiplexed. Demultiplexing is the process of converting serial data into parallel data. The D/A converter is then used to change the digitized

audio into an analog signal. The analog audio signal is then sent to the passenger headphones. It should be noted that multiplexers and demultiplexers, as well as A/D and D/A converters are typically circuits found within an LRU. These circuits are not stand-alone components.

Architecture

As shown in Figure 7-5-2, ACESS is comprised of 17 different LRUs. ACESS can be thought of as a pyramid of control units which are used to receive inputs, process data, coordinate activities, and send audio and control signals to their proper destination (Figure 7-5-3). Please refer to this pyramid periodically during the following discussion on ACESS.

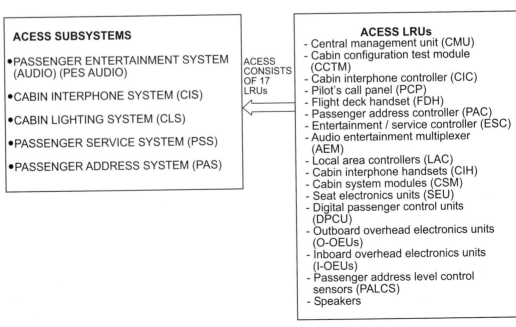

Figure 7-5-2. ACESS system and subsystem interface

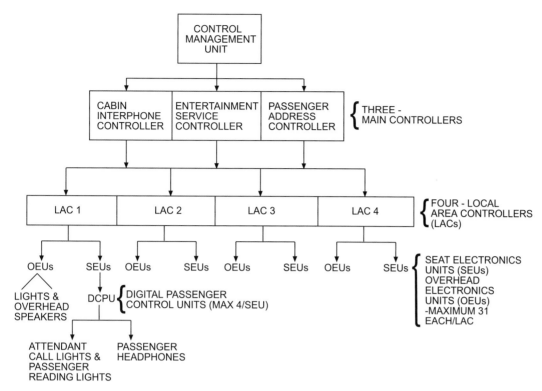

Figure 7-5-3. The hierarchy of ACESS control units

Three of the ACESS LRUs are called main controllers, they are the:

1. Cabin Interphone Controller (CIC) supervising the Cabin Interphone System (CIS) through the cabin/flight deck handsets, and the flight deck interphone system.

2. Entertainment/Service Controller (ESC) controlling the Passenger Entertainment System (PES) audio to the passenger headphones, Cabin Lighting System (CLS), and the Passenger Service System (PSS).

3. Passenger Address Controller (PAC) controlling the Passenger Address System (PAS) audio output to the cabin speakers and passenger headphones.

Each of the three main ACESS controllers interfaces with four local area controllers and the control management unit (Figure 7-5-4). The Control Management Unit (CMU) is a digital interface unit used to coordinate programming, testing, and monitoring of ACESS. The CMU receives inputs from the cabin configuration test module, the software data loader, and the EFIS/EICAS interface units. The Cabin Configuration Test Module (CCTM) is used to control portions of ACESS and to perform system tests. The CCTM will be discussed later in this chapter. The software data loader is used to program ACESS in the event of an aircraft configuration change. The EFIS/EICAS Interface Units (EIU) are used to interface ACESS with the central maintenance computer system and the integrated display units.

Figure 7-5-5. shows the interface between the three main ACESS controllers and the local area controllers. Remember the three main ACESS controllers are the CIC, ESC, and the PAC. Each Local Area Controller (LAC) is comprised of three independent circuits, one dedicated to each main controller. As seen in Figure 7-5-6, each LAC controls a given zone of the aircraft; LAC 1 controls zones A & B, LAC 2 controls zones C & D, LAC 3 controls zone E and LAC 4 controls the upper deck.

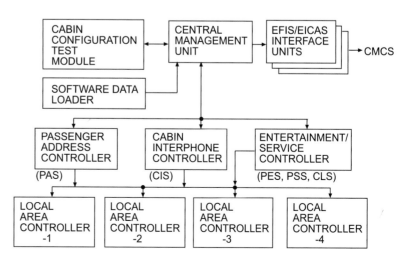

Figure 7-5-4. Block diagram of the ACESS controller interface

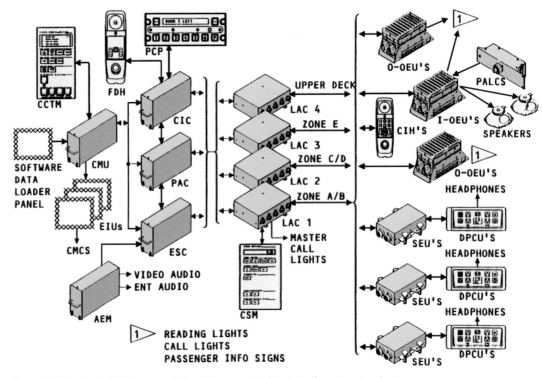

Figure 7-5-5. Pictorial diagram if the Advanced Cabin Interface Service System

Courtesy of Northwest Airlines, Inc.

The main controllers are located in the aircraft's main equipment rack; the LACs are located in the cabin area. LACs are typically located above various ceiling panels. The LACs act as a distribution unit for each of the three main controllers and send lighting commands and digitized audio signals to the overhead electronic units and the seat electronic units for distribution to the individual passenger stations and related lighting.

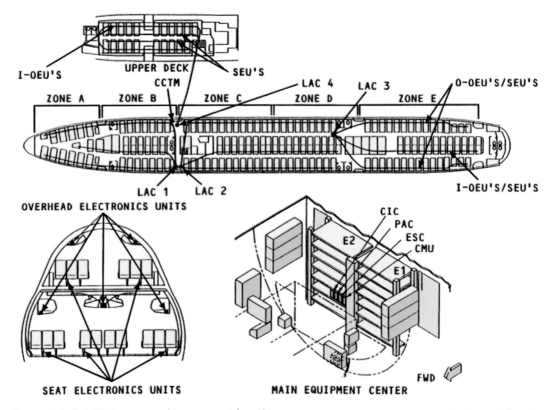

Figure 7-5-6. ACESS zones and component locations

Courtesy of Northwest Airlines, Inc.

The ACESS Overhead Electronic Units (OEU) control the overhead speakers, cabin lighting, passenger reading lights, flight attendant call lights, and passenger information lights (*no smoking* and *fasten seat belts*). A maximum of 31 OEUs can be connected to one LAC. The Seat Electronic Units (SEU) are used to interface with each seat control for selecting various audio channels, volume, reading lights, and attendant call lights. The SEU also distributes the requested audio to each passenger headphone connection. One SEU is located under each group of seats, and can interface with up to four seats in a group. A maximum of 31 SEUs can connect to one OEU. For locations of OEUs and SEUs refer to Figure 7-5-6.

During the preceding discussions, the major LRUs of ACESS have been introduced. The ultimate goal of ACESS is for these LRUs to interface with the aircraft crew and passengers. There are various control panels, handsets, headphones, and speakers all of which are part of ACESS. These items will be discussed as the specific systems are presented in upcoming sections of this chapter.

Configuration Database

The configuration database is the software program used to inform ACESS of the current cabin layout. There are several LRUs within ACESS that contain the configuration database stored in a nonvolatile memory. The Central Management Unit (CMU) stores the configu-

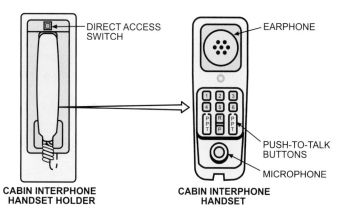

Figure 7-5-7. Cabin interphone handset

ration database. The CMU also coordinates database loading and storage for other ACESS LRUs.

The database can be used to identify items, such as seating configurations, cabin interphone dial codes, entertainment audio channels, passenger address areas, and the passenger address volume levels. This system provides flexibility to the airline for easy reconfiguration of the passenger compartment. A shop technician or the engineering department using a personal computer typically modifies the configuration database. The database is then stored on a CD-ROM or USB/Flash Drive (solid state memory stick) and transferred to the CMU using the aircraft's data loader. Aircraft that have not been updated may still use a 3.5" floppy disk. After the database has been downloaded to the CMU,

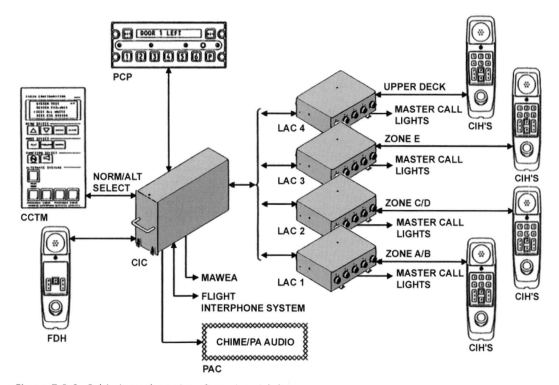

Figure 7-5-8. Cabin interphone interface pictorial diagram

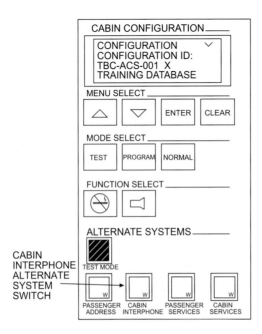

Figure 7-5-9. Cabin configuration test module.

the ACESS Cabin Configuration Test Module (CCTM) is used to download the database from the CMU to the various ACESS LRUs.

Cabin Interphone System

The Cabin Interphone System (CIS) provides communications between different flight attendant stations, and between flight attendant stations and the flight deck. A handset, resembling a telephone receiver, is located at each flight attendants station (Figure 7-5-7). From the flight deck, the pilots can interface with the CIS using the flight deck, handset, or the flight interphone system.

Architecture

A CIS interface diagram is shown in Figure 7-5-8. Please refer to this diagram during the following discussion on CIS architecture. Up to 25 Cabin Interphone Handsets (CIH) can be connected to the CIS. Each cabin handset is wired to one of the four Local Area Controllers (LAC). The LACs connect to the cabin interphone controller via a digital data bus. The Cabin Interphone Controller (CIC) is the central multiplexer/demultiplexer for the CIS. All CIS audio goes through the CIC.

The CIC has two fully redundant circuits: normal and alternate. Each redundant circuit is capable of operating the complete CIS. If the normal circuit fails, the alternate is activated using the Cabin Configuration Test Module (CCTM) (Figure 7-5-9). The CIC performs all cabin interphone coordination activities and

interfaces with other necessary LRUs. The CIC must be programmed with the current configuration database for proper operation. The CCTM interfaces with the CIC to download the database from the CMU when updating the system. The various CIC interfaces and their related functions are listed in Table 7-5-1.

Operation

To describe the operation of the cabin interphone system, three examples will be presented. The sequences of events described in these examples are typical of those which take place during cabin interphone use. The event sequence becomes very important when troubleshooting the system. The sequence of events can help a technician determine which LRUs are used during the various phases of an interphone call.

Example 1

When the flight deck wishes to place a call to a flight attendant's station, the following occurs:

1. The process begins with entering the correct code into the PCP.

2. The CIC transmits a signal to the PAC to sound a chime at the attendant's station. The CIC also sends a signal to the appropriate LAC which turns on the master call light at the attendant's station.

3. When the attendant picks up the handset, the CIC connects the pilot's and attendant's station and extinguishes the master call light. The pilot can use either the flight deck handset or flight interphone system for the communication. The proper selection must be made on the pilot's Audio Control Panel (ACP). The first officer's ACP is used to control the first officer's audio. The ACP was discussed under the flight interphone section of this chapter.

Example 2

If a flight attendant initiates a call to the flight deck, the following would occur:

1. The flight attendant enters the interphone station code on the handset.

2. The station code signal is sent through the LAC to the CIC.

3. The CIC sends a signal to MAWEA to sound the flight deck chime, tells the PCP to display an incoming message, and sends a signal to the ACP to illuminate the CAB INT light.

ACESS CIC INTERFACE LIST	
CIC INTERFACE	PURPOSE OF INTERFACE
Cabin Configuration Test Module (CCTM)	To download configuration database files, and to activate alternate CIC circuitry
Pilot Control Panel (PCP)	Used to place an interphone call from the flight deck to the cabin
Modularized Avionics and Warning Electronics Assembly (MAWEA)	To sound a flight deck chime for interphone calls from the cabin to the flight crew
Flight Interphone System (FIS)	To allow flight crew personnel to communicate to CIS using the FIS
Passenger Address Controller (PAC)	To sound the appropriate cabin chime when a request is made

Table 7-5-1. ACESS CIC interfaces

4. When the pilot activates his/her handset, the CIC connects the call and turns off all call lights.

Example 3

If a flight attendant calls another attendant station the following series of events would take place:

1. The station code is entered into the handset.

2. The station code goes to the CIC through the LAC.

3. The CIC sends a signal to the PAC that sounds a chime at the appropriate attendant's station. The CIC also initiates a signal to the appropriate LAC to light the master call light at the attendant's station being called.

4. When the called station handset is removed from its holder, the CIC connects the two stations and turns off the master call light via the LAC.

Call Priority

The cabin interphone system has a priority program that routes calls according to their importance. The order of CIS call priority is:

1. **Pilot Alert Call.** A pilot alert call can be made from any flight attendant's station

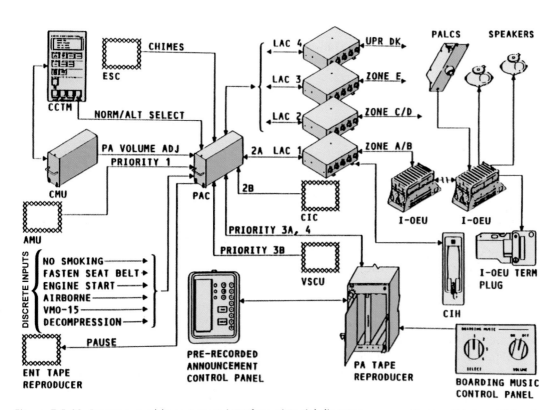

Figure 7-5-10. Passenger address system interface pictorial diagram   *Courtesy of Northwest Airlines, Inc.*

PASSENGER ADDRESS CONTROLLER (PAC) ARINC 429 DATA INPUTS	
PAC INTERFACES WITH	PURPOSE OF INTERFACE
Cabin Configuration Test Module (CCTM)	To select normal or alternate PAC circuits to control database updates from CMU controls
Central Management Unit (CMU)	To recieve configurations database information, and for fault monitoring
Audio Management Unit (AMU)	For connection to the flight interphone system
Video System Control Unit (VSCU)	For inputs of video system audio
PA Tape Reproducer	For inputs of prerecorded audio tape signals
Cabin Interphone Controller (CIC)	To coordinate calls from all cabin handsets (CIH) to the PAC for PA announcements
Local Area Controller (LAC) 1	Used for direct connection of 2 cabin handsets to the PAC
Local Area Controller (LAC) 1-4	Multiplexes and digitizes audio and control signal to connect all CIHs to the PAC for announcements from any attendants station
Entertainment/Service Controller (ESC)	Provides signals to PAC for operations of cabin chimes

Table 7-5-2. Passenger Address Controller ARINC 429 data inputs

using the code PP. Pilot alert call overrides all other interphone operations except another pilot alert call.

2. **All Call.** An All Call can be placed from any station using code 55. All call contacts all interphone stations.

3. **Priority Line Calls.** One of the cabin interphone stations can be selected as a priority station. This station takes priority over other cabin stations.

4. **Attendants' All Call.** Using code 54 any cabin station will call all other cabin stations. The flight deck is not included in the attendants all call.

5. **Normal station-to-station calls.** In this mode, one statin connects to another in the order received by the CIC. The CIC can store up to four call requests. Each call is processed in order of reception. The waiting call is processed after the preceding call is finished.

Passenger Address System

The Passenger Address System (PAS) transmits audio through a series of cabin speakers to announce messages from the:

1. Flight crew

2. Flight attendants

3. Prerecorded tape reproducer (messages and boarding music)

4. Video tape reproducer (audio only)

5. To the sound cabin chimes

The PAS on the B-747-400 is segmented into various zones, so that different cabin sections can receive specific announcements, or so all sections of the cabin can receive the same announcement.

Architecture

Figure 7-5-10 shows a PAS interface diagram and will be used for the following discussion on the PAS. The Passenger Address Controller (PAC), one of the three main ACESS controllers,

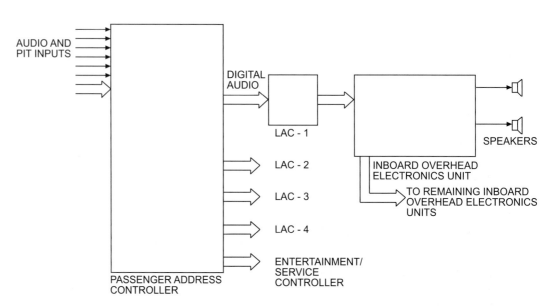

Figure 7-5-11. Passenger address control audio output diagram   *Courtesy of Northwest Airlines, Inc.*

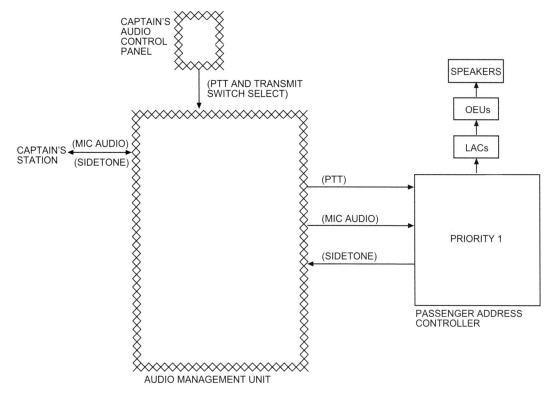

Figure 7-5-12. Passenger address system interface with the flight interphone system

Courtesy of Northwest Airlines, Inc.

provides the main coordinating functions for the PAS. The PAC contains two complete circuits, normal and alternate. A list of the PAC input interfaces and their basic functions are shown in Table 7-5-2.

There are six discrete inputs shown on the interface PAS diagram (upper left of Figure 7-5-11). The no smoking and fasten seat belt discrete are used to signal the need for a cabin chime. The cabin chime system is controlled through the PAC. The engine start, airborne, Vmo-15 KTS (maximum speed less 15 knots), and decompression discrete are used to inform the PAC of different aircraft configurations. This configuration information is used for automatic PA volume adjustments.

The main PAC output signals include:

1. Data to the CMU for system fault monitoring

2. Control signals to the entertainment, and PA tape reproducers

3. Multiplexed and digitized audio signals to the Local Area Controllers (LAC)

The LACs demultiplexes the signals and send the digitized audio to the Overhead Electronics Units (OEU). The OEUs convert the audio to an analog signal and send that signal to two cabin speakers (Figure 7-5-11).

Each OEU contains two separate amplifiers, one for each speaker.

Operations

The PAS has six operating modes related to the following functions:

1. Flight deck interphone PA announcements

2. Direct access PA messages

3. Cabin interphone PA transmissions

4. Broadcasting PAS tape reproducer music/messages

5. Broadcasting video system audio

6. Cabin chime activation

The following three examples will describe some of the operational functions of the PAS.

Example 1

Placing a PA announcement from the flight interphone system (Figure 7-5-12):

1. The pilot selects the PA system for transmission and presses the PTT switch on the audio control panel (ACP). This signal goes to the AMU.

2. The pilot announces the message into the headset or oxygen mask MIC. This signal goes to the AMU.

3. The AMU sends the press-to-talk (PTT) discrete and audio signals to the PAC.

4. The PAC multiplexes and digitizes the signals, sending them to the LACs (reference the PAS interface diagram, Figure 7-5-11).

5. The LACs demultiplexes the data and sends the signals to the OEUs.

6. The OEUs converts the digital audio to analog signals, amplifies, and sends the signals to the cabin speakers.

Example 2

Placing a PA announcement from a direct access handset. Direct access PA stations are located at each interphone station, the number one, and number four doors. PA announcements made from these stations have "direct access" to the PAC, bypassing the LAC circuitry.

1. Direct access is initiated by pressing the **DIRECT ACCESS** switch on the handset station (Figure 7-5-13). This sends a direct access signal to the LACs. The PTT switch is pressed to begin PA announcement.

2. When receiving the direct access signal, the appropriate LAC closes an internal switch and connects the handset directly to the PAC.

3. The PAC digitizes, multiplexes, and transmits the audio/control signal to the LACs.

4. The LACs demultiplexes the data and sends the audio signal to the OEUs.

5. The OEUs converts the digital audio to an analog signal, amplifies the signal, and sends audio to the cabin speakers.

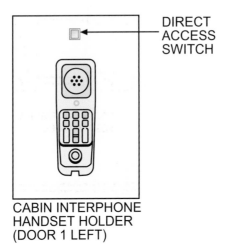

DIRECT ACCESS SWITCH

CABIN INTERPHONE HANDSET HOLDER (DOOR 1 LEFT)

Figure 7-5-13. Handset holder showing the direct access switch

Example 3

Using the cabin interphone system to place a PA announcement from any cabin interphone station (Figure 7-5-14):

1. The flight attendant presses the **PTT** switch at any flight interphone handset. (Note: The PTT switch is used for PA announcements only.)

2. Pressing the PTT signals the appropriate LAC that a PA announcement is to begin from that handset station. The LAC receives a PTT discrete, MIC audio, and a handset station identification signal from the handset.

3. The LAC digitizes and mutiplexes the input data. The LAC then transmits the data to the cabin interphone controller (CIC).

4. If more than one transmission is placed at a time, the CIC determines handset priority. The CIC then sends the appropriate multiplexed signals to the PAC.

5. The PAC then distributes the multiplexed audio and control signals to the LACs.

6. The LACs demultiplexes the data and sends the audio signal to the OEUs.

7. The OEUs converts the audio to analog, amplifies, and sends the signals to the cabin speakers.

Understanding the sequence of events for each activity conducted by the PAS will help a technician during troubleshooting. The three preceding examples contained a brief outline of the activities. Be sure to become familiar with the system and which components are being used during a given PA announcement prior to troubleshooting.

Automatic Volume Control

The PAS automatic volume adjustment system is used to set PA volume levels according to the amount of cabin noise. The automatic volume adjustment can be accomplished by the OEUs or PAC (Figure 7-5-15). The OEUs are the primary means for adjusting PA volume. Each OEU receives inputs from a Passenger Address Level Control Sensor (PALCS) which monitors local cabin noise levels. The OEUs adjust their audio output level according to the noise level signal received from the PALCSs.

The PAC software contains automatic volume adjustment circuitry to increase PA volume as cabin noise increases. The PAC determines cabin noise indirectly by monitoring aircraft status: engine on, airborne, Vmo - 15 KTS (max-

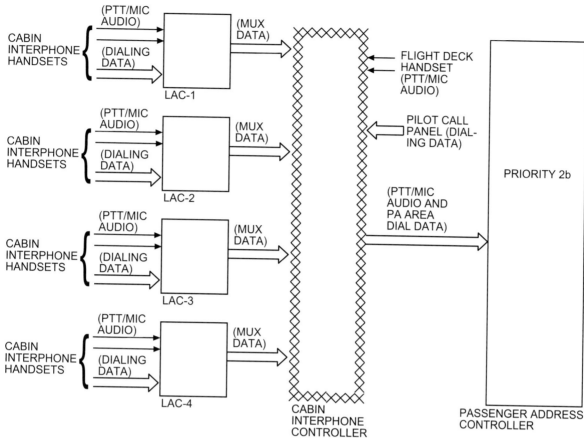

Figure 7-5-14. Cabin interphone system input diagram

Courtesy of Northwest Airlines, Inc.

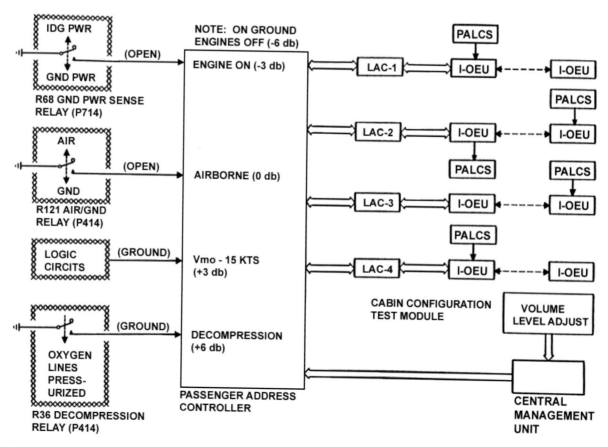

Figure 7-5-15. Passenger address controller automatic audio level control diagram

Courtesy of Northwest Airlines, Inc.

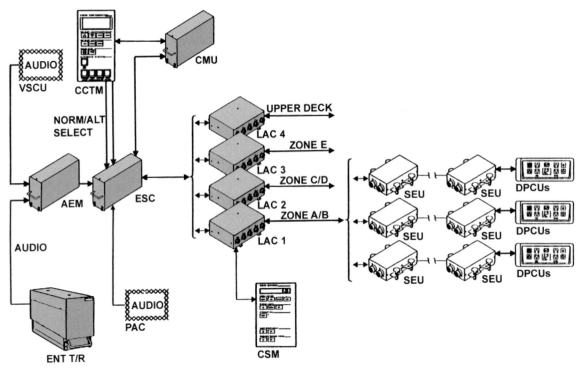

Figure 7-5-16. Passenger entertainment system pictorial interface diagram

Courtesy of Northwest Airlines, Inc.

imum speed less 15 knots), and decompression. The PAC is the backup adjustment source and is used only if the PALCSs are turned off or malfunctioning. The PAC also receives a volume level adjustment command from the Central Management Unit (CMU). The CMU receives volume level commands from the Cabin Configuration Test Module (CCTM).

Passenger Entertainment System

The Passenger Entertainment System (PES) audio provides prerecorded audio signals to each passenger seat interface unit. The PES typically contains several channels of music or other entertainment audio, as well as the audio portion of any in-flight video presentation. Each passenger can select a given channel for his or her listening pleasure through the individual seat mounted control unit. The selected audio is transmitted through headphones to the passenger.

Architecture

During the following discussion on the PES audio, refer to the interface diagram shown in Figure 7-5-16. The Entertainment/Service Controller (ESC) is the main ACESS controller that coordinates the distribution of the audio-tape reproducer and the passenger video system audio signals. The system uses the basic ACESS philosophy, which digitizes and multiplexes the audio signals and sends them to local areas for distribution to the passenger headsets. The ESC can operate on one of two completely independent circuits, normal or alternate. The LRU input/output interfaces with the ESC. The purposes for these interfaces are listed in Table 7-5-3.

The ESC can receive audio signals from three different sources: the Video System Control

PASSENGER ENTERTAINMENT SYSTEM (AUDIO) MAIN CONTROLLER	
INTERFACE LIST	
LRU WITH INPUTS TO ESC	PURPOSE OF INTERFACE
Cabin Configuration Test Module (CCTM)	Controls normal/alternate circuit selection, used to install the configuration database onto the ESC, runs PES test function
Audio Entertainment Multiplexer (AEM)	Supplies the multiplexed audio signal to the ESC for control and distribution (the AEM received the audio signals both the audio and video reproducers)
Passenger Address Controller (PAC)	Sends passenger address announcements to the ESC for distribution to passenger headphones during PA announcements
Local Area Controllers (LAC)	Receives multiplexed audio signals from the ESC for distribution to the seat electronic units (SEU)

Table 7-5-3. Passenger entertainment system main controller interfaces

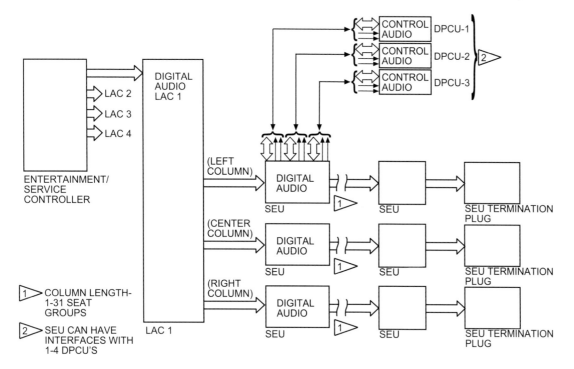

Figure 7-5-17. Passenger entertainment system audio interface diagram   *Courtesy of Northwest Airlines, Inc.*

Unit (VSCU), the Audio Entertainment Multiplexer (AEM), or the Passenger Address Controller (PAC). Each of these three sources send multiplexed digital audio to the ESC. The ESC then controls the audio for distribution. One of the major functions of the ESC is to replace entertainment audio with passenger address audio when needed. The passenger address audio system will always have priority over any entertainment audio.

As seen in Figure 7-5-17, the ESC sends multiplexed audio to the local area controllers (LAC). Each LAC has three output channels: left, center, and right. Each LAC output channel is connected to a maximum of 31 Seat Electronic Units (SEU). The SEUs receive passenger audio selection and volume commands from the Digital Passenger Control Units (DPCU) located on the armrest of each seat (Figure 7-5-18). The SEUs demultiplexes and amplifies the ESC signals. The SEUs also converts the digital audio to an analog signal. Using the DPCU command data, the SEU sends the selected audio to the passenger headphone jack.

A termination plug must be installed at the end of each series of seat electronic units (Figure 7-5-19). The output from the LAC connects to the J1 (Jack 1) plug of the SEU, the J2 connection of that SEU connects to the J1 plug of the next SEU in the series, and so on (Figure 7-5-19). On the last SEU in the series, the J2 plug must be connected to a 75-ohm termination plug. The termination plug must connect to the aircraft electrical ground. The termination plug

is needed to ensure the transmission cable's impedance is correct.

Operation

The following example will cover the operation of the entertainment audiotape system as it applies to the PES. Other audio subsystems of the PES operate in a similar manner.

The following are the series of events for entertainment audiotape use:

1. The Cabin System Module (CSM) is used to turn on/off audio entertainment. The

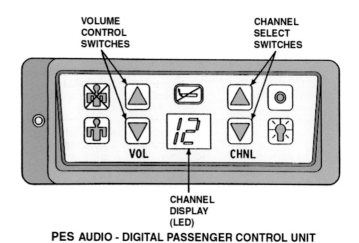

PES AUDIO - DIGITAL PASSENGER CONTROL UNIT

Figure 7-5-18. Digital passenger entertainment control unit located in the armrest of each passenger seat   *Courtesy of Northwest Airlines, Inc.*

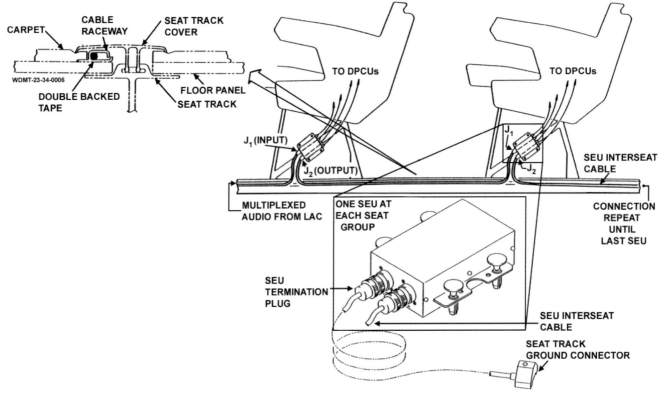

Figure 7-5-19. SEU cable interconnect diagram

Courtesy of Northwest Airlines, Inc.

CSM is a small switch panel activated by the flight attendant. The CSM sends a digital signal to LAC 1 to turn on/off the Entertainment Tape Reproducer (ENT T/R).

2. As seen in Figure 7-5-20, the ENT T/R sends up to 12 channels of analog audio to the Audio Entertainment Multiplexer (AEM). The AEM also receives six ana-

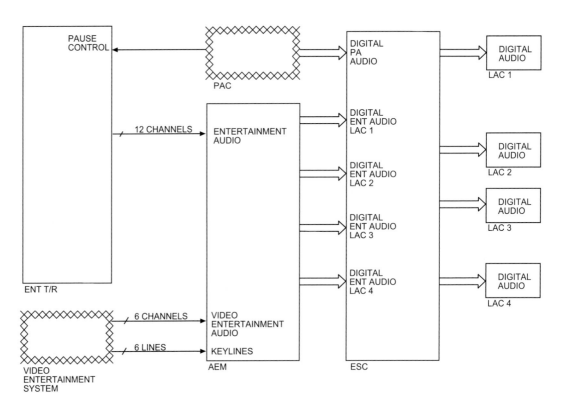

Figure 7-5-20. Passenger entertainment system audio interface diagram   *Courtesy of Northwest Airlines, Inc.*

log audio signals from the Video System Control Unit (VSCU).

3. The AEM digitizes and multiplexes the audio channels and transmits the signals to the Entertainment/Service Controller (ESC).

4. The ESC prioritizes entertainment and passenger address audio. The ESC sends multiplexed signals to the LACs.

5. The LACs distribute the signals to the SEUs.

6. The SEUs convert the signals to analog and distribute them to the passenger headphone jacks. The specific audio channel and volume is controlled by the individual DPCUs.

Passenger Service System

The Passenger Service System (PSS) is used by passengers to control reading lights and attendant call functions. The PSS is also used to control passenger information signs and attendant call functions from the lavatory. Passenger information signs are *no smoking*, *fasten seatbelts*, and *lavatory occupied* (Figure 7-5-21).

Architecture

During operation of the PSS most of the functions are controlled by the LACs. On the audio system just discussed, most of the functions are controlled by the system's main controller. Refer to the PSS interface diagram in Figure

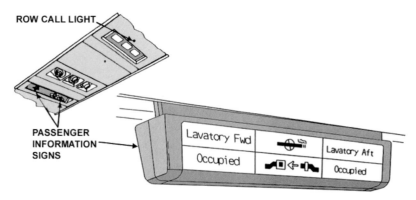

Figure 7-5-21. Passenger information signs

7-5-22 during the following discussions on PSS architecture. Table 7-5-4 gives a list of LRUs, which interface with the ESC and a description of their function. The ESC is the main controller for the system. The ESC provides coordination functions and sends multiplexed control signals to the LACs for distribution.

Operation

This section will present two common functions performed by the PSS: operation of the passenger reading lights and the passenger attendant call.

Example 1

Passenger turns on the reading light using the seat-mounted control:

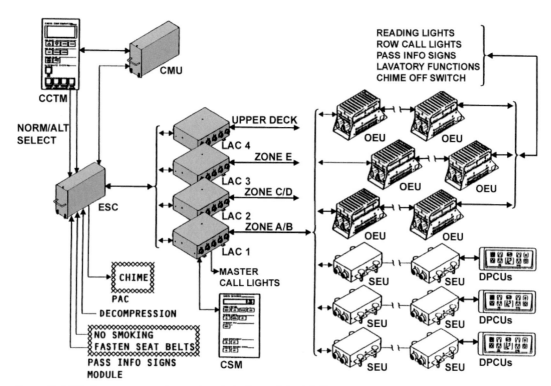

Figure 7-5-22. Passenger service system interface pictorial diagram   *Courtesy of Northwest Airlines, Inc.*

PASSENGER SERVICE SYSTEM MAIN CONTROLLER INTERFACE LIST	
ESC Interface	Purpose of Interface
Cabin Configuration Test Module (CCTM)	Controls normal/alternate circuit selection, used to install the configuration database onto the ESC, to configure No Smoking sign locations, and to run test functions
Passenger Address Controller (PAC)	Used to sound attendant station chimes when needed for attendant call functions
Passenger Information Sign Module	Supplies discrete inputs to the ESC for passenger information signs
Local Area Controller (LAC)	Transmits attendant call signals to ESC for sounding chime
Central Management Unit (CMU)	Controls data from CCTM to ESC

Table 7-5-4. Passenger service system main controller interfaces

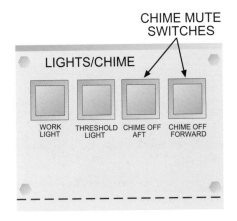

Figure 7-5-23. Chime light control module

Courtesy of Northwest Airlines, Inc.

1. The passenger pushes the **LIGHT ON** button on the Digital Passenger Control Unit (DPCU).

2. The DPCU sends a signal to the Seat Electronic Unit (SEU).

3. The SEU sends a digital multiplexed request signal to the Local Area Controller (LAC).

4. The LAC sends a command signal to the appropriate Overhead Electronics Unit (OEU).

5. The OEU turns on the reading light as requested by the passenger.

Example 2

Passenger activates the attendant call function:

1. The passenger pushes the **ATTENDANT CALL** button on the DPCU.

2. The DPCU sends a command signal to the LAC.

3. The LAC controls three functions: (A) the LAC turns on the appropriate master attendant call light, (B) sends a command to the inboard OEU to turn on the call light for the appropriate seat row, and (C) sends a signal to the ESC to sound the attendant call chime.

4. The ESC then sends the chime request to the PAC.

5. The PAC activates the chime.

A switch on the chime/light control module can turn off the attendant call chime. The chime/light control module is located at the attendants' station. The threshold and workstation lights are also controlled at this module (Figure 7-5-23).

Cabin Lighting

The Cabin Lighting System (CLS) is a part of ACESS that controls the following cabin lights:

1) Indirect ceiling lights

2) Sidewall wash lights

3) Direct ceiling lights

4) Night lights

Figure 7-5-24 shows the various types of cabin lighting. Fluorescent type lights are used for wall wash, indirect ceiling lights, and some night lighting. Incandescent lamps are used for direct lighting. Flight Attendants control cabin lights using the Cabin System Modules (CSM) (Figure 7-5-25). A lighting command entered into the CSM is sent to the Local Area Controllers (LAC). The LACs processes the information and transmits a digital signal to the ESC and the Overhead Electronics Units (OEU). If the ESC agrees with the command the OEUs will turn on the appropriate lights.

ACESS Maintenance and Troubleshooting

The maintenance and repair of ACESS is similar to multiplexed passenger address/entertainment/interphone systems found on other aircraft. In general, these multiplexed systems are somewhat high maintenance simply due to their complexity. Miles of wires controlled by dozens of electronic units link hundreds of light bulbs and switches. With this large number of components comes a massive number of connector plugs. Connector assemblies are often a weak link in any electronic system and the large number required for ACESS makes the system vulnerable to failures. Be sure to

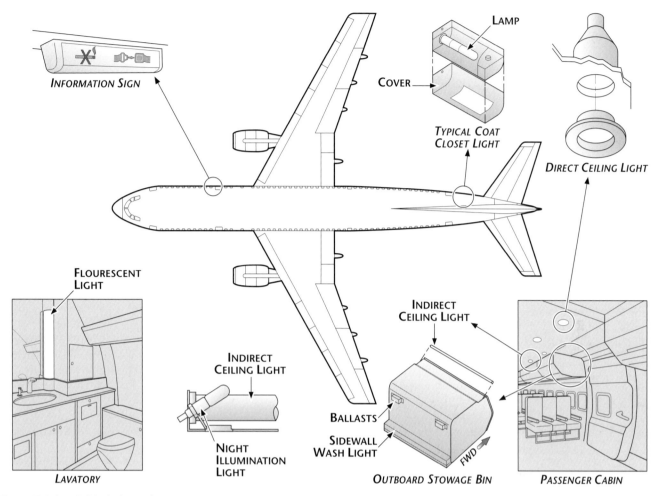

Figure 7-5-24. Cabin lighting locations

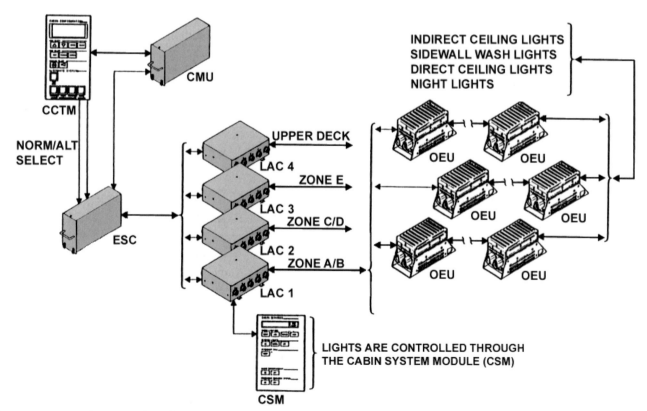

Figure 7-5-25. ESC/LAC interface diagram

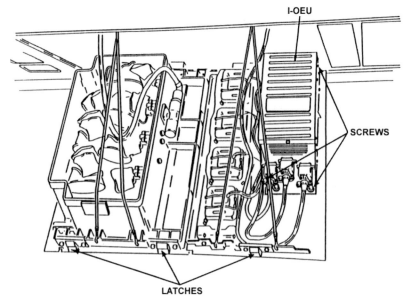

Figure 7-5-26. Access to the overhead electronic unit

Courtesy of Northwest Airlines, Inc.

carefully inspect all suspect connectors whenever troubleshooting the system.

Most of the ACESS LRUs are located in the cabin behind decorative panels or mounted to seat structures. For example, Overhead Electronic Units (OEU) are located behind ceiling panels directly above the passenger seating. The OEUs are easily replaced by depressing the three latches and removing the screws holding the unit in place (Figure 7-5-26). The four Local Area Controllers (LAC) are located above the cabin ceiling in three different areas.

The aircraft service manual will provide diagrams that can be used to locate various ACESS components.

On the B-747-400, one electronic LRU is located under each group of passenger seats. As seen in Figure 7-5-27, these Seat Electronic Units (SEU) are located in the structure of the seat that is mounted to the seat track on the aircraft floor. Since passengers often rest their feet, or store baggage in this area, SEUs are often prone to wiring failures. As passengers move items or their feet near the SEUs, the wiring may get snagged. This causes connector problems or wiring failures due to the excess stress placed on the wiring. Whenever working on an SEU, be sure that all of the wiring is secured correctly and that all plastic covers are in place. This will help to prevent wire/connector damage.

To aid in troubleshooting, ACESS employs built-in-test equipment that detects faulty LRUs, interface connections, and lighting components. The various ACESS LRUs continuously monitor system status and report any failures to the Central Management Unit (CMU). The CMU sends fault data to the EFIS/EICAS Interface Units (EIU), which sends all fault data to the Central Maintenance Computer (CMC) and the Integrated Display System (IDS). EICAS will display seven different messages related to ACESS (Table 7-5-5). Since ACESS is a relatively non-flight critical system, all EICAS messages for ACESS are status messages.

When requested by the technician, the ACESS CMU sends all fault data to the Cabin

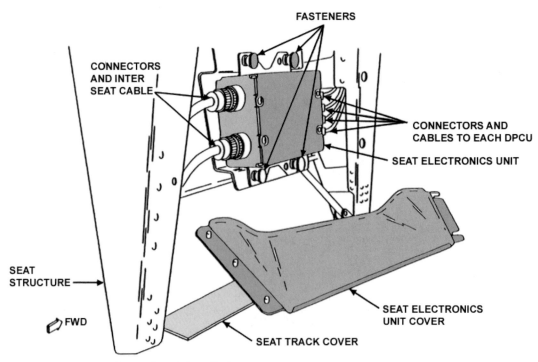

Figure 7-5-27. A seat electronic unit installation

Courtesy of Northwest Airlines, Inc.

ACESS MAINTENANCE MESSAGE AVAILABLE ON EICAS		
MESSAGE DISPLAYED	TYPE OF MESSAGE	DESCRIPTION
Acess MGT Unit	Status	Failure of the central management unit (CMU)
Pass Address 1	Status	Failure of the passenger address controllers (PAC) normal controller circuits
Pass Address 2	Status	Failure of the passenger address controller (PAC) alternate controller circuits
Cabin INT 1	Status	Failure of the cabin interphone controllers (CIC) normal controller circuits
Cabin INT 2	Status	Failure of the cabin interphone controllers (CIC) alternate controller circuits
Pass Service 1	Status	Failure of the entertainment/service controllers (ESC) normal controller circuits
Pass Service 2	Status	Failure of the entertainment/service controllers (ESC) alternate controller circuits

Table 7-5-5. Possible ACESS messages displayed by EICAS

Configuration Test Module (CCTM). The CCTM provides a means to monitor fault activity for the various ACESS LRUs. The CCTM is located in the aircraft cabin and can be used to perform system status checks, system tests, and software configuration updates, and verifications.

System Status Checks

The ACESS system status check can be performed during flight or on the ground to retrieve failure data stored in the CMU memory. During operation, the CMU constantly monitors the main ACESS controllers for fault data. Any faults reported to the CMU are stored in memory. This memory is accessed using the CCTM located at attendants' station door No. 2 right. The operation of the ACESS components is unaffected by the system status check.

The following steps are needed to retrieve fault data stored in the CMU (Figure 7-5-28):

1. Press the **NORMAL** button on the CCTM twice. Pressing the **NORMAL** button the first time activates the ACESS display. Pressing it a second time initiates the normal ACESS operating mode. While in the normal mode, the CMU will access all fault data from the reporting LRUs.

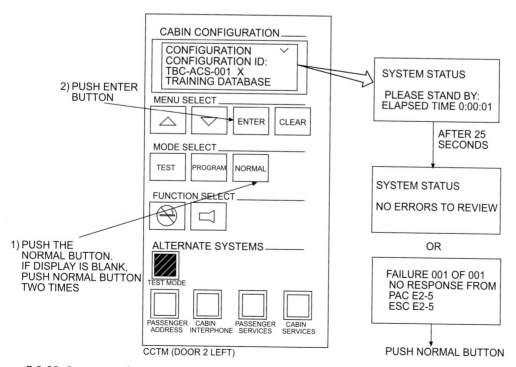

Figure 7-5-28. Sequence of events to perform an ACESS status check

Courtesy of Northwest Airlines, Inc.

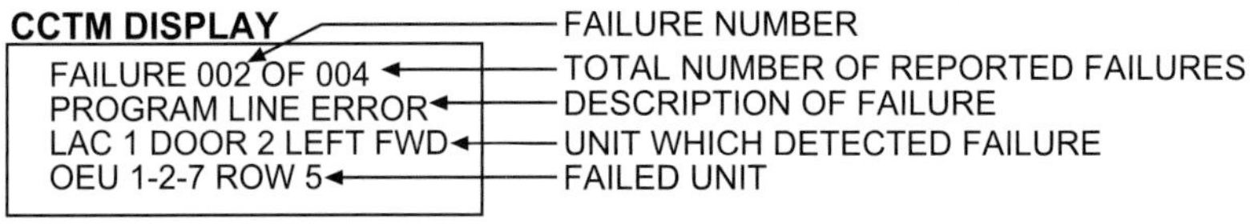

CCTM DISPLAY ———————— FAILURE NUMBER

FAILURE 002 OF 004 ◄——— TOTAL NUMBER OF REPORTED FAILURES
PROGRAM LINE ERROR ◄——— DESCRIPTION OF FAILURE
LAC 1 DOOR 2 LEFT FWD ◄——— UNIT WHICH DETECTED FAILURE
OEU 1-2-7 ROW 5 ◄——— FAILED UNIT

Figure 7-5-29. CCTM display if a failed ACESS status check

2. Press the **ENTER** button on the CCTM. This will cause the CMU to send a list of active fault data to the CCTM display. One of two messages can be displayed: *NO ERRORS TO REVIEW*, or the last reported failure. If more than one fault is present, each failure will then be displayed in succession.

3. Press the **NORMAL** button to exit the system status mode.

Any failure presented on the CCTM display will always contain five types of information, the:

1. Failure number

2. Total number of failures

3. Description of the failure

4. Unit which detected the failure

5. Failed component or interface

Figure 7-5-29 shows an example of failure data. In this case, the fault (2nd in a total of 4) is a program error, which was detected by LAC 1, and the fault lies in OEU 1-2-7 row 5. This same format is used for other types of fault data presented by the CCTM.

System Tests

The ACESS system tests feature allows technicians to run real time tests of the various ACESS LRUs. Since the system test function disables ACESS functions, the test can only be performed on the ground. Five ACESS subsystems are tested using the systems test feature, these are the:

1. Passenger Service System (PSS)

2. Passenger Entertainment System audio (PES audio)

3. Cabin Lighting System (CLS)

4. Cabin Interphone System (CIS)

5. Passenger Address System (PAS).

The attendant master call lights, the row call lights, and the passenger information sign lights can also be tested. A test all units function is included in this feature, which performs a test on each of the five systems and records their faults.

The system tests are initiated using the CCTM. First, press the **TEST MODE** button (Figure 7-5-30).

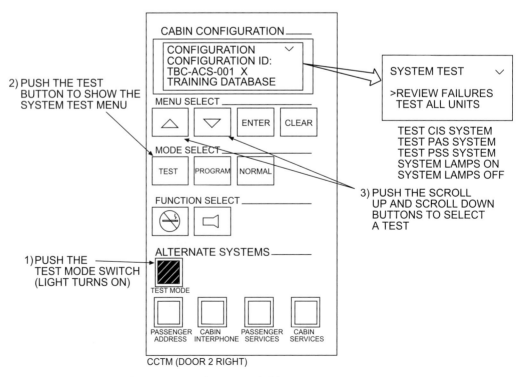

Figure 7-5-30. Sequence of events to perform an ACESS system test   *Courtesy of Northwest Airlines, Inc.*

Second, press the **TEST** button and the CCTM display will show the system test menu. Third, use the **SCROLL DOWN** button to move to the desired system test. The system test menu is shown in Table 7-5-6. After a test is complete, the CCTM display will list the number of errors found.

If one or more errors are found, the technician can press the **TEST** button to return to the system test menu. From the menu, the review failures function can be accessed to retrieve a list of the detected faults. The CCTM will display a list of the failures in a format as described in Figure 7-5-29. Additional failures are displayed by pressing the **SCROLL UP/DOWN** buttons on the CCTM.

Lamp Test

The lamp test initiated through the CCTM is used to test the attendant master call lights, the row call lights, the lavatory call lights, and the passenger information sign lights (Figure 7-5-31). The lamp test simply turns on all call lights and information signs. The technician walks through the aircraft to identify if the lights are on. If a lamp is off the bulb is most likely defective. If an entire section of lamps are off, it is most likely caused by a defective Overhead Electronic Unit (OEU) or associated wiring.

To turn on the lights for testing, press the **TEST MODE** switch on the CCTM. Next press the **TEST** button and the system test menu will appear in the CCTM display. Scroll down to the system lamps on selection from the menu and press **ENTER** (Figure 7-5-32). To turn off the lamp test function, simply scroll down the menu to system lamps off and press **ENTER**. The Cabin System Module (CSM) lights can also be tested through the CCTM. As mentioned earlier, there are two CSMs used to control cabin lighting: PES audio and the passenger service system. The two CSMs are located at Door No. 2 left in the main cabin and in the upper deck at the attendant's station.

ACESS Software

ACESS is a software driven system designed to be reprogrammed any time the cabin configuration is changed. When software modifications are made, several ACESS LRUs are reprogrammed. If an ACESS LRU is replaced for maintenance, be sure to update the software. There are two categories of software that must be loaded into ACESS: operational software and the configuration database. Although the exact procedures for downloading LRU software is beyond the scope of this text, the basic process is described in the following paragraphs.

SYSTEM TEST MENU SELECTION	
TEST SELECTION	FUNCTION OF TEST
Review failure	Used to review failures which were found during previous system tests
Test all units	Performs functional test of all ACESS subsystems (CIC, PAS, PSS, PES audio, and CLS)
Test CIS system	Used to test the cabin interphone system
Test PAS system	Used to test the passenger address system
Test PSS system	Used to test passenger service system, PES audio, and cabin lighting system
System lamps on	Used to turn on the master call lights, row call lights, and passenger information signs
System lamps off	Used to turn off the master call lights and passenger information signs

Table 7-5-6. System test menu selections

The operational software is used to instruct ACESS LRUs how to function in the system. Operational software is typically installed in the shop prior to the LRU reaching the aircraft. In some cases, however, operational software can be installed with the LRU on the aircraft. Be sure to consult the manual for proper software upgrade procedures.

The configuration database is used by the LRUs to determine the specific layout of the aircraft's cabin. Configuration software is

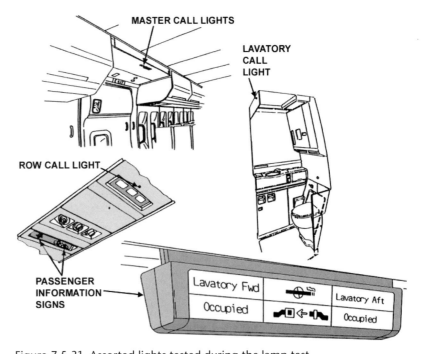

Figure 7-5-31. Assorted lights tested during the lamp test

Courtesy of Northwest Airlines, Inc.

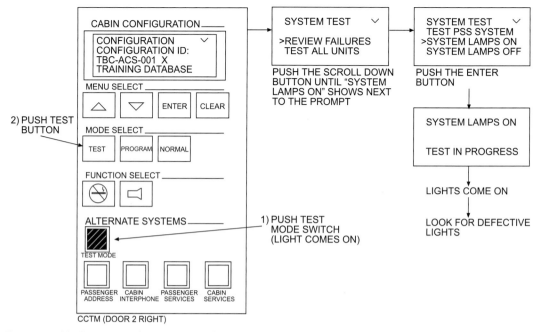

Figure 7-5-32. Sequence of events to perform an ACESS lamps test   *Courtesy of Northwest Airlines, Inc.*

typically installed into an LRU's memory after the LRU has been installed in the aircraft. Using the CCTM, the configuration software is downloaded from the Central Management Unit (CMU) to other LRUs in the system. This is required only if the aircraft's cabin configuration is changed or a new LRU, requiring the database is installed. If the aircraft's CMU is changed, both the operational and configuration software can be downloaded using the aircraft's data loader. The data loader can also be used to update the CMU.

Passenger Entertainment Video

The Passenger Entertainment System Video (PES video) is used to display video entertainment on various LCD flat panel display units located throughout the cabin. Many 747 aircraft have been upgraded with a modern video system. This system includes the replacement of older CRT and projector video units with modern flat panel displays. Some displays are located in the seat backs directly in front of each passenger,

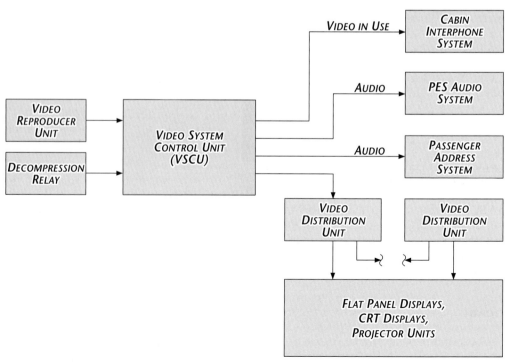

Figure 7-5-33. Passenger entertainment system video interface pictorial diagram

or the display can fold away into the armrest of each seat. Those aircraft that have not undergone a retrofit with new video displays will still employ traditional CRT's located in various spots in the cabin. Some aircraft employed video projector units designed to project video images on a screen lowered from the ceiling. When these video systems were modified, the original video tape unit was also replaced with a DVD player or a solid state memory system which stored all video information. The system uses prerecorded digitized video/audio stored on DVD's of a solid state memory. The audio data is transmitted to the passenger address controller for broadcast over cabin speakers, or to the Entertainment/ Service Controller (ESC) for distribution to passenger headphones.

Architecture

Reference the interface diagram in Figure 7-5-33 during the following discussions on the PES video. The Video System Control Unit (VSCU) provides the major control and distribution unit for the entertainment video signals. The VSCU receives video and audio inputs from the Video Reproducer Unit (VRU) and a discrete signal from the decompression relay. The decompression relay monitors for sudden cabin decompression. In the event that cabin pressure is lost, all video is automatically turned off.

The VSCU sends control and video outputs in digital format to the video distribution units. The Video Distribution Units (VDU) are located throughout the cabin. Up to 24 VDUs can interface with the VSCU; although most aircraft have 10 to 12 VDUs. The VSCU also sends a video in use discrete signal to the cabin interphone system to inform the crew that the video is in operation. The audio signals from the video reproducer are also sent through the VSCU to the PES (audio) and PAS. The VDUs send an on signal to the VSCU.

The video distribution units are each connected to the VSCU via two bidirectional data bus cables (Figure 7-5-34). A 75-ohm termination plug must be installed at the end of the data bus cables to provide proper bus impedance.

The VSCU is located under the stairway to the upper deck in the Video Control Center (VCC). A maximum of two video reproducer units will also be located in the VCC. Storage for videodiscs and other materials is also located in the VCC.

Control

All video system-operating controls are located on the front panel of the VSCU (Figure 7-5-35). The VSCU contains controls for:

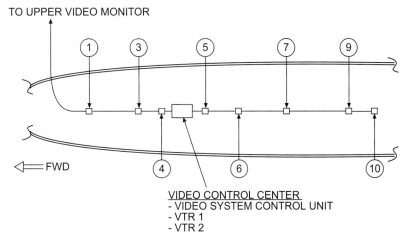

Figure 7-5-34. Passenger entertainment system video component locations

1. *MASTER POWER*

2. *ANTENNA* selection - used for systems receiving ground-based video programming (this system is typically not used on US aircraft)

3. *PREVIEW MON* (monitor) selection - to select the video source for the display monitor and audio source for VSCU monitor

4. *AREA ON/OFF* controls - used to select the specific area where the video/audio entertainment will be distributed (area 1, 2, and 3 are in the main cabin, UD is the upper deck)

5. *SOURCE SEL* - used to select the video source (up to three input sources can feed the VSCU)

6. *PA SELECT* - used to select the primary or secondary audio source for output to the PAC, also controls volume

7. *INDIVIDUAL PROJECTOR/MONITOR CONT* (control) - used to turn on/off individual monitors

8. *MANUAL OVERRIDE MODE* - these controls are used to set the VSCU selections for all video zones simultaneously

The VSCU system monitor is located in the lower right corner of the VSCU front panel. The system monitor is yellow during normal operation. The monitor illuminates orange if the VSCU BITE system detects an internal failure.

Maintenance and Troubleshooting

The PES video system is a relatively simple system to maintain. Whenever troubleshooting the system, always remember which LRUs control and/or feed data to the unit in

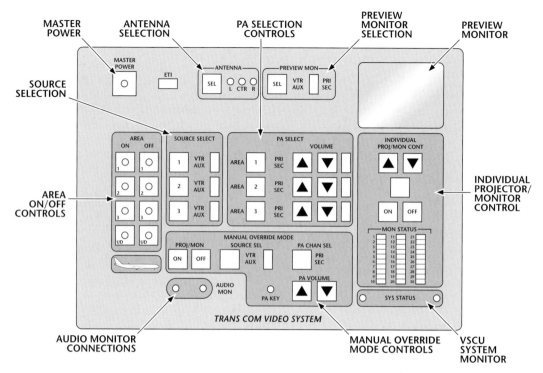

Figure 7-5-35. Video system control unit panel

question. For example, if several monitors are inoperative, it is most likely that the area VDU which feeds those monitors is defective. Remember the VSCU controls the signals to the VDU. Therefore, if the VDU does not repair the system, the associated wiring or the VCDU should be suspected. In most cases, the LRUs are simply swapped with other units to troubleshoot the system. Connector pins are also likely fault areas. Be sure to inspect all electrical connections related to the failed system(s).

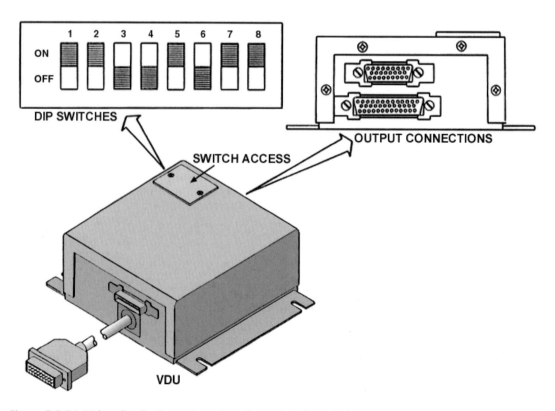

Figure 7-5-36. Video distribution unit and configuration dip switches   *Courtesy of Northwest Airlines, Inc.*

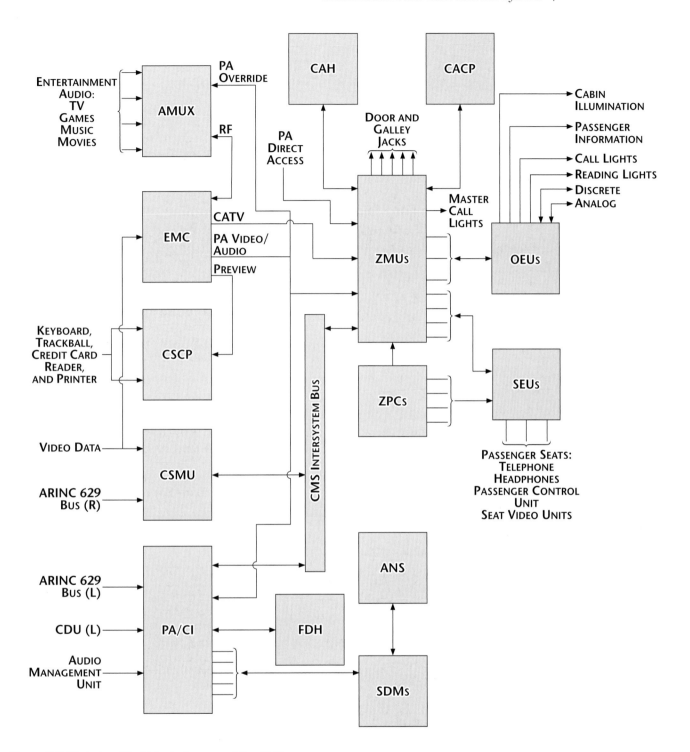

Figure 7-5-37. A simplified block diagram of the CMS

The VCSU and VDUs each have a series of dip switches, which are used to define the configuration of the video system (Figure 7-5-36). Dip switches are small two position switch that are set to binary 1 or 0 for system/component configuration. The VSCU dip switches define the number of VDUs, VRUs, and video sources in the system, the audio configuration, the software configuration, the BITE enable, and the upper deck configuration. The VDU dip switches define the location of the VDU, the VDU data bus address, and the number of monitors and projectors controlled by the VDU.

The Boeing 777

The B-777 employs a Cabin Management System (CMS) for operations control and monitoring various passenger compartment communication and entertainment systems. Similar to the B-747 ACESS, the CMS is an integrated digital,

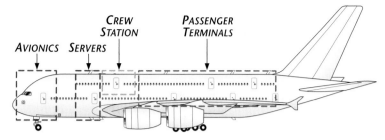

FLIGHT OPERATIONS	CABIN CREW	PASSENGERS	MAINTENANCE
Aircraft Documentation	PAX Database	Email	Maintenance Documentation (TSM, AMM)
Electronic Logbook	Crew Email	Intranet	Maintenance Improvements
Weight and Balance	Cabin Logbook	News/Sports	Aircraft Condition Monitoring
Performance	Aircraft Documentation	Live Television	Electronic Logbook
Operational Checklists	Credit Card Validation	Internet	Data Loading
Access to Flight Information Services (Weather, NOTAM, Etc.)	Cabin Inventory	E-commerce	Operational Software and Database Storage
Charts and Maps	Quality Monitoring		Email
Crew Email	Reservations		Equipment List
FOQA Download			
Airline Specific Applications			

Figure 7-5-38. The four domains of the On-board Information System (OIS)

multiplexed system that integrates the following functions:

1. Passenger addresses
2. Cabin interphone
3. Passenger entertainment
4. Passenger service
5. Cabin lighting

Since these systems were described in the last section of this chapter only a brief description follows here.

The integration of the five subsystems of the CMS allows for centralized controls, monitoring, and testing of all CMS components. The CMS is a software-based system, which can easily be reconfigured as needed whenever the cabin configuration changes or system updates are necessary. The system Configuration Database Generator (CDG) is a menu-driven database editor that runs on a personal computer (PC). The CDG program is used to make

changes and then download the information into the aircraft CMS. The Cabin System Control Panel (CSCP), located in the main flight attendants' station, is used to initiate configuration downloads. Maintenance personnel us the CSCP for testing functions.

A simplified block diagram of the CMS is shown in Figure 7-5-37. Please refer to this diagram during the following discussions on CMS. The system operates using a dedicated intersystem data bus for all audio and video signals. The five main LRUs needed for operation of the system are the:

1. AMUX (audio multiplexers) which receives individual passenger entertainment request signals and sends them to the EMC

2. EMC (entertainment multiplexer controller) is used to control and transmit all entertainment signals (audio and video) to the individual passenger seat

3. CSCP (cabin system control panel) previously discussed

4. CSMU (cabin system management unit) which processes data to/from the CSCP

5. PA/CI (passenger address/cabin interphone) controller used for management of PA/CI audio

The aircraft also employs three ZMUs (zone management units). The ZMUs are located in the forward, middle, and aft portions of the aircraft. Each ZMU sends audio and video data to the overhead electronics units and seat electronics units, which in turn deliver the signals to each passenger seat. The ZMUs also controls any passenger request for changes in seat lighting, audio volume and channel, or video selections. The ZMUs also contain analog to digital and digital to analog conversion circuitry for system control audio signals. Digital signals are multiplexed and transferred on an ARINC 629 data bus dedicated to the CMS.

A-380 Cabin Systems

The Airbus A-380, one of the newest airliner in production at the time this text was written, incorporates the concepts of IMA (Integrated Modularized Avionics) and high-speed Ethernet-type data bus systems (AFDX). These advanced concepts, which are also employed on the Boeing 787 provide better cabin communications and passenger entertainment than available on previous aircraft. The system also employs fiber optic cable to reduce weight.

On the A-380 each seat back contains a flat panel "smart display". Each unit incorporates a liquid crystal display and the processing circuitry. The smart display simply connects to the digital data bus and performs all necessary operations to create the video image. This combination is part of the IMA design philosophy and allows for easy modifications when needed. The system is software reliant and updates can be installed through the central control unit.

The cabin communications and in-flight entertainment systems are part of the A-380 OIS (Onboard Information System). As discussed earlier, the OIS is an integrated system, which coordinates and processes four distinct segments. As seen in Figure 7-5-38, the OIS is separated into four domains, which are heavily partitioned through software firewalls. This allows the four domains, Flight Operations, Cabin Crew, Passenger, and Maintenance systems; to communicate and operate collectively and yet each system is isolated in the event of a failure. This type of fail operational concept is vital to flight safety.

Another advancement found on the A-380 is the ability for IFE (In-Flight Entertainment) systems to report faults through the OIS. The OIS has the ability to communicate fault data using wireless technologies and send data to the Airbus global network (Figure 7-5-39). This allows IFE faults to be reported prior to landing. The maintenance grew can diagnose the fault and have all materials ready to initiate the repair as soon as the aircraft reaches the gate.

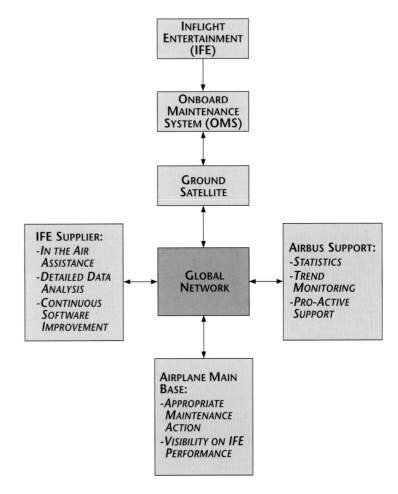

Figure 7-5-39. Communication network for the OIS

Chapter 8

GLOBAL
positioning system

Section 1

Introduction

The *Global Positioning System* (GPS) is a satellite-based system capable of providing position and navigational data for ground-based and airborne receivers. GPS actually determines the position of a given receiver, or a receiver's antenna to be precise. The navigational capabilities of GPS are determined through multiple calculations and comparison to the World Geodetic System (WGD) map. The WGD map is an extremely accurate reference to the earth's latitude and longitude lines. The global positioning system, therefore, provides location coordinates based on a latitude/longitude reference, not with reference to the known location of a ground-based transmitter. Conventional forms of aircraft navigation rely on ground-based transmitters that employ limited range and relatively low frequency signals. GPS employs ultra high frequency transmitters and requires only 21 transmitters for worldwide navigation.

The global positioning system was first implemented by the military in the late 1970s. Through the early 80s, the system was tested and refined. During the 1980s and early 90s, the initial satellites were replaced by more accurate and powerful units. The FAA had granted Technical Standard Orders (TSOs) to several GPS units, permitting installation on civilian aircraft, making GPS a navigation aid option on many aircraft. Most newer aircraft make GPS standard equipment.

In the United States, the formal name for the global positioning system is the NAVSTAR GPS. The U.S. deputy secretary of defense first

Left. Very long range aircraft, like this Airbus A340, can fly non-stop for more than 18 hours. Satellite navigation makes these flights safe and efficient.

GPS SYSTEM

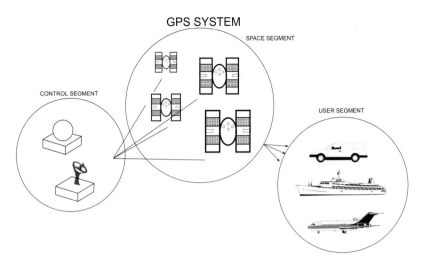

Figure 8-1-1. The three segments of the Global Positioning System (GPS)

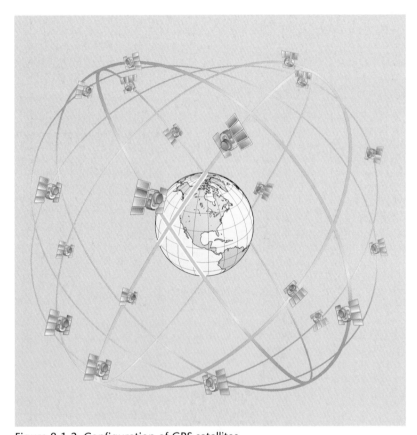

Figure 8-1-2. Configuration of GPS satellites

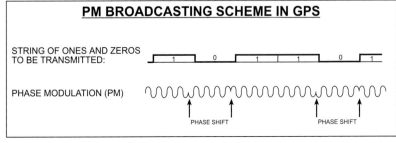

Figure 8-1-3. Example of phase modulated RF signal

assigned this name in 1973 and the name has stuck ever since. The United States is not the only country that has developed a usable GPS. The former Soviet Union began the implementation of their satellite navigation system called Glanoss several years before the country's demise. Today the Commonwealth of Independent States (made up of former Soviet Union member nations) is continuing to develop their GPS. Both nations have explored the possibility of combining efforts to reduce cost and increase worldwide coverage.

GPS System Elements

The NAVSTAR GPS consists of three distinct elements: the space segment, the control segment, and the user segment (Figure 8-1-1). The space segment consists of a "constellation" of 24 orbiting satellites. Twenty-one of the satellites are active leaving three as spares, which can be moved into position in the event an active unit fails. The satellites are in a near geosynchronous orbit approximately 10,900 nm (20,200km) above the earth. As seen in Figure 8-1-2, the satellites are equally spaced around six different orbits to provide worldwide coverage. Each satellite completes one orbit approximately every 12 hours. The system is designed so that a minimum of five satellites should be in view of a ground-based user at any given time, at any location on earth.

Each satellite in the system transmits position and precise time information on two frequencies known as L_1 and L_2. L_1 operates at 1,575.42 MHz and L_2 has an operating frequency of 1,227.6 MHz. The signals are digitally modulated and have a bandwidth of 20MHz or 2MHz depending on the type of information that is being transmitted. The individual satellites are identified by the information broadcast by the modulated carrier waves. *Phase modulation* (PM) is the process that modulates the carrier wave. As seen in Figure 8-1-3 phase modulation is achieved by shifting the phase of the carrier wave to represent a change in the digital "information" signal. To improve receiver reception of faint GPS signals, each satellite employs 12 helical antennas arranged in a tight circular pattern (Figure 8-1-4).

The control segment of the GPS consists of five ground-based monitor stations, one master control station and three ground antennas located at different sites throughout the world (Figure 8-1-5). The five monitor stations receive/transmit time and range data from various satellites within its region. The raw data received by these unmanned stations is sent to the Master Control Station in Colorado Springs, Colorado. The Master Control Station computers analyze the data and provide correction signals as needed to the three ground antennas. The signals are periodically uplinked to the satellites and the necessary corrections are made by the system's software.

When GPS was first conceived, it was intended that the entire user segment would be restricted to high accuracy military applications. Today, it would be difficult to determine if there are more civilian or military GPS receivers in operation. Once released by the Department of Defense, the civilian use of GPS has skyrocketed. GPS is currently being used and researched for future use by farmers, surveyors, archaeologists, miners, recreational hikers, and of course, aircraft owners and pilots. Anyone who needs to navigate, measure position, or determine velocity can become part of the GPS user segment.

The user segment typically receives time and position data from four or more satellites and processes that data into the desired output. The specific equipment needed for these operations is a function of the equipment installation (if any), the desired accuracy, and specific output data. Output data ranges from simple display of latitude and longitude, to moving maps and displays of local airports or ground terrain.

All GPS receivers must have at least three major elements: the control/display unit, the receiver/processor circuitry, and the antenna. These elements may all be combined in one unit or consist of three individual elements. Modern hand held receivers are available for just a few hundred dollars. These units are completely self-contained but may not be approved for aircraft use.

The Bendix/King model KLX 100 shown in Figure 8-1-6 is one of the first popular FAA approved hand-held GPS. This unit contains both a GPS receiver/processor and communications transceiver in one. The KLX 100 is certified under TSO-C129. Many aircraft systems are designed to be permanently installed on the aircraft and often interface with other navigational equipment. On aircraft GPS equipment, the display and receiver processor are often combined into one unit; the antenna is always mounted on the top of the aircraft fuselage to ensure proper satellite reception.

Theory of Operation

To begin this discussion, two assumptions are made: 1) the exact location of each satellite is known at any given time, and 2) the distance to each satellite can be calculated by the GPS receiver/processor. In reality, the exact location of a moving satellite is easy to predict. Once in orbit, each satellite will follow a consistent path. Exact satellite position data is transmitted

Figure 8-1-4. A typical NAVSTAR GPS satellite

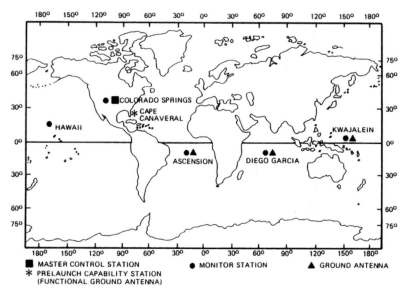

Figure 8-1-5. Locations of GPS ground facilities

Figure 8-1-6. A Bendix/King KLX 100 hand held GPS/COMM transceiver

Courtesy of AlliedSignal General Aviation Avionics

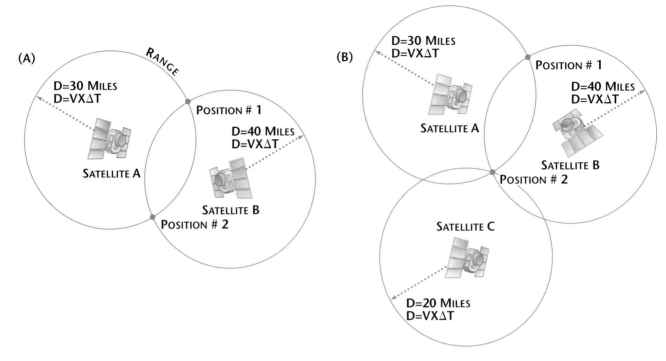

Figure 8-1-7. (A) Using two satellites to determine position, (B) Using three satellites to determine position

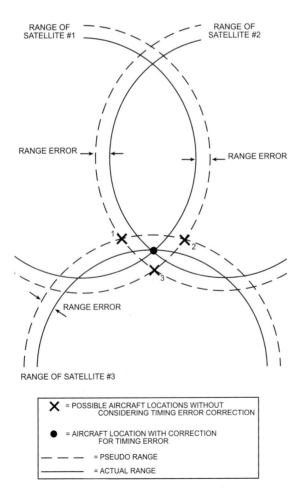

Figure 8-1-8. The aircraft's exact location can only be determined after corrections for range error have been mathematically removed by the receiver/processor

to the receiver as part of the satellite message. Calculating distance from the satellite is also a simple matter. Since distance equals velocity multiplied by time, the receiver/processor need only measure the time it took for the GPS signal to reach the receiver. The speed at which the signal traveled to the receiver is a constant 186,000 miles/second (the speed of light). Using time and velocity to derive distance (range) is known as the *time of arrival* (TOA) ranging concept.

For simplicity, a two-dimensional model will be presented first. Assume the GPS user segment and all satellites are located in one geometric plane. In this case, knowing the distance (range) from just two satellites would provide the location of your aircraft (Figure 8-1-7A). In this example, the aircraft must be located somewhere on a circle with a radius of 30 miles from satellite A, and somewhere on a circle with a radius 40 miles from satellite B. In this two-dimensional model, the aircraft can be in one of two positions. To determine the correct location of our aircraft in the two-dimensional model, a third GPS satellite must be added. As seen in Figure 8-1-7B, if the aircraft was 30 miles from satellite A, 40 miles from satellite B, and 20 miles from satellite C, the aircraft must be in position #2. Aircraft can travel in three dimensions. To determine the exact location of an aircraft in three dimensions, the aircraft must monitor at least four GPS satellites.

Timing errors. In the previous example, the ability to establish distance is based on accurate time measurements. Because the satellite radio signals travel at the speed of light, even the slightest timing error will result in significant range errors. For example, if a timing error of just 1/1000th of a second is present, the distance calculated from the satellite would be off by 186 miles. This error is obviously too large for aircraft navigation.

There are two ways to ensure accurate timing for all space and user segments of the system. One is to install very expensive atomic clocks in each GPS receiver. Each satellite in the constellation already employs four such atomic clocks. This solution, although relatively simple, would put the cost of the user segment out of practical reach for almost everyone, even for the military.

Another solution to ensure accurate timing is through using mathematical formulas. Figure 8-1-8 represents a two-dimensional model for GPS positioning. The dotted lines represent the range from all three satellites calculated with the time bias error. This distance is the *pseudorange*. Since all satellites are synchronized by their atomic clocks, the same time bias error exists between each satellite and the receiver. The solid lines in Figure 8-1-8 represent the actual range from each satellite. Since the same range error (based on the same timing error) exists for all three satellites, the receiver/processor can simply add or subtract any error until it "finds" a common intersection from all three satellites. As before, in a three-dimensional model this is slightly more difficult; however, if four satellites are being used the GPS processor has no difficulty canceling out time bias errors.

Achieving accurate time and range measurements. In most cases, the GPS receiver/processor incorporates a low-cost free-running crystal-controlled oscillator. A free-running oscillator will establish a given frequency without being compensated for frequency drift. That is, the accuracy of the receiver's oscillator is not extremely critical. On the other hand, the frequencies established by the oscillator within the satellites are accurate to within 0.003 seconds per one thousand years.

Satellites transmit a digital code that repeats periodically. The transmission of this code is synchronized for each satellite in the constellation. In other words, every satellite will begin and end the digital code at the same exact time. The GPS receiver also generates the same digital code; however, the receiver code is not synchronized to the satellites. Figure 8-1-9 shows digital codes of three satellites and one receiver. When the GPS receiver first begins to establish a position fix, the time bias between the free-running clock and the satellites is calculated by the processor circuitry. In effect, the

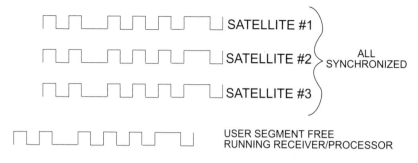

Figure 8-1-9. Synchronized pseudorandom digital codes transmitted by three satellites compared to the pseudorandom code generated the receiver/processor

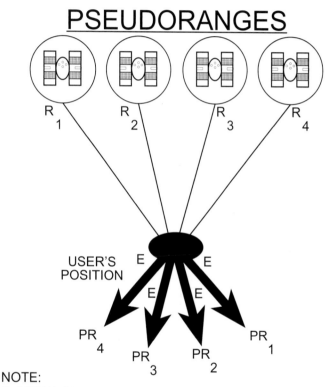

NOTE:
R=ACTUAL GEOMETRIC RANGE (FROM EACH SATELLITE TO USER)
E=RANGE ERROR CAUSED BY GPS RECEIVER CLOCK BIAS
PR=PSEUDORANGE=R+E=OBSERVED TOA VALUE x SPEED OF LIGHT

Figure 8-1-10. Measurement of pseudorange from four satellites

user segment now knows the correction factor to apply to the range calculations for each satellite. To establish the range to a given satellite, the receiver measures the time of arrival of the digital signal. Next, the processor calculates the time it took for the signal to travel from the satellite to the receiver (Δt). Lastly, the receiver multiplies the Δt by the speed of light.

Another way to look at the range and time corrections is shown in Figure 8-1-10. In this example, the receiver actually measures pseudorange from four satellites. The aircraft's GPS processor mathematically determines the

A. Data processor obtains pseudorange measurements (PR$_1$, PR$_2$, PR$_3$, PR$_4$) from four satellites

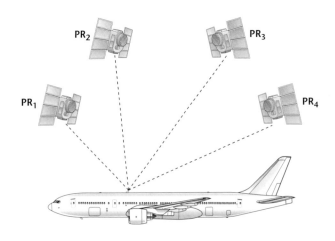

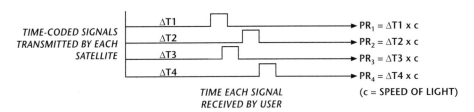

TIME-CODED SIGNALS TRANSMITTED BY EACH SATELLITE

ΔT1 → PR$_1$ = ΔT1 x c
ΔT2 → PR$_2$ = ΔT2 x c
ΔT3 → PR$_3$ = ΔT3 x c
ΔT4 → PR$_4$ = ΔT4 x c

TIME EACH SIGNAL RECEIVED BY USER

(c = SPEED OF LIGHT)

B. Data processor applies deterministic corrections

PR$_I$ = PSEUDORANGE (I = 1, 2, 3, 4)

1. Pseudorange includes actual distance between satellite and user plus satellite clock bias, atmospheric distortions, relativity effects, receiver noise, and receiver clock bias

2. Satellite clock bias, atmospheric distortions, relativity effects are compensated for by incorporation of deterministic adjustments to pseudoranges prior to inclusion into position/time solution process

C. Data processor performs the position/time solution

FOUR RANGING EQUATIONS

$$(X_1 - U_x)^2 + (Y_1 - U_y)^2 + (Z_1 - U_z)^2 = (PR_1 - CB \times c)^2$$
$$(X_2 - U_x)^2 + (Y_2 - U_y)^2 + (Z_2 - U_z)^2 = (PR_2 - CB \times c)^2$$
$$(X_3 - U_x)^2 + (Y_3 - U_y)^2 + (Z_3 - U_z)^2 = (PR_3 - CB \times c)^2$$
$$(X_4 - U_x)^2 + (Y_4 - U_y)^2 + (Z_4 - U_z)^2 = (PR_4 - CB \times c)^2$$

X$_I$ x Y$_I$ x Z$_I$ = SATELLITE POSITION (I = 1, 2, 3, 4)

1. Satellite position broadcast in 50Hz navigation message

DATA PROCESSOR SOLVES FOR:

1. U$_X$, U$_Y$, U$_Z$ = USER POSITION

2. CB = GPS RECEIVER CLOCK BIAS

Figure 8-1-11. Calculations used to determine an aircraft's location using the pseudorange measurements from four satellites

actual range and position. The actual range (R) is equal to the pseudorange (PR) less the range error (E). Figure 8-1-11 shows a simplified version of the calculations used by the receiver/processor to calculate the aircraft's position.

Other errors. Errors other than time bias exist in the system. The four most common errors are:

1. *Satellite clock errors*, which exist due to minor inaccuracies in synchronizing the satellite atomic clocks,

2. *Ephemeris errors*, which are caused by slight variations in the satellite's position as it orbits the earth,

3. *Atmospheric propagation errors*, which are caused by the distortion of the transmit-

ted signal as it travels from the satellite to the receiver through the ionosphere, and

4. *Receiver errors,* which are caused by local electrical noise (interference), computational errors, and errors in matching the pseudorandom digital codes.

As seen in Table 8-1-1, the errors mentioned above are relatively minor when considering the distance from satellite to receiver. In some cases, these errors are correctable. The control segment of the GPS monitors the satellite clock errors and ephemeris errors with periodic adjustments made to correct these errors. Using two signals of different frequencies and comparing the difference in time delays can minimize the amount of atmospheric error. The receiver error is often a function of equipment quality. In general, higher quality receivers incur less error.

C/A- and P-codes. As mentioned earlier, the satellites transmit a pseudorandom digital code on two L-band frequencies. The digital codes transmitted by the GPS satellites are known as the course/acquisition (C/A) code and the precision (P) code. The C/A- and P-codes contain the timing and satellite position information required by the receiver to calculate navigational data. The C/A- and P-code information is often referred to as the NAV message.

The NAV message is decoded using a combination of the C/A- and P-code data, which transmits at 50 bits/second and contains 1,500 bits/frame. Twenty-five data frames are transmitted before the information repeats. Since much of the data is repeated during subsequent frames, the GPS receiver can typically "lock on" to the satellite within approximately 30 seconds.

In order to ensure full reception of the CA- and P-codes, the GPS receiver must be in line of sight to four satellites at least 5° above the horizon (Figure 8-1-12). Satellite position is also important to achieve the proper geometry for position calculations. The ideal situation is for all four satellites to be more than 5° above the horizon and equally spaced above the aircraft. This provides the receiver with the best triangulation for position calculation.

GPS and Aircraft Navigation

GPS provides an excellent means of general navigation; however, there are several limitations that must be addressed for civilian aircraft use. Aviation GPS service must meet the four following basic criteria:

1. *Accuracy* — the difference between the measured position and the aircraft's actual position. GPS accuracy is adequate

SOURCE OF ERROR	APPROXIMATE ERROR DISTANCE
Satellite clock errors	2 ft.
Ephemeris errors	2 ft.
Receiver errors	4 ft.
Atmospheric propagation errors	2 ft.

Table 8-1-1. Range error values

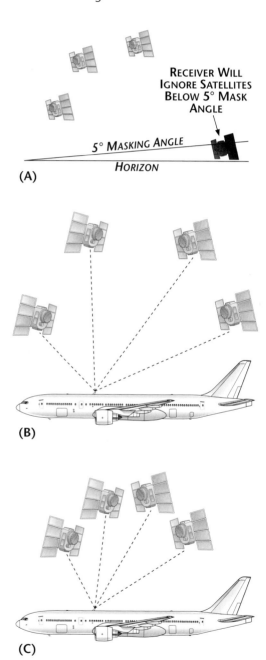

(A)

(B)

(C)

Figure 8-1-12. Correct satellite location is important to accurate positioning: (A) Satellites must be a least 5° above the horizon to provide accurate data, (B) The use of four satellites with good separation provides accurate position data, (C) The use of four satellites clustered together provides less accurate position data

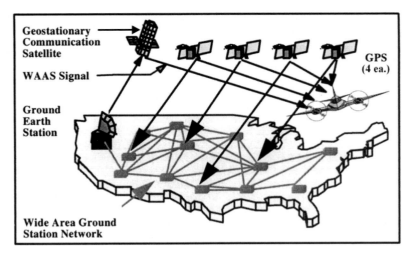

Figure 8-1-13. The basic system elements needed for differential GPS (DGPS)

for enroute navigation, but fails to meet approach and landing requirements.

2. *Availability* — the ability of the system to be used for navigation whenever it is needed, and the ability to provide that service throughout the entire flight.

3. *Integrity* — the ability of the system to shut itself down when it is unsuited for navigation or to provide timely warnings to the pilot in the event of a system failure.

4. *Continuity* — the probability that GPS service will continue to be available for a period of time necessary to complete the navigation requirements of the flight.

While aircraft are flying on an IFR flight plan, their en route separation is typically maintained at five miles or greater. The accuracy of GPS is sufficient for civilian en route navigation. GPS can easily provide navigational signals capable of the five mile separation ±4 percent. However, while flying an approach to land, the accuracy level must be significantly higher. The future of GPS aircraft navigation depends on improved accuracy. As for availability, the U.S. government has stated that GPS will be consistently available for civilian use. Since systems do periodically fail, continuity may become a problem. Integrity, therefore, becomes a very important issue for aircraft use of GPS. Remember, integrity in this instance is the ability of the system to provide a timely warning in the event of a system failure.

In general, many of the GPS limitations can be solved using navigation with more than four satellites. If one satellite fails, the system can simply navigate using a fix from another satellite. When using five satellites to navigate, the airborne equipment can detect a system malfunction, but cannot detect which satellite has failed. Six or more satellites are required to provide enough information to determine which

satellite has failed and use the remaining satellites for accurate navigation. Modern receivers are capable of tracking several satellites at once to eliminate this problem.

The best solution is for the airborne GPS equipment to be notified immediately if a satellite fails. This is especially important for aircraft using GPS for approach and landing navigation. Each GPS satellite broadcasts an integrity message to assure the health of the system; however, one-half hour or longer may elapse from the time that a fault occurs to the time that the aircraft is notified. This time delay is obviously too long.

Differential GPS

To overcome the previously mentioned limitations, the FAA has developed a concept known as *Differential GPS* (DGPS). DGPS was designed to increase the accuracy, availability, integrity, and continuity of basic GPS to a level sufficient for complete aircraft navigation. DGPS uses a series of ground-based stations to compensate for the inherent shortcomings of GPS. There are two basic subcategories of Differential GPS, the Wide Area Augmentation System and the Local Area Augmentation System.

Wide area augmentation system. The Wide Area Augmentation System (WAAS) is an enhanced GPS designed to improve the integrity, accuracy, availability, and continuity of the basic satellite navigation system. At the time that this text was written, WAAS was being developed to provide wide area coverage for aircraft navigation through all phases of flight, including the Category I precision approach.

Figure 8-1-13 depicts the WAAS network, which consists of approximately 35 ground-based reference stations that receive and monitor GPS signals. Data from the reference stations is transmitted to a master station. The master station contains data on the precise location of each reference station, in addition to monitoring and detecting any GPS satellite errors or failures. The master station then uplinks all correction data to one or more geostationary communications satellites. The communication satellites broadcast the correction data to the aircraft on the same frequency as GPS (L1, 1575.42 MHz), providing instantaneous correction and fault data to any aircraft in the WAAS coverage area.

WAAS will improve basic GPS accuracy to approximately seven meters vertically and horizontally. WAAS accuracy is also considered to be within three meters or less 95 percent of the time. This is more than adequate for en route navigation and for a Category

I precision approach. A Category II or III approach requires even more rigid standards. In 2006, WAAS was approved for non-precision approach procedures for all appropriately equipped aircraft. This allows aircraft to make a full instrument approach with vertical guidance down to a maximum of 200 feet above ground level. At the time this text was written, WAAS/GPS satellite coverage is only available in North America and Hawaii. Currently countries in South America, Europe, and Asia are developing or considering the implementation of a WAAS-type system.

Local area augmentation system. The Local Area Augmentation System (LAAS) is a GPS upgrade designed to be used during a Category II or Category III precision approach. This type of approach and landing requires extremely high accuracy, availability, and integrity. LAAS relies on a ground-based station located on or near each airport, to monitor the GPS satellites in a localized area. The geographic position of the ground-based station is accurately determined and loaded into the memory of the station's processor. After receiving GPS satellite data, the ground station can calculate all errors and determine the health of the GPS system. As seen in Figure 8-1-4, the error data is then transmitted to the aircraft. LAAS will broadcast the correction message via a very high frequency (VHF) radio signal to the airborne receiver. The aircraft's receiver/processor then makes the necessary computations to compensate for all GPS errors. It is anticipated that LAAS will provide accuracy to less than one meter both vertically and horizontally.

GPS Past, Present, and Future

Currently LAAS is under development and testing and WAAS is in limited use only in North America. Although the exact outcome of Differential GPS is difficult to predict; most experts agree that in the future DGPS will become a primary navigation source for both precision and non-precision landings worldwide.

All TSO equipment is currently approved for en route navigation. IFR approach-to-land situations are another story. In 1993, the FAA approved a non-precision approach, commonly called the overlay approach. This IFR approach required relatively good visibility and high altitude minimums. The overlay approach also required that GPS equipment could only be used if the appropriate conventional ground-based navigation equipment was operating onboard the aircraft during the approach. Specifically, it required that the pilot monitor the conventional systems (VOR, ILS, glide slope, and/or DME) during the approach.

As of 1996, all equipment certified under TSO-C129a allowed pilots to utilize the GPS as supplemental area navigation equipment. When using TSO-C129a receivers, there is no need to monitor conventional navigational systems. The equipment can only be used for non-precision approaches and conventional navigational equipment is still required onboard the aircraft. Equipment certified under TSO-C145, released in 1998, meet the standards for GPS navigation augmented by the Wide Area Augmentation System. TSO-C146a, released in 2002, contains the minimum performance standards required for stand-alone airborne navigation equipment using GPS/WASS. Stand-alone systems do not require pilots to monitor traditional navigation systems while using GPS for en route or approach navigation. GPS systems on modern aircraft should meet TSO-C146a. Additional TSOs will be introduced as new GPS systems are developed.

Aircraft GPS Equipment

All aircraft GPS equipment must meet a minimum certification standard through a TSO. There are, however, several units currently available that are not TSO certified. Most of these GPS receivers were intended for non-aircraft uses such as hiking, marine, or automobile navigation. In some cases, pilots use non-TSO equipment as a secondary reference during VFR flights. This is perfectly legal, but only equipment approved by a TSO may be used as a source for IFR or VFR navigation. The TSO ensures a minimum quality and accuracy standard set by the FAA; hence, it provides approval for aircraft use. In addition, any GPS equipment installed in an aircraft must meet all FAA standards. In general, the major difference between an aircraft GPS receiver and one designed for marine or other use is the database; aircraft GPS receivers contain a database of various airports and standard navigational waypoints.

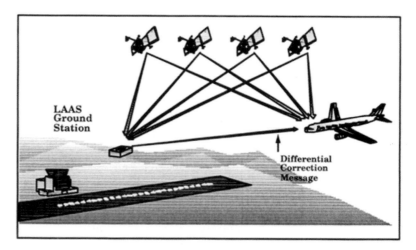

Figure 8-1-14. Example of the Local Area Augmentation System (LAAS)

Figure 8-1-15. A typical portable (hand-held) GPS unit

Courtesy of Bendix/King by Honeywell

As mentioned earlier, GPS transmits position information to the user segment. The position data is not relative to a transmitter located on the earth's surface like other forms of aircraft navigation. The GPS user must therefore, navigate via waypoints. Waypoints can be any location in two dimensions (latitude and longitude). By some means, the waypoints are entered into the airborne GPS equipment. On older systems waypoints were manually entered using the GPS control panel; new systems contain waypoint information in the airborne GPS database.

Portable GPS Equipment

There are several GPS receiver/processors currently available, which are portable or "hand-held" units. The hand-held receivers are designed for hikers, hobbyists, surveyors, boaters, and even pilots. Some modern portable units are designed specifically for aircraft use and contain an extensive database with airport information as well as the ability to display SiriusXM satellite weather. A modern portable unit is completely self-contained; including the receiver/processor, control panel/display, antenna, and batteries in a compact unit as seen in Figure 8-1-15. Portable GPS units have become very popular and are being used extensively by general aviation pilots. Some advantages to the hand-held equipment include: 1) portability, the units can be moved between various aircraft, or taken to a convenient location to change waypoint or other database information, and 2) cost, there is virtually no installation cost and these units are traditionally less expensive than panel mounted systems. The major disadvantage of portable GPS receivers is that they are not FAA certified for use as a primary navigation source. Another major problem of portable units is the difficulty in achieving appropriate antenna placement. Since the aircraft structure blocks the signal transmitted from the satellites, any antenna located inside the aircraft will have poor reception at best. Many portable receivers incorporate antennas that clip onto the aircraft windshield. This solution is less than optimal and may still result in temporary signal loss. One solution includes connecting the portable unit to an antenna permanently mounted on top of the aircraft. Of course, any equipment permanently installed on the aircraft must meet all FAA requirements.

Panel-Mounted GPS Equipment

The alternative to the hand-held GPS receiver is the panel-mounted system found on many aircraft. With the increased use of GPS for en route and approach navigation inevitable, many pilots and aircraft owners are selecting panel-mounted GPS receivers. These systems, permanently installed in the aircraft, are similar to conventional navigation radios. Currently produced aircraft may contain one or more permanently installed GPS units; and most of the newest systems are integrated with a host of other navigation systems as discussed earlier. Figure 8-1-16 shows two panel-mounted GPS units, a "stand-alone" GPS and an integrated system with GPS capabilities. The early GPS unit contains the receiver/processor and control/display in one unit. The integrated system contains a receiver/processor in a remote location on the aircraft and all GPS information is displayed on one or more large integrated displays. The antenna on all panel-mounted GPS equipment is remotely located on top of the aircraft.

Figure 8-1-16. Typical Panel Mounted GPS Receivers: (A) A stand-alone unit, (B) A navigational display with integrated GPS capabilities

Figure 8-1-17. Bendix/King AFD-840, a retrofit PDF with GPS capabilities

Courtesy of Bendix/King by Honeywell

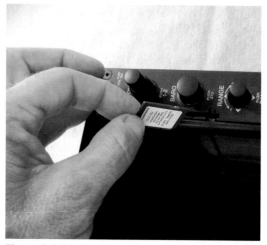

Figure 8-1-18. An SD card used to update a GPS database

Many panel-mounted units are designed to fit the standard radio rack configuration. The GPS receiver is typically mounted in the center of the instrument panel, which provides good visibility of the display for both pilot and co-pilot. Many panel-mounted GPS receiver/processors have the capability to interface with other navigational displays, such as, PFDs, NDs, MFDs, or EHSIs. For example, the Bendix/King AFD-840 is a retrofit system designed to receive input signals from a separate GPS receiver/processor (Figure 8-1-17).

Two of the most common airborne GPS features are the extended database and the moving map display. Any GPS receiver must incorporate some type of database; extended databases are found on almost all aircraft GPS receivers. An *extended database* may include an encyclopedia of waypoints, airport locations, runway headings, runway lengths, and other pertinent data. Many units now include an option that allows for the entry of checklist data. Once entered, the pilot can use the GPS unit to display aircraft checklist procedures.

Since navigational data can change periodically, the GPS database must be updated at regular intervals to maintain accuracy. This update responsibility falls on the GPS user. Some units require that the database be updated manually. Some GPS systems use an advanced technology memory chip (SD card) installed into the receiver/processor to update the database (Figure 8-1-18). Another means of updating a database employs an RS-232 or USB connection between the GPS processor and a personal computer. The updated information can be loaded onto the computer typically from CD-ROM or the Internet and downloaded onto the receiver/processor. On some integrated systems, a wireless downloading system might also be used.

Figure 8-1-19. Representation of a typical digital display showing bearing, range, track, and estimated time of arrival

Courtesy of Garmin International, Inc.

Earlier aircraft GPS receivers provided a digital readout of navigation data as seen in Figure 8-1-19. This type of display is good for position data, but it is difficult to interpret for continuous navigation. Many GPS receivers incorporate some type of electronic CDI (course deviation indicator). For example, the Garmin GPS 150 employs a CDI at the lower portion of the LCD display (Figure 8-1-20). The CDI provides a display of the aircraft's position relative to the desired course.

Figure 8-1-20. Panel mounted GPS with CDI located at bottom of display

Courtesy of Garmin International, Inc.

Figure 8-1-21. A panel mounted GPS with full color LCD moving map display

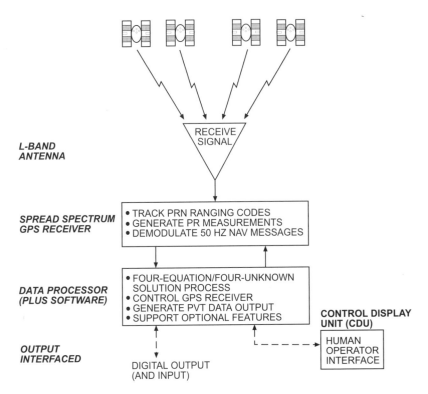

Figure 8-1-22. The four functional blocks of a GPS receiver

Moving map displays provide a horizontal picture of the aircraft and surrounding navigational reference points. Figure 8-1-21 shows a panel-mounted GPS unit with a typical moving map LCD (liquid crystal display). As the name "moving map" implies, the ground reference moves in relation to the aircraft as the flight progresses. For most pilots, the moving map is the display of choice and some type of moving map is employed in the newest GPS units integrated with navigational or multifunction displays.

Many airborne GPS receivers are now combined into one unit with a conventional NAV/ COM transceiver. Figure 8-1-21 shows the Garmin GNS 430 and installation on a typical light aircraft panel. This unit is capable of VHF communications, VHF navigation (VOR and localizer) as well as GPS navigation. Pilot inputs are made on the GNS 430 panel and the color LCD is used to display radio tuning frequencies, VOR/LOC course deviation, and GPS moving maps. The unit is slightly taller than a typical NAV/COM radio, which allows for a larger LCD display. The unit is a standard width, which allows for easy installation into a standard radio rack on light aircraft.

GPS airborne components. In general, all civilian GPS receivers/processors operate in a very similar manner. The specific circuitry, types of controls, displays, and interface options may vary among models of equipment. The generic GPS receiver diagrammed in Figure 8-1-22 includes the four function areas that each unit must contain:

1. The L-band antenna

2. A spread spectrum receiver

3. A data processor (plus software)

4. An output interface

The antenna. The L-band antenna receives the signal transmitted from the space segment satellites. The antenna must pick up the phase-modulated signal, convert the electromagnetic energy to electrical signals, amplify those signals and send them to the receiver through the coaxial cable. This is a difficult task since the space vehicle (SV) signal is extremely weak. In fact, the signal's strength is actually less than the sky's normal background noise. This makes the reception strictly line of sight. Although rain and clouds do not effect reception, a thin layer of ice or snow on the receiving antenna will limit signal reception.

Since the L-band antenna receives such a weak signal, almost all units contain a signal amplifier. The amplifier gets electrical power from the receiver circuitry through the coaxial cable (coax). The antenna provides amplification to overcome the anticipated signal loss through the coax. Antennas mounted less than two feet from the receiver circuitry often do not contain an amplifier.

The receiver. There are several different types of receivers available for the GPS user segment. The type of circuitry (digital or analog) can categorize the receivers, or the number of receiving channels. Digital receivers are more popular on aircraft and multi-channel receivers provide the greatest accuracy and reliability, especially for high-speed aircraft.

All receivers must perform three functions:

1. Acquire and track the satellite range code information

2. Determine the pseudorange (PR) from each satellite

3. Demodulate the 50Hz navigation message

The main input to the receiver is the antenna signal; the outputs include data to the processor unit, and in some cases to other systems that interface with the GPS. To accomplish these three functions, the generic receiver will typically include a down converter, a quartz clock/frequency synthesizer, and one or more PM (phase modulation) signal tracking channels.

The *down converter* filters out any extraneous signals and converts the RF (radio frequency) input from the antenna down to a given intermediate frequency (IF). The IF is a much lower frequency than the RF and easier to control while processing the transmitted information. The down converter must lower the frequencies of both L_1 and L_2 signals received by the antenna. While this occurs, the individual signals must remain completely independent to ensure signal integrity.

The *quartz clock/frequency synthesizer* provides the master time pulse and is the source of all reference frequencies within the receiver/processor. An accurate time pulse is critical to determine PR (pseudorange) measurements. The quartz clock is a very low noise crystal oscillator placed inside an oven to guard against frequency and time fluctuations caused by temperature change. The oven is typically a vacuum bottle containing an electric heater to keep the crystal at a constant 100° C. Although the quartz clock is not as accurate as the atomic clock in the satellites, it is far less expensive and is adequate for the job.

The *PM signal tracking channels* use the internal master time pulse along with the reference fre-

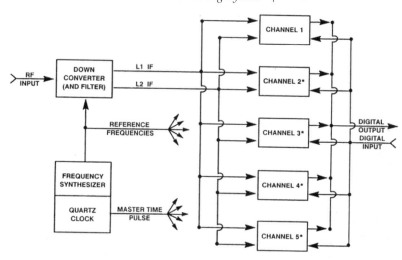

Figure 8-1-23. A typical five-channel receiver

quencies from the quartz clock to process the down-converted L_1 and L_2 signals at their IF. Each channel can track only one component of the transmission at any given time (the C/A- or P-code from L_1 or L_2). To simultaneously track all signal components from a satellite would require at least three channels. Most receivers contain multiple channels. Each channel operates independently under the control of a common processor circuitry. A generic five-channel receiver is shown in Figure 8-1-23.

A five-channel receiver has four channels dedicated to making range measurements continuously with four different satellites. This will satisfy the processor's need for a four-equation/four-unknown position solution as shown earlier in Figure 8-1-11. The choice of which satellites to track is made by the processor circuitry. Since only four channels are needed to solve the position equation, the fifth channel is free to handle other necessary functions, including making range measurements on the other frequency needed to calculate the errors induced by the ionosphere. Advanced planning can also be done with the fifth channel to verify data from satellites that are just rising on the horizon.

A one-channel receiver must still perform most of the functions of the five-channel receiver using a time-sharing process. Since a single channel can only track one satellite at a time, the data processor uses the single channel to perform sequential tracking. With a one-per-second operating rate, the channel will be sequenced among the four best satellites. Using this process, the receiver can process one range measurement per second, completing the same operations of the five channel unit in five seconds. This situation works well for receivers that are relatively stationary; however, for aircraft, this update rate is typically insufficient.

A two-channel receiver provides a low cost alternative to the five-channel unit and still provides the accuracy required by many users. The two-channel receiver can alternate between two satellites and minimize the update rate as compared to a single channel receiver. Most modern receivers contain multiple receiver channels. Multiple channels allow the GPS receiver to track several satellites simultaneously, which improves system performance.

The data processor. In addition to providing the control functions for each PM signal-tracking channel in the receiver, a data processor uses the resulting range measurements and a 50 Hz NAV message to solve for position, velocity, and time (PVT). The four-equation/four-unknown position solution process described previously is useful from a conceptual point of view, but very few processors actually operate this way. Instead, the processors perform their calculations using an advanced software algorithm known as the Kalman Filter. The mathematical theory behind Kalman Filters goes beyond the scope of this text; however, with advanced software, there is little need for a flightline technician to be concerned with this process.

The output interface. Aircraft GPS receivers contain some type of display panel, either a portable or panel mounted unit, used to provide the pilot with the desired navigational data. Many panel-mounted units contain outputs, which interface with other aircraft systems, allowing for GPS data to be integrated with other navigational displays, the flight data recorder, and even the autopilot system. Whenever troubleshooting any aircraft GPS, be sure to examine all interface wiring carefully. A loose pin connection or broken wire should always be considered a possible cause of the problem.

Integrated GPS systems. The modern GPS system is often designed to be part of an integrated navigation and display system. In many cases, the only dedicated GPS components might be the antenna and antenna cable. The GPS receiver/processor is part of the software which operates a variety of systems. The GPS information is displayed on only part of a large multifunction (MFD) or navigational display (ND). The GPS controls are also typically integrated into a control panel, which operates a variety of systems. The Garmin G-1000 is typical of an integrated system that employs one or more GPS receivers/processors.

There are currently several corporations manufacturing light aircraft electronic instrument systems. Many of the traditional manufactures, such as Honeywell and Bendix/King, produce systems for corporate and general aviation aircraft. In general, these systems can be found as retrofit packages for existing aircraft or as factory installed systems sold as an integral part of a new aircraft. In either case the system contains an integrated GPS similar to the one discussed here.

As seen in Figure 8-1-24, the system utilizes two integrated avionics units (IAUs), which contain the main circuitry for all communication and navigation systems (including GPS). The IAUs also provide the processing capabilities for three large LCD flat panel displays. Each IAU contains the receiver/processor circuitry and software needed for GPS operation. In order to provide redundancy, the G-1000 system contains two complete GPS systems, including two separate antennas. Since these are integrated systems, the IAUs process the GPS data and send all necessary signals to the video display units. The system contains a common control panel used for the GPS settings. There are typically two GPS antennas located on the top of the aircraft. These antennas could be external mounted on metal aircraft or internally mounted on composite aircraft.

Modern GPS equipment is extremely reliant on software for troubleshooting, operational updates, and system configuration. In order to allow for flexibility in design, the system has the ability to change configuration through modifications to the system software. For example, if one aircraft employs only one navigational display showing GPS data and another aircraft uses two, the IAUs can be modified to configure the systems accordingly. Each aircraft could have the exact same hardware; only the software would be different. With this philosophy of software configurations, the manufacturer can produce limited variations of the hardware and still offer a variety of options to an aircraft owner. In the case of the IAU found in the Garmin system discussed here, the inputs and outputs are configured using software. The GPS database is also contained in the system software and can be changed as needed. Since the GPS is part of a complete navigation system, updates to airports, waypoints, and other flight related information is critical and updated at regular intervals.

If the configuration of the system needs to be changed due to a change in aircraft equipment or a periodic software modification, the software update would most likely be downloaded from the manufacturer's website and loaded onto the aircraft equipment. In the case of the Garmin system, the update is loaded onto a solid-state memory known as a secure digital (SD) card. The SD card is then installed into a slot on the bezel of the PFD or ND and uploaded into the Garmin IAU.

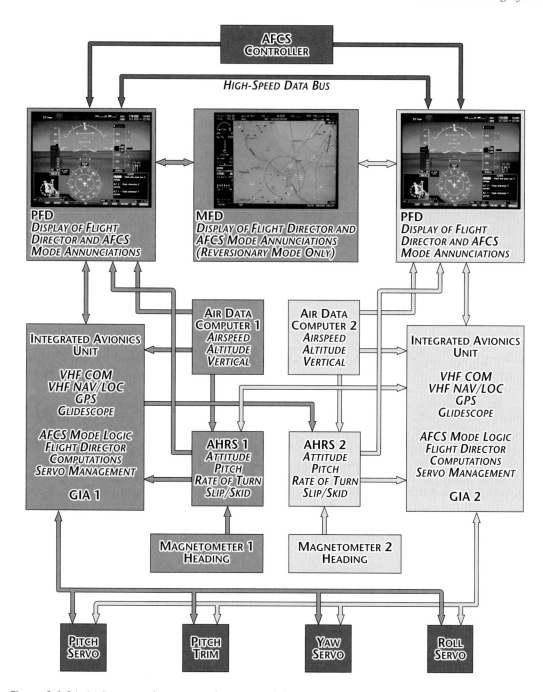

Figure 8-1-24. An integrated system with GPS capabilities

Whenever troubleshooting any system of this type it is important to identify the exact depth of the fault. The technician must understand the system architecture completely and how data is transmitted and received (i.e., through which sensors or probes, through which bus or busses, and processed by which LRUs). If the aircraft experiences a GPS failure, a red *X* appears over all GPS indications. There will also be an alert message of a GPS failure. Since the system is extremely software reliant, the BITE program can perform a variety of troubleshooting and diagnostic functions. This is very important since most of the communication between LRUs in the GPS is completed through various data busses and troubleshooting can become extremely complex.

The system software continually monitors each data bus for activity and ensures data validity through a series of cross checks with parallel systems. The maintenance manual for this system contains a troubleshooting section detailing the various steps to help isolate defective components. In the event of an inoperative system, the PFD or ND will display a red *X* in place of the lost data. The technician can then refer to the system troubleshooting guide in the maintenance manuals and follow the recommendations to find the failed component.

Section 2

The KLN90

Bendix/King avionic systems, has been viewed as an industry standard for several decades. The KLN90 was one of the first widely accepted panel mounted GPS receivers produced by Bendix/King (Figure 8-2-1). The KLN90 was designed for general aviation use and is often installed in corporate-type aircraft, such as the Beechcraft King Air. The following discussion on the KLN90 will focus on a general description and installation of the system. Most of the information presented here is general and will apply to most panel mounted GPS receivers. Although this GPS unit was very popular, the KLN90 is now considered old technology and has been replaced in most aircraft with a more advanced model. The following information is for training purposes only and is not to be used as a guide for installation of an actual GPS receiver.

General Description

The KLN90 was designed for installation in a standard radio rack configuration with dimensions of 6.3 inches (16.03cm) wide, 2.0 inches (5.08cm) tall, and 12.5 inches (31.8cm) deep (Figure 8-2-2). The receiver processor and display/control panel are contained in one unit. Containing an extensive database, the system provides real time position information with respect to a flight plan defined by the pilot.

The KLN90 can present position and course deviation information on its own LCD display or the unit can interface with a CDI, EFIS, or radar graphics unit. The internal database contains worldwide information on airports, waypoints, VORs, NDBs, intersections, and outer markers. The database memory is stored in a cartridge, which plugs into the back of the KLN90 receiver/processor. Database informa-

tion can also be updated using an interface with a personal computer. Many of the input/output signals for the processor unit are formatted according to the ARINC 429 digital data specification. Gray code inputs are accepted for altitude information, and several analog output signals are used for interface message annunciators, such as the CDI and navigation flags. If needed to communicate with other equipment, an interface adaptor kit is available to convert ARINC 429 signals to analog data.

This system can be installed for VFR or IFR certification. If the KLN90 is used for IFR flights, it must meet certain installation criteria that are not required for VFR use. In other words, additional equipment or installation procedures may be required for IFR certification. For example, when the KLN90B is used for IFR, additional annunciators must be installed, an altitude source or encoding altimeter must be interfaced with the receiver/processor, and a specific antenna must be used. An approved pilot's guide must be accessible to the pilot(s) during flight.

Basic Installation

As shown in Figure 8-2-2, the KLN90 was designed for a standard panel mount configuration. The receiver/processor/display unit is one component and designed to fit into a typical slide-in radio rack. The connectors for both the antenna and related wiring are mounted at the rear of the rack. A jack for the remote data loader can be installed on the aircraft panel using a 0.693-inch hole. As seen in Figure 8-2-3, a special cable can be made to connect a PC to the data loader jack.

As with any avionics, equipment cooling is extremely important to the long-term reliability of the product. Compared to older units, modern systems are in some cases more vulnerable to heat problems since they contain a large number of electronic components in a compact case. For this reason, most panel

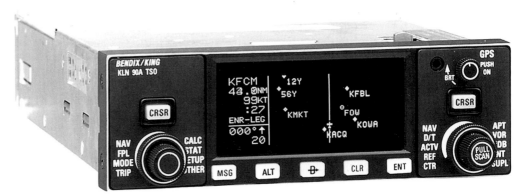

Figure 8-2-1. Bendix/King KLN90 GPS receiver

Courtesy of AlliedSignal General Aviation Avionics

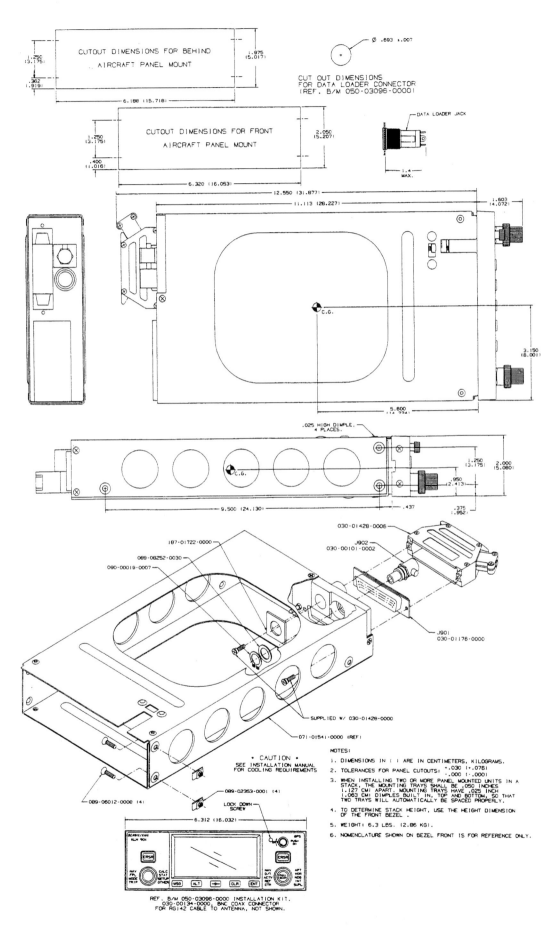

Figure 8-2-2. Installation drawing for the KLN90B

Courtesy of AlliedSignal General Aviation Avionics

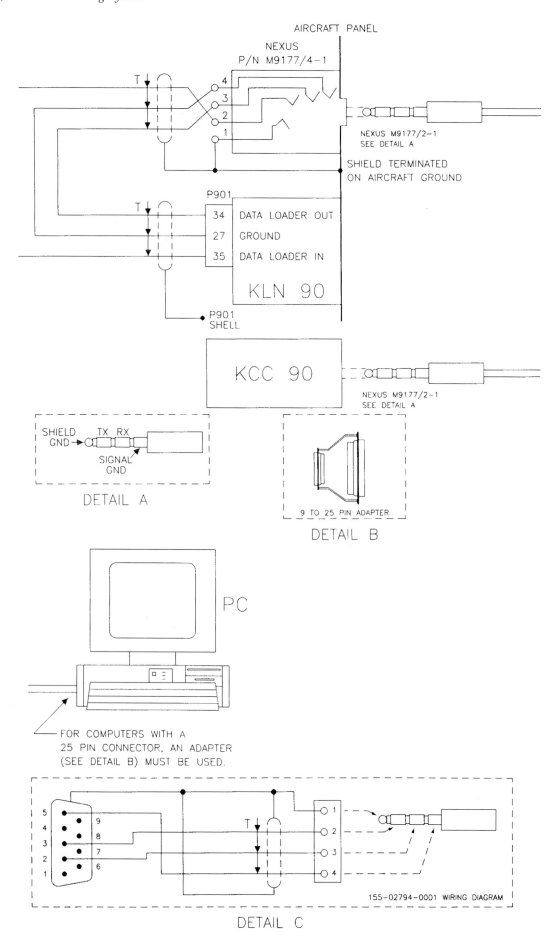

Figure 8-2-3. Data loader interface cable for the KLN90B

Courtesy of AlliedSignal General Aviation Avionics

mounted GPS receiver/processors should be cooled by forced air blowers. Forced air-cooling is most important when several avionics units are installed, one on top of the next or side-by-side.

The KLN90 antenna is a relatively small unit that is installed on top of the aircraft. As shown in Figure 8-2-4, the antenna is a streamline shape mounted with four 10-32 machine screws. The antenna should be mounted to the aircraft skin using a backer plate. It is also important that the antenna be mounted in a location so it is level ±5° when the aircraft is in level flight. This will help to ensure capture of satellites that are close to the horizon. To ensure communication signals do not interfere with satellite reception, the antenna must be mounted at least three feet from any communications antenna. As some GPS antennas are directional, be sure to install the unit facing the correct direction.

Proper installation of the antenna coaxial cable is critical for system operation. The maximum allowable loss through the coaxial cable is 8.0 dB. At the time of installation, the specific coax wiring kit should be ordered. Different kits are available for aircraft requiring between 0 to 40, 0 to 80 or 0 to 100 feet of cable. The coax must be run using smooth curves and must not be crushed or pinched. It is recommended that a right angle coax connector similar to the one shown in Figure 8-2-5, be used for certain installations between the cable and the antenna (or between the cable and the processor). This type connector will ensure the coax is not kinked at this location.

The system wiring for the KLN90B is relatively simple. If the unit is installed as a stand-alone system, the only wiring required is the antenna coax, the electrical power, ground, and the display lighting wires. If the receiver/processor is to interface with other aircraft systems, such as the EFIS or an HSI, additional wiring will be required. When connecting the receiver/processor to other aircraft systems, pin programming may also be required. Program pins are used to configure the GPS to your particular installation. For example, if pin number 1 is connected to high (positive voltage), the KLN90B is set to use an external analog OBS (omnibearing selector). If pin number 1 is connected to ground, no external OBS will be used.

Operational tests

After installation, the system must be thoroughly tested for proper operation of the receiver/processor, the display, and any unit that interfaces with the GPS. Proper antenna and coax installation is verified by operation of

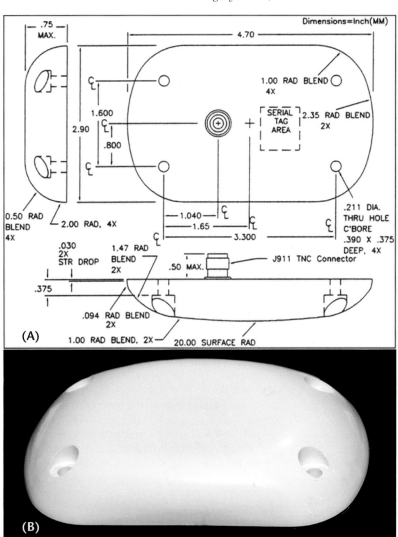

Figure 8-2-4. A typical aircraft mounted GPS receiver antenna

Courtesy of AlliedSignal General Aviation Avionics

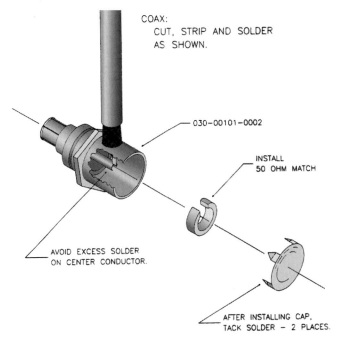

Figure 8-2-5. A right angle coaxial cable connector used to eliminate cable kinks at the base of the antenna *Courtesy of AlliedSignal General Aviation Avionics*

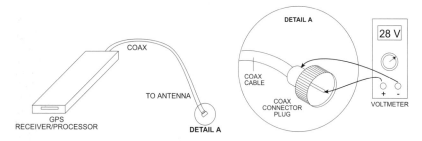

Figure 8-3-1. Measuring the DC voltage sent from the GPS receiver to the antenna

the receiver/processor. Before applying power to the system, verify that all wiring is installed properly and that adequate equipment cooling will be supplied during the test. Move the aircraft to a location outside of the hanger away from any tall structures that may block satellites low on the horizon.

On some equipment, the initialization requires entering the current time and position of the aircraft. This is typically completed through the GPS display/control unit. To test the system, simply follow the procedures outlined in the pilot's guide. Once the receiver/processor has "locked on" to four satellites, the GPS display should provide your current location in latitude and longitude. To verify system operation, simply compare this to an accurate value of your known position. A receiver may require several minutes to provide accurate position data, although typically 30 seconds is adequate.

The GPS must also be tested for interference caused by other equipment on the aircraft. Two common sources of interference are the communication (comm) radio and the emergency locator transmitter (ELT). To ensure proper GPS operations, the verification check outlined in the last paragraph should be repeated while transmitting on the communication radio frequency 121.50 MHz. The test should be repeated for all comm radios installed on the aircraft and at any frequency recommended by the manufacturer. Common frequencies that can interfere with GPS signals include 121.175, 121.20, 131.250, 131.275, and 131.30 MHz. The 12th and 13th harmonics of these frequencies are often strong enough to interfere with GPS reception. On some comm radios, a simple in-line filter may solve the interference problem. Other situations may require that the comm and GPS antennas be moved further apart. It is also wise to test the ELT as a source of interference. The comm radio can excite a tank circuit in the ELT that can radiate GPS interference. Grounding the ELT antenna should verify interference from the ELT. (i.e., if grounding the antenna stops the interference, the ELT is responsible for

generating the interference.) In most cases, the ELT manufacturer should be consulted for the proper repair.

Other systems that interface with the GPS must also be tested for proper operation. This includes the data loaders, CDI/HSIs, EFIS displays, encoding altimeters, and any discrete annunciators. Each of these units requires a specific test, which is beyond the scope of this text; however, each system must be thoroughly tested and its proper operation verified prior to completing the GPS installation.

Section 3

Troubleshooting Aircraft GPS Equipment

The troubleshooting of a failed GPS receiver is typically a relatively simple task. If the unit is a portable device, there is not much that can be done outside of returning the unit to a repair facility. A line technician may look closely at the battery, battery connections, any external power supply, and the antenna and coax if applicable. If the unit is a panel-mounted unit, once again, the power supply and antenna system are the primary suspects for a failed unit.

Panel-mounted GPS receivers often interface with other aircraft systems. If a problem exists between one of these interfaces, be sure to check the system wiring. Connector pins and sockets, which have become corroded or bent can often cause intermittent operation or complete system failure. If the GPS fails to operate correctly on a display device, such as the HSI, be sure to verify that the HSI is operational when driven by another source.

The antenna system of the GPS airborne equipment is often overlooked during troubleshooting. Be sure that the coaxial cable is in good condition, with no kinks or tight bends. Poor coaxial cable can cause poor or no reception of satellites low on the horizon. It should also be noted that any remotely mounted antenna can contain an amplifier used to boost signal strength. The power source for the antenna amplifier is located in the receiver circuitry. If the power source fails, the antenna will be inoperative even though it is in perfect working condition. As seen in Figure 8-3-1, a DC voltage is sent to the antenna unit through the coaxial cable. In most cases, a voltmeter connected to the coax cable at the antenna can be used to test the power supplied to the antenna amplifier.

Of course, the receiver/processor unit must be operating for this test.

Many GPS simulators, test units, are commercially available. One such system, produced by IFR Systems Inc. is shown in Figure 8-3-2. The GPS-101 is a portable unit that simulates the satellite output and is used to test the antenna, coax and connectors, and the receiver/processor. In general, testing with this unit is a simple process of elimination. For example, in a system with a defective coaxial cable between the receiver and the antenna, a technician would complete the following steps:

1. Place the GPS-101 transmitting antenna (the triangular shaped unit in Figure 8-3-2) over the GPS receiver antenna on the aircraft. This will permit testing of the entire airborne system. For this example, the unit fails the test.

2. To eliminate the antenna as a possible fault, plug the GPS-101 simulator into the coax cable at the antenna and run another test. Once again, for this example the airborne equipment fails the test.

3. Move the test unit inside the aircraft and connect the simulator output directly to the receiver/processor. If the coaxial cable is faulty, the GPS receiver will pick up the signals transmitted by the test unit.

In the above example, the GPS-101 simulator was used to eliminate different sections of the airborne GPS equipment (the antenna, the coax, and the receiver). When troubleshooting, always keep in mind the power supply to the antenna amplifier. If the receiver had worked properly with the antenna bypassed (step 2 above), the fault lies in the antenna or the power sent to the antenna for the amplifier. Check for the proper voltage to the antenna before condemning the antenna assembly.

As previously discussed, modern integrated systems employ extensive diagnostics and often pinpoint defects to a faulty LRU. The built in test equipment will most likely test power, data bus connections, and processor circuitry. The diagnostics can also test the antenna and related cable. Typically, if the antenna and wiring have been eliminated as possible faults, the receiver/processor is defective. The receiver/

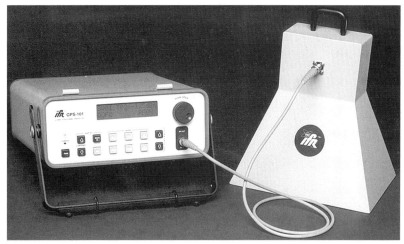

Figure 8-3-2. Model GPS-101 GPS simulator *Courtesy of IFR Systems, Inc.*

processor cannot be repaired on the flight line and must be sent to the proper repair facility.

Interference created by other avionics systems can also cause GPS problems. Most GPS displays will allow the technician to view the signal strength of the received signal. If signal strength level is erratic or disappears and reappears rapidly, suspect an external electronics device as a possible cause. If the GPS is experiencing intermittent signal loss, be sure to question the pilot as to when signal loss occurs. Ground-based and aircraft radio transmitters, cell phones, and PDAs may interfere with the GPS reception. Although interference is extremely limited for ground-based transmitters, airborne VHF communication can easily cause interference. Ask the pilot where and when the GPS signal was lost. If the communication radio was in use at the time, be sure to suspect interference from the communication radio.

Be sure all cellular devices, which might be in the cabin, are turned off. These devices can interfere with GPS units even in the monitoring mode; they must be off. In some cases, the system may have become corrupt and the technician should reset the aircraft GPS equipment according to the maintenance manual. The next step would most likely be to cycle power, allowing the system to restart, making sure the aircraft is clear of all metal structures (outside the hanger) and allow three to five minutes for the system to acquire the necessary satellite signals.

Appendix A

Excerpt from ATA iSpec 2200 (Formerly ATA Specification 2200)

(Includes ATA 100 and ATA 2100 specifications, Revision 2001.1)
NOTE 1: *Only assigned numbers shown in this table*
NOTE 2: *Subsystem/Section set to zero are all general information and are not shown in this table*

SYS/ CHAP	SUBSYS/ SECTION	TITLE
1		RESERVED FOR AIRLINE USE
2		RESERVED FOR AIRLINE USE
3		RESERVED FOR AIRLINE USE
4		RESERVED FOR AIRLINE USE
5		TIME LIMITS AND MAINTENANCE CHECKS
	-10	Time limits
	-20	Scheduled maintenance checks
	-50	Unscheduled maintenance checks
6		DIMENSIONS AND AREAS
7		LIFTING AND SHORING
	-10	Jacking
	-20	Shoring
8		LEVELING AND WEIGHING
	-10	Weighing and balancing
	-20	Leveling
9		TOWING AND TAXIING
	-10	Towing
	-20	Parking
10		PARKING, MOORING, STORAGE, AND RETURN TO SERVICE
	-10	Parking and storage
	-20	Mooring
	-30	Return to service
11		PLACARDS AND MARKINGS
	-10	Exterior color schemes and markings
	-20	Exterior placards and markings
	-30	Interior placards
12		SERVICING
	-10	Replenishing
	-20	Scheduled servicing
	-30	Unscheduled servicing
18		VIBRATION AND NOISE ANALYSIS (HELICOPTER ONLY)
	-10	Vibration analysis
	-20	Noise analysis
20		STANDARD PRACTICES - AIRFRAME
21		AIR CONDITIONING

SYS/ CHAP	SUBSYS/ SECTION	TITLE
	-10	Compression
	-20	Distribution
	-30	Pressurized control
	-40	Heating
	-50	Cooling
	-60	Temperature control
	-70	Moisture and contaminant control
22		AUTOFLIGHT
	-10	Autopilot
	-20	Apeed and attitude correction
	-30	Autothrottle
	-40	System monitor
	-50	Aerodynamic load alleviating
23		COMMUNICATIONS
	-10	Speech communications
	-20	Data transmission and automatic calling
	-30	Passenger address, entertainment, and comfort
	-40	Interphone
	-50	Audio integrating
	-60	Static discharging
	-70	Audio and video monitoring
	-80	Integrated audio tuning
24		ELECTRICAL POWER
	-10	Generator drive
	-20	AC generation
	-30	DC generation
	-40	External power
	-50	AC electrical load distribution
	-60	DC electrical load distribution
25		EQUIPMENT AND FURNISHINGS
	-10	Flight Compartment
	-20	Passenger
	-30	Gallery
	-40	Lavatories
	-50	Additional compartments
	-60	Emergency
	-70	Available

SYS/ CHAP	SUBSYS/ SECTION	TITLE
	-80	Insulation
26		FIRE PROTECTION
	-10	Detection
	-20	Extinguishing
	-30	Explosion suppression
27		FLIGHT CONTROLS
	-10	Aileron and tab
	-20	Rudder and tab
	-30	Elevator and tab
	-40	Horizontal stabilizer
	-50	Flaps
	-60	Spoiler, drag devices, and variable aerodynamic farings
	-70	Gust lock and dampener
	-80	Lift augmenting
28		FUEL
	-10	Storage
	-20	Distribution
	-30	Dump
	-40	Indicating
29		HYDRAULIC POWER
	-10	Main
	-20	Auxiliary
	-30	Indicating
30		ICE AND RAIN PROTECTION
	-10	Airfoil
	-20	Air intakes
	-30	Pitot and static
	-40	Windows, windshields, and doors
	-50	Antennas and radomes
	-60	Propeller and rotors
	-70	Water lines
	-80	Detection
31		INDICATING AND RECORDING SYSTEMS
	-10	Instrument and control panels
	-20	Independent instruments
	-30	Recorders
	-40	Central computers
32		LANDING GEAR
	-10	Main gear and doors
	-20	Nose gear and doors
	-30	Extension and retraction
	-40	Wheels and brakes
	-50	Steering
	-60	Position and warning
	-70	Supplementary gear

SYS/ CHAP	SUBSYS/ SECTION	TITLE
33		LIGHTS
	-10	Flight compartment
	-20	Passenger compartment
	-30	Cargo and service compartments
	-40	Exterior
	-50	Emergency lighting
34		NAVIGATION
	-10	Flight environment data
	-20	Attitude and direction
	-30	Landing and taxiing aids
	-40	Independent position determining
	-50	Dependent position determining
	-60	Flight management computing
35		OXYGEN
	-10	Crew
	-20	Passenger
	-30	Portable
36		PNEUMATIC
	-10	Distribution
	-20	Indicating
37		VACUUM
	-10	Distribution
	-20	Indicating
38		WATER AND WASTE
	-10	Potable
	-20	Wash
	-30	Waste and disposal
	-40	Air supply
41		WATER BALLAST
	-10	Storage
	-20	Dump
	-30	Indication
44		CABIN SYSTEMS
	-10	Cabin core system
	-20	Inflight entertainment system
	-30	External communication systems
	-40	Cabin mass memory system
	-50	Cabin monitoring system
	-60	Miscellaneous cabin system
45		CENTRAL MAINTENANCE SYSTEM (CMS)
	-5 thru -19	CMS and aircraft general
	-20 thru -49	CMS and airframe system
	-45	Central maintenance system
	-50 thru -59	CMS and structures
	-60 thru -69	CMS and propellers
	-70 thru -89	CMS and power plant

SYS/ CHAP	SUBSYS/ SECTION	TITLE
46		**INFORMATION SYSTEMS**
	-10	Airplane general information systems
	-20	Flight deck information systems
	-30	Maintenance information systems
	-40	Passenger cabin information systems
	-50	Miscellaneous information systems
49		**AIRBORNE AUXILIARY POWER**
	-10	Power plant
	-20	Engine
	-30	Engine fuel and control
	-40	Ignition and starting
	-50	Air
	-60	Engine controls
	-70	Indicating
	-80	Exhaust
	-90	Oil
50		**CARGO AND ACCESSORY COMPARTMENTS**
	-10	Cargo compartment
	-20	Cargo loading systems
	-30	Cargo related systems
	-40	Available
	-50	Accessory compartments
	-60	Insulation
51		**STANRDARD PRACTICES AND STRUCTURES – GENERAL**
	-10	Investigation, clean up, and aerodynamic smoothness
	-20	Processes
	-30	Materials
	-40	Fasteners
	-50	Support of airplanes for repair and alignment check procedures
	-60	Control surface balancing
	-70	Repairs
	-80	Electrical bonding
52		**DOORS**
	-10	Passenger and crew
	-20	Emergency exit
	-30	Cargo
	-40	Service and miscellaneous
	-50	Fixed interior
	-60	Entrance stairs
	-70	Monitoring and operation
	-80	Landing gear
53		**FUSELAGE**
	-10 thru -90 as req.	Fuselage sections
54		**NACELLES AND PYLONS**

SYS/ CHAP	SUBSYS/ SECTION	TITLE
	-10 thru -40 as req.	Nacelle section
	-50 thru -80 as req.	Pylon
55		**STABILIZERS**
	-10	Horizontal stabilizer or canard
	-20	Elevator
	-30	Vertical stabilizer
	-40	Rudder
56		**WINDOWS**
	-10	Flight compartment
	-20	Passenger compartment
	-30	Door
	-40	Inspection and observation
57		**WINGS**
	-10	Center wing
	-20	Outer wing
	-30	Wing tip
	-40	Leading edge and leading edge devices
	-50	Trailing edge and trailing edge devices
	-60	Ailerons and elevons
	-70	Spoilers
	-80	(as required)
	-90	Wing folding system
58		**UNASSIGNED**
59		**RESERVED FOR AIRLINE USE**
60		**STANDARD PRACTICES – PROPELLER AND ROTOR**
61		**PROPELLERS AND PROPULSION**
	-10	Propeller assembly
	-20	Controlling
	-30	Braking
	-40	Indicating
	-50	Propulsion duct
62		**ROTORS**
	-10	Rotor blades
	-20	Rotor heads
	-30	Rotor shafts and swashplate assemblies
	-40	Indicating
63		**ROTOR DRIVES**
	-10	Engine and gearbox couplings
	-20	Gearboxes
	-30	Mount attachments
	-40	Indicating
64		**TAIL ROTOR**
	-10	Rotor blades
	-20	Rotor head

SYS/CHAP	SUBSYS/SECTION	TITLE
	-30	Available
	-40	Indicating
65		TAIL ROTOR DRIVE
	-10	Shafts
	-20	Gearboxes
	-30	Available
	-40	Indicating
66		FOLDING BLADES AND PYLON
	-10	Rotor blades
	-20	Tail pylon
	-30	Controls and indicating
67		ROTORS FLIGHT CONTROL
	-10	Rotor control
	-20	Anti-torque rotor control (yaw control)
	-30	Servo-control system
70		STANDARD PRACTICES – ENGINES
71		POWER PLANT
	-10	Cowling
	-20	Mounts
	-30	Fireseals
	-40	Attach fittings
	-50	Electrical harness
	-60	Air intakes
	-70	Engine drains
72		ENGINE TURBINE AND TURBO-PROP DUCTED AND UNDUCTED FAN
	-10	Reduction gear, shaft section (turbo-prop/front mounted gear-driven propulsor)
	-20	Air inlet section
	-30	Compressor section
	-40	Combustion section
	-50	Turbine section
	-60	Accessory drives
	-70	Bypass Section
	-80	Propulsor Section (rear-mounted)
73		ENGINE FUEL AND CONTROL
	-10	Distribution
	-20	Controlling
	-30	Indicating
74		IGNITION
	-10	Electrical power supply
	-20	Distribution
	-30	Indicating
75		AIR
	-10	Engine anti-icing
	-20	Cooling
	-30	Compressor control

SYS/CHAP	SUBSYS/SECTION	TITLE
	-40	Indicating
76		ENGINE CONTROLS
	-10	Power control
	-20	Emergency shutdown
77		ENGINE INDICATING
	-10	Power
	-20	Temperature
	-30	Analyzers
	-40	Integrated engine instrument systems
78		EXHAUST
	-10	Collector and nozzle
	-20	Noise suppressor
	-30	Thrust reverser
	-40	Supplementary air
79		OIL
	-10	Storage
	-20	Distribution
	-30	Indicating
80		STARTING
	-10	Cranking
81		TURBINES
	-10	Power recovery
	-20	Turbo-supercharger
	-30	Dumping and purging
	-40	Indicating
83		ACCESSORY GEARBOXES
	-10	Drive shaft section
	-20	Gearbox section
84		PROPULSION AUGMENTATION
	-10	Jet assist takeoff
91		CHARTS
95		RESERVED FOR AIRLINE USE
96		RESERVED FOR AIRLINE USE
97		RESERVED FOR AIRLINE USE
98		RESERVED FOR AIRLINE USE
115		FLIGHT SIMULATION SYSTEMS
116		FLIGHT SIMULATOR CUING SYSTEMS

Appendix B

List of Common Acronyms Used in this Text

AAP - Altitude Awareness Panel

ACARS – Airborne Communications Addressing and Reporting System

ACESS - Advanced Cabin Entertainment Service System

ADC - Air Data Computer

AFDS - Autopilot Flight Director System

AHC - Attitude Heading Computer

AIM - Acknowledgement, Iso-alphabet, Maintenance

AIMS - Airplane Information Management System

APP - Autopilot Panel

ARP - Air Data Reference Panel

ASCB - Avionics Standard Communication Bus

ASM - Aircraft Schematic Manual

ATA - Air Transport Association

ATC - Automatic Trim Coupler

AWL - Aircraft Wiring List

AWM - Aircraft Wiring Manual

BCD - Binary Coded Decimal

BITE - Built In Test Equipment

BNR - Binary Data

CCTM - Cabin Configuration Test Module

CDC - Control Display Coupler

CDU - Control Display Unit

CFDIU - Centralized Fault Display Interface Unit

CFDS - Centralized Fault Display System

CHP - Course Heading Panel

CIC - Cabin Interphone Controller

CIS - Cabin Interphone System

CLS - Cabin Lighting System

CMC - Central Maintenance Computer

CMC - Current Mode Coupler

CMCS - Central Maintenance Computer System

CMU - Control Management Unit

CSDB - Commercial Standard Digital Bus

DAU - Data Acquisition Unit

DCP - Display Control Panel

DEMUX - Demultiplexer

DEU - Display Electronic Unit

DGPS - Differential Global Positioning System

DMC - Display Management Computer

DMU - Data Management Unit

DSDL - Dedicated Serial Data Line

DSP - Display Select Panel

EAD - Engine Alert Display

EADI - Electronic Attitude and Director Indicator

ECAM - Engine Condition and Monitoring

ECS - Environmental Control System

ED - Electronic Display

EEC - Electronic Engine Control

EFD - Electronic Flight Display

EFIS - Electronic Flight Instrument System

EHSI - Electronic Horizontal Situation Indicator

EICAS - Engine Indicating and Crew Alerting System

EIS - Electronic Instrument System

EIU - Electronic Interface Unit

ES - Elevator Station

ESC - Entertainment Service Controller

FAC - Flight Augmentation Computer

FCC - Flight Control Computer

FDE - Flight Deck Effect

FIM - Fault Isolation Manual

FIN - Functional Item Number

FMC - Flight Management Computer

FMS - Flight Management System

FPD - Flat Panel Display

FRM - Fault Reporting Manual

FS - Fuselage Station

FWC - Flight Warning Computer

GAMA - General Aviation Manufacturers Association

GMT - Greenwich Mean Time

GPS - Global Positioning System

GPWC - Ground Proximity Warning Computer

IAPS - Integrated Avionics Processor System

IDS - Integrated Display System

IDU - Integrated Display Unit

INS - Inertial Navigation Systems

IRS - Inertial Reference System

IRU - Inertial Reference Unit

LAAS - Local Area Augmentation System

LASER - Light Amplification by Stimulated Emission of Radiation

LCD - Liquid Crystal Display

LSK - Line Select Keys

LVDT - Linear Voltage Differential Transducer

MAWEA - Modularized Avionics and Warning Electronics Assembly

MCDU - Maintenance Control Display Unit

MFD - Multifunction Display

MCDU - Multipurpose Control and Display Unit

MPU - Multifunction Processor Unit

MSP - Mode Select Panel

MU - Management Unit

MUX - Multiplexer

NBPT - No Break Power Transfer

ND - Navigational Display

NRZ - Non-Return to Zero

NS - Nacelle Station

PAC - Passenger Address Controller

PAS - Passenger Address System

PES - Passenger Entertainment System

PFD - Primary Flight Display

PM - Phase Modulation

PSS - Passenger Service Systems

PWR - Power module

RAT - Ram Air Turbine

RCP - Radio Communications Panel

RLG - Ring Laser Gyro

RMM - Ramp Maintenance Manual

RVDT - Rotary Voltage Differential Transducer

RZ - Return to Zero

SAT - Static Air Temperature

SD - System Display

SDD - Sensor Display Driver

SDI - Source Destination Identifier

SDU - Sensor Display Unit

SG - Synchronization Gap

SIM - Serial Interface Module

SSM - Sign Status Matrix

TAT - True Air Temperature

TC - Terminal Controller

TG - Terminal Gap

TI - Transmit Interval

TR - Transformer Rectifier

TSM - Trouble Shooting Manual

VCC - Video Control Center

VDU - Video Distribution Unit

VSCU - Video System Control Unit

WAAS - Wide Area Augmentation System

WGD - World Geodetic System

WS - Wing Station

Glossary

Accelerometer - device that senses aircraft acceleration

Acknowledgement, ISO-alphabet, Maintenance (AIM) - an ARINC 429 data word format used for systems that require large amounts of data transfer

Active Matrix LCD - a type of LCD with a transistor located at each pixel intersection

Activity Monitor - monitors the data bus signals to ensure data transmission at regular intervals

Air Data Computer (ADC) - a system that monitors pitot pressure, static pressure and air temperature to determine various parameters, such as airspeed, altitude, and vertical climb

Air Transport Association (ATA) – an organization that represents airlines, aircraft manufacturers, and various system manufactures in an effort to ensure uniformity in various facets of the aviation industry

Airborne Communications Addressing and Reporting System (ACARS) - a digital air/ground communications service designed to reduce the amount of voice communications on the increasingly crowded VHF frequencies

ARINC 629 (also called 629) - a digital data bus format that permits up to 120 receiver/transmitters to share a bidirectional serial data bus

ARINC Incorporated - a global corporation made up of various U.S. and international airlines, aircraft operators and their subsidiaries. The company provides services related to a variety of aviation communication and navigation systems

Attenuation - occurs when a digital data signal becomes weak or the transmitted voltage becomes too low

Availability - the capability of the GPS system to be used for navigation whenever it is needed, and the ability to provide that service throughout the entire flight

Avionics Standard Communication Bus (ASCB) - a bidirectional data bus operating at 0.667 MHz

Binary Coded Decimal (BCD) - a specific ARINC 429 data word format

Bipolar - a digital data format that reverses polarity (two-polarity) when it changes from binary 1 to binary 0

BITS Mode - used to access real time data from the RAM memory of the DPU/MPU on the Collins EFIS 85/86

Block Diagram - defines electrical circuits by separating a system or subsystem into functional blocks

Built-In Test Equipment (BITE) - systems used in conjunction with many digital circuits to aid in system troubleshooting

Bus Hierarchy - a design concept to ensure that the most critical electrical systems are the least likely to fail

Butt Line - a vertical reference plane, which divides the aircraft front to rear through the center of the fuselage

Cabin Interphone System (CIS) - provides communications between different flight attendant stations, and between flight attendant stations and the flight deck on the B-747-400

Caution - a notation that will call attention to any methods, materials, or procedures, which must be followed to avoid damage to equipment on the aircraft

Central Maintenance Computer System (CMCS) - the advanced built-in troubleshooting systems found on Boeing aircraft

Centralized Fault Display Interface Unit (CFDIU) - the main computer that manages the CFDS information

Centralized Fault Display System (CFDS) - is the advance diagnostic system used on Airbus Industries transport category aircraft

Character Generator - processor circuitry used to produce letters, numbers, and symbols

Cold Cathode Fluorescent Tube - a highly efficient fluorescent tube often used to illuminate a transmissive LCD

Commercial Standard Digital Bus (CSDB) - a one way data bus system between one transmitter and a maximum of 10 receivers

Compensator Unit - used to correct for magnetic errors and flux detector misalignment

Configuration Strapping - *(also known as pin programming)* specific electrical connections used to determine various display, input/output signal formats and other system parameters

Continuity - the probability that a GPS service will continue to be available for a period of time necessary to complete the navigation requirements of the flight

Control Display Unit (CDU) - an LRU containing a display and alphanumeric keyboard typically used to control the auto-flight and other system functions

Current Mode Coupler (CMC) - an inductive coupling device, which connects the LRU to the 629 data bus

Data Bus Analyzer - test equipment used to troubleshoot digital systems

Datum - a vertical reference plane for aircraft, typically located toward the front of the fuselage

Dedicated Serial Data Line (DSDL) - a digital bus structure unique to Airbus aircraft

Demultiplexer (DEMUX) - a circuit that converts serial data into parallel data

Differential GPS (DGPS) - a design concept to increase the accuracy, availability, integrity and continuity of the basic GPS to a level sufficient for complete aircraft navigation

Differential Transducer - a solid-state device used to generate an electrical signal, typically used to identify the position of moving components

Discrete Word Format - an ARINC 429 data word format used to transmit the status of several individual components

Display Electronic Unit (DEU) - used on the B-747-400 to provide the processing power for the six display units and other aircraft systems

Display Management Computer (DMC) - the main processor for the CRT displays found on the Airbus A-320

Doublet Signal - a short positive and negative pulse (spike) on the bus whenever the data value changes from binary one to binary zero or back

Down Converter - a circuit found in GPS receivers used to filter out any extraneous signals and converts the RF input to a given intermediate frequency (IF)

Dutch Roll - a slow oscillation of the aircraft about its longitudinal axis

Dynamic Pressure - the difference between pitot pressure and static pressure

Effectivity - used to determine if the aircraft is covered by the information stated in that section (page) of a manual

Electric Servo - utilizes an electric motor and clutch assembly to move the aircraft's control surface according to autopilot commands

Electronic Flight Display (EFD) - an LRU used to provide primary flight and navigation data on a LCD or CRT display

Electronic Flight Instrument System (EFIS) - employs 2 or more CRTs or LCDs to present alphanumeric data and graphical representations of aircraft flight instruments

Electronic Instrument System (EIS) - found on the B-747-400 to display flight and navigational data, as well as engine parameters and warning information

Electrostatic Discharge - the discharge (movement of electrons) created when any material containing a static charge is exposed to a material containing a different or neutral charge

EICAS - Engine Indicating and Crew Alerting System

Ephemeris Errors - caused by slight variations in the GPS satellites positions as they orbit the earth

Equipment Identifier - part of the ARINC 429 data word used to further define the word label

Fault Flag - any indication made on one or more displays related to a complete or partial system failure

Fault Isolation Manual (FIM) - a part of the aircraft maintenance manuals, which contains various repair strategies for given faults

Flag Note - used to bring attention to a specific effectivity on a diagram or schematic

Flat Panel Display (FPD) - a solid-state device used to display various formats of video information in modern aircraft flight displays

Flight Deck Effects - any EFIS or EICAS display, or discrete annunciator used to inform the flight crew of a system fault

Flight Director - a visual aid to the pilot, present on the ADI, EADI or PFD used for reference or manual control of the aircraft

Flight Interphone - a system used to provide communications between flight crew members and/or other aircraft operations personnel

Flight Management System (FMS) - computer-based systems that reduce pilot workload by providing: automatic radio tuning, lateral and vertical navigation, thrust management, and the display of flight plan maps

Flight Phase - a time period of a given flight often recorded by the CMC to help the technician determine the airplane's configuration at the time of a system fault

Flight phase Screening - a software function used to help eliminate nuisance messages from entering the central maintenance system

Flux Detector - remote sensor used in a Flux Gate Compass system

Flux Gate Compass - a system that employs one or more remote sensors to produce an electric signal that can be used to determine the aircraft's position relative to magnetic North

Free Running Oscillator - a circuit to establish a given frequency

Functional Item Number (FIN) - a unique number given to each line replaceable unit on Airbus aircraft

Fuselage Station (FS) - used to indicate locations longitudinally along the aircraft fuselage

Glanoss - the former Soviet Union's satellite navigation system

Global Positioning System (GPS) - a satellite-based system that is capable of providing position and navigational data for ground based and airborne receivers

GLOBALink - the name of the ACARS service provided by ARINC Incorporated

Home Diagram - the diagram / schematic where the component is shown in full detail

Hydraulic Servo - the most powerful type of servo actuator; hence, these units are typically used on transport category aircraft

Inertial Navigation Systems (INS) - an aircraft and navigation system that does not rely on external radio signals; laser gyros and accelerometers provide three dimensional navigation capabilities

Inertial Reference System (IRS) - a combination of laser gyros and accelerometers used to sense the aircraft's angular rates and accelerations

Inertial Reference Unit (IRU) - an LRU containing laser gyros and accelerometers used for aircraft navigation

Integrated Avionics Processor System (IAPS) - provides integration functions for various avionics systems

Integrated Display System (IDS) - a state of the art instrument system, which employs integrated display and processor circuitry

Integrity - the ability of the GPS system to shut itself down when it is unsuited for navigation or to provide timely warnings to the pilot of the system failure

Interphone System - found on transport category aircraft to provide communications between flight crew, ground crew, flight attendants, and maintenance personnel

Ionization Air Blower - a special air blower that will safely delete any static charge formed on nonconductors

Kalman Filter - an advanced software algorithm used in GPS receivers to process the complex position calculations necessary to determine aircraft position

Line Select Keys (LSK) - push button switches found on the control display units used to select items for display or activate functions available on the CDU

Link Test - used to test ACARS link between airborne equipment and ground based equipment

Liquid Crystal Display (LCD) - a type of flat panel display that employs the use of liquid crystals to control light

Local Area Augmentation System (LAAS) - a GPS upgrade designed to be used during a Category II or Category III precision approach

Maintenance Control Display Unit (MCDU) - used on some Boeing aircraft to monitor and test the flight control computers, flight management computers, and the thrust management computers

Maintenance Manual - provides specific information for flight line and hangar maintenance activities

Manchester II Code - (often simply referred to as Manchester) is a serial digital data format that incorporates a voltage change in each data bit

Microfiche - a 4 by 6 inch filmcard, which is capable of storing up to 288 pages of data

Microfilm Strips - referred to as film, and used mainly by airlines to store maintenance and parts

Multifunction Display (MFD) - an EFIS display used to display weather radar data, course and flight plan information, system checklist, and provide back-up functions in the event of a partial system failure

Multifunction Processor Unit (MPU) - a processor used with the Collins EFIS 85/86 that provides signal processing and switching to the multifunction display

Multiplexer (MUX) - converts parallel data into serial data

Multipurpose Control and Display Unit (MCDU) - LRU containing an alpha numeric keyboard and CRT display found on Airbus aircraft used to input data to the CMCS and the CFDS

Nacelle Station (NS) or Elevator Station (ES) - used on larger aircraft to help technicians find components in specific areas of the aircraft

NAVSTAR GPS - the formal name for the global positioning system in the United States

No Break Power Transfer (NBPT) - an automated system that allows the aircraft to switch AC power supplies without a momentary interruption of electrical power

Non-Return to Zero (NRZ) - a self-clocking data bus format that does not return to zero at the end of each data bit

Note – found in various service manuals, used to draw your attention to a particular procedure, that will make the task easier to perform

Nuisance Message - any FDE or CMC message that is not caused by an actual fault

Pad Bits - fill in any portion of the data field not used for the transmission of data

Page Block - used to categorize a given chapter-section-subject of the maintenance manual

Parity Bit - a binary bit set by the transmitter that permits error checks by each receiver connected to the bus

Passenger Address System (PAS) - system for transmitting audio through a series of cabin speakers to announce messages

Passive Matrix LCD - employs a grid of conductors with pixels located at each intersection of the grid

Phase Modulation (PM) - digital modulation of the carrier for GPS and satellite

Pin Programming - *(also known as configuration strapping)* specific electrical connections used to determine various display, input/output signal formats and other system parameters

Pitot Pressure - the absolute pressure of the air that is *pressed* into the front of the aircraft

Pneumatic Servo - vacuum actuated autopilot component used to move control surfaces; typically found on light aircraft

Power Distribution System - one or more electrical distribution points used to connect the main power supply for the aircraft

Ram Air Turbine (RAT) - a propeller driven system which is deployed from the fuselage to provide emergency power in the event of a catastrophic hydraulic or electrical system failure

Ramp Maintenance Manual (RMM) – an abbreviated maintenance manual designed specifically for line maintenance and minor troubleshooting

Resolution - the smallest unit of data that can be measured or transmitted by a given system

Return to Zero (RZ) - a self-clocking data bus format that returns to zero during the second part of each bit

Ring Laser Gyro (RLG) - an angular rate sensor that employs a laser to detect aircraft motion.

Rotating Mass Gyro and **Ring Laser Gyro** - used by many autopilots to detect movement of the aircraft, gyro outputs are also used for reference on certain navigation systems

Sensitive Wires - considered critical to flight safety and must not be modified without specific manufacturer approval

Serial - transmission of data one binary digit or bit at a time

Serial Interface Module (SIM) - an ARINC 629 subsystem that changes the Manchester current signal from the LRU into an analog voltage doublet signal

Servo - a device used to move the flight control surface in accordance with autopilot command signals

Sign Status Matrix (SSM) - a portion of the ARINC 429 data word provides information, which might be common to several peripherals

Simplified Schematic - wiring illustration with intermediate depth and scope

Single Wire Electrical System - uses the airframe to distribute negative voltage

Split Parallel - power distribution systems used on some modern four-engine aircraft such as the Boeing 747-400

Standard Practices - the procedures and practices used repeatedly during aircraft maintenance, troubleshooting, and repair

Static Air Temperature (SAT) - the temperature of the undisturbed air surrounding the aircraft

Structural Modal Oscillations - an undesired effect created by turbulence, which causes bending of the fuselage around the wing area

Stub Cable - a 4-wire cable used to connect the TC to the current mode coupler on an ARINC 629 subsystem

Synchronization Gap (SG) - a time period common to all transmitters on the ARINC 629 data bus

Tach Generator - used in electric servo systems as rate sensors

Terminal Controller (TC) - an ARINC 629 subsystem that moves data to and from the LRU memory

Terminal Gap (TG) - period unique for each transmitter connected to the ARINC 629 data bus

Timing Tolerance Fault - occurs when the rise or fall of a digital signal responds too slowly to be within specifications of the data bus standard

Transformer Rectifier (TR) - A device to change 115 VAC to 26 VDC

Transmissive Display - LCD containing a dedicated light source mounted to the rear of the display

Transmit Interval (TI) - a common period for all transmitters on the ARINC 629 data bus

Troubleshooting Manual (TSM) - designed specifically for system troubleshooting

True Air Temperature (TAT) - a measure of the air temperature as it is compressed by the moving aircraft

Vector Generator - processor circuitry used to draw lines on a CRT display

Vertical Navigation - an autopilot mode used to fly the selected altitude or glide path

VU Number – specific number used to designate all panels and racks on Airbus aircraft

Warning - calls attention to any methods, materials, or procedures that must be followed to avoid injury or death

Water Line - a horizontal reference plane, which runs the length of the aircraft from nose to tail

Wide Area Augmentation System (WAAS) - a GPS enhancement designed to improve the integrity, accuracy, availability, and continuity of the basic satellite navigation system over a large area of coverage

Wing Leveler - a simple autopilot system used to provide guidance along only the longitudinal axis of the aircraft. These systems, often found on light single engine aircraft, were called because they were used to keep the wings level

Wing Station (WS) - used to indicate locations longitudinally along the wings of the aircraft

Wire Identification Number – a label placed on all wires three inches or longer

Wire Routing Chart - used to locate wire bundles that run through the aircraft

Wiring Diagram - very specific diagrams with details on wires, connectors, and pin numbers for a given system

Wiring Manual - contains various diagrams, charts, lists and schematics needed to maintain the various electrical/electronic systems on the aircraft

Yaw Damper - system designed to control rudder and eliminate dutch roll

Zoning – a reference system designed by the Air Transport Association (ATA) to further identify the location of components on large aircraft

Index

Corrections, Suggestions for Improvement, Request for Additional Information

It is Avotek's goal to provide quality aviation maintenance resources to help you succeed in your career, and we appreciate your assistance in helping.

Please complete the following information to report a correction, suggestion for improvement, or to request additional information.

REFERENCE NUMBER *(To be assigned by Avotek)*	
CONTACT INFORMATION*	
Date	
Name	
Email	
Daytime Phone	
BOOK INFORMATION	
Title	
Edition	
Page number	
Figure/Table Number	
Discrepancy/Correction *(You may also attach a copy of the discrepancy/correction)*	
Suggestion(s) for Improvement *(Attach additional documentation as needed)*	
Request for Additional Information	

FOR AVOTEK USE ONLY	Date Received	
	Reference Number Issued By	
	Receipt Notification Sent	
	Action Taken/By	
	Completed Notification Sent	

Contact information will only be used to provide updates to your submission or if there is a question regarding your submission.

Send your corrections to:

Email: comments@avotek.com
Fax: 1-540-234-9399
Mail: Corrections: Avotek Information Resources
P.O. Box 219
Weyers Cave, VA 24486 USA